Statistics

A Bayesian Perspective

Donald A. Berry
Institute of Statistics and Decision Sciences, Duke University

An Alexander Kugushev Book

Duxbury Press
An Imprint of Wadsworth Publishing Company
I(T)P® An International Thomson Publishing Company

Belmont Albany Bonn Boston Cincinnati Detroit London Madrid Melbourne
Mexico City New York Paris San Francisco Singapore Tokyo Toronto Washington

Editorial Assistant: Cynthia Mazow
Production Editor: Susan L. Reiland
Print Buyer: Stacey Weinberger
Permissions Editor: Peggy Meehan
Designer: Andrew P. Zutis
Copy Editor: Adela C. Whitten
Cover Design: Gary Head
Compositor: G & S Typesetters, Inc.
Printer: R. R. Donnelley, Crawfordsville

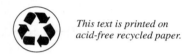 *This text is printed on acid-free recycled paper.*

COPYRIGHT © 1996 by Wadsworth Publishing Company
A Division of International Thomson Publishing Inc.
I(T)P   The ITP logo is a registered trademark under license.

Printed in the United States of America
2 3 4 5 6 7 8 9 10

For more information, contact Wadsworth Publishing Company:

Wadsworth Publishing Company
10 Davis Drive
Belmont, California 94002, USA

International Thomson Publishing Europe
Berkshire House 168-173
High Holborn
London, WC1V 7AA, England

Thomas Nelson Australia
102 Dodds Street
South Melbourne 3205
Victoria, Australia

Nelson Canada
1120 Birchmount Road
Scarborough, Ontario
Canada M1K 5G4

International Thomson Editores
Campos Eliseos 385, Piso 7
Col. Polanco
11560 México D.F. México

International Thomson Publishing GmbH
Königswinterer Strasse 418
53227 Bonn, Germany

International Thomson Publishing Asia
221 Henderson Road
#05-10 Henderson Building
Singapore 0315

International Thomson Publishing Japan
Hirakawacho Kyowa Building, 3F
2-2-1 Hirakawacho
Chiyoda-ku, Tokyo 102, Japan

All rights reserved. No part of this work covered by the copyright hereon may be reproduced or used in any form or by any means—graphic, electronic, or mechanical, including photocopying, recording, taping, or information storage and retrieval systems—without the written permission of the publisher.

Library of Congress Cataloging-in-Publication Data

Berry, Donald A.
    Statistics : A Bayesian perspective / by Donald A. Berry.
      p.   cm.
    Includes index.
    ISBN 0-534-23472-0
    1. Bayesian statistical decision theory.   I. Title.
QA279.5.B48   1996
519.5—dc20

95-34996

Preface

THIS IS an introduction to statistics for general students. It differs from standard texts in that it takes a Bayesian perspective. It views statistics as a critical tool of science and so it has a strong scientific overtone. While my outlook is conventional in many ways, its foundation is Bayesian.

The Bayesian Perspective

There are several advantages of the Bayesian perspective:

- It allows for direct probability statements, such as the probability that an experimental procedure is more effective than a standard procedure.
- It allows for calculating probabilities of future observations.
- It allows for incorporating evidence from previous experience and previous experiments into overall conclusions.
- It is subjective. This is a standard objection to the Bayesian approach; different people reach different conclusions from the same experiment results. There would be comfort in giving an answer that others would also give. But differences of opinion are the norm in science and an approach that explicitly recognizes such differences is realistic.

Despite differences in focus between the standard and Bayesian approaches, there are more similarities than differences. Many of the principles illustrated in the examples and exercises of this text are not peculiar to either approach.

Other Distinguishing Features

In addition to those mentioned above, this text has several other distinguishing features:

- *It is about statistical ideas,* not solely about methods. Ideas are developed using examples, many of which are case studies in which data from experiments are brought to bear on scientific questions. However, statistics is not merely a set of methods for analyzing data. *It is also a way for integrating data into the scientific process.*

- My emphasis is on sound principles of experimentation and on learning through observation. The process of learning requires calculation. This in turn leads to a secondary emphasis on methodology. Calculations involved in developing and motivating general settings can be tedious. I have endeavored to keep difficulties with this aspect of the learning process to a minimum.
- Statistics involves learning from data. However, the development in this text is not one of "data analysis" per se. Not all data sets are fodder for statistical methods. There must be a well-defined question being addressed.
- Most of the examples and exercises deal with substantive scientific issues as well as with the scientific method. In my view, applying statistical ideas requires familiarity with the substantive scientific issues and should not deal merely with numbers. So it is inevitable that readers will learn some science as well as some statistics.
- In keeping with a subjective nature of the Bayesian approach, I write in the first person and draw my own conclusions in the various examples.
- I write with the student in mind, for example, giving detailed discussions and redundant examples when the concepts are difficult for students. And I sacrifice numerical accuracy for understanding.

Examples, Exercises, and Real Data

Almost all examples and exercises in this text involve real settings and real data. The examples come from the sciences and from sports, with many involving medicine. A reason for the number of medical examples is that health issues are usually interesting for students. Also, medical studies tend to be well designed and easy to report. Finally, data in medicine are readily available. Nearly all the examples and exercises that involve artificial settings occur in the two chapters on probability.

The examples and exercises are chosen to be interesting and relevant for students. They are also chosen to display the effective and fruitful use of statistical thinking and methods.

Many examples and exercises appear more than once in the text. The second time usually addresses an improvement on the first—sometimes the analysis is more appropriate and sometimes the calculations are easier. Occasionally, an analysis is wrong; my intention is to show you what some people would do, saying why it is wrong. When the same problem appears a second time I repeat the information given earlier to minimize your having to leaf back through the book. But I give the earlier reference so you will know where you worked on the problem before.

Using the Computer

This text does not require the use of computers. However, computer programs that make many of the calculations described in this text are included on a 3.5″ disk that comes with this text. These were written by James Albert, who has also written appendices for many of the chapters; these demonstrate the use of the computer programs for the chapter in question. For instructors who wish to use computers even more intensively in their course, a companion Bayesian software package and a guide to its use—*Bayesian Computation Using Minitab,* by James Albert, 1996—have been published separately and are available from Duxbury Press. This software refers specifically to examples and exercises in this text. It uses some methodology that is more advanced than the level of this text.

The enclosed 3.5″ disk also contains in electronic form all the datasets used in the text, in both ASCII format and Minitab format.

Mathematical Prerequisites

The mathematical level of this book is minimal with only an exposure to high school algebra expected. Exercises in several chapters require formulas. The latter are usually motivated by calculations and arguments that require some familiarity with numerical operations.

Coverage

This book has five parts:

- Part I: Chapters 1, 2, and 3
- Part II: Chapters 4 and 5
- Part III: Chapters 6, 7, 8, and 9
- Part IV: Chapters 10, 11, 12, and 13
- Part V: Chapter 14

Part I presents basic concepts. Chapter 1 is an introduction to scientific inference and sampling. Chapter 2 describes data displays and data summaries. It also serves to provide insights into statistical problems and to introduce statistical ideas that will be developed later in the book. Chapter 3 discusses experimental design and the limitations that poor designs place on inference. Part II introduces probability (Chapter 4) and conditional probability (Chapter 5). The ideas in this part are used extensively in the remainder of the text. Part III considers statistical problems dealing with proportions. Chapters 6 and 7 deal with a single proportion, and Chapters 8 and 9 address the comparison of two proportions. Part IV deals with general populations. Chapters 10 and 11 consider a single population and Chapter 12 addresses the comparison of means of two or more populations. Chapters 11 and 12 make assumptions about the population shape, and Chapter 13 relaxes those assumptions. Part V describes regression analysis and fitting by least squares.

Chapters 1–14 contain enough material for a one-semester course. A shorter course might skip Chapter 13 and the various optional sections. Chapters 4 and 5 are essential (except for betting odds in Section 4.3), and exposure to some of the material in Chapters 6–9 is required for Chapters 10–13.

Chapters 6, 8, and 10 proceed from first principles and Bayes' rule from Chapter 5 to develop the basic ideas and methods of inference. These are used to motivate the simpler and yet more sophisticated and practical methods of Chapters 7, 9, 11, and 12.

These latter chapters address the same types of problems as do Chapters 6, 8, and 10, but they handle calculations much more smoothly. A course that focuses on statistical methods might place less emphasis on the material in Chapters 6, 8, and 10.

References

There are several advanced Bayesian texts that are excellent future references for students in this course. All require more mathematical background than is assumed in this book, and in particular, all require the calculus. They are listed below approximately in increasing order of mathematical level.

>P. M. LEE. *Bayesian Statistics: An Introduction.* London: Charles Griffin, 1989.
>M. H. DEGROOT. *Probability and Statistics,* 2d ed. Reading, Massachusetts: Addison-Wesley, 1986.
>S. J. PRESS. *Bayesian Statistics.* New York: Wiley, 1989.
>G. E. P. BOX and G. C. TIAO. *Bayesian Inference in Statistical Analysis.* New York: Wiley, 1973.
>J. M. BERNARDO and A. F. M. SMITH. *Bayesian Theory.* Chichester: Wiley, 1994.
>J. O. BERGER. *Statistical Decision Theory and Bayesian Analysis.* New York: Springer-Verlag, 1985.

Acknowledgments

John Andersen, David Banks, George Barnard, James Berger, David Blackwell, Kathryn Chaloner, Hong Chang, Ronald Christensen, Murray Clayton, George Diamond, David Draper, Ruth Etzioni, Ian Evett, Prem Goel, Ram Gopalan, Lawrence Joseph, Jay Kadane, Michael Lavine, Roger Lewis, Chengchang Li, John Monahan, James Press, Richard Savage, Tom Short, Jonathan Skinner, Mike West, Robert Winkler, Jeff Witmer, and Arnold Zellner provided advice and help in the development of this book. James Albert gave important advice and provided the Minitab programs and computing appendices that follow many of the chapters. Jon Stroud gave the solutions of the exercises a diligent and thorough checking. I want to acknowledge the advice and suggestions of all my colleagues at Duke University and, formerly, at the University of Minnesota, especially Bernard Lindgren, who taught me that not all data deserve to be analyzed. I particularly want to recognize Frank Anscombe and Dennis Lindley—both have been inspirations to me in many ways. The views of Leonard J. Savage have been of immeasurable influence on me generally and while writing this book in particular, though I could never set them down as cleanly and elegantly as he enunciated them. I am grateful for the thorough readings and excellent suggestions of the reviewers: George T. Duncan, Carnegie Mellon University; Daniel Heitjan, Pennsylvania State University; S. Rao Jammalamadaka, University of California, Santa Barbara; Thomas A. Louis, University of Minnesota; Carl N. Morris, Harvard University; and Alan Zaslavsky, Harvard University. Cheryl McGhee tracked down many original references. Finally and most importantly, Donna, Donald, Michael, Timothy, Scott, Jennifer, and Erin helped me see many of the examples more clearly, sometimes challenging my initial interpretations and conclusions.

Contents

1 Statistics and the Scientific Method — 1

1.1 The Scientific Method 2
1.2 Statistics As Learning 4
1.3 Samples, Populations, and Predictions 8
1.4 Organization of This Book 8

2 Displaying and Summarizing Data — 10

2.1 Dot Plots 11
2.2 Bar Charts 14
2.3 Scatterplots and Line Plots 18
*2.4 Number Plots, Area Plots, and Star Plots 31
2.5 Stem-and-Leaf Diagrams and Data Histograms 33
2.6 Means, Standard Deviations, and Medians 40
Appendix Using Minitab for Displaying Data 54

3 Designing Experiments — 59

3.1 Controlled Studies 60
3.2 Randomized vs Observational Studies 62
3.3 Taking Samples 72
3.4 Multiplicities and Design 79
3.5 Paired vs Parallel Designs 85

*These sections are not used in the sequel and are optional.

3.6 Crossover Designs   91
***3.7** Factorial Designs   97

4

Probability and Uncertainty   105

4.1 Properties of Probability   106
4.2 Probability as Long-Run Frequency   114
***4.3** Betting Odds and Probability   116
4.4 Probability as Degree of Belief   120

5

Conditional Probability and Bayes' Rule   124

5.1 Joint Probabilities   125
5.2 Conditional Probability   130
5.3 Law of Total Probability   140
5.4 Bayes' Rule   147
Appendix   Using Minitab for Conditional Probability and Bayes' Rule   161

6

Models for Proportions   165

6.1 Sampling from Populations Having Two Types of Members   167
6.2 Law of Large Numbers   168
6.3 Assessing Information about Models and Finding Means   169
6.4 Likelihoods, Posterior Probabilities, and Predictions   174
6.5 Calculations in an Example   182
Appendix   Using Minitab with Models for Proportions   193

7

Densities for Proportions   196

7.1 Many Models   196
7.2 Infinite Numbers of Models and Densities   198
7.3 Updating Rule for Beta Densities   205
7.4 Choosing Beta Densities as Priors   211
7.5 Finding Areas and Probabilities of Intervals   217
7.6 Percentiles and Probability Intervals   221
7.7 Prediction   227
Appendix   Using Minitab with Densities for Proportions   233

8 Comparing Two Proportions — 238

8.1 Two Population Models 239
***8.2** Testing a Null Hypothesis 257
Appendix Using Minitab for Comparing Two Proportions 270

9 Densities for Two Proportions — 277

9.1 Multiplying Likelihoods for Two Proportions 277
9.2 Applications of Densities: PdAL 282
9.3 Probability Intervals for Differences 291
Appendix Using Minitab with Densities for Two Proportions 305

10 General Samples and Population Means — 310

10.1 General Population Models 310
10.2 Normal Models 318
Appendix Using Minitab with General Samples and Population Means 334

11 Densities for Means — 336

11.1 Prior Densities and Normal Models 336
11.2 Choosing a Normal Density As a Prior 347
11.3 Rule of Means 349
11.4 Normal Densities for Means of Large Samples 355
11.5 Probability Intervals for Population Means 359
Appendix Using Minitab with Densities for Means 368

12 Comparing Two or More Means — 370

12.1 Normal Densities for Differences 371
12.2 PdAL and Probability Intervals for Differences 378
12.3 Comparing Several Means 392
Appendix Using Minitab for Comparing Two or More Means 405

13 Data Transformations and Nonparametric Methods — 409

13.1 Outliers   410
13.2 Transforming Data: Taking Logarithms   416
13.3 Categorizing Data   426
13.4 Ranking Data   440
13.5 Probabilities of Intervals and the Sign Test   449

14 Regression Analysis — 459

14.1 Least Squares Line   461
14.2 Relating Two Measurements: Regression   479
14.3 Regression Effect   494
Appendix   Using Minitab for Regression Analysis   505

Appendix Using the Bayesian Minitab Macros — 509

Short Answers to Selected Odd-Numbered Exercises — 513

Index — 515

1 Statistics and the Scientific Method

YOU meet a nice couple from Lake Wobegon, Minnesota. Does that mean most couples from Lake Wobegon are nice? Mozart, Beethoven, and Schubert were born in the winter. Does this mean that someone born in the winter is more likely to be a great composer than someone born in another season? A study of 50 coffee drinkers and 50 non-coffee drinkers finds that the coffee drinkers died an average of 3 years sooner. Does this mean coffee is detrimental to one's health? Such questions are statistical: They deal with how we learn from evidence. The following is typical of the scenarios that we will deal with in this book.

Can fish oil prolong pregnancy? An article[1] with this title describes a study of 533 healthy Danish women who were in their last 3 months of pregnancy. Half of the women (266) received daily capsules containing fish oil, about half of the rest (136) received daily capsules containing olive oil, and the remainder (131) received no capsules. The women were assigned to these three treatment groups in a random fashion. In effect, the investigators tossed a coin. If the coin came up heads, the woman was given fish oil; if it came up tails, the coin was tossed again. If heads resulted on the second toss, then she was given olive oil; if tails resulted again, then she received no capsules. Pregnancies lasted "an average of 4 days longer" in the fish-oil group than in the olive-oil group. Pregnancies in the no-capsule group lasted "about 2.4 days longer" than in the olive-oil group. It is not clear why the author gives the latter comparison instead of no-capsule vs fish-oil since fish oil is the focus of the article. But no matter, we can subtract to find that the fish-oil group's average gestation was only 1.6 days longer than that of the no-capsule group. According to the article, "Nobody understands the mechanism behind fish-oil's action." Based on my reading of the data, it is far from clear that fish oil *has* an action! Teaching you to draw conclusions in similar circumstances is a major goal of this book.

Inferences from data are statistical. Statistical inferences have two characteristics:

1. Experimental or observational evidence is available or can be gathered.
2. Conclusions are uncertain.

This text provides a way to make statistical inferences.

> **Statistical conclusions are based on observational evidence and involve uncertainty.**

Experimentation and quantifying uncertainty are fundamental in science. Scientists learn using the *scientific method*. This term is sometimes understood to mean what scientists use. But different scientists use different methods, and some of their methods do not meet very high standards of science. It is important to have a precise definition, and that is the purpose of the next section.

1.1 The Scientific Method

According to physicist Stephen Hawking,[2] the ultimate goal of science "is nothing less than a complete description of the universe we live in." Scientists strive for this goal by building theories and checking the theories' predictions—this is the heart of the scientific method. Hawking does not view science as a way of approaching reality (whatever that may be) but as a way of *thinking* about reality. Such a view is central to the approach in this text.

Hawking defines a scientific theory as "a model of the universe, or a restricted part of it, and a set of rules that relate quantities in the model to observations that we make." For example, "the earth is round" is a scientific theory. Hawking relates how Aristotle and other early Greeks used the scientific method to come to believe in the correctness of this theory. They made observations and compared them with the theory's predictions. First, if the earth were round it would always cast a round image on the moon during a lunar eclipse (it does), whereas if it were a flat disk the image would sometimes be elliptical. Second, if the earth were round the North Star would appear lower in the sky when viewed further to the south (it does). Third, if the earth were round the sails of approaching ships would appear before their hulls (they do).

Expanding on the above interpretation, the scientific method is a process of devising experiments and updating knowledge using evidence from the experiments. Better experiments are more informative. But experiments are costly, where costs are measured in time, money, and experimental resources. While costs may seem to be a consideration that is separate from the methodology of science, they are central to that methodology and should be considered explicitly.

Scientific Method

1. Ask a question or pose a problem.
2. Assemble and evaluate the relevant information.
3. Based on current information, design an investigation or an experiment (or perhaps no experiment) to address the question posed in step 1. Consider costs and benefits of the available experiments, including the value of any information they may contain. Recognize that step 6 is coming.
4. Carry out the investigation or experiment.
5. Use the evidence from step 4 to update the previously available information; draw conclusions, if only tentative ones.
6. Repeat steps 3 through 5 as necessary.

The following are questions scientists might pose at step 1. Some are examples considered in this course. Their answers are known with varying levels of precision. Moreover, you will get different answers, depending on whom you ask!

- What is the distance between the Sun and Sirius?
- Is an attractive human face merely average?
- What killed the dinosaurs?
- When will the next avalanche occur on this mountain road?
- Has intelligent life developed on celestial bodies other than Earth?
- Is the drug AZT effective in delaying the symptoms of AIDS?
- What is the atomic weight of nitrogen?
- Was Ted Williams a better baseball hitter than Joe DiMaggio?
- Why do burrowing owls line their nests with cattle dung?
- Is exposure to the sun unhealthy?
- Is this machine still producing acceptable items?
- What is the melting point of villiaumite?
- How long will this lung cancer patient live?
- Can fish oil prolong pregnancy?
- What is the diameter of Pluto?
- When will the next big earthquake occur?
- Is Al Edged the father of Suzy Smith?
- Is cold fusion possible?
- How does a particular kind of acid rain affect an ecosystem?

Asking the right question is important. A reliable answer to the wrong question may be worthless. The scientist formulates the question (step 1). This step is not really statistical, but an effective use of statistics can help show that a question being addressed is inappropriate. Statistics can also show that answering the question will be difficult or impossible. Also, while a statistician may help in carrying out an experiment (step 4), that is usually the scientist's responsibility. Statistics plays a central role in all other aspects of the scientific method.

It is difficult to overemphasize the importance of designing good experiments (step 3). Designing experiments is the subject of Chapter 3. While good design will be important in subsequent chapters as well, they focus on steps 2 and 5.

> **Statistics plays important roles in designing and drawing conclusions from experiments.**

Statistics As Learning

The word *statistics* has several meanings. *Statistics are numbers.* They serve to *summarize* results of a study. In its singular form, *statistics is a scientific discipline.* It is the study of the way people learn as they make observations—from experiments or otherwise. It answers the question: What does the study mean? In many applications the observations are numbers that are related to the question posed in step 1 of the scientific method. Hence, the same word is used for numbers and the discipline that deals with them.

Learning is only part of statistics. The rest is using what is learned. Consider a simple example. You buy a can opener that does not work. You return it to the store. Should you replace it with one of the same type or one of another, more expensive type? Since you have tested but a single opener, you do not have much information to make an intelligent decision. But you do have some information—more now than when you bought the faulty opener. So it may be appropriate to change your strategy and try the other type, despite the extra cost.

Statistical inference formalizes the process of learning through observation—such as finding that the can opener did not work. One makes decisions as learning occurs. A possible decision in the can opener problem is to buy another opener of the same type; a second is to switch types; a third is to try to get more data before you commit to buying either; and there are others. Statistical thinking guides the decision making process.

The following example is more important than the can opener example, but it too involves a single observation.

EXAMPLE 1.1

▷ **What caused milk sickness?** Abraham Lincoln's mother, Nancy Hanks Lincoln, died of milk sickness when Abe was only 7 years old.[3] The disease got its name from an apparent association with drinking milk. It was virtually unknown in populated areas, but it claimed many victims among the American pioneers who settled west of the Appalachians in the early 19th century. It wiped out up to half of some settlements.

It was generally understood that milk sickness occurred in areas where cattle suffered from a fatal disease called trembles. But it was not known whether there was a relationship between milk sickness and trembles: While the only people who died were milk drinkers, cows that gave milk never died of trembles. (We now know that milking cows got rid of the toxin responsible for the disease in humans by passing it out of their bodies in their milk.) Trembles seemed to occur among cattle that had grazed in woodlands. Various plants found in woodlands were suspected, but doctors were unable to determine the specific source.

Enter Anna Pierce. In those days women were not allowed to attend medical school. Pierce had taken midwife and nursing courses and had settled in Illinois as a doctor. She made important observations about milk sickness on her own, but only when she befriended a Shawnee woman known as Aunt Shawnee did she learn the cause. Aunt Shawnee took the doctor into the woods and pointed to a plant called white snakeroot, which her people used to treat snakebite and other maladies. Aunt Shawnee claimed the plant was responsible for trembles and in turn for milk sickness.

Being a good scientist, Pierce was skeptical. The scientific question (step 1) is, Does white snakeroot cause trembles? (The corollary question is, Does it cause milk sickness?) She assembled information (step 2), including that from Aunt Shawnee. She designed and carried out an experiment (steps 3 and 4): She fed white snakeroot to a calf and the calf soon developed trembles. In view of Aunt Shawnee's claims, this single observation was conclusive to Pierce (step 5) and she managed to convince others. But it was not conclusive to everyone, and some doctors refused even to accept that the disease existed. To convince one dubious doctor, she repeated the experiment (step 6)—on one of the doubter's calves! ◁

If you are thinking that a single observation provides compelling evidence, consider the next example.

EXAMPLE 1.2
▷ **Is making rain a lot of hot air?** According to the August 12, 1990 *Santa Barbara News-Press,* a balloonist named Sam Carter asked a fee of $60,000 to bring rain to that drought-stricken area: "The idea is you have a massive formation of clouds stopped by the coastal mountains from California to Alaska. By taking two to 10 hot air balloons in close formation at sunrise, it will cause clouds to come apart and when the sun shines through there is a chain reaction which turns the (stratus) clouds to cumulus clouds." The only evidence he cited was a time in June 1974 when two balloons took to the skies and broke the Los Angeles basin temperature inversion—2 days later a good rain fell on Albuquerque! ◁

Examples 1.1 and 1.2 are the same in that both involve a single observation. But if you are like me, you will draw opposite conclusions. Anna Pierce's conclusion seems quite reasonable in view of the circumstances—something was causing trembles, and other evidence suggested that it was something that the cows ate in the woods. The result of the second experiment, in which the other doctor's calf was fed white snakeroot, seems quite predictable. However, one would have to be pretty desperate and foolish to pay the balloonist $60,000 to bring rain. It would take much more evidence to lead me to put even a little credence in the balloonist's theory. The next time balloonists "take to the skies," it may or may not rain—in Albuquerque or elsewhere.

As this course develops, you will see that it is reasonable to come to different conclusions in different scientific settings even though the available experimental evidence is the same. You will also see that it is reasonable for different people to come to different conclusions about the same question, even though both have the same experimental evidence.

> **The same data can lead to different conclusions depending on other available information.**

The two principal parts of the learning process are designing experiments (What data should be collected and how should they be collected?) and analyzing data from experiments (How do we learn from the data collected?). The second part is the main focus of this text. Learning from numerical information is not an intuitive process. People sometimes give too much credibility to a few observations and sometimes give too little credence to data that are inconclusive.

Scientific studies involving a single observation are rare. When there are many observations the learning process is more difficult. Repeated observations frequently give different results: Changes in blood pressure after taking a drug vary from one patient to the next; one white rat may take 10 seconds to learn a task while the next takes 110 seconds. Variability does not invalidate experimental results, but it is important to know how to deal with it. Indeed, recognizing and handling variation in data is a critical aspect of the discipline of statistics.

We need a formalism for learning, for incorporating the results of an experiment into what we already know. This is step 5 of the scientific method. Such a formalism should allow for describing knowledge available to us at any time. Knowledge can then be converted into inferences (such as a certain type of can opener tends to be faulty), decisions (do/do not buy another opener of this type), and designs for additional experiments (take a can to the store and test some openers before you buy one).

The problem of choosing from among available actions occurs in business, medicine, science, and everyday life. The appropriate choice depends on one's objectives, the costs of the various consequences of one's actions, and on one's state of knowledge concerning what the consequences will be.

The next two examples convey the flavor of the problem of *quantitative inference*. The researchers collected data to help address particular questions. I will indicate how the examples relate to steps 1–4 of the scientific method and will return to both examples in later chapters to carry out step 5.

EXAMPLE 1.3

▷ **Malaria and sickle cells.** Step 1: Are carriers of the sickle-cell gene more resistant to malaria than noncarriers? Step 2: In the 1950s, scientists decided that they did not have much information concerning this question. Steps 3 and 4: Some scientists conducted a study. They injected 30 African volunteers with malaria—15 were sickle-cell carriers (heterozygotes, meaning that they had exactly one copy of the sickle-cell gene) and 15 were noncarriers. Fourteen of the 15 noncarriers came down with malaria, whereas only 2 of the 15 carriers got the disease.

These scientists had a perverted notion of costs (step 3): This study was clearly unethical. Of the 30 subjects exposed to the disease for the purpose of getting information, 16 contracted the disease. This is a high price to pay for information.

Statistical thinking shows how to use the information obtained to conclude whether or not sickle-cell carriers tend to be protected from malaria. You have learned something from these data—besides the fact that some scientific research is unethical. Statistics helps you express what you have learned. The process of learning is not always obvious. Commenting in *Natural History*,[4] scientist Jared Diamond not only deplored the study as unethical, but claimed that the sample size was too small for it to be conclusive. In Chapter 8, however, we will see that the study *was* quite conclusive. ◁

Statistics is not simply a set of techniques for analyzing data, although some statisticians take this limited view. Consider an experiment with the same data as Example 1.3. But now suppose that each of the 30 subjects tossed a coin, calling it while it was in the air. Fourteen of the 15 noncarriers called their toss correctly and 2 of the 15 carriers called theirs correctly. The data are the same as in the example: 14/15 vs 2/15. But my conclusion would be different. A relationship in the original experiment is plausible and so these data are more convincing than they are in the coin-toss experiment. That is, I am more willing to believe that sickle-cell carriers are resistant to malaria than that they are worse at calling coin tosses.

The next example shows that even a lot of data—in this case, 628 putts—can strain one's credibility.

EXAMPLE 1.4
▷ **PGA putting.** Step 1: How well do professional golfers putt, depending on the length of the putt? Step 2: Recognizing that recollections and impressions may be biased, researchers set out to get objective information. Steps 3 and 4: In the latter half of 1988, the Professional Golfers Association (PGA) collected data on the putting success of PGA tour players. They selected one relatively flat green at each of 15 tournaments. For every putt attempted on that green during the tournament, a PGA staff member used triangulation to measure the distance to the hole (to the nearest foot) and recorded whether the golfer made it. Here are the results for three distances:[5]

Length of putt (ft)	Number tried	Number made	Percentage success
9	217	69	31.8
11	237	75	31.6
14	174	54	31.0

How much should you trust these percentages? Certainly not completely. They are essentially equal. But from 9 feet away the hole is only $\frac{9}{14}$ as far as it is from 14 feet away. So it should be easier to hit from a distance of 9 feet. Why isn't the success percentage from 14 feet only $\frac{9}{14}$ of that from 9 feet?

On the other hand, the success percentages in the table do convey information. Suppose a professional golfer is about to try a 10-foot putt on a flat green. If you are going to predict whether he makes it, you would rather have seen these data than not. So they do have some value. ◁

In this course you will learn how to interpret and weigh observational data. The next section sets the stage for this and presents some of the language used in statistical problems.

1.3 Samples, Populations, and Predictions

Statistics deals with observations. Observations made in a specified way (for example, carrying out a particular experiment repeatedly) form a **sample**. The central problem of statistics is learning from samples.

> **A sample is a collection of observations.**

The sample should be tied in some way to the question that is being addressed (step 1 of the scientific method). In particular, if we had a sufficient number of observations of the same type, we should be able to give a firm answer to the scientific question. The question must deal with the process that produces the observations. The set of all possible observations (real or imagined) is called a **population**. A population may be finite or infinite.

> **A population is the collection of potential observations of which the sample is a part.**

A typical statistics problem is drawing a conclusion about some characteristic of a population on the basis of a sample from that population. A conclusion may entail making a decision based on the sample. Conclusions almost always involve uncertainty. For example, having observed 37 females and 63 males, one might conclude that *about* 37% of the population is female. Eventually, we will address meanings for the "about."

> **Statistical inferences extrapolate from a sample to the population being sampled.**

A particular type of inference is making conclusions or **predictions** about the next observation from the population. Predictions are not usually specific, such as "the next observation will be a 7." Rather, they are subject to error; developing methods for addressing such errors is a major focus in this text.

> **A prediction is an inference about the next sample observation or set of observations.**

1.4 Organization of this Book

Chapter 2 deals with data displays and summaries. Data displays are pictorial methods for communicating the information in a sample. Sometimes a data display is all that is needed in drawing conclusions. Sometimes supplementary methods are required for interpreting a display. Chapter 2 serves several purposes beyond teaching you the mechanics of making graphs and drawing pictures. One is to introduce some statistical ideas. Another is to provide insights into scientific problems. You will learn about sampling variability. Also, you will deal with samples from populations in which there are several measurements of interest—say, height and weight. Chapter 2 also gives a glimpse into designing experiments.

Chapter 3 presents some fundamental principles for designing experiments. Because some researchers give little attention to design issues, they end up with worthless results. Some experiments convey faulty information and thus are detrimental to science.

Chapters 4 and 5 introduce probability. Probability is the language of uncertainty and provides a mechanism for learning. Probability will be used extensively in the remainder of the text.

Chapters 6 and 7 consider a special and rather simple type of population: The measurement of interest has only two possible values (success and failure, say). For such a population, a complete description is the proportion of one or the other value. The main statistical questions involve inferences about these proportions and making predictions about future observations. Chapters 8 and 9 consider *two* populations in which the measurement of interest has two values—the same two values in both populations. For example, one population is that of experimental units exposed to a treatment and each unit either succeeds or fails; the other population is that of units exposed to a control and each of these either succeeds or fails. Of interest is which success proportion is larger, and by how much.

Chapters 10 and 11 deal with more general types of populations: The quantity or measurement of interest can take any number of values. Chapter 12 addresses the comparison of two or more such populations. In Chapters 10–12 we will make specific assumptions about the shape of populations being sampled; in Chapter 13 we will relax those assumptions.

Chapter 14 considers a common type of applied problem, but one that is more complicated than those treated in earlier chapters. It considers a single population from which two measurements are made on individuals in a sample from the population. The goal is to understand the relationship between the two measurements in the population, again based on the sample observations, which are now pairs of measurements.

Chapter Notes

1. *Science News* 141 (May 16, 1992): 334.
2. S.W. Hawking, *A Brief History of Time: From the Big Bang to Black Holes* (New York: Bantam Books, 1988).
3. This example is taken from an article by ecologist D. C. Duffy in *Natural History* (July 1990): 4–8.
4. Jared Diamond, *Natural History* (February 1989): 8–18.
5. J. Diaz, "Perils of putting," *Sports Illustrated* 70 (April 13, 1989): 76–79.

2 Displaying and Summarizing Data

STATISTICS deals with data. A set of data that is collected in a particular way (carrying out an experiment, for example) is a **sample.** The central issue in this course is how we learn from samples. Most of the course deals with formalizing the learning process. This chapter addresses a rather primitive aspect of that process: Presenting and summarizing sample information. The emphasis in the first five sections is on pictorial displays; the last section considers numerical summaries.

Reporting information is an exercise in communication. The way sample information is displayed can be more important than any words that accompany the displays. This includes words describing the statistical methods you will learn later in this book. Indeed, a good technique for displaying information may make further statistical description unnecessary. On the other hand, all data displays are subject to overinterpretation—what you see may not be real. Most of the descriptive methods presented in this chapter are related to quantitative methods presented in later chapters. The latter methods do not obviate the need for data displays but rather serve to bolster them by giving a way to measure the strength of the evidence present in the display.

We will describe several methods for displaying data in Sections 2.1–2.5. Most are easy to learn and use. Some can be made using computer programs [see Albert's[1] book] and others have to be made by hand. Many methods that are in use are not included here; see the book by Cleveland[2] for examples. When faced with an application you may be able to use a method that is presented in this chapter. But since each problem has its own peculiarities, you may have to devise a new method. There are several (related) principles that you should follow:

○ Choose a mode of display that is understandable; displays should be pleasing to the eye and not overly complicated.

○ Faithfully represent the information in the experiment. When grouping and summarizing data (which will be necessary with large data sets), some information will be lost. Convey as much information as possible.

○ Some displays hide an aspect of the experiment or of the data; this should be made clear to the viewer either in the display or in a footnote to the display.

There is a conflict between not hiding information and making displays understandable. In practice, compromise is required.

Section 2.6 describes numerical ways for summarizing data.

2.1 Dot Plots

A **dot plot** (or dot diagram) is the simplest way to display data. Each observation is represented by a dot that is placed on a line to show the observation's position relative to the other data.

EXAMPLE 2.1
▷ **Viscosity of dimethylaniline.** To find the viscosity of dimethylaniline at 20°C, a scientist makes the following 12 measurements (in centipoises or cP):[3]

    146   154   141   140   136   132   147   140   147   139   140   140

The measurements vary because the scientist cannot control all aspects of the procedure. A dot plot will convey the accuracy and variability of the measuring process and show how the 12 observations relate to each other. A dot plot is a simple picture that shows each observation as a dot on a scale. The symbol used for the "dot" does not matter. In the figure shown for this example (Figure 2.1), it is a ◆, but subsequent examples will use ■ and ● and other symbols as well. Using small dots is necessary when there are many observations. When there are replicates, as in this example, one can stack them up in some way. The extreme viscosities are easy to see in this dot plot. Something else you can see is that the measurements seem to be centered near 140 cP.

Figure 2.1 Dot plot for Example 2.1

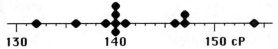

◁

A typical question in science is whether there is a relationship between two measurements. Dot plots (and other devices) can be used to investigate this question. They are especially useful when two or more of them are used to compare different groups, as in the next example.

EXAMPLE 2.2
▷ **Alcohol vs caffeine consumption.** Do college students who consume large amounts of alcohol also consume large amounts of caffeine? Three Duke students[4] who thought so took a survey in April 1993. They asked 97 undergraduates about their daily consumption of coffee, tea, types of soft drinks, and other drinks containing caffeine. They converted these to amounts of caffeine (in mg). They also asked students how many alcoholic drinks they have at "one sitting." ("Drinks per sitting" may not be the best indicator of alcohol consumption. But it is related to total amount consumed and, in any case, the researchers thought this measure was important.)

The researchers queried students concerning alcohol consumption using the following categories of number of drinks: 0, 1–2, 3–5, 6+. The results are shown (by category) in Table 2.1 (page 12). For example, nine students said they never drank and these students' daily caffeine amounts were 68, 180, 180, etc.

Table 2.1
Caffeine intake per day (in mg), listed by four different categories of alcoholic drinks per sitting

0	1–2	3–5			6+		
68	0	0	210	505	68	260	390
180	0	68	260	598	68	260	390
180	68	68	260	805	68	260	390
210	93	68	278	810	98	260	398
210	98	98	373	975	98	260	525
323	165	98	390		98	260	570
368	210	98	390		98	270	665
368	210	98	398		113	323	860
698	210	113	405		165	323	
	260	165	405		180	328	
	293	180	405		180	373	
	323	180	435		210	373	
	405	180	435		210	373	
	405	210	480		225	373	
	435	210	495		260	373	

To investigate the possibility of a relationship, the data can be plotted in four dot plots, one for each of the four categories. These are shown in Figure 2.2. Each dot plot shows caffeine consumption for one of the four categories of alcohol consumption.

Figure 2.2 Dot plots for Example 2.2

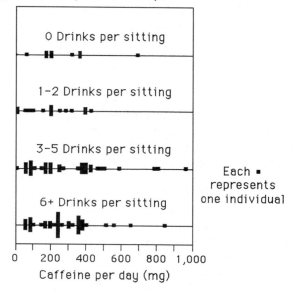

The scale at the bottom applies for all the plots. A single individual is represented by a square dot: ■. So there are 97 dots overall. Duplicates (and values too close together to show separately) are shown as two dots, one on top of the other: ▮. Triplicates are shown by three dots combined (▌), etc.

The four dot plots are different, of course. But they are not *very* different. They seem consistent enough to suggest that the study provides little evidence of a relationship between caffeine intake and number of drinks per sitting. ◁

> **Use dot plots when the observations are numerical and the sample size is small enough to show one dot per observation.**

When making dot plots, try to place the center of the dot on the appropriate number. Accuracy is not crucial. Depending on the size of the dots and the range of the data, it is usually possible to pay attention only to two significant digits. For example, if the data vary from 0.00 to 10.00, you can ignore the hundredths digit.

If the dots start to pile up and you have a hard time fitting them in, consider bar charts (Section 2.2) or histograms (Section 2.5).

EXERCISES

2.1 Darwin once carried out an experiment to learn whether cross- or self-pollination would produce superior plants.[5] (For a description of the experiment see Example 3.18.) He paired 30 *zea mays* (Indian corn) plants genetically (both from the same seed) and assigned them to be cross- or self-pollinated. When the plants matured he measured their heights (in eighths of an inch) as follows, with the last row giving the difference between the cross- and self-pollinated plants in each pair.

Pair	1	2	3	4	5	6	7	8	9	10	11	12	13	14	15
Cross	188	96	168	176	153	172	177	163	146	173	186	168	177	184	96
Self	139	163	160	160	147	149	149	122	132	144	130	144	102	124	144
Diff.	49	−67	8	16	6	23	28	41	14	29	56	24	75	60	−48

Make a dot plot of the differences. (The pairs with negative differences are interesting because Darwin tried to explain them. For example, other plants in the pot with pair 2 died or were stunted, "probably by some larva gnawing at their roots.")

2.2 An experimental drug designed to ease the symptoms of congestive heart failure was given to 32 patients.[6] One of the measures of interest was whether the patients could walk farther while on the drug than while off. All 32 patients were encouraged to walk as much as possible for two consecutive weeks; during the second week they were given the drug. They were outfitted with a pedometer that measured distances walked. The differences in miles walked while on the drug as compared with the previous week are shown. (The patients with negative differences walked less in the second week than in the first week.) Make a dot plot of these data.

```
  0.00    +.56    +3.27   −2.55   +8.42   +1.07   −1.31   +3.19
  −.59   +10.75  +11.73    −.05   +1.65   −3.42   +1.73   −1.44
 +6.04   +12.21   +4.97   +1.68   +2.28   −6.57   −2.11    +.75
  −.96    +1.68   +8.85   +7.45    −.59   +2.91    0.00   +5.40
```

2.3 Cannulae are tubes that are inserted into body cavities during various surgical procedures. It is important that they do not come out accidentally. An experiment was designed to compare the holding ability of two types of cannulae, A and B. Cannulae were inserted into animal hearts. Measurements of the pressure required to remove them were recorded, as follows:

Cannula A: .30 .30 .32 .43 .44 .47 .52 .59 .70 .77 .79 .81 .95 1.33 1.43 1.54
Cannula B: .04 .14 .19 .21 .25 .26 .32 .35 .40 .40 .42 .53 .53 .56 .57 .68

Using a single scale of pressure, make dot plots for both cannula types and compare them. In your estimation do the data suggest that one or the other cannula type requires more pressure to remove it?

2.2 Bar charts

There are various ways to make a dot plot. We could put all the dots above the line instead of having some above and some below. If we use square or rectangular dots they will stack to form bars. The result is a **bar chart.**

EXAMPLE 2.3
▷ **Viscosity of dimethylaniline (revisited).** Example 2.1 presented a dot plot for 12 measurements of the viscosity of dimethylaniline. Replacing the diamond-shaped dots with square dots or rectangular dots forms a bar chart. Three possible charts are considered in Figure 2.3, the first with square dots and the other two with rectangular dots. Narrow rectangles in the third chart give very narrow bars; this is also called a **spike chart.**

Figure 2.3 Alternative bar charts for Example 2.3

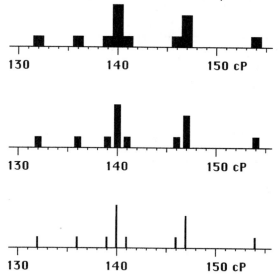

Of the three charts, the first one has the benefit of conveying the fact that the value 132, say, is really an interval extending from 131.5 to 132.5 (see the discussion of histograms in Section 2.5). It is also more visually impressive. ◁

The next example is more complicated and more interesting than the previous one because it shows four bar charts; this allows for comparison of the four groups that they represent.

EXAMPLE 2.4
▷ **Alcohol vs caffeine consumption (revisited).** Example 2.2 presented dot plots for daily caffeine consumption by 97 students, one for each of four categories of alcohol consumption. Sliding the vertical bars in the dot plots to align their bases on the corresponding scale forms bar charts, as shown in Figure 2.4.

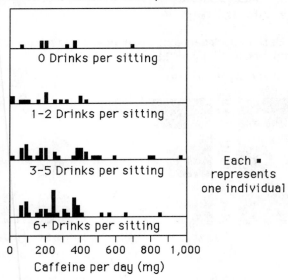

Figure 2.4 Bar charts for Example 2.4

◁

The measurements in the preceding two examples are numerical. Bar charts are commonly and appropriately used when the possible observations are not numerical. If they are naturally ordered (such as letter grades in a course), then it is clear how to arrange the categories before constructing the bars. When they are not naturally ordered (such as country of citizenship), then the presenter has some freedom in choosing an order. The following is an example in which the categories are ordered.

EXAMPLE 2.5
▷ **Arrowhead fracture.** Archaeologists[7] found 77 broken arrowheads in central Tennessee. They were fractured either at the base, the middle, or the tip, with frequencies indicated by the bars in Figure 2.5 (page 16). The investigators were interested in

whether the location of the fracture might be related to whether fire caused the fracture. The back-to-back bar charts in Figure 2.6(a) show both cause and fracture. (The sums of the heights of pairs of bars equal the heights of the bars in the first chart.) Still another way to show the relationship is to use a split-bar chart, as in Figure 2.6(b).

Figure 2.5 Bar chart for Example 2.5

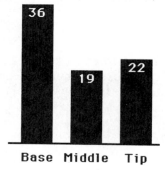

Figure 2.6 Back-to-back bars and split bars showing relationship between cause and location of fracture for Example 2.5

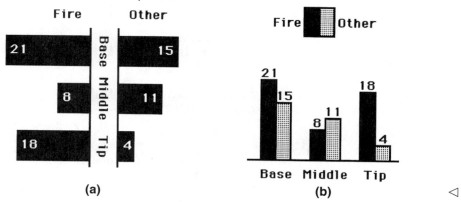

The next example is simpler than the earlier examples in that there are only two possible outcomes. Pictorial representations such as dot plots and bar charts have less benefit in this simple setting, but they still help to show the data.

EXAMPLE 2.6

▷ **Treating leukemia.** A study[8] reported in 1963 was designed to evaluate the effectiveness of a chemotherapeutic agent, 6-mercaptopurine (6-MP), for the treatment of acute leukemia. Such an evaluation requires a comparison group (see Chapter 3). Patients were randomized to therapy groups by coin tosses. The first patient was assigned to the 6-MP group if the coin came up heads and to the placebo group if it came up tails; the second patient received the other therapy. This process was repeated for the third and fourth patients, fifth and sixth patients, and so on. For each pair of patients, the investigators recorded whether the 6-MP patient or the placebo patient stayed in remission longer. There were 21 pairs of patients in the study. The results were as follows (where B means that the 6-MP patient fared **B**etter than the placebo patient in that pair and W means that the 6-MP patient fared **W**orse):

BWBBB   WBBBB   BBBWB   BBBBB   B

So 6-MP was more effective in 18 of the 21 pairs of patients. Thus, the proportion of pairs in which it was better is $18/21 = 85.7\%$. The dot plot and bar chart are shown in Figure 2.7.

This study will be analyzed in Chapter 6.

Figure 2.7 Dot plot and bar chart for Example 2.6

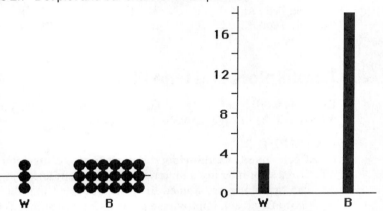

> **Use bar charts to show distributions on a small or moderate number of categories.**

If the categories are next to each other and there are many of them, then combining nearby observations into a histogram (Section 2.5) may be a better choice than a bar chart.

EXERCISES

2.4 Redo Exercise 2.1 (Darwin's pollination study) but make a bar chart instead of a dot plot. Also, use the rounded-off differences given here instead of the differences you used in the earlier exercise.

Pair	1	2	3	4	5	6	7	8	9	10	11	12	13	14	15
Diff.	49	−67	8	16	6	23	28	41	14	29	56	24	75	60	−48
Rnd. Diff.	50	−70	10	20	10	20	30	40	10	30	60	20	80	60	−50

2.5 Redo Exercise 2.2 (miles walked) but use a bar chart instead of a dot plot. Also, use the mileage differences given here, which are rounded off from the earlier example.

0	+1	+3	−3	+8	+1	−1	+3
−1	+11	+12	0	+2	−3	+2	−1
+6	+12	+5	+2	+2	−7	−2	+1
−1	+2	+9	+7	−1	+3	0	+5

2.6 Reconsider Exercise 2.5. Now consider only whether there was an improvement (+1) or not (−1), and include 0's with the second category. Make a bar chart of the resulting data:

```
−1  +1  +1  −1  +1  +1  −1  +1
−1  +1  +1  −1  +1  −1  +1  −1
+1  +1  +1  +1  +1  −1  −1  +1
−1  +1  +1  +1  −1  +1  −1  +1
```

2.7 Redo Exercise 2.3 (pressures to remove cannulae) but use bar charts instead of dot plots. Present them in such a way that they can be easily compared. Also, use the pressures given here, which are rounded off from the earlier exercise.

```
Cannula Type A:  .3  .3  .3  .4  .4  .5  .5  .6  .7  .8  .8  .8  1.0  1.3  1.4  1.5
Cannula Type B:  .0  .1  .2  .2  .3  .3  .3  .4  .4  .4  .4  .5  .5   .6   .6   .7
```

2.3 Scatterplots and Line Plots

Dot plots work nicely in two dimensions. In the next example there are two measurements, and the linkage between them is of interest.

EXAMPLE 2.7

▷ **Effectiveness of amiloride; dot plots in two dimensions.** A study[9] evaluated the effectiveness of inhaling a diuretic drug called amiloride in patients with cystic fibrosis. The experimenters wanted to know whether using the drug in an aerosol spray would be effective in improving a patient's lung function, as compared with an aerosol spray lacking the drug. The 14 patients in the study used an aerosol spray four times a day for two 6-month periods; in one period the spray contained the drug and in the other period it did not ("vehicle only"). Half the patients were on the drug in the first period and the other half were on the drug in the second. At the end of each 6-month period, each patient's lung function was ascertained by measuring "forced vital capacity." The measurements are given in Table 2.2. The two dot plots in Figure 2.8 show the two sets of measurements separately.

Table 2.2
Data for Example 2.7: Forced vital capacity (in ml)

Patient Number	Vehicle Only	Amiloride	Patient Number	Vehicle Only	Amiloride
1	2,925	2,760	8	4,108	3,880
2	4,190	4,490	9	2,646	2,732
3	5,067	5,617	10	3,635	3,758
4	2,588	2,543	11	2,890	2,960
5	3,934	3,810	12	3,125	3,387
6	3,952	3,985	13	3,805	4,048
7	2,547	2,392	14	1,741	1,787

Much of the story in the data does not come through when comparing the two dot plots because it is impossible to tell which dot in one plot goes with which dot in the other. Thus, a critical aspect of the design and of the resulting data is lost. A way to make the connection is simply to connect the dots across the two plots, as shown in the "connected-dot" plot in Figure 2.9. The connecting lines are nearly vertical because the two measurements—off and on drug—are close to each other.

Figure 2.8 Dot plots for the data of Example 2.7

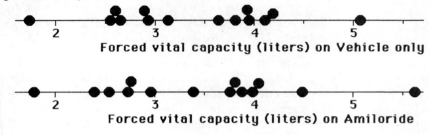

Figure 2.9 Connected-dot plot for the data of Example 2.7

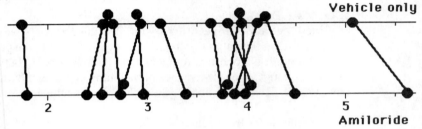

The picture with connected dots tells the whole story, but it is difficult to read. An alternative is a **scatterplot.** This is a dot plot in two dimensions—one response on amiloride and the other response on vehicle only. The scatterplot for the data of this example is shown in Figure 2.10. The dotted 45° line indicates where the two measurements are equal.

Figure 2.10 Scatterplot for Example 2.7

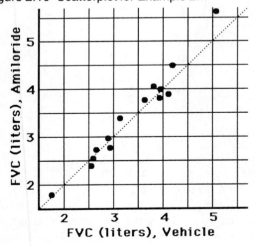

The scatterplot makes it clear that there is a lot of variability in the data. However, the data points hug the 45° line. This is because the two measurements are taken on the same patient and a patient's forced vital capacity tends to be similar in different periods. An advantage of a paired analysis is described in Section 3.5. ◁

> **Use scatterplots to show the relationship between two measurements.**

In the scatterplot of the previous example, none of the 14 dots lies on top of another dot. Should replicates occur it is not clear how to handle them. The next section deals with several alternatives.

There are various modifications of scatterplots that can be helpful, depending on the problem. The next example contains a typical scatterplot. But enhancing it using a *third* measurement shows an interesting feature.

EXAMPLE 2.8
▷ **Blood factor vs temperature by lot.** During cardiac bypass surgery, a mechanical pump circulates the patient's blood. Over a period of time the pump can have an impact on the blood. It is important to understand the relationship between a particular blood factor that gauges the performance of the pump and other characteristics such as blood temperature. The scatterplot in Figure 2.11 shows temperature and blood factor for 64 pumps.

The plot is far from conclusive, but it suggests that temperature has little effect on blood factor. If anything, the effect seems negative in the sense that higher temperatures are associated with smaller blood factor. Also, there is a hint of more variability in blood factor when the temperature is lower.

Figure 2.11 Scatterplot for Example 2.8

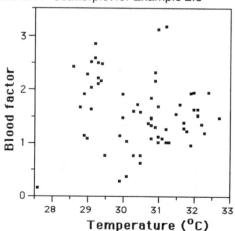

There are other characteristics of the data points. One is the lot of pump manufacture. Pumps in the same lot may behave similarly. The 64 pumps were selected from four lots. The scatterplot in Figure 2.12 is identical to that of Figure 2.11, except that in place of dots I have used different symbols for each of the four lots: ■, X, Y, Z. This second plot shows that the pumps cluster with other pumps in the same lot. The lot with the symbol ■ ran at a higher temperature and had less variability in blood factor—possibly because of temperature, but also possibly because of other characteristics of the lot.

Figure 2.12 Scatterplot for Example 2.8, with symbols indicating lots

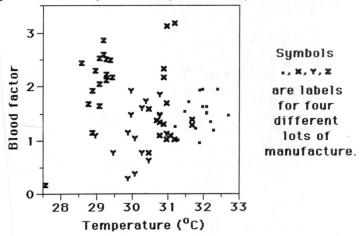

One can draw various conclusions from the scatterplot when lot is taken into account. One is that temperature may have an effect after all—a positive effect! Here is why. Focus only within a lot: first lot 1, then lot 2, and so on. You will see that, within each lot, higher temperatures seem to be associated with higher blood factor. This effect is not dramatic, but there is at least some evidence in its favor. And it would be missed completely without considering lot. ◁

In the next example the scatterplot conveys an important message from the data.

EXAMPLE 2.9
▷ **Forecasting tornadoes.** Meteorologists measure various characteristics of a storm to predict whether it will spawn tornadoes.[10] Two of them are **umax,** the rate of outflow (meters per second) from the storm's anvil, and **mda,** the maximum deviation angle (degrees) of the anvil to the ambient wind. These measurements were made for 175 storms and are shown in the scatterplot in Figure 2.13[11] (page 22). Each dot represents a storm and its position shows the values of the two measurements for that storm. There are three types of dots corresponding to three types of storms: those that produced no tornadoes, those that produced a weak tornado, and those that produced a strong tornado.

The scatterplot provides a neat method of showing the clustering and orderly arrangement of these three types of storms. Storms with large values of both umax and mda are more likely to spawn tornadoes. It is sometimes helpful to reduce several

Figure 2.13 Scatterplot for Example 2.9 showing three types of storms

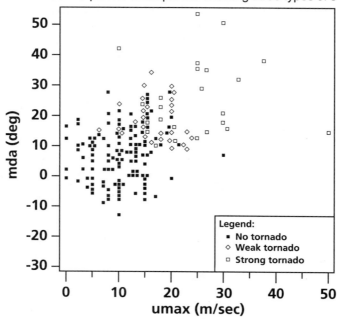

measurements into one, thus lowering the problem's dimensionality. While there are more sophisticated and better ways of combining these two measurements into one measurement for the purposes of predicting tornadoes,[11] let's simply add umax and mda together. (The first is measured in meters per second and the second is in degrees, so adding them gives a quantity without units.) The result is quite remarkable. The dot plots (with dots stacked to form bar charts) in Figure 2.14 show the distribution of umax + mda for the three types of storms. Since all 80 storms having umax + mda no bigger than 20 were tornadoless, it seems safe not to forecast a tornado should the next storm meet this condition. Similarly, forecasting a tornado—a strong one—for umax + mda greater than 50 is clearly in order. Whether one should forecast a tornado if umax + mda = 25, say, depends on the consequences of incorrectly forecasting one or of incorrectly forecasting that there will not be one.

Figure 2.14 Dot plots of umax + mda for Example 2.9 for three types of storms

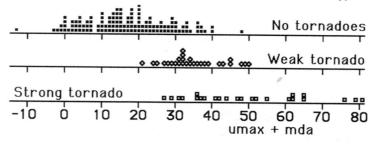

Sec. 2.3 / Scatterplots and Line Plots

> **Use different symbols or markers in a scatterplot to show other dimensions or aspects of the data.**

The next example shows ways of dressing up a scatterplot to address additional issues or to make a point more convincingly.

EXAMPLE 2.10
▷ **Home vs away performance in basketball.** Most sports fans are aware of the *home-team advantage,* in which teams are more likely to win a home game than an away game. This effect is especially pronounced in basketball. Three Duke students[12] collected the data from the 1992 National Basketball Association (NBA) season shown in Table 2.3. Home and away records are given as wins–losses.

For data such as these, it is best to compare proportions. But in this case all the denominators are equal (41). So plotting number of wins is the same as plotting proportion of wins—in the sense that one is just a change of scale from the other. Numbers of wins for home and away games are plotted in Figure 2.15 (page 24). Points on the diagonal line are those in which home wins equal away wins.

Several modifications of this basic plot are also presented. The first modification (Figure 2.16) shows that points below the diagonal—in the shaded region—have fewer away wins than home wins. This is an example in which a conclusion stands out clearly from a picture: There are good basketball teams and not-so-good basketball teams, but all win more games at home. The home-team advantage applies to all teams, good and not-so-good.

Table 2.3
Data for Example 2.10: Numbers of wins–losses by home and away, 1992 NBA season

Team	Home	Away	Team	Home	Away
Atlanta	23–18	15–26	Milwaukee	25–16	6–35
Boston	34–7	17–24	Minnesota	9–32	6–35
Charlotte	22–19	9–32	New Jersey	25–16	15–26
Chicago	36–5	31–10	New York	30–11	21–20
Cleveland	36–5	22–19	Orlando	13–28	8–33
Dallas	15–26	7–34	Philadelphia	23–18	12–29
Denver	18–23	6–35	Phoenix	36–5	17–24
Detroit	25–16	23–18	Portland	33–8	24–17
Golden State	31–10	24–17	Sacramento	21–20	8–33
Houston	28–13	14–27	San Antonio	31–10	16–25
Indiana	26–15	14–27	Seattle	28–13	19–22
LA Clippers	29–12	16–25	Utah	37–4	18–23
LA Lakers	24–17	19–22	Washington	14–27	11–30
Miami	28–13	10–31			

24　　　Ch. 2 / Displaying and Summarizing Data

Figure 2.15 Scatterplot for Example 2.10

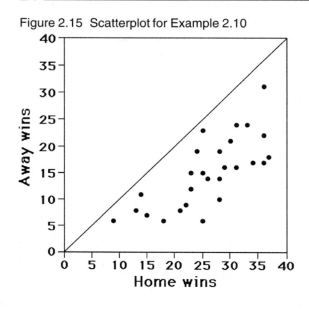

Figure 2.16 Scatterplot for Example 2.10, with "Away < Home" region shaded

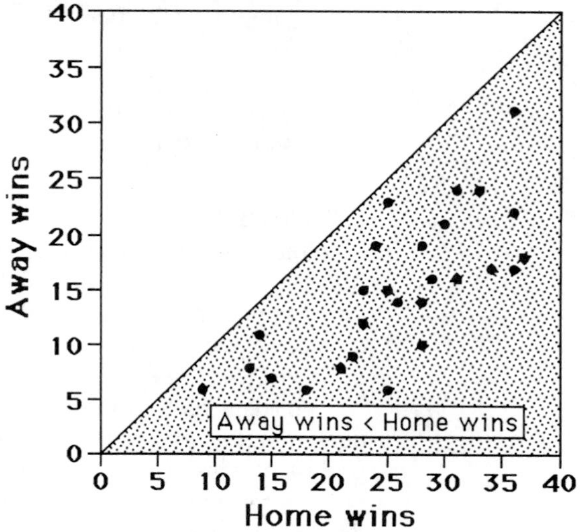

The second modification (Figure 2.17) shows a line corresponding to away wins equal to 60% of home wins. New Jersey falls on this line exactly since 60% of 25 home wins equals 15, their number of away wins. This line straddles the data points, with about half falling on each side. The third modification (Figure 2.18) shows the points

Figure 2.17 Scatterplot for Example 2.10, with 60% line added

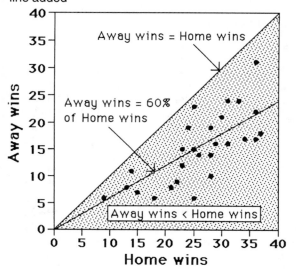

Figure 2.18 Scatterplot for Example 2.10, with points labeled

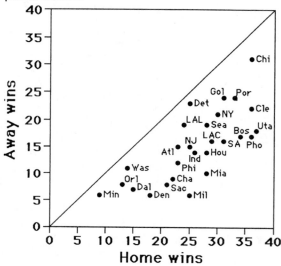

labeled by the corresponding team city. The fourth (Figure 2.19) adds a dimension showing total wins—the sum of home and away wins. These can be read from the boxes labeling the diagonals. For example, Milwaukee had 25 home wins and 6 away wins for a total of 31. The dot for Milwaukee lies just above the diagonal line labeled $\boxed{30}$. Milwaukee's total number of wins is the same as Charlotte's (22+9), and so they lie in the same diagonal position.

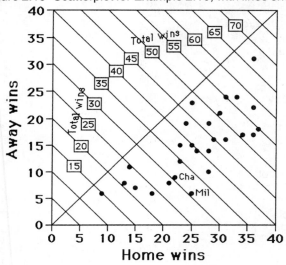

Figure 2.19 Scatterplot for Example 2.10, with lines showing total number of wins

EXAMPLE 2.11

▷ **In baseball, do you get what you pay for?** Sports salaries in the United States have increased to very high levels. Three Duke students[13] were interested in whether team owners who pay large salaries are getting their money's worth relative to other team owners. Table 2.4 (page 26) gives major league baseball teams' payrolls (in $100,000s) and win percentages (number of games won divided by number of games played) for the years 1988 and 1992. We will consider only the latter year in this example.

The scatterplot for 1992 payroll and win percentage is shown in Figure 2.20. [Interesting pattern! All data displays have their own quirks and regularities. But this one looks especially regular. It is easy to see how the ancients were able to perceive forms of animals in the (essentially random) arrangements of the stars.] The issue of whether there is a relationship here seems pretty clear: There is not. Some teams that pay a lot get a lot, but others get little—whether because they are lucky, or skillful in deciding which players to hire. The scatterplot in Figure 2.21 shows the points with team labels. You may notice that California and New York teams tend to have high payrolls—but some win and some do not.

Table 2.4
Data for Example 2.11

Team	1988 Payroll[a] ($100,000s)	Win %	1992 Payroll ($100,000s)	Win %
Atlanta	118	.338	330	.605
Baltimore	135	.335	210	.549
Boston	139	.546	422	.451
California	119	.463	335	.444
Chicago Cubs	131	.475	294	.481
Chicago White Sox	62	.441	256	.531
Cincinnati	89	.540	352	.556
Cleveland	79	.461	81	.469
Detroit	129	.543	256	.463
Houston	123	.506	134	.500
Kansas City	141	.532	318	.444
Los Angeles	169	.584	438	.389
Milwaukee	85	.537	303	.568
Minnesota	125	.562	274	.556
Montreal	97	.500	159	.537
New York Mets	153	.625	445	.444
New York Yankees	194	.528	345	.469
Oakland	101	.642	397	.593
Philadelphia	138	.424	238	.432
Pittsburgh	60	.531	326	.593
St Louis	129	.469	266	.512
San Diego	93	.516	274	.506
San Francisco	124	.512	325	.444
Seattle	73	.402	222	.395
Texas	54	.435	282	.475
Toronto	121	.537	427	.593

[a] From *USA Today,* April 2, 1993.

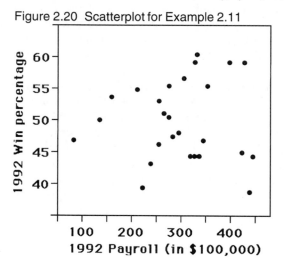

Figure 2.20 Scatterplot for Example 2.11

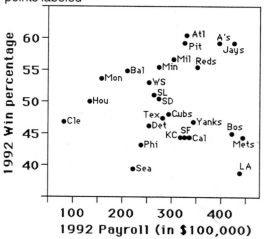

Figure 2.21 Scatterplot for Example 2.11, with points labeled

Earlier in this section I suggested scatterplots as alternatives to connected-dot plots. Scatterplots are easier to work with and to understand, and they are more versatile. But the reverse is true when there are more than two dimensions. The next example shows six connected-dot plots or **line plots.** While showing the connections is somewhat cumbersome, a six-dimensional scatterplot is out of the question.

EXAMPLE 2.12
▷ **Digoxin levels over time—Line plots.** Many people who suffer from congestive heart failure take the drug digoxin (digitalis). If they also suffer from ventricular arrhythmia, they may be taking other drugs. A concern is that a second drug will change a patient's metabolism of digoxin or otherwise affect the amount of digoxin in the patient's blood plasma. If the second drug decreases the amount of digoxin in the plasma, then the patient may not receive the full benefit of the digoxin; if it increases the digoxin level, then the patient may suffer ill effects from this potent drug—perhaps even death.

Investigators set up a study[14] to examine any effect of an experimental antiarrhythmic drug on digoxin in the plasma. Fifteen patients took digoxin for several weeks and the experimental drug for part of the time. Periodically (approximately every 4 days), investigators measured the patients' digoxin levels in the plasma. The results are shown in Table 2.5. On days A and B and again on days E and F, the patients had been taking only digoxin. On days C and D (shown in boldface type), they had also been taking the experimental drug. The question was whether the plasma digoxin levels were changed on days C and D as compared with the other days, thus implicating

Table 2.5
Digoxin levels over time for Example 2.12

Patient Number	Plasma Digoxin (ng/ml/100)					
	Day A	Day B	**Day C**	**Day D**	Day E	Day F
1	34	40	**56**	**45**	121	34
2	74	15	**44**	**49**	38	37
3	31	49	**52**	**58**	41	23
4	40	22	**38**	**25**	29	32
5	143	33	**57**	**51**	49	46
6	54	58	**59**	**43**	41	35
7	31	48	**65**	**53**	44	30
8	52	61	**70**	**62**	66	38
9	51	63	**76**	**50**	32	16
10	21	36	**56**	**40**	39	38
11	42	38	**68**	**52**	55	51
12	20	30	**36**	**40**	31	40
13	51	47	**47**	**38**	46	31
14	27	28	**61**	**67**	92	75
15	52	43	**73**	**56**	56	44

the experimental drug. If so, patients taking the experimental drug may have to have their digoxin dose modified.

If there had been only two days of testing—say, days A and C—we could make a scatterplot and draw conclusions. But the investigators wisely tested over more days and, in particular, the two days *after* the patients had stopped taking the experimental drug. Any display that shows only part of the information is flawed. Line plots do capture all the information in the table, as shown in Figure 2.22. In the figure, the data have been split into two parts, with the first eight patients' levels in the left-hand plot and the last seven patients' levels in the right-hand plot. The days are ordered from A to F and each patient's levels are connected across days.

The plots are difficult to read even with the split, but the important points come through. Every patient's level increased between days B and C, but the increases were not dramatic. There is a lot of within-patient variability as well as between-patient variability. The levels for patient 5 on day A and patient 1 on day E are out of line with the other levels in the plots. These patients' levels on other days are quite ordinary. Their presence is challenging to any analysis. (In Example 13.1 we will return to these data and reconsider these two interesting points. But I want to indicate something that should *not* be done in analysis. A standard procedure in some statistical computer packages is to discard unusual observations. Don't! They sometimes contain more information than the rest of the observations combined. In this particular study, discarding these two outlying points gives a false notion of inherent variability in the data collection and measurement processes.)

Figure 2.22 Line plots for Example 2.12; labels are patient numbers

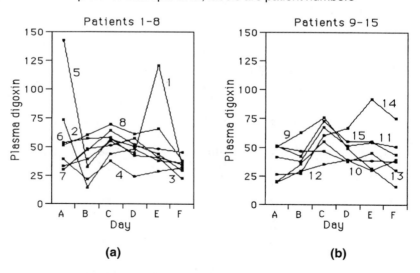

(a)　　　　　　　　　　　(b)

Use line plots to connect individual sets of observations over time.

Sec. 2.3 / Scatterplots and Line Plots   29

EXERCISES

Use graph paper for scatterplots; make your own if you do not have any.

2.8 Exercise 2.1 considered an experiment designed by Darwin to learn whether cross- or self-pollinated plants matured sooner. His observations are repeated here. Make a scatterplot of heights on cross- (horizontal axis) and self-pollinated plants (vertical axis).

Pair	1	2	3	4	5	6	7	8	9	10	11	12	13	14	15
Cross	188	96	168	176	153	172	177	163	146	173	186	168	177	184	96
Self	139	163	160	160	147	149	149	122	132	144	130	144	102	124	144

2.9 In half of the cases in Exercise 2.3, there was another measure of retention of the cannula. For these 16 hearts, both measures are listed here as meas1 and meas2. Make a scatterplot of these data, putting meas1 on the horizontal axis.

meas1	meas2	meas1	meas2	meas1	meas2	meas1	meas2
.04	.06	.25	.21	.40	.10	.47	.58
.14	.00	.30	.47	.40	.39	.52	.67
.19	.22	.32	.56	.43	.60	.59	.45
.21	.17	.32	1.04	.44	1.17	.95	1.34

2.10 The table given here gives home and away win–loss records for 1992 major league baseball teams. Make a scatterplot of number of away wins (vertical axis) vs number of home wins (horizontal axis). Label the points on your plot by the team name. Point out on your plot the team(s) with the most total wins and the team(s) with the least total wins.

Numbers of wins–losses by home and away for 1992 baseball season

Team	Home	Away	Team	Home	Away
Atlanta	51–30	47–34	Minnesota	48–33	42–39
Baltimore	43–38	46–35	Montreal	43–38	44–37
Boston	44–37	29–52	New York Mets	41–40	31–50
California	41–40	31–50	New York Yankees	41–40	35–46
Chicago Cubs	43–38	35–46	Oakland	51–30	45–36
Chicago White Sox	50–32	36–44	Philadelphia	41–40	29–52
Cincinnati	53–28	37–44	Pittsburgh	53–28	43–38
Cleveland	41–40	35–46	St. Louis	45–36	38–43
Detroit	38–42	37–45	San Diego	45–36	37–44
Houston	47–34	34–47	San Francisco	42–39	30–51
Kansas City	44–37	28–53	Seattle	38–43	26–55
Los Angeles	37–44	26–55	Texas	36–45	41–40
Milwaukee	53–28	39–42	Toronto	53–28	43–38

2.11 The pressures generated by two heart pumps similar to those described in Example 2.8 were checked on numerous runs and under various conditions. Both diastolic and systolic pressures were measured (in mmHg). The data are shown in the table, ordered by diastolic reading within each pump type. Make a scatterplot of these data—only *one* scatterplot. In place of a dot, use an "A" for pump A and a "B" for pump B. Plot diastolic readings on the horizontal axis and systolic on the vertical axis. (If your plot is very crowded, you may have to shift the labels a bit to fit them in.)

Pump A								Pump B					
Dia	Sys	Dia	Sys	Dia	Sys	Dia	Sys	Dia	Sys	Dia	Sys	Dia	Sys
6	17	15	45	23	64	44	116	1	26	17	54	44	109
7	24	16	45	23	59	44	82	2	29	18	57	45	118
9	32	16	29	24	56	45	107	3	26	20	53	54	130
9	28	16	52	30	86	53	120	4	40	26	84	60	107
10	38	17	42	30	70	55	136	8	32	29	84	63	140
11	34	17	52	31	83	59	140	11	50	30	77	64	123
11	36	18	55	31	71	62	115	12	43	31	80		
12	30	19	64	32	85	65	102	14	49	31	80		
12	20	19	54	33	79	66	120	15	53	32	91		
13	27	21	60	39	93	69	123	16	52	35	91		

2.12 Refer to Example 2.11 on page 25. Consider the 1988 season and plot win percentage (vertical axis) by payroll (horizontal axis). Label the data points using team names (see Figure 2.21). (As was true in 1992, there appears to be little relationship between money spent and winning percentage. If you combine your picture with that in Example 2.11, you will see an interesting characteristic: There was more variability in payroll in 1992 but less variability in winning percentage in 1992.)

2.13 Refer to Example 2.11 on page 25. Teams that were good in 1988 might be good in 1992. Address this issue by making a scatterplot of the winning percentages in the two years (1988 on horizontal axis). Label the data points using team names.

(You will see little relationship between winning percentages in the two years. Either personnel changes that take place over a short period of time really matter or the element of chance dominates winning at the major league level. In my view, the latter is more important. An old saw says baseball is 90% pitching—with the balanced competition of the major leagues, winning is 90% luck!)

2.14 An experiment[15] was conducted to assess the amount of ascorbic acid (AA, mg/100 g dry weight) in five varieties of lima beans. Five observations for each variety were taken at different stages of maturity. The percentage of dry matter (DM) was measured as an index of maturity. Results for the five varieties are shown in the table given here.

Variety 1		Variety 2		Variety 3		Variety 4		Variety 5	
DM	AA	DM	AA	DM	AA	DM	AA	DM	AA
34.0	93.0	39.6	47.3	31.7	81.4	34.5	61.5	31.4	80.5
33.4	94.8	39.8	51.5	30.1	109.0	31.5	83.4	30.5	106.5
34.7	91.7	51.2	33.3	33.8	71.6	31.1	93.9	34.6	76.7
38.9	80.8	52.0	27.2	39.6	57.5	36.1	69.0	30.9	91.8
36.1	80.2	56.2	20.6	47.8	30.1	38.5	46.9	36.8	68.2

(a) Plot all 25 points with ascorbic acid on the horizontal axis and dry matter on the vertical axis. Use the treatment number to label the location of the point. (Since there are five points per treatment, your plot will contain five each of the numbers 1, 2, 3, 4, and 5.) The data points fall roughly along a curve, with some variation among treatments.

(b) One of the treatments seems to have less variability in both dry matter and ascorbic acid than do the others. Which treatment is it?

(c) One of the treatments seems to have consistently more dry matter for a given amount of ascorbic acid than do the others. Which treatment is it? (This question would be difficult to answer without a plot.)

2.15 The following table gives measurements of plasma citrate concentrations of 10 subjects taken at five times during the day (units not given).[16] Make a line plot with time of day on the horizontal scale and a line for each subject. Your plot will show (at least) one point that seems unusual. Note which one it is and why it is unusual. (There is no single correct answer to this question.)

Subject	8 A.M.	11 A.M.	2 P.M.	5 P.M.	8 P.M.
1	93	121	112	117	121
2	116	135	114	98	135
3	125	137	119	105	102
4	144	173	148	124	122
5	105	119	125	91	116
6	109	83	109	80	104
7	89	95	88	91	116
8	116	128	127	107	119
9	151	149	141	126	138
10	137	139	125	109	107

2.4 Number Plots, Area Plots, and Star Plots*

When points lie on top of each other, showing a single point in that place misrepresents the data. There are several remedies. One is to stack the points atop each other as was done in the dot plots of Example 2.2 on page 11. This remedy does not work when there are thousands of points. Still another is to show the multiplicity by plotting the number of points that occur in that position, as in the next example.

EXAMPLE 2.13
▷ **Alcohol vs caffeine consumption (revisited)—Number plots and area plots.** In Examples 2.2 and 2.4 we plotted caffeine consumption for each category of alcohol consumption. There are up to five points lying on top of each other; when trying to plot nearby points, the duplication is even greater. A remedy is to make a **number plot** by putting a number showing the number of points at that location, as in Figure 2.23 (page 32). I have interchanged the axes from the figures in the earlier examples. This figure represents category 1–2 as 1.5, 3–5 as 4, and 6⁺ as 8. This more accurately reflects distances on the drinks-per-sitting scale; the 3–5 category is shown further from the 1–2 category than from the 6⁺ category.

Another type of data display is an **area plot:** Show the points using dots, as usual, but make the dots bigger when there are more data points at that position. Make the areas of dots proportional to the multiplicities, as shown in Figure 2.24.

*Optional section; not used in the sequel.

Figure 2.23 Number plot for Example 2.13 showing numbers of observations

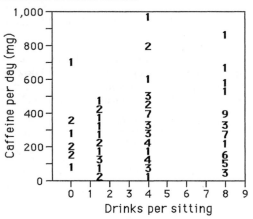

Figure 2.24 Area plot for Example 2.13; areas of dots equal number of observations

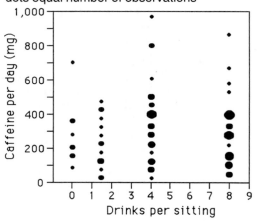

Still another way to represent data multiplicities is to show a ray emanating from a point for each repeat of the point. This is called a **star plot,** as shown in the next example.

EXAMPLE 2.14

▷ **Alcohol vs caffeine consumption (revisited)—Star plots.** Figure 2.25 shows a star plot for the data in Example 2.2. It shows a ray (❘) instead of a dot for a single observation. If there are two observations occurring at the same point, then a second ray is added to the first (—). Also, ⋎ (and rotations such as ⊱) means three observations at that location, and ✶ means five.

Figure 2.25 Star plot for Example 2.14; rays indicate multiplicities

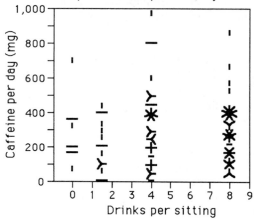

The eye sees area. The fundamental tenet—here, as generally—is to represent more points with more area. And area should be proportional to multiplicity. So, for

example, the amount of blackness in ✶ is about five times that in !. This is not the case for number plots because, for example, the area occupied by the numeral 9 is not three times that of the numeral 3. ◁

> **Use number plots, area plots, and star plots to show multiple observations at points in a two-dimensional grid when the number of points is not very large.**

EXERCISE

2.16* The table shows the data from Exercise 2.11 (page 29), but with pressure rounded off to the nearest 10 (I rounded numbers smaller than 10 up to 10). Make a star plot of these data, ignoring the matter of pump lot—simply regard these as 66 data points, all of which contribute equally to the star plot.

Pump A								Pump B							
Dia	Sys	Dia	Sys	Dia	Sys	Dia	Sys	Dia	Sys	Dia	Sys	Dia	Sys
10	20	20	50	20	60	40	120	10	30	20	50	40	110
10	20	20	50	20	60	40	80	10	30	20	60	50	120
10	30	20	30	20	60	50	110	10	30	20	50	50	130
10	30	20	50	30	90	50	120	10	40	30	80	60	110
10	40	20	40	30	70	60	140	10	30	30	80	60	140
10	30	20	50	30	80	60	140	10	50	30	80	60	120
10	40	20	60	30	70	60	120	10	40	30	80		
10	30	20	60	30	90	70	100	10	50	30	80		
10	20	20	50	30	80	70	120	20	50	20	90		
10	30	20	60	40	90	70	120	20	50	40	90		

Stem-and-Leaf Diagrams and Data Histograms

Dot plots, bar charts, and star plots are useful as pictorial representations of data. For some purposes, it is better to see the actual data. Stem-and-leaf diagrams convey the data, but they do it in such a way that they preserve some of the visual perspective of bar charts and dot plots.

EXAMPLE 2.15
▷ **Weight gains of piglets.** One hundred young pigs were weighed when they reached a certain age; then they were fed a particular diet and weighed again 20 days

---

*Exercise for optional section.

later.[17] The following table is a complete list of their weight gains (in pounds). The second table gives the gains from smallest to largest.

Weight gains of 100 piglets for Example 2.15

33	25	22	34	40	42	44	46	32	20	31	24	24	23	30	12	39	22	31	37
21	28	38	18	15	31	3	27	33	30	35	26	43	7	29	43	25	19	42	28
30	17	47	21	33	36	16	49	57	28	41	53	41	29	27	33	27	24	30	38
35	30	31	21	36	20	29	42	39	18	26	34	13	18	40	41	32	30	26	34
33	30	39	26	48	45	23	30	30	30	17	35	25	37	11	14	19	19	36	29

Reordered weight gains of 100 piglets for Example 2.15

3	7	11	12	13	14	15	16	17	17	18	18	18	19	19	19	20	20	21	21
21	22	22	23	23	24	24	24	25	25	25	26	26	26	26	27	27	27	28	28
28	29	29	29	29	30	30	30	30	30	30	30	30	30	30	31	31	31	31	32
32	33	33	33	33	33	34	34	34	35	35	35	36	36	36	37	37	38	38	39
39	39	40	40	41	41	41	42	42	42	43	43	44	45	46	47	48	49	53	57

A **stem-and-leaf diagram** is an alternative to dot plots for showing where the data are concentrated and how they are spread out. In the diagram that follows, the *stem* of a weight gain is its tens digit and the *leaf* is its units digit. There is one leaf in the diagram for each weight gain—100 leaves in all—and all leaves for weight gains with the same tens digit grow on the same stem. For example, a weight gain of 31 has a stem of 3 and a leaf of 1; since there are four 31's among the 100 weight gains, the diagram shows four 1's at stem 3. The two 2's to their right correspond to the two 32's in the sample, and so on.

Stem-and-leaf diagram for piglet weight gains of Example 2.15

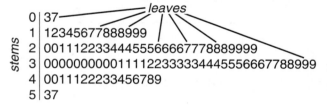

```
       0 | 37
       1 | 12345677888999
stems  2 | 00111223344455566667778889999
       3 | 000000000011112233334445556667788999
       4 | 0011122233456789
       5 | 37
```

There are variations of stem-and-leaf diagrams that may be used to make the diagram more readable. One such variation has two stems for each tens digit, as in the next diagram. If the unit digit is 0, 1, 2, 3, or 4, then the leaf is attached to the first stem and if the unit digit is 5, 6, 7, 8, or 9, then the leaf is attached to the second stem. This variation better shows the shape of the distribution of data.

Both diagrams show that the data are concentrated near 30 lb. For example, about 65% of the data lie between 20 and 40 lb, and almost 90% lie between 15 and 45 lb. Also, the data are reasonably symmetric about 30 lb; that is, each data point tends to be matched by another data point that is about the same distance from 30, but on the other side of 30.

Stretched-out stem-and-leaf diagram for Example 2.15

```
0 | 3
0 | 7
1 | 1234
1 | 5677888999
2 | 001112233444
2 | 55566667778889999
3 | 000000000011112233333444
3 | 5556667788999
4 | 00111222334
4 | 56789
5 | 3
5 | 7
```

Using too few or too many stems may not adequately show the distribution in the data. As a rule of thumb, use between 5 and 20 stems. The next example shows that you may have to drop digits to achieve this. Another aspect of the example that is different from the previous one is that there are several stem-and-leaf diagrams.

EXAMPLE 2.16

▷ **Alcohol vs caffeine consumption (revisited).** In Examples 2.2, 2.4, 2.13, and 2.14, we considered caffeine consumption by categories of number of alcoholic drinks per sitting. Table 2.6 shows four different stem-and-leaf diagrams, one for each drinking category. If the first two digits of caffeine consumption were used as stems, there would be nearly 100 stems. This is too many. Instead, I have rounded off the caffeine consumption to the tens digits, using them as leaves and the hundreds digits as the stems.

Table 2.6
Stem-and-leaf diagrams; caffeine by number of drinks for Example 2.16

0 Drinks	1–2 Drinks	3–5 Drinks	6+ Drinks
0 | 7	0 | 0079	0 | 0777	0 | 777
1 | 88	1 | 07	1 | 000017888	1 | 00001788
2 | 11	2 | 11169	2 | 111668	2 | 11366666667
3 | 277	3 | 2	3 | 799	3 | 22377777999
4 |	4 | 114	4 | 0111448	4 | 0
5 |	5 |	5 | 01	5 | 37
6 |	6 |	6 | 0	6 | 7
7 | 0	7 |	7 |	7 |
8 |	8 |	8 | 11	8 | 6
9 |	9 |	9 | 8	9 |

Use stem-and-leaf diagrams for small and moderate samples to show individual observations and their location in the sample distribution.

A **data histogram** is easy to relate to a stem-and-leaf diagram. It is also closely related to a bar chart. It loses information as compared to both of these, but it is not constrained by the size of the sample—it works well even for millions of data points.

EXAMPLE 2.17
▷ **Weight gains of piglets (revisited).** Example 2.15 contains a stem-and-leaf diagram (see page 35) of weight gains of piglets. This is repeated in Figure 2.26(a), with lines drawn around those leaves having the same stem. These lines form bars, better seen by erasing the leaves, as shown in Figure 2.26(b). The bar heights are the frequencies of intervals of values and not of single points. Such a special type of bar chart is called a **histogram.** An option is how to label the heights of the bars; one possibility is as the number of observations in each interval [Figure 2.27(a)] and the other is as the proportion of the observations [Figure 2.27(b)]. The sum of the heights of the bars

Figure 2.26 Constructing a histogram for Example 2.17

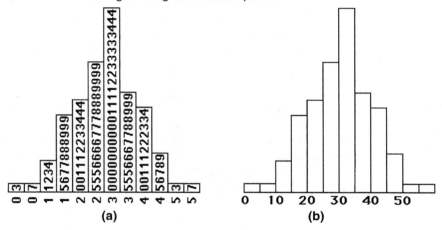

Figure 2.27 Histogram for Example 2.17, with bars shaded and heights labeled

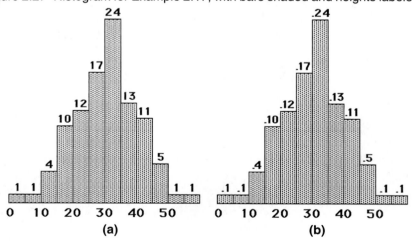

of the former is 100, the total number of observations, and the sum of the latter heights is 1. Shading the bars is a cosmetic touch. ◁

A histogram is just the shell of a stem-and-leaf diagram. As such, it is a reduction of the information available. In the previous example, with the stem-and-leaf in hand, you know that there are exactly four piglets that gained 31 lb; but with only the histogram, you do not know if *any* piglets gained 31 lb, or if 10 did. You do know that 24 piglets gained 30–34 lb.

A histogram is simply a special type of bar graph in which the categories are intervals, as shown in Example 2.17. Even though the data points included in the categories are not all the same (30 lb is different from 34 lb), they are not very different in that the distances between them are small.

It is possible (and usually easier) to construct a histogram without first making a stem-and-leaf diagram. Simply decide on the intervals—making all intervals the same width—and count the number of data points in each. The next example demonstrates this method.

> **Use histograms for any size sample to show the sample distribution of a numerical measurement.**

EXAMPLE 2.18
▷ **Mothers' smoking habits and birth weights.** Information was collected on 72 mothers enrolled in a health maintenance organization (HMO) and their newborn children.[18] Two pieces of information obtained were the child's birth weight (in lb) and whether the mother smoked. Table 2.7 gives this information.

Table 2.7
Birth weights of children by mother's smoking category in Example 2.18

Mother smokes			Quit	Mother never smoked			
4.5	6.6	7.2	5.4	3.3	6.6	7.3	7.9	9.2
5.4	6.6	7.5	6.6	5.3	6.6	7.4	8.3	9.2
5.6	6.6	7.6	6.8	5.6	6.6	7.4	8.3	10.9
5.9	6.9	7.6	6.8	5.6	6.7	7.6	8.3	
6.0	6.9	7.8	6.9	5.6	6.7	7.8	8.4	
6.1	7.1	8.0	7.2	6.1	6.9	7.8	8.5	
6.4	7.1	9.9	7.3	6.1	7.1	7.8	8.6	
			7.4	6.1	7.1	7.8	8.6	
				6.5	7.1	7.8	8.6	
				6.6	7.1	7.9	8.8	

We will make two histograms—one combining mothers who smoke with those who once smoked but have quit, and the other for those who never smoked. For ease in comparing the two, we will use the same intervals and put them back to back. (In terms of the orientation of the histograms of the previous example, think of turning them 90° to the left, laying one on top of the other, and then flipping the top one over so the two are joined at their bases.) The smallest birth weight is 3.3 lb and the largest is 10.9 lb. If we have a category for each number of pounds from 3 through 10, then there are eight categories. Having "nice" intervals—instead of intervals such as 3.3 to 3.8, 3.9 to 4.4, and so on—makes life easier.

Consider the nonsmokers. Of the 43 mothers who did not smoke, one's baby weighed between 10.0 and 10.9 lb, two between 9.0 and 9.9, nine between 8.0 and 8.9, and so on. The next table completes this list and also gives the corresponding list for smokers ("mother smokes" and "quit" combined).

Smoked	Category	Never Smoked
0	10.0 to 10.9	1
1	9.0 to 9.9	2
1	8.0 to 8.9	9
10	7.0 to 7.9	15
12	6.0 to 6.9	11
4	5.0 to 5.9	4
1	4.0 to 4.9	0
0	3.0 to 3.9	1

The ranges of the categories in the previous table correspond to a particular stem-and-leaf diagram. You will get a different histogram if you use different categories—say, 3.3 to 3.8, 3.9 to 4.4, and so on.

All that remains is to construct bars with these heights, and we can follow the scheme of the table. Figure 2.28 gives the histograms, where the order of categories is the same as in the table. (You can choose between having large or small numbers at the top; I chose large.) The bars have been shaded. The second set of histograms (Figure 2.29) are versions of the first, with counts labeled on the bars. Some of the counts are inside the bars and some are outside. This alternative is more cluttered, but it properly conveys the sample sizes—the first version could apply to 72,000 births as well as to 72.

Figure 2.28 Histograms for data of Example 2.18

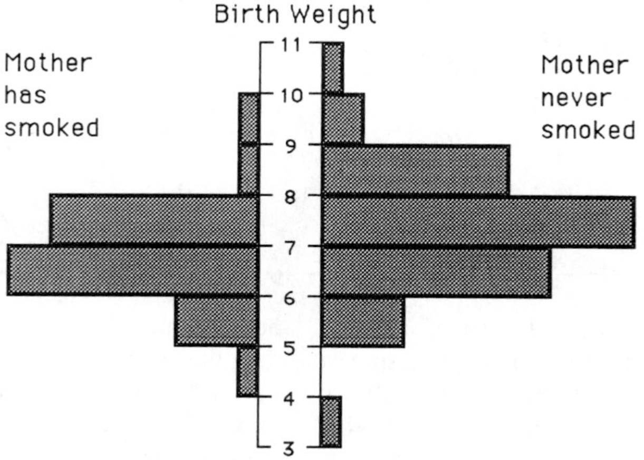

Figure 2.29 Alternative histograms for data of Example 2.18; counts added

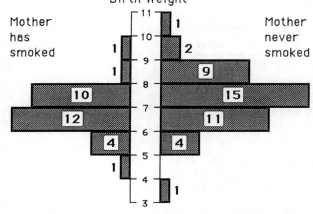

The histograms suggest that birth weights tend to be greater in the "never smoked" group. In later chapters we'll address the strength of this suggestion. ◁

> **Show histograms back-to-back to compare the two sample distributions.**

EXERCISES

2.17 Strengths of spot welds vary even though they are made by the same welding tool. The following are the strengths (in psi) of a sample of 50 welds:[19]

400	395	398	421	445	389	372	408	398	401
399	386	423	364	394	414	390	412	398	363
388	431	392	438	411	399	399	408	390	420
400	389	430	426	388	406	431	411	404	424
450	416	397	404	388	405	392	405	379	419

(a) Make a stem-and-leaf diagram of these data using the first two digits as stems (that is, use stems 36, 37, ..., 45).

(b) Make a histogram using 360–369, 370–379, up to 450–459 as intervals of spot-weld strengths.

2.18 The data given in Exercise 2.2 on page 13 are repeated here, but rounded to tenths:

0.0	+.6	+3.3	−2.6	+8.4	+1.1	−1.3	+3.2
−.6	+10.8	+11.7	−.1	+1.7	−3.4	+1.7	−1.4
+6.0	+12.2	+5.0	+1.7	+2.3	−6.6	−2.1	+.8
−1.0	+1.7	+8.9	+7.5	−.6	+2.9	0.0	+5.4

(a) Make a stem-and-leaf diagram using miles as stems: −6, −5, −4, −3, −2, −1, 0, +1, ..., +12.

(b) Make a histogram of these data; start from −7 and make your intervals 2.5 miles wide. (The leftmost interval goes from −7 to −4.5, the next goes from −4.5 to −2.0, and so on.)

2.19 Make stem-and-leaf diagrams of the diastolic readings for each of the pumps in Exercise 2.11; these are repeated here. Use the same scheme for the two diagrams so they can be compared easily.

Diastolic Pressures for Pump A				Diastolic Pressures for Pump B		
6	15	23	44	1	17	44
7	16	23	44	2	18	45
9	16	24	45	3	20	54
9	16	30	53	4	26	60
10	17	30	55	8	29	63
11	17	31	59	11	30	64
11	18	31	62	12	31	
12	19	32	65	14	31	
12	19	33	66	15	32	
13	21	39	69	16	35	

2.20 Make histograms of the systolic readings for each of the two pumps in Exercise 2.11; these are repeated here. Use the same scale and interval scheme for both histograms so they can be compared easily. Use at least 10, but no more than 15, intervals.

Systolic Pressures for Pump A				Systolic Pressures for Pump B		
17	45	64	116	26	54	109
24	45	59	82	29	57	118
32	29	56	107	26	53	130
28	52	86	120	40	84	107
38	42	70	136	32	84	140
34	52	83	140	50	77	123
36	55	71	115	43	80	
30	64	85	102	49	80	
20	54	79	120	53	91	
27	60	93	123	52	91	

2.6 Means, Standard Deviations, and Medians

The earlier sections of this chapter deal only with pictorial displays. Displays are powerful communication tools and good displays may not require further exposition. But in extrapolating from a sample to a population, it sometimes helps to reduce the data into forms more manageable than displays. In particular, it can be helpful to use one or two numbers to represent an entire sample—a number calculated from a sample is called a **statistic.** This is the approach of Chapters 10–13. One commonly used statistic is the **sample average**—also called the **sample mean.** This is distinct from the *population mean* to be introduced in Chapter 10. The word *average* means common, typical, or middle-sized. An average is a center of the sample.

A second statistic of importance in Chapters 10–13 is the **sample standard de-**

viation. This is a typical deviation from the mean and is a measure of spread or variability in the data.

The usual notation for a sample mean is $\bar{x}$ and for a standard deviation is s. These notations are so pervasive that they appear on buttons of many scientific calculators. Whether you have calculators or not you should know how to calculate them. The next example shows how.

EXAMPLE 2.19

▷ **Increases in corn yield.** Fourteen fields that were planted with corn in Iowa were each divided into two strips; one strip in each field was sprayed for corn borers and the other was not. The increase in yields of corn (in tenths of bushels per acre) for the sprayed over the unsprayed strips are as follows:[20]

$-57 \quad 37 \quad 64 \quad 15 \quad 43 \quad 48 \quad 33 \quad 36 \quad 5 \quad 50 \quad 240 \quad 88 \quad 45 \quad 11$

The sample mean is their simple arithmetic average: First add them up and then divide by the sample size, 14. Their sum is

$-57 + 37 + 64 + 15 + 43 + 48 + 33 + 36 + 5 + 50 + 240 + 88 + 45 + 11 = 658$

So the sample mean is

$$\bar{x} = \frac{658}{14} = 47$$

The standard deviation is a little more complicated. There are two methods for calculating s. The first is more instructive and the second is easier to use. The first method is **SSADR**: **S**ubtract, **S**quare, **A**dd, **D**ivide, square **R**oot. First, we **S**ubtract $\bar{x} = 47$ from each observation to get these 14 *deviations from the mean:*

$$
\begin{array}{ll}
-57 - 47 = -104 & 36 - 47 = -11 \\
37 - 47 = -10 & 5 - 47 = -42 \\
64 - 47 = +17 & 50 - 47 = +3 \\
15 - 47 = -32 & 240 - 47 = +193 \\
43 - 47 = -4 & 88 - 47 = +41 \\
48 - 47 = +1 & 45 - 47 = -2 \\
33 - 47 = -14 & 11 - 47 = -36 \\
\end{array}
$$

The average of these 14 deviations is 0, as the average deviation always must be. The 14 original data points are shown in the dot plot in Figure 2.30 (page 42), along with a dot plot showing the deviations from the mean. The two pictures are identical except that the scale is shifted to the right by $\bar{x} = 47$ units in the second one.

The next step in calculating the standard deviation is to **S**quare the deviations:

$$
\begin{array}{ll}
(-104)^2 = 10{,}816 & (-11)^2 = 121 \\
(-10)^2 = 100 & (-42)^2 = 1{,}764 \\
(+17)^2 = 289 & (+3)^2 = 9 \\
(-32)^2 = 1{,}024 & (+193)^2 = 37{,}249 \\
(-4)^2 = 16 & (+41)^2 = 1{,}681 \\
(+1)^2 = 1 & (-2)^2 = 4 \\
(-14)^2 = 196 & (-36)^2 = 1{,}296 \\
\end{array}
$$

Figure 2.30 Dot plots for Example 2.19

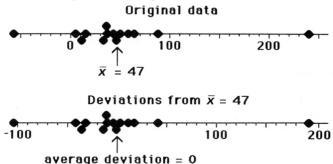

Next **A**dd these squared deviations:

$$10{,}816 + 100 + 289 + 1{,}024 + 16 + 1 + 196 \\ + 121 + 1{,}764 + 9 + 37{,}249 + 1{,}681 + 4 + 1{,}296 = 54{,}566$$

To find s^2, **D**ivide by the number of observations:

$$s^2 = \frac{54{,}566}{14} = 3{,}897.6$$

Finally, the square **R**oot of s^2 is the sample standard deviation:

$$s = \sqrt{\frac{54{,}566}{14}} = 62.43 \quad \text{or about } 62$$

A standard deviation is a "typical" deviation. Some deviations are bigger than s and some are smaller. The dot plot in Figure 2.31 shows that, for the data at hand, all but the two extreme observations (the biggest and the smallest) have deviations smaller than s.

Figure 2.31 Dot plot for Example 2.19 indicating $\pm s$

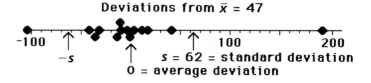

The second method of calculating the sample standard deviation is **SADSR**: **S**quare, **A**dd, **D**ivide, **S**ubtract, square **R**oot. First **S**quare the observations (without subtracting $\bar{x}$ from each), then **A**dd the squares:

$$(-57)^2 + (37)^2 + (64)^2 + (15)^2 + (43)^2 + (48)^2 + (33)^2 \\ + (36)^2 + (5)^2 + (50)^2 + (240)^2 + (88)^2 + (45)^2 + (11)^2 = 85{,}492$$

Sec. 2.6 / Means, Standard Deviations, and Medians

Divide this sum of squares by the number of observations and then **S**ubtract the square of the sample mean $\bar{x}$, giving s^2:

$$s^2 = \frac{85{,}492}{14} - 47^2 = \frac{54{,}566}{14} = 3{,}897.6$$

This agrees with the value of s^2 we got using the first method. So taking the square **R**oot of s^2 again gives

$$s = \sqrt{\frac{54{,}566}{14}} = 62.43$$

Seeing the relationship between $\bar{x}$ (and also s) and the sample distribution is easier for larger samples, as in the next example.

EXAMPLE 2.20

▷ **Weight gains of piglets (revisited).** Examples 2.15 and 2.17 consider the weight gains of 100 piglets. Calculating $\bar{x}$ is easy for this sample:

$$\bar{x} = \frac{3 + 7 + 11 + 12 + \cdots + 57}{100} = \frac{3{,}000}{100} = 30$$

The mean is the "center of gravity" of the sample distribution. So 30 is the balance point for the histogram of these data shown in Figure 2.32.

I will spare you the tedium in calculating s—it turns out to be 10.[21] One standard deviation to the left of $\bar{x}$ is 20 and one standard deviation to the right is 40. These three points are labeled on the histogram.

Figure 2.32 The value $\bar{x}$ balances the histogram for Example 2.20; split histogram shows $\bar{x} - s$ and $\bar{x} + s$ balancing each half

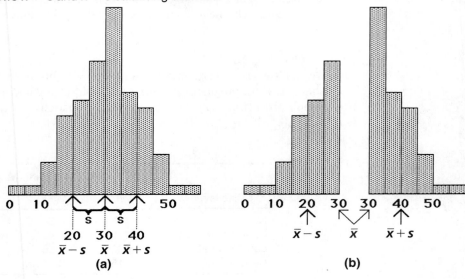

To show an interpretation of s, I have split the histogram at $\bar{x}$, as shown in Figure 2.32(b). The points $\bar{x} - s$ and $\bar{x} + s$ roughly balance these two halves. ◁

> **A sample mean, $\bar{x}$, is a center of a distribution of sample values.**

> **A sample standard deviation, s, is a typical deviation from the sample mean.**

EXAMPLE 2.21

▷ The following list gives volumes of 17 head and neck cancers (in cm^3):

| 15.75 | 11.25 | 1.12 | 3.38 | 4.50 | .80 | 1.42 | 22.00 | .06 |
| 1.12 | .03 | .42 | .28 | 16.00 | 2.50 | 11.81 | 1.66 | |

Summing these 17 values gives 94.10. Dividing by 17 gives $\bar{x} = 5.54$ (two-decimal accuracy).

Using the **SADSR** method for calculating the standard deviation, first square the observations:

| 248.06 | 126.56 | 1.25 | 11.42 | 20.25 | .64 | 2.02 | 484.00 | 0.00 |
| 1.25 | 0.00 | .18 | .08 | 256.00 | 6.25 | 139.48 | 2.76 | |

Adding up these 17 numbers gives 1,300.21. Dividing by 17 gives 76.48. The square of $\bar{x}$ is $5.54^2 = 30.64$. Subtracting this gives $76.48 - 30.64 = 45.84$. Taking the square root gives $s = 6.77$. ◁

Published reports seldom give the actual data but report sample means and standard deviations. The next example is typical. It also shows a common use of bar charts.

EXAMPLE 2.22

▷ **Mother's exercise level and child's birth weight.** A study[22] addressed the relationship between a pregnant woman's exercise level and her child's birth weight. Only the sample sizes (n), means ($\bar{x}$), and standard deviations (s) were published, as shown in Table 2.8.

Table 2.8

		Birth weight (g)	
Exercise level	n	$\bar{x}$	s
None	185	3,389.3	487.9
Changing pattern	213	3,451.4	502.4
Low/Moderate	49	3,554.4	382.1
Heavy	15	3,713.8	298.4

The category "changing pattern" means that women reported different levels of exercise in different trimesters.

Authors of research articles frequently use bar charts to show means as heights of bars, as in Figure 2.33. They are different from the bar charts discussed in Section 2.2 in which the heights of the bars were frequencies. The effect is to show in the vertical direction what we have previously shown in the horizontal direction. Such bar charts show only one aspect of a sample distribution (the mean) and so are not as informative as histograms, for example, which show entire distributions.

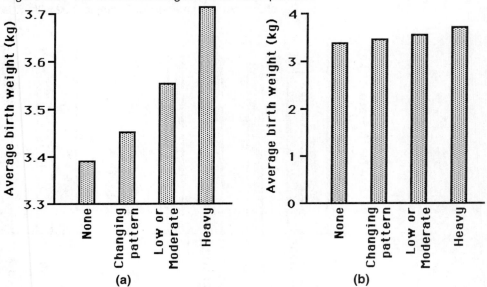

Figure 2.33 Bar charts showing means in Example 2.22

The bar chart in Figure 2.33(a) starts at 3.3 kg. The same amount (0 to 3.3) is cut off the bottom of each bar. Such bar charts are commonly used, and they *seem* reasonable—it seems wasteful to show parts of the bars where they are the same. But this practice exaggerates the differences among the heights of the bars. The first bar is only 35% as high as the third one, whereas the mean for the no-exercise group is actually more than 95% of the mean for the low-exercise group. The vertical scale tells the truth, but the viewer sees the bars first and the scale second—if at all! Not showing the 0 on the vertical scale can be deceptive. People who do so may be trying to influence the reader by exaggerating differences. Perhaps the author simply wants to convey things as being interesting. Differences are catchy; sameness is dull.

The bars in the chart of Figure 2.33(b) start at 0 and so this chart does not deceive from this perspective. But this chart has another type of flaw, one that is unavoidable when there is no natural ordering of the categories. Because the mean birth weight for the "changing pattern" group is between "none" and "low/moderate," showing it between them lends more credence to the importance of exercise. In terms of total effort or effort that counts toward lowering birth weights, I do not know that "changing pattern" is between "none" and "low/moderate." Consider the bar chart in Figure 2.34(a). Any relationship between mother's exercise level and child's birth weight is less clear. If you arrange a set of groups so the bars go from low to high or high to low, the eye picks this up and associates a trend: There are several ways of displaying data and the displayer can choose the one that makes a point best. The problem is one of "multiplicities" (to be discussed in Section 3.4). There may be no way to tell a story—any story!—without deceiving. As a reader, you have to be aware of the options available to authors and make suitable adjustments. When you are the author, be aware of the power of charts and graphs and of the inevitability of deception.

Figure 2.34 Bar charts showing means for Example 2.22 with order of groups changed

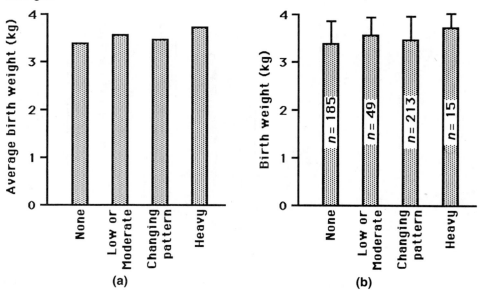

Another common practice is to show sample standard deviations on the tops of the bars using a T-shaped extension—see the bar chart in Figure 2.34(b). The height of the T is the size of the standard deviation in that group. So each bar with its T shows a second aspect of the corresponding distribution. In effect, this bar chart repeats the information given in Table 2.8 at the beginning of this example (including sample sizes). Such a picture conveys the variability in the sample. However, while showing standard deviations in addition to means is more informative than means alone, such charts are still less informative than histograms of birth weights within the exercise groups. ◁

Real estate groups report mean selling prices of homes by location. The calculation is the same as for $\bar{x}$—namely, add up the prices of homes sold and divide by the number sold. But these groups realize that this can be misleading. Suppose the prices (in thousands of dollars) of the last nine homes sold are as follows:

$$56 \quad 68 \quad 45 \quad 6{,}200 \quad 59 \quad 70 \quad 49 \quad 52 \quad 61$$

The mean is

$$\bar{x} = \frac{56 + 68 + 45 + 6{,}200 + 59 + 70 + 49 + 52 + 61}{9} = 740$$

So the mean selling price is $740,000. But this mean is dominated by a single home—or rather, an estate—selling for $6.2 million, and $740,000 is hardly a typical selling price. So real estate groups often give another statistic: the **median** selling price. This is the middle selling price. First, order the list from smallest to largest:

45 49 52 56 **59** 61 68 70 6,200

The median is the middle one in the list: $59,000, which is shown in boldface type. This is a far cry from the mean of $740,000.

Suppose a sample size, n, is even. Then the middle number is not unique. Suppose the next home sells for $68,000; then the list becomes:

45 49 52 56 **59** **61** 68 68 70 6,200

The median is defined to be the average of the two middle numbers: $(59 + 61)/2 = 60$.

A further division into fourths defines **quartiles.** The median divides the sample in two. Consider the two halves, dropping the median when n is odd. For the original sample the two halves are as given here:

45 **49** **52** 56   and   61 **68** **70** 6,200

The median of the smaller half is the **first quartile:** $(49 + 52)/2 = 50.5$. The median of the larger half is the **third quartile:** $(68 + 70)/2 = 69$. The quartiles (including the median which is also the second quartile) divides the sample into four parts:

45 49 **(50.5)** 52 56 **(59)** 61 68 **(69)** 70 6,200

For the above sample with an additional $68,000, the first quartile is 52, the median of the smaller half:

45 49 **52** 56 59

The third quartile is 68, the median of the larger half:

61 68 **68** 70 6,200

The median is little influenced by the 6.2-million-dollar home. For example, in the original sample, changing 6,200 to any selling price bigger than 59 leaves the median unchanged at 59. In this sense the median is a better measure of center. On the other hand, if one is interested in the tax base, then the sample mean $\bar{x}$ is better since $\bar{x}$ can be used to find the total selling price (multiply by n) while the median cannot.

EXAMPLE 2.23
▷ **Increases in corn yield (revisited).** Reorder the 14 increases in corn yield from Example 2.19:

−57 5 11 15 33 36 **37** **43** 45 48 50 64 88 240

The median is (37 + 43)/2 = 40. Compare this with $\bar{x} = 47$, calculated in the earlier example. Both values are shown on the dot plot in Figure 2.35. The first quartile is 15 and the third quartile is 50:

−57   5   11   **15**   33   36   37   43   45   48   **50**   64   88   240

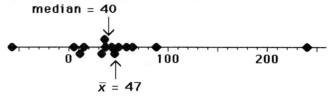

Figure 2.35 Median vs mean in Example 2.23

The median of a histogram divides the histogram into two equal parts: 50% of the observations and therefore 50% of the histogram's area are in each half. For example, in the piglet weight gains of Examples 2.15 and 2.20, the median is 30 (see the ordered data listing for Example 2.15). The mean and the median are equal, which is true when the histogram is symmetric about its midpoint (the midpoint being both $\bar{x}$ and the median). (A histogram is symmetric when looking at it backwards—such as by turning the page and holding it to the light—leaves it unchanged.)

> **A median is the middle number (in the ordered sample).**

The median is less susceptible to extreme observations (such as a 6.2-million-dollar house, or the corn yield of 240 in the previous example). In this sense, it is more *robust* than is the sample mean, in the sense of Chapter 13. Nevertheless, most statisticians prefer working with the mean and we will do so in this text.

The quartiles of a histogram divide it into four equal parts, with 25% of the area in each part. The first quartile has 25% of the area to its left and the third quartile has 25% of the area to its right.

> **Quartiles divide the ordered sample into four pieces; the first quartile has one quarter of the sample to its left and the third quartile has one quarter of the sample to its right.**

EXAMPLE 2.24
▷ **Molecular weights of fragments of DNA.** In some criminal and paternity cases, laboratories use measurements of molecular weights of DNA fragments from blood or semen. The histogram in Figure 2.36 shows measured molecular weights of a particular segment (locus D2S44) for 1,546 Caucasians.[23] Since each individual has two frag-

ments (one maternal and the other paternal), the sample size is $n = 3,092$, which is the sum of the frequencies shown on the tops of the bars in the histogram.

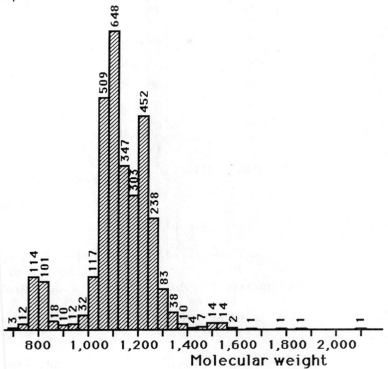

Figure 2.36 Histogram of molecular weights of DNA, locus D2S44 (Caucasians) for Example 2.24

The median has half the area—or equivalently, half the total frequency—to its left. Since $3,092/2 = 1,546$, we seek the molecular weight with this frequency to its left. Adding the frequencies to the left of 1,120 gives $3 + 12 + 114 + 101 + \cdots + 648$, or 1,576, which is 30 too much. Interpolate between molecular weights of 1,080 and 1,120. The total frequency in this interval is 648, about $648/40 = 16.2$ per unit of molecular weight. So the median is $30/16.2$ or about 2 units down from 1,120—that is, 1,118.

The first quartile has total frequency $3,092/4 = 773$ to its left. The frequency to the left of 1,040 is 419 and that to the left of 1,080 is 928. Interpolating between these two points gives the first quartile to be about 1,068. Similarly, the third quartile is about 1,208. These are shown on the version of the histogram in Figure 2.37 (page 50), where the four quarters of the data are shown using different shadings. The middle two quarters are narrower than the outer two because more than half the data is rather tightly packed near the median in comparison with the smallest and largest molecular weights.

Figure 2.37 Histogram for Example 2.24 showing quartiles with bars in different quarters shaded differently

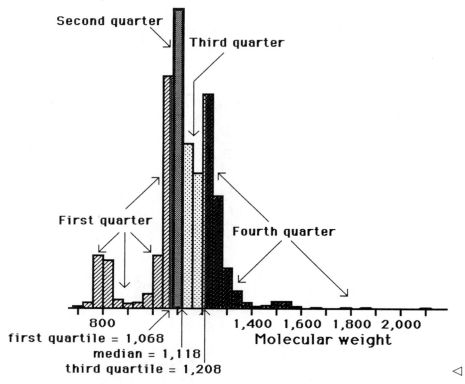

Just because you know how to calculate something does not mean you should do it. The following is an example in which the median and the mean are both irrelevant.

EXAMPLE 2.25
▷ **Chicago engineer test scores.** In 1966, 223 prospective Chicago city engineers took an examination to compete for 15 jobs.[25] The scores are given in Table 2.9 and presented in the histogram in Figure 2.38. What is the median and what is the sample mean? My answer is that they are irrelevant and the mere act of calculating them is therefore useless. What matters is the score required to get one of the 15 jobs. The histogram shows that the top score was 95 and there were two such scores. This was followed by 93 (4), 92 (3), 91 (3), and 90 (3). So a score of 90 was required to get a job.

The histogram shows something else, and shows it dramatically. There is a big gap of five possible scores between 84 and 90. The clear suggestion is that the test was fixed and the 15 winners had been picked in advance. Apparently, on the basis of the histogram alone, the examiners were accused of fraud for rigging the exam.

Sec. 2.6 / Means, Standard Deviations, and Medians

Table 2.9
Test scores of 223 applicants for 15 jobs; for Example 2.25

61	48	58	27	57	39	37	57	47	60	55	67	51	42	50
84	61	53	91	93	33	49	80	74	52	71	74	55	45	83
68	35	43	30	45	83	60	36	63	61	37	82	58	51	58
62	64	26	37	84	41	40	54	56	46	43	93	45	39	69
60	37	53	84	29	54	80	34	47	45	31	44	56	66	95
59	49	59	33	31	44	46	76	43	69	47	61	95	60	32
90	33	42	37	69	39	56	58	48	27	49	81	84	46	47
50	48	60	57	27	27	32	42	91	54	65	58	82	36	61
57	62	69	34	51	55	44	62	61	43	51	42	93	80	56
37	36	39	58	47	30	30	67	59	54	34	63	51	73	52
66	52	42	69	84	43	67	83	44	83	80	44	56	78	33
47	92	39	48	58	58	30	75	48	76	53	48	54	44	46
46	66	45	37	69	72	46	33	93	84	92	90	84	53	81
53	90	52	45	31	43	43	45	75	68	39	43	71	81	48
59	91	35	60	74	67	48	52	39	69	69	92	49		

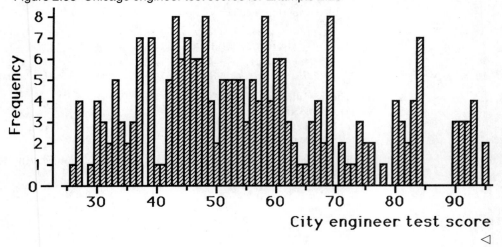

Figure 2.38 Chicago engineer test scores for Example 2.25

EXERCISES

2.21 Consider the viscosity of dimethylaniline at 20°C (in centipoises), as given in Example 2.1:

146   154   141   140   136   132   147   140   147   139   140   140

Calculate the following:
  (a) Sample mean
  (b) Sample standard deviation
  (c) Sample median
  (d) First and third quartiles

2.22 Exercise 2.1 on page 13 described an experiment that Darwin carried out to learn whether cross- or self-pollination would produce more vigorous seeds. The differences between the cross- and self-pollinated plants in each pot are repeated here:

Pair	1	2	3	4	5	6	7	8	9	10	11	12	13	14	15
Diff.	49	−67	8	16	6	23	28	41	14	29	56	24	75	60	−48

For these 15 differences, calculate the following:
(a) Sample mean
(b) Sample standard deviation
(c) Sample median
(d) First and third quartiles

2.23 A study[25] of the effects of dumping wastes into the sea gives the following measurements of zinc concentration (mg/kg) at 13 points within 1 mile of a dump site:

13.5   23.8   23.3   20.9   23.8   29.0   20.9   24.4   16.4   18.3   17.6   25.4   23.3

(a) Draw a histogram of these data, using categories 12–14, 14–16, and so on.
(b) Calculate the median.
(c) Calculate the sample mean and standard deviation.

2.24 Exercise 2.2 on page 13 gives the following differences in miles walked for patients while on a drug vs off a drug (repeated here).

```
  0.00    +.56    +3.27   −2.55   +8.42   +1.07   −1.31   +3.19
  −.59   +10.75  +11.73   −.05   +1.65   −3.42   +1.73   −1.44
 +6.04  +12.21   +4.97   +1.68   +2.28   −6.57   −2.11   +.75
  −.96   +1.68   +8.85   +7.45   −.59    +2.91   0.00   +5.40
```

Calculate the following:
(a) Sample mean
(b) Sample standard deviation
(c) Sample median

2.25 Exercise 2.5 on page 17 gives the rounded-off differences in miles walked that you considered in Exercise 2.24 (repeated here).

```
  0    +1    +3    −3    +8    +1    −1    +3
 −1   +11   +12     0    +2    −3    +2    −1
 +6   +12    +5    +2    +2    −7    −2    +1
 −1    +2    +9    +7    −1    +3     0    +5
```

Calculate the following:
(a) Sample mean
(b) Sample standard deviation
(c) Sample median
[Your answers should be close to those of the previous exercise.]

2.26 The strengths of 50 spot welds given in Exercise 2.17 are repeated here:

```
400   395   398   421   445   389   372   408   398   401
399   386   423   364   394   414   390   412   398   363
388   431   392   438   411   399   399   408   390   420
400   389   430   426   388   406   431   411   404   424
450   416   397   404   388   405   392   405   379   419
```

Sec. 2.6 / Means, Standard Deviations, and Medians

Calculate the following:
(a) Sample mean
(b) Sample standard deviation
(c) Sample median
(d) First and third quartiles

2.27 Exercise 2.3 gives the following pressures required to remove cannulae from animal hearts:

Cannula A: .30 .30 .32 .43 .44 .47 .52 .59 .70 .77 .79 .81 .95 1.33 1.43 1.54

Cannula B: .04 .14 .19 .21 .25 .26 .32 .35 .40 .40 .42 .53 .53 .56 .57 .68

For cannula A, calculate the following:
(a) Sample mean
(b) Sample standard deviation
(c) Sample median
(d) First and third quartiles

2.28 For cannula B in Exercise 2.27, calculate the following:
(a) Sample mean
(b) Sample standard deviation
(c) Sample median
(d) First and third quartiles

2.29 Example 2.2 presented daily caffeine intake for individuals, depending on alcohol consumption. The data are repeated here. Calculate the sample means, standard deviations, and medians of each of the four groups (with n's of 9, 15, 35, and 38).

Caffeine intake per day, listed by number of alcoholic drinks per sitting

0	1–2	3–5		6+	
68	0	0	210	505	68	260	390
180	0	68	260	598	68	260	390
180	68	68	260	805	68	260	390
210	93	68	278	810	98	260	398
210	98	98	373	975	98	260	525
323	165	98	390		98	260	570
368	210	98	390		98	270	665
368	210	98	398		113	323	860
698	210	113	405		165	323	
	260	165	405		180	328	
	293	180	405		180	373	
	323	180	435		210	373	
	405	180	435		210	373	
	405	210	480		225	373	
	435	210	495		260	373	

2.30 Example 2.11 presented payrolls and win percentages for various major league baseball teams. The payrolls (in $100,000s) for 1988 and 1992 are repeated at the top of page 54. Calculate the means, standard deviations, and medians for both years.

1988:
```
118   135   139   119   131    62    89    79   129   123   141   169    85
125    97   153   194   101   138    60   129    93   124    73    54   121
```

1992:
```
330   210   422   335   294   256   352    81   256   134   318   438   303
274   159   445   345   397   238   326   266   274   325   222   282   427
```

Appendix: Using Minitab for Displaying Data

Minitab is a system for performing statistical calculations. General information about starting Minitab and executing the programs on the disk is contained in the Appendix at the end of the text. In this section, we briefly introduce Minitab by illustrating some basic commands to graph and summarize data.

Minitab performs calculations on groups of numbers that are organized into a worksheet or spreadsheet. One uses Minitab by first placing data into the columns of the worksheet and then executing commands which perform calculations on the worksheet. In Release 10 of Minitab, one can execute commands either by typing commands in a "session window" or using menus. Here we illustrate the typing method since all of the programs included in this text use this method.

Suppose that you are interested in displaying and summarizing the viscosity data of Example 2.1. We first place the data into the worksheet. Columns of the worksheet are referred to by the symbols C1, C2, etc. and it is often helpful to give the columns meaningful names. We name the column C1 'dimeth' by typing the command

```
MTB > name c1 'dimeth'
```

We then place the 12 measurements into the column 'dimeth' using the set command:

```
MTB > set 'dimeth'
DATA> 146 154 141 140 136 132 147 140
DATA> 147 139 140 140
DATA> end
```

Note that entries are separated by one or more spaces. You can place the values on different lines and you indicate that you have finished entering data by typing 'end' at the last line. Now you can type commands which work on this column of data. You can construct a dot plot by the 'dotplot' command:

```
MTB > dotplot 'dimeth'
```

```
                         :
          .        . :  .       . :              .
       +---------+---------+---------+---------+---------+-------dimeth
     132.0     136.0     140.0     144.0     148.0     152.0
```

A histogram can be constructed using the command 'hist':

```
MTB > hist 'dimeth'
```

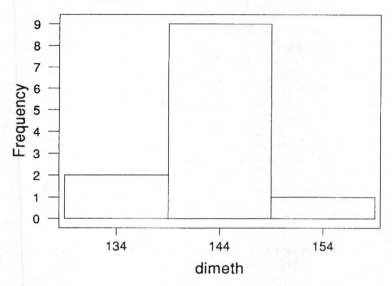

You can obtain some summary statistics of this data set by the 'describe' command:

```
MTB > describe 'dimeth'

                N      MEAN    MEDIAN    TRMEAN     STDEV    SEMEAN
dimeth         12    141.83    140.00    141.60      5.81      1.68

              MIN       MAX        Q1        Q3
dimeth     132.00    154.00    139.25    146.75
```

The variable N gives the number of observations and MEAN, MEDIAN, STDEV correspond to the sample mean, the sample median, and the sample standard deviation, respectively. MIN and MAX give the smallest and largest observations, and Q1 and Q3 denote the first and third quartiles. (Minitab uses a formula for Q1 and Q2 that is slightly different from the one given in Section 2.6.) The remaining summary numbers TRMEAN and SEMEAN will not be discussed in this text.

Minitab can also be used to understand the relationship between two measurements. To illustrate, consider the data in Example 2.2, which gives the caffeine consumption and number of drinks for 97 undergraduate students. We enter this data into two columns of the worksheet where a row corresponds to the caffeine amount and number of drinks for a particular student. Since we do not know the exact number of drinks for each student, we will represent the drink numbers for the categories using the numbers 0, 2, 5, and 6. We call column C3 'caffeine' and column C4 'drinks' by the 'name' command. We place the data into the two columns by two 'set' commands.

```
MTB > name c3 'caffeine' and c4 'drinks'

MTB > set 'caffeine'
DATA> 68 180 180 210 210 323 368 368 698
DATA> 0 0 68 93 98 165 210 210 210 260 293 323 405 405 435
DATA> 0 68 68 68 98 98 98 98 113 165 180 180 180 210 210
DATA> 210 260 260 278 373 390 390 398 405 405 405 435 435 480 495
DATA> 505 598 805 810 975
DATA> 68 68 68 98 98 98 98 113 165 180 180 210 210 225 260
DATA> 260 260 260 260 260 260 270 323 323 328 373 373 373 373 373
DATA> 390 390 390 398 525 570 665 860
DATA> end

MTB > set 'drinks'
DATA> 0 0 0 0 0 0 0 0 2 2 2 2 2 2 2 2 2 2 2 2 2 2 2
DATA> 5 5 5 5 5 5 5 5 5 5 5 5 5 5 5 5 5 5 5
DATA> 5 5 5 5 5 5 5 5 5 5 5 5 5 5
DATA> 6 6 6 6 6 6 6 6 6 6 6 6 6 6 6 6 6 6 6
DATA> 6 6 6 6 6 6 6 6 6 6 6 6 6 6 6 6
DATA> end
```

There are different Minitab commands that can be used to help understand the relationship between alcohol and caffeine consumption. One can construct a dot plot of the caffeine numbers for each drinking category. This is done on Minitab by the 'dot plot' command with the 'by' subcommand. (It is important to place a semicolon at the end of the first line, so Minitab expects the 'by' subcommand.)

```
MTB > dotplot 'caffeine';
SUBC> by 'drinks'.

drinks
0
                        .       : :     . :              .
         +---------+---------+---------+---------+---------+-------caffeine
drinks                  .
2
         : . :  . :  . ... :.
         +---------+---------+---------+---------+---------+-------caffeine
                                    :
drinks          . :    . .         :
  5        . : :. .: : :.      .: : ..:    .          ..         .
         +---------+---------+---------+---------+---------+-------caffeine
                           .
drinks                    :     .
  6            . :     . :  . ::
            : :. .: : :. : ::         . . .           .
         +---------+---------+---------+---------+---------+-------caffeine
         0        200       400       600       800      1000
```

A second graph that can be constructed is a scatterplot with caffeine consumption on the vertical axis and number of drinks on the horizontal axis. This graph is obtained by the Minitab 'plot' command. The accompanying scatterplot will appear in a separate graph window.

```
MTB > plot 'caffeine'*'drinks'
```

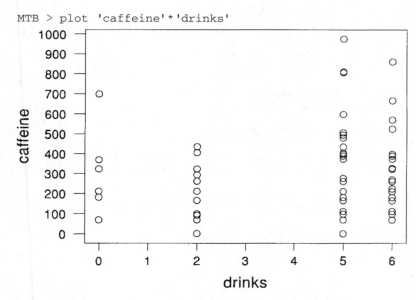

Although only a few basic commands have been illustrated here, Minitab is actually a very powerful program that can be used to perform sophisticated statistical analysis. One nice feature of the software is that you can write programs or macros to perform special types of calculations. The use of these Minitab macros will be illustrated throughout this text.

Chapter Notes

1. J. Albert, *Bayesian Computation Using Minitab* (Belmont, Calif.: Duxbury Press, 1996).

2. W. S. Cleveland, *The Elements of Graphing Data* (Pacific Grove, Calif.: Wadsworth Advanced Books, 1985).

3. S. A. Schmitt, *Measuring Uncertainty* (Reading, Mass.: Addison-Wesley, 1969), 199.

4. I am indebted to David Niccolini, Peter Pavlacka, and Nan Stillinger for these data.

5. C. Darwin, *The Effects of Cross- and Self-Fertilisation in the Vegetable Kingdom*, 2nd ed., John Murray (London, 1878), Ch. 1. See also G. A. Barnard, "Darwin's data on growth rate of plants," in *Data*, D. F. Andrews and A. M. Herzberg (New York: Springer-Verlag, 1979), 9–12.

6. D. Salsburg, unpublished manuscript, 1987. [Data referred to and analyzed further in D. A. Berry, *Statistical Methodology in the Pharmaceutical Sciences* (New York: Marcel Dekker, 1989), 40–44.]

7. J. L. Hofman. "Eva projectile point breakage at Cave Spring: Pattern recognition and interpretive possibilities," *Midcontinental Journal of Archaeology* 11 (1986): 79–95.

8. E. J. Freireich et al., *Blood* 21 (1963), 699–716.

9. M. R. Knowles et al., *New England Journal of Medicine* 322 (1990): 1189–1194.

10. K. J. Schrab, C. E. Anderson, and J. F. Monahan. "Techniques used to identify tornado producing thunderstorms using geosynchronous satellite data." *Sixth Conference on Satellite Meteorology and Oceanography.* AMS (1992) 159–162.

11. J. F. Monahan, K. J. Schrab, and C. E. Anderson. "Statistical models for forecasting tornado intensity." *Statistical Sciences and Data Analysis: Proceedings of the Third Pacific Area Statistical Conference,* K. Matushita ed., 1993. I am indebted to John Monahan for providing the raw data.

12. I am indebted to John Adams, Julie Anderson, and David Becker for these data.

13. I am indebted to David McIntosh, Edward Marblestone, and Andrea Kirshenbaum for collecting these data.

14. D. A. Berry. "Logarithmic transformations in ANOVA," *Biometrics* 43 (1987): 439–456.

15. R. G. D. Steel and J. H. Torrie, *Principles and Procedures of Statistics: A Biometrical Approach,* 2nd ed. (New York: McGraw-Hill, 1980), 411–412.

16. C. E. Lunneborg, *Modeling Experimental and Observational Data* (Belmont, Calif.: Duxbury Press, 1994), 398.

17. G. W. Snedecor and W. G. Cochran, *Statistical Methods,* 6th ed. (Ames, Iowa: Iowa State University Press, 1967), 67.

18. J. L. Hodges, D. Krech, and R. S. Crutchfield, *Stat Lab* (New York: McGraw-Hill, 1975).

19. D. A. Berry and B. W. Lindgren, *Statistics: Theory and Methods* (Pacific Grove, Calif.: Brooks/Cole Publishing Co., 1990), 323.

20. G. W. Snedecor and W. G. Cochran, *Statistical Methods,* 6th ed. (Ames, Iowa: Iowa State University Press, 1967), 98.

21. In the reference for these data, the authors say they "slightly modified" the experimental data so the sample mean would be exactly 30 and sample standard deviation, s, would be exactly 10.

22. M. C. Hatch, X-O. Shu, D. E. McLean, et al., "Maternal exercise during pregnancy, physical fitness, and fetal growth," *American Journal of Epidemiology* 137 (1993): 1105–1114.

23. I thank Lifecodes Corporation, Valhalla, NY, for providing these data.

24. K. Bemesderfer and J. D. May, *Social and Political Inquiry* (Belmont, Calif.: Duxbury Press, 1972). Discussed by D. Draper, D. P. Gaver, Jr., P. K. Goel, et al., *Combining Information* (Washington, D.C.: National Academy Press, 1992), 109–112.

25. K. J. Borwell, D. M. G. Kingston, and J. Webster, "Sludge disposal at sea—the Lothian experience," *Water Pollution Control* 85 (1986): 269–276.

3 Designing Experiments

CHAPTER 2 focused on displaying and summarizing data. Many of the examples consider the results of experiments. Broadly speaking, an experiment is doing something and observing the result. The objective is to learn about some aspect of the world. Some experiments involve multiple observations. The simplest observation is made on an **experimental unit.** Examples of experimental units are plots of land, patients, petri dishes, households, ants, engines, and so on. As examples and exercises in this chapter will show, identifying the experimental unit is important in drawing conclusions from an experiment and extrapolating to some larger population of units. Sometimes there is more than one type of experimental unit. For example, to judge customer satisfaction with a product, a company may select 10 grocery stores and survey 20 customers at each store. Responses may vary from store to store and they may vary from customer to customer within the stores. At one level, the experimental unit is the store, but at another level it is the customer.

Setting up an experiment that conveys information is the subject of this chapter. Designing experiments plays a fundamental role in the scientific method (see Chapter 1). The focus is on characteristics of good experiments. I will not address the costs and benefits indicated in step 3 of the scientific method. The approach of this text is ideally suited to these considerations, but the subject matter is large enough to comprise a separate course. The interested student is referred to the book by Clemen.[1] For an insightful and more detailed discussion of some of the topics addressed in this chapter and for related topics as well, see the book by Wilson.[2]

Before describing types of designs, I want to address the question of how closely the experimenter should follow the design. Designing an experiment and analyzing data from the experiment are not and cannot be kept distinct. Sometimes information becomes available during the course of an experiment that suggests modifying the original design. Changing a design can markedly affect the analysis of the results. Changes are consistent with the scientific method, but all aspects of the experiment's design must be recorded. Such a record includes the data collected before any design switch, as well as when and why the design was changed.

> **While carrying out any experiment, record any deviations from its original design.**

A related problem of utmost importance can occur in any study, however pure its design. Engineers, social scientists, physicians, and others have fallen prey to the **data-fixer problem**. It is easy to do. An observation turns out to be "bad." After counting bacteria, a laboratory technician finds evidence of dirt on the petri dish; a psychologist suspects that the subject who made outlandish responses was lying; a physician now believes that the patient who died during the procedure may have had a heart problem. Or, perhaps the observation was simply unexpected, with no explanation available. People do not like to tell messy stories and so they clean up their data before the telling. They may mean well, but their action is bad science. It is essential to report everything, including and especially peccadillos that the experimenter may not like to have others see. Such peccadillos can be more informative than the rest of the experiment. True, they may cast a shadow over the experiment, and that is the worry of the data fixer, but this should be addressed by the entire scientific community and not just by the experimenter. Darwin's cross- vs self-fertilization experiment[3] (see Exercise 2.1) is a good example. Darwin clearly believed that cross-fertilization would yield heartier plants. He was disappointed with the cross-fertilized plants that had "some larva gnawing at their roots." But he still reported *all* the data while offering his explanations.

Sometimes experimenters make a second measurement if the first one seemed strange. That is alright provided they record both. Unlike golfers who have to take a two-stroke penalty for dropping a second ball to improve their lie, scientists are lucky. They can "drop a second ball" without penalty—if they are open about why they did it and tell what happened to the first ball.

> **In carrying out any experiment, report every aspect of the results, including any data thought to be flawed.**

Some statistical methods (see Chapter 13) are aimed at minimizing the effect of unusual observations. These methods are especially relevant when you do not know whether the observations are defective or simply unusual.

3.1 Controlled Studies

Most scientific questions are relative, or comparative. "How big is it?" Compared with what? "How effective is it?" Compared with what? "How fast does it travel?" Compared with what? Most of the examples and exercises of Chapter 2 deal with comparisons. Examples 2.6, 2.8, and 2.15 deal with comparisons of two groups and others deal with comparisons of many groups or levels. The comparison in Example 2.11, for instance, is winning percentage for various payrolls.

For an example in which controls are necessary, consider a simple scientific hypothesis: Dark colors absorb more heat than do light colors. What experiment would you perform to check this hypothesis? You might paint a rock black, put it in the sun for an hour or so, and then take its temperature. Suppose it is 95°F. The statement "dark colors absorb heat" is comparative—dark colors are compared with colors that are not dark. It is not possible to check such statements unless you either (1) have an idea about how much heat light colors absorb or (2) include light colors in your experiment. Perhaps a white rock would also register 95°F.

A second color gives a frame of reference. Suppose you take two identical looking rocks, paint one black and the other white and place them both in the sun for an hour. Suppose the black rock's temperature is 95°F and the white rock's is 78°F. Even though it involves only two rocks, this is a very informative experiment because it allows for a comparison. If there were only a slight difference in temperatures, then we would have to increase the number of rocks (not a bad idea in any case) and perhaps also the types and sizes of rocks. In some problems, moderate to large sample sizes are essential for drawing conclusions.

A fundamental requirement in scientific investigations is the need for a **control**. The following examples show this.

EXAMPLE 3.1

▷ **Vitamin C and the common cold.** A study addressed whether vitamin C helps prevent the common cold. Of 208 university students who were especially susceptible to colds and took vitamin C regularly, 65.5% experienced a reduction as compared with their recollection of colds that they had had in the previous year. So vitamin C seems to be effective. But how would these students have fared had they not received vitamin C? The study also contained a *control* group of 155 students. Students in the study were assigned either vitamin C or a **placebo**, an inert tablet that had the same appearance and taste as the vitamin C tablets. (Supposedly, the assignment alternated between vitamin C and placebo—how it happened that 53 more students received vitamin C I do not know.) The study participants were *blinded* as to their assignment in the sense that no one knew which pill they were taking. The placebo group reported a 62.7% reduction in colds. According to E. B. Wilson,[4] "Presumably the reduction in both groups was a purely psychological effect," also called a **placebo effect.** (I doubt it. It was more likely the result of the subjects' poor recollection of the previous year's colds or the regression effect that you will learn about in Section 14.3.) ◁

EXAMPLE 3.2

▷ **Amiloride for cystic fibrosis (revisited).** Consider the data in Example 2.7. But suppose now that only the amiloride results were available (see Table 3.1 on page 62). Is amiloride effective? You cannot say and neither a dot diagram nor anything else will help. You have no point of reference. You need to know the forced vital capacities for these patients or for similar patients who had not received amiloride. You need *controls*.

Table 3.1
Forced vital capacity (FVC in ml) on amiloride for Example 3.2

Patient Number	FVC	Patient Number	FVC	Patient Number	FVC
1	2,760	6	3,985	11	2,960
2	4,490	7	2,392	12	3,387
3	5,617	8	3,880	13	4,048
4	2,543	9	2,732	14	1,787
5	3,810	10	3,758		

These two examples use different types of controls. In Example 3.1 a different set of subjects received a placebo than received vitamin C. This is a **parallel** design—see Section 3.5. In standard terminology, one population is the **treatment group** and the other is the **comparison group** or **control group.** The scientific goal is to compare treatment and control groups on the basis of samples from the corresponding population. In Examples 2.7 and 3.2, the patients themselves were the controls; that is, the comparison was of the *same* patients but during a period in which they received the vehicle only, instead of vehicle plus amiloride. Such observations are **paired**—see Section 3.5. The scientific question is the change or difference between the two measurements in a single population.

> **Controls are essential for interpreting results of an experiment.**

There are two meanings for the phrase *controlled study:*

1. A study containing a control group
2. A study in which the investigator exercises "control" by assigning treatment or control to experimental units in some fashion.

Both interpretations are legitimate. In the remainder of this text I will use *controlled study* in the former sense.

3.2 Randomized vs Observational Studies

There are two principal types of studies, randomized and observational. Randomization minimizes bias in treatment assignment. While observational studies are problematic, they are essential in science.

Randomized Studies

An important way to assign treatment is to **randomize.** In effect, the experimenter tosses a coin (or uses another type of randomization device) and assigns treatment if heads, and control if tails. A formal definition of *randomization device* involves the notion of probability and so will be delayed until Chapter 4. However, an intuitive understanding of the term is sufficient:

> **A randomization device is one for which the outcomes are symmetric in the sense that none seems more or less likely to result than the others.**

Randomization can be employed in various ways. The most common is in assigning experimental units (patients, plots, machines, etc) to treatment or control. Another is that both treatment and control are used on the same experimental unit, but at different times, with the order decided by a coin toss. In the amiloride study, Example 2.7, half the patients were administered amiloride first followed by vehicle only; for the other half, the order was reversed, with those receiving amiloride first decided randomly.

> **In a randomized study the researcher assigns treatment using a randomization device.**

Another important way to use randomization is to assign experimental units in pairs. The first unit is assigned treatment randomly and the second is assigned the other treatment (or control); the third is assigned randomly and the fourth is assigned the other treatment; and so on. This has the effect of balancing units over time: Whenever the experiment stops, the numbers of units in the two treatments will differ by no more than one. There is a disadvantage associated with the researcher knowing that even-numbered units will receive the opposite treatment from the previous (odd-numbered) unit—namely, the researcher may favor one of the treatments and may subconsciously bias the results by choosing "good" or "bad" experimental units accordingly.

A closely related method is assignment in blocks of four, or six, and so on. Within each block, half the units are assigned to one treatment and the other half are assigned to the other. This assignment is random. Consider blocks of four. For treatments A and B these are the six possibilities:

$$\begin{array}{cccccc} \text{AABB} & \text{ABAB} & \text{ABBA} & \text{BAAB} & \text{BABA} & \text{BBAA} \\ 1 & 2 & 3 & 4 & 5 & 6 \end{array}$$

These are selected randomly (by rolling a die, say) and placed in the resulting sequence. For example, suppose the die rolls are 4, 5, 4, 1, 5, ... The resulting assignment sequence is

$$\text{BAAB} \quad \text{BABA} \quad \text{BAAB} \quad \text{AABB} \quad \text{BABA} \quad \ldots$$

As in the case of assignment in pairs, this method tends to balance the two treatments over time.

Randomization in blocks is a special case of **stratification.** It balances treatment assignment for units next to each other in case they happen to be similar in some important respect. Other types of stratification are commonly used in experimentation to balance other patient characteristics. For example, in a clinical trial some patients may be sicker than others. We would like to ensure that the competing treatments are used equally for very sick patients as well as for patients who are relatively well off. The strata in question are the levels of sickness. Such balance can be achieved using separate randomization schemes—perhaps randomization in pairs—within each patient stratum. First decide which stratum a patient falls into and use the randomization scheme for that stratum.

Observational Studies

A study in which the investigator does not assign treatment or control to the experimental units is an **observational study.** The researcher notes a naturally occurring difference (for example, pregnant women who smoke or not), and observes a response or makes a measurement (birth weights of the children). In observational studies, treatment has a slightly different meaning from the usual since there is no active intervention in the "treatment" group. For example, it is a bit strained to call smokers the treatment group.

> **In an observational study, a researcher observes differences between two groups—treatment and control—but does not assign treatment.**

Observational studies can be of enormous value, especially when randomized studies do not exist and are impossible to carry out. But they are always difficult to interpret and evaluate. By their nature they come with built-in biases. The following examples describe some of them. The first two address the impossibility of attributing cause from an observational study.

EXAMPLE 3.3
▷ **"Pet a day keeps doctor away."** This headline appeared in the August 2, 1990, issue of the *Santa Barbara News-Press*. It was referring to a study[5] of 938 Medicare enrollees in a health maintenance organization. Participants were followed for 1 year and the frequency of doctor contacts noted. Of these, 345 owned pets and 593 did not. Those who owned pets had contact with doctors an average of 8.42 times while those without pets had an average of 9.49 contacts. The study's conclusion is that pet ownership has a moderating role in helping the elderly through times of stress.

I would very much like to believe that pets decrease doctor visits. Perhaps the effect is physical and not just psychological. People who own some pets—dogs, for example—usually get exercise walking the pet. Such exercise may be healthy, both physically and psychologically.

Unfortunately, this study does not convince me. People who are healthy or who have a healthy outlook may be more predisposed to owning a pet: Perhaps the good health caused the pet rather than the pet causing the good health. The researcher controlled for differences in health among patients at the start of the study. But health is impossible to measure exactly. (Indeed, owning a pet may be a better indication of overall health than any other known measure!) Also, the researcher did not try to control for attitude. People who are disinclined to care for a pet may be more likely to visit a doctor. And providing elderly people with pets may in fact be detrimental to their health. ◁

The next example is similar to the previous one in suggesting that attributing cause is often impossible in an observational study. But it also addresses problems present when interpreting survey data.

EXAMPLE 3.4
▷ **Marriage, sex, and mental health.** In a health examination survey, 636 women and 642 men aged 25 to 34 were asked: "Have you ever felt you were going to have a nervous breakdown?" A statistics text[6] gives the responses to the survey (Table 3.2) and, as an exercise, asks students to address for men and women separately whether the population proportions of "yes" answers depend on marital status. The authors conclude that it does not for men, but "For the women, marital status does seem to be related to mental health, and the married ones seem to be under much more stress." Also, responses to other survey questions regarding mental health were consistent with the answers to this question. Interestingly, married women's responses were similar whether or not they had children.

The differences are striking—but they are difficult to interpret. Many readers will attribute cause: Being married drives women crazy but not so for men! Could be. But marriage has not been assigned. Married people are self-selected. It is possible that women (but not men) who are prone to nervous breakdowns work harder at finding a mate. As in the previous example, we cannot tell which is the cause and which is the effect.

Table 3.2
Data for Example 3.4: Felt going to have nervous breakdown?

	Women		Men	
	Married	Never married	Married	Never married
Yes	143 (25%)	5 (8%)	40 (7%)	9 (9%)
No	431 (75%)	57 (92%)	503 (93%)	90 (91%)
Total	574	62	543	99

Surveys have problems of interpretation beyond those of other types of observational studies. People's responses to questions depend on their truthfulness (perhaps on what they think the interviewer would like to hear and on whether they think their answers will influence policy), their understanding of the questions, and the acceptability of answers (depending on socio-economic status, sex, marital status, and so on). Regarding their understanding of the question, presumably there are degrees of proximity to nervous breakdowns; how one decides that a breakdown is imminent may depend on one's sex or marital status. Regarding the acceptability of certain answers, some married women may volunteer that they are going to have a nervous breakdown, almost as a commentary on marriage, but men do not accede easily to such a statement of vulnerability. So it is quite possible that mental health is the same in both sexes. ◁

> **Showing relationship is not the same as showing cause and effect.**

The next example is similar to the previous two in addressing problems with observational studies, but it also shows that there may be no alternative.

EXAMPLE 3.5
▷ **Suntans and health.** There is abundant evidence that prolonged exposure to the sun causes skin cancer. But the sun must have beneficial health effects as well. For if it did not, evolution would have awarded our place in the sun to another species. But what are the beneficial effects, how beneficial are they, and do they result from direct exposure to the sun?

Health benefits of the sun are difficult to assess. Before drugs were available for treating tuberculosis, doctors advised their patients to get plenty of sunshine. However, I do not think the evidence of benefit was strong and I do not know of any randomized studies concerning this question. Exposure to the sun has been associated with lower rates of breast and colon cancers, which are more serious than many sun-induced skin cancers. This association is from observational studies: Cities receiving more sunlight tend to have lower rates of these cancers. The evidence is scanty at best since there are differences among cities other than the amount of sunlight they receive. It would be better to address the issue more rigorously and at the level of the individual.

Consider designing a more detailed type of observational study. Identify two sets of individuals: those who have spent lots of time in the sun and those who have not. Then assess the health of both groups. This is a **retrospective** study because the data already exist. Not all retrospective studies are bad, but this one has an obvious flaw: Individuals who have had cancer or other serious health problems may no longer be around for observation.

To repair this flaw we might include all individuals born in a particular year in a certain large New York City hospital, say. But we do not want to wait 60 years for a conclusion, so we do not use the current year. Instead, let's use 1940. In 1940 this hospital had a heterogeneous and highly transient patient population, so not all babies born there will continue to live in New York City.

The plan is to find out what happened to these individuals and assess their health and level of exposure to the sun—a mammoth job! We will lose track of many as soon as they leave the hospital and many others in later years. We could hire many detectives and have a modicum of success finding people. Some people will have died. Death certificates will help us determine the cause of death in many, but not all, cases. In some, but not in all, cases we will be able to locate relatives who will talk with us; they may or may not be able to provide accurate information about their deceased relative's level of exposure to the sun. If we are lucky and sufficiently diligent, we will get usable data for 50% of the individuals in our sample. But the ones lost may be lost for reasons related to their health. All in all, any conclusions will be questionable.

In deciding whether to undertake any such venture, one should consider the possible outcomes and then weigh these outcomes by their probabilities—the subject of the next chapter. In the case of this study, the answer is clear. Imagine the best possible results: Superhuman efforts have netted us information for our entire sample! It turns out that those with greatest exposure to the sun had the lowest rates of breast and colon cancer. Would this be enough evidence to recommend doses of sunshine to prevent cancer? No. The problem is that the subjects chose their own treatment. So it is possible that exposure to the sun has no benefit whatever. Perhaps a developing cancer makes people prefer cool temperatures or, for some other reason, shy away from the sun. Perhaps people who are not genetically disposed to cancer tend to seek out vitamin D from sunshine. Perhaps some types of suntan lotion prevent cancer in some people. There are many other possible explanations. To show that there is a tendency for two characteristics to coexist is not to show that one causes the other. ◁

> **A retrospective study is an examination of existing data. In a prospective study, treatments are assigned in advance and the subjects followed thereafter.**

These examples indicate problems with observational studies. What about randomized studies? Suppose we get a large number of volunteers and divide them into two groups. One group receives heavy doses of sunshine and the other is forever shielded from the sun. Assign each volunteer to one of the two groups by flipping a coin: heads means sun and tails means shade. Then watch to see how many cancers develop in each group. Sounds absurd. Even if we could get the volunteers (by paying them a lot of money, say) and even if we could spend the many years required to get results, I would have no confidence that the participants behaved according to their assignment.

> **When subjects choose their own treatments, comparisons are inherently flawed (although perhaps not fatally so).**

Reported studies commonly succumb to the fallacies mentioned in the previous example. The next is one of many.

EXAMPLE 3.6
▷ **Parkinson's disease and country living.** A study reported on television newscasts in January 1994 claimed a higher incidence of Parkinson's disease for people who live in rural areas. (Could this be another reason to avoid sun?) The study suggested a culprit, namely, fertilizer. I put little credence in the study's conclusion. This was a *retrospective* study. Researchers identified people with the disease and decided whether they were rural or urban. It is not possible to get a sample of Parkinson's patients without looking for them. You have to look in either urban or rural areas. You can try to adjust, perhaps using the number of patients at the various reporting institu-

tions who have other diseases. But you cannot adjust for differing abilities at these institutions in diagnosing various diseases, or for attitudes at the institutions or in the communities toward the diseases.

Moreover, urban folk may tend to be poorer and less well educated than rural folk. Such people with the disease may not be as inclined to seek treatment. Or their culture may be different and they regard this neurological disorder of people over 50 years old as the way some people age. Or, city folk may contract more serious forms of the disease and tend to die early, thus eliminating themselves from the eventual sample.

What would it take to make me a believer? A plausible connection—a chemical that occurs predominantly in rural areas, say, and that is shown to cause nerve damage of a type implicated in Parkinsonism. ◁

A randomized study is not subject to the biases that so complicate conclusions from observational studies. In the previous example, a prospective randomized study could proceed as follows. Identify a (very) large number of children at birth and randomly assign them to live in the country or city. Then follow them to observe the proportion in each group that contracts the disease. As in the suntanning example, such a study is impossible. Even if it were possible, answering the question would not be worth the enormous effort and expense. The prevalence of the disease is not sufficient to have an impact on where people choose to live. Also, concluding that a particular environment is associated with a higher incidence of disease is of little help in identifying a cause.

Randomization has obvious virtues in medical research. If clinicians are allowed to assign patients to treatment, they may subconsciously assign the sickest patients to the treatment they perceive to be better or, perhaps, to their least favorite treatment. Treatment comparisons would be flawed. Randomization minimizes the possibility of assignment bias. There are many illnesses in which accepted medical techniques or therapies have been shown by randomized trials to be worthless or essentially worthless. The following is a standard example.[7]

EXAMPLE 3.7

▷ **Gastric freezing.** In 1958, surgeon Owen Wangensteen introduced a procedure called gastric freezing for treating ulcers. The procedure is as follows. The surgeon places a balloon in the patient's stomach and pumps coolant through it. The goal is to stop the digestive process and allow the ulcer to heal. Wangensteen treated 24 patients and all reported that their symptoms were improved. He used no controls. Not all doctors were convinced and some set up a randomized study[8] to evaluate the procedure. The patients agreed to have their treatment assigned randomly, receiving either the gastric freezing procedure or a control procedure. Moreover, they would not be told which they received, and so were **blinded** as to treatment. The study was **double-blind** in that the surgeon was also not told. It is not difficult to keep this information from the patient, but it is difficult to blind the surgeon. They used a sham procedure as a control. One of two types of balloons—selected randomly—was placed in the patient's stomach. Balloons assigned to control patients had a shunt that returned the coolant before reaching the stomach. The surgeon could not tell which type of balloon was used.

Table 3.3
Example 3.7 data: Randomized trial for treating ulcers

After	Gastric Freezing		Control	
	Total	Success	Total	Success
6 weeks	82	64 (78%)	78	53 (68%)
24 weeks	71	39 (55%)	70	43 (61%)

Table 3.3 gives the results of the randomized trial: numbers (and percentages) of patients who were clinically improved or stable (success) at 6 and 24 weeks after surgery. [I give counts because they provide the basis later for Exercise 9.21. I had to estimate some of the counts because the original article gives only percentages. Also, for the 24-week data, the authors give only the total number of patients (141) in the two groups. To provide numbers I assumed equal drop-out rates in the two groups.]

It is not clear how seriously to take the 10-percentage-point improvement in success rate at 6 weeks. Wangensteen had not claimed a long-lasting benefit and indeed had proposed that the procedure could be repeated with relative ease as necessary. Results from any study can be deceiving, but since the control patients were actually better at week 24 (61% vs 55% rates of success), it seems likely that any improvement had disappeared by then. The procedure has at best a slight, temporary benefit—hardly worth the trouble. The authors of this study credited the patient-reported benefit in early studies to a psychological effect. ◁

Randomizing treatment assignment eliminates selection biases.

Randomization eliminates selection biases and allows for neat, uncomplicated comparisons. It is great for agricultural and industrial experiments, as well as for medical experiments. But it must be used with care in medicine. In clinical studies conducted in many countries, patients have to sign informed consent forms indicating that they have been informed about all aspects of the study. In particular, they have to agree to be randomly assigned to therapy. (Whether patients really understand what they are signing is another matter.) A possible problem with randomization is that patients who agree to be randomized may be different in a way that is related to the effectiveness of the treatments. Attempts to address this question have not been uniformly successful.

Participation by clinicians in randomized studies is problematic. They usually have information about the effectivenesses of the therapies and so, while they may be far from certain about relative benefits, they have a preference. It is then difficult to encourage patients to accept randomization. It is a delicate trade-off: The truth about treatment benefits may not become known without randomization, but a randomized patient may not receive what the clinician thinks may be the best therapy.

A common type of design is to compare a sample having a particular outcome with retrospective controls who do not have the outcome. For example, a sample of lung cancer patients is compared with a *normal* sample to see whether there are differences that suggest causation. Now the outcome plays the role of the treatment. The advantage is one of convenience: It is easy to identify people having lung cancer. But finding a good comparison group is not so easy. One approach is to consider each lung cancer patient in turn and find an individual who matches that patient in particular ways—such as age and sex—but does not have lung cancer. Then the individuals are analyzed as pairs (see Section 3.5) in comparing other factors. For example, one might be interested in differences in smoking habits in the two groups. This is a **matched case–control study.** While it does not have the credibility of a randomized, controlled study, it can be a useful research tool.

EXAMPLE 3.8

▷ **Caffeine and miscarriage: A case–control study.** Investigators[9] were interested in whether caffeine consumption increased the incidence of miscarriage. They identified 331 women who had experienced a miscarriage. For each of these women, they found three women who had experienced a normal pregnancy and who had similar demographic characteristics. So the total number of controls was 993. Then they compared caffeine consumption in the two groups.

Like other retrospective studies, case-controlled studies are easy to carry out. Like other retrospective studies, they are subject to biases. There may be factors not accounted for by the investigators (such as having to work overtime or holding down two jobs) that increase one's risk of miscarriage and also increase one's caffeine consumption. Changing coffee consumption would not have an effect on such a factor and, therefore, it would not have an effect on propensity to miscarry. ◁

Sometimes people believe observational studies over randomized studies because they reinforce their prejudices. The next example is a case in point.

EXAMPLE 3.9

▷ **"Mammogram best for early detection."** This is the way one paper titled a syndicated column by Lawrence Lamb, MD, that appeared in U.S. newspapers on March 8, 1994. In the column, Dr. Lamb suggested that a 42-year-old letter writer should get mammograms. He writes, "There have been two reports, one from a Canadian study and the other from New Zealand, saying that mammograms before the age of 40 [he meant 50] did not decrease the death rate from breast cancer in these age groups." I do not know the New Zealand study. The Canadian study was highly publicized and resulted in the American Cancer Society changing its recommendations for screening mammograms for women under 50. In the Canadian study, more than 50,000 women were randomized (only a portion of them under 50), with patients assigned to receive regular mammograms or not. The women were then followed for a minimum of 7 years and incidences of breast cancer were noted and compared between the two groups. There was a lower incidence of breast cancer in the group that did *not* receive mammograms.

Dr. Lamb continued: "However, a study from Massachusetts General Hospital and Harvard of 117 women under the age of 50 found to have breast cancer with a mammogram suggests otherwise. They were compared to 928 women in the same below-age-50 group who were found to have palpable breast cancer by physical examination.

The 5-year survival [rate] for women whose cancer was detected by mammography was 92% compared to 74% for women whose cancer was physically palpable."

It may be that mammography has a benefit in this age group. But the Massachusetts study provides no evidence for it. Mammograms identify breast cancer sooner than do physical exams. Perhaps the difference between 92% and 74% at 5 years reflects the fact that the second group of women had already survived a number of years longer than the first group and this time is not counted in the 5-year survival. But there are other important differences between the two groups. Not all breast tumors grow fast and not all are lethal. Some tumors discovered by mammography are not life-threatening, some would never be discovered using palpation, and some may never affect the woman's well-being.

Dr. Lamb puts more credence in a small *retrospective* study than in a much larger *prospective,* randomized study. His advice may be correct, but it is whim and not science. ◁

We all have prejudices. Do not suppress yours, but recognize that they exist. Also recognize that observational studies have flaws, even though they play important roles in science.

EXERCISES

3.1 There are two routes, A and B, that you can take to work in the morning. In the interest of finding the shortest route timewise, you set aside a 2-week period for experimentation, planning to time your trip on each of the five work days in each week (Monday through Friday). You will take route A on five of the days and route B on the other five days. There may be daily variation with, say, heavier rush hour traffic on Monday than on Friday. There might also be a weekly variation—perhaps there is a convention in town one of the two weeks. So you would not want to use route A all the first week and route B all the second.

(a) Design an experiment; that is, tell how you will decide which route to take on each of the days. Use randomization, but ensure that you will use routes A and B at least twice each in both weeks.

(b) Describe what data from the experiment you plan to analyze. It is enough to say how many numbers you will use and whether they are travel times or differences in travel times.

3.2 You want to compare users' reactions to two types of hand soap. You plan to sample 80 undergraduates at your university: 40 will get type A soap and 40 will get type B. You will give a bar of soap to each person selected and ask them to fill out questionnaires with their reactions. (You will get their telephone numbers so you can call them in case they forget to comply.) You anticipate that men and women will react differently and that students in different undergraduate classes may also react differently. So you will stratify on both these factors. You have complete lists of the four undergraduate classes. Explain how you would take your sample.

3.3 Suppose you work for a company that makes heart valves and 17 of the valves have been returned to you in the last 2 months because they leaked excessively. They come from several different lots; the other valves in these lots did not leak. You have inspected the leaky valves and found nothing that is clearly wrong. You know which lots they were in, their time of manufacture, which quality engineers were on duty at the time, and you can make six different size and function measurements on the valves. Design a matched case–control experiment with the goal of identifying the problem and, perhaps, fixing it. Use the lot of manufacture for matching and the six measurements as outcomes.

3.4 Burrowing owls sometimes build their nests in holes dug by prairie dogs, coyotes, or badgers and since abandoned. They sometimes line their nests with cattle or horse dung. Why? Their motivation may be to insulate the nest from temperature extremes. Another possibility (suggested by biologist Dennis Martin) is that the owls use the dung to keep predators away. To test this theory, biologist Gregory Green observed lined and unlined owl nests in the Columbia River basin.[10] Of the 25 nests lined with dung, two were raided by badgers. Recognizing the need for a control, Green also observed 24 unlined owl nests. He found that 13 of these 24 were raided by badgers. Is this an observational study or a randomized study? Might there be biases that account for the differences between these two samples other than the treatment—dung or no dung?

3.5 Is there a relationship between having experienced physical or sexual abuse as a child and developing post-traumatic stress disorder (PTSD) as a soldier? A study[11] suggests that there is. Of 38 Vietnam combat veterans seeking help for PTSD, 11 reported experiencing childhood physical or sexual abuse. Of 28 Vietnam combat veterans seeking medical care unrelated to PTSD, only two reported incidences of abuse. (You will analyze these data in Exercise 9.17.) According to *Science News,* "The new findings indicate that child abuse can foster the link between dissociation and PTSD." This goes too far. Why? [Consider the appropriate populations for this observational study and address cause-and-effect conclusions in the context of such studies.]

3.6 Suppose someone does a study of major league baseball players and finds that those who developed potentially fatal diseases (such as cancer or Lou Gehrig's disease) during their playing careers tended to have higher lifetime batting averages than other players. (I have never seen such a study, but I suspect that this would be the result.) One conclusion is that having a fatal disease makes for a better baseball player. But there are others that are more plausible. Give one.

3.7 In October 1994, *The New York Times* ran a front-page article indicating that "men from traditional families, in which the wives stay at home to care for their children, earn more and get higher raises than men from two-career families." The statement is based on studies of salary histories of male managers. These studies suggest cause and effect: To make more money, have a traditional family. But, to my mind, there are other more plausible conclusions. Name one.

3.3 Taking Samples

Carrying out an experiment means taking a sample from some population. Sometimes the only part of the population that exists is the sample. An example is a laboratory experiment in which, say, 1 gram of chemical A is added to 3 grams of chemical B and we measure the resulting amount of chemical C. This sample may not be representative of any larger population. Results may depend on factors such as temperature and humidity that may be related to time. If so, results from that part of the experiment conducted tomorrow may be different from the one conducted today. And days missing from the experiment—perhaps weekends—will not be represented at all. An experimenter might try to control for such factors, but even if they are all controlled, there may still be a time trend. The experimenter should assess the trend and judge its impact on the conclusions from the experiment, but in the end may have to live with it.

In other types of experiments, there exists an actual population. If we have the time and resources we could select the entire population and make the required measurements. But usually we can obtain observations from only a fraction of the population. We would like the sample to be as much like the population as possible. How to make the selections? The following are examples of how *not* to make selections.

EXAMPLE 3.10
▷ **The *Cosmo* Girl's Sex Survey.** An article with this title appeared in the April 1993 issue of *Cosmopolitan* magazine. It reported results of a survey that had been published in a previous issue. Thousands of readers had responded. One question was: "If married, have you ever been unfaithful to your husband?" to which 39% of the responders said "Yes."

The responders constitute a sample. But what population do they represent? Here are a few of the many possibilities:

○ Adult women in the U.S.
○ *Cosmo* readers
○ Women who respond to surveys
○ Women who respond to surveys about their sex lives
○ *Cosmo* readers who respond to surveys
○ *Cosmo* readers who respond to surveys about their sex lives

The last population is clearly appropriate. The others are not. People who respond to surveys are different from those who do not. People willing to tell others (even anonymously) about their sex lives are very different from those who are not. People who read magazines are different from those who do not. People who read *Cosmopolitan* magazine are very different from those who do not.

So the population *is* the sample! Everyone who qualifies for the population is in the sample. The response of 39% "Yes" is irrelevant, except in this very narrow population. (*Cosmopolitan*'s editors would not be concerned by this statement. They are interested in selling magazines and not in making inferences about larger populations.) For any other use, this figure is biased in ways that are unpredictable. Perhaps women who are unfaithful to their husbands are more likely to read *Cosmo* and respond, but perhaps they are less likely. The same is true for women who are faithful to their husbands.

Finally, we have no check on the accuracy of the responses. People who respond to sensitive questions may well lie, and the lying may be in either direction. So the 39% figure is untrustworthy even for the narrowest of populations. ◁

Self-selected samples are practically worthless for scientific inference.

EXAMPLE 3.11
▷ **The *Literary Digest* poll.** In 1936 a now defunct magazine, *Literary Digest*, published results of a poll of 2,400,000 people: 57% favored Republican Alf Landon over

Democrat Franklin Roosevelt in the upcoming U.S. presidential election. They predicted a Landon victory and were off by 19% when Landon received only 38% of the vote. (Roosevelt carried 46 of the 48 states in an electoral college landslide.)

Modern polls use relatively tiny samples of about 1,000 people and claim to be accurate to within 4%. If you have followed Gallup, Harris, and other reputable polls in recent years you will not disagree with this claim. Why couldn't *Literary Digest* do as well with a sample thousands of times larger? There are several reasons, all related to the way they took the sample. Using telephone directories and existing mailing lists, they mailed out 10 million questionnaires. If you were on a mailing list in 1936, you were likely to be reasonably well off financially and tended to vote Republican. This is a clear bias but its magnitude is difficult to assess. Moreover, the response rate of 24% is paltry. As was indicated in the previous example, nonresponders may well be very different from responders. The magnitude and directions of these differences are impossible to predict. Such a low response rate is sufficient in itself to make the poll worthless, regardless of the *number* of responders. ◁

> **A low rate of response in surveys can weaken and even invalidate conclusions.**

Random Sampling

A way to minimize the biases discussed in the preceding two examples is to use **random sampling.** Taking a random sample means using a randomization device (Section 3.2). One way is to place tokens labeled with the various members of the population into a container, stir it up, and select one. Repeat this process as necessary or as otherwise indicated in the study's design.

> **To obtain a random sample from a population of interest, use a coin, die or other such randomization device.**

A consideration in this process is whether previously drawn tokens are replaced for the subsequent selection. If they are, then the sampling is called **with replacement** and, if they are not, then the sampling is called **without replacement.** If the population is very large in comparison with the sample, then the two types of sampling are essentially the same—when duplicates are not likely when sampling with replacement, then the sample is as though one were sampling without replacement. Even though samples are usually taken without replacement, we will frequently assume it is done with replacement because results obtained by sampling with replacement are much easier to analyze.

There are other ways to obtain a random sample. One is to consider each member of the population in turn and decide to include that member depending on the result of a coin toss, die roll, spinning pointer, or the like. For example, suppose a 10% sample is desired. A computer random number generator is used to decide which member numbers to include in the sample:

6, 29, 39, 72, 73, 74, 99, 132, 138, 150, 153, 160, 162, 163, 171, 185, 191, 204, 210, 211, 221, 235, 239, 244, 259, 268, 317, ...

This tells us to include the sixth member of the population, the 29th, and so on, but none of the others. As is characteristic of random samples, some members selected are next to each other on the original list and then there are some large gaps in which no members are selected.

Still another possibility is to list all possible samples of a predetermined size and choose one of these randomly, taking the sample indicated. Because the number of samples is very large even when the population is small, this method is not very useful.

A way to select a random sample that is useful even for large populations is to randomize at increasingly finer levels until getting a single population member. For example, consider selecting a head of household randomly from the United States, and consider only those with listed telephone numbers. (Selecting randomly from the entire United States population is effectively impossible!) First select a state, including the District of Columbia. The randomization device must be employed in such a way that each state's chance of being selected is proportional to the state's population (assumed to be proportional to the number of heads of households). One way to do this is to use a calculator's random number key; this gives a number between 0 and 1 that is supposed to be random. List the states with their populations and calculate the corresponding proportion of the U.S. population. Suppose the first state considered has 1% of the U.S. population, the second state has 5%, the third state has 2%, and so on. Press the random number key. If the number is between 0 and .01, then the first state is selected; if it is between .01 and .06 (1% + 5%), then the second state; if it is between .06 and .08 (1% + 5% + 2%), then it is the third state; and so on. Once the state has been identified, select a city (or other municipality with a phone book) from the state in the same way. Then select randomly from the households in the phone book. (If the phone number you get is not that of a household, then there are various possible fixes—one is to start over.) This gives one member of the sample; repeat as necessary.

Other Sampling Methods

Sometimes—such as when selecting a head of household randomly from the United States—random samples are difficult to obtain. Usually, they can be approximated reasonably well by sampling procedures that are easier to use and that are designed to yield representative samples. The two types of sampling considered next are examples.

Systematic sampling either does not use randomization or uses it in a more limited way. Suppose you have a list of the population (say, a telephone book) and want a 10% sample. Then you can select members who occur at the following locations in your list:

10, 20, 30, 40, ...

Or you could randomize only for the first name and take every 10th name thereafter. Compare the regularity in this sequence with that of the previous random sequence

(6, 29, 39, 72, 73, . . .). This regularity is good and bad. It is good because there are no large gaps in the sequence—which contrasts with random sequences. It is bad because if there happened to be some regularity in the list (or in parts of the list), you would oversample some categories of members and undersample others. An obvious advantage of systematic sampling is that it is easier to carry out.

Stratified sampling uses randomization within subpopulations. Again consider taking a sample of the U.S. population. Suppose you want a sample of size 100. Focus on each state in turn. The first state has 1% of the population, so take exactly one person (randomly) from that state. The second state has 5% of the population, so take five people from that state. The third state has 2%, so take two people from the third state, and so on. Such methods are common because they are easy to use and they help deliver representative samples.

The modern political polls mentioned in Example 3.11 employ stratification and randomization within strata.

Convenience sampling means that items are selected because they are handy. There is no intentional selection bias, but being handy may be related to response.

EXAMPLE 3.12

▷ **Too much violence on TV?** I wanted to know how Duke undergraduates feel about the amount of violence on TV. Taking a random sample (or a random sample stratified by undergraduate class) would be difficult. It would be relatively easy to identify a sample from class lists—the hard part is to find the individuals so identified. I could try the telephone, but I would be able to contact fewer than half at the first call. Then a prolonged endeavor to locate the others would follow. I could not just forget those not contacted because of the problems of nonresponse already indicated—perhaps students not at home when I called tend to watch less TV and have different opinions about what is on TV.

So I asked the students in my statistics class! I asked men and women separately. For future reference, these were the responses:

	Too much TV violence?	
	Yes	No
Men	8	12
Women	17	5

So I know how students who are in this particular class feel. But can this apply to any larger population? No one really knows the answer, and that is the problem with convenience sampling. Consider U.S. college students. I do not expect Duke students to have typical political ideologies and I would not be willing to extrapolate to U.S. college students. It would not surprise me if a national poll showed quite different attitudes.

Consider undergraduates at Duke. The question is whether students in my class are representative of Duke students. They are not representative in every respect, but they may be representative regarding their attitudes toward violence on TV. They represent all four undergraduate classes and in roughly the same proportions. I see no reason that statistics students would feel differently about this issue, at least not very differently. Sticklers will put no credence in this sample for any inference beyond this particular class. A concern is that students in the same class are more likely to be friendly with each other than students selected randomly and, thus, have similar attitudes. But I give it some credence and am willing to accept it as approximating opinion at Duke at the time. In particular, I suspect that the evident difference between men and women holds more generally. ◁

Convenience sampling is important because of its efficiency. Savings in time and resources are important. But the loss may be difficult to assess. Convenience can easily lead to samples that are not representative. The reason has to do with why they are convenient. Convenience—time, place, or other factor—leads to similarities among the units in the sample. So, convenience samples tend to be more homogeneous and, therefore, less representative than random samples. Perhaps the sentiment expressed in the sample of the previous example is representative, but I would not weigh it as heavily as I would a random sample of size 42. Perhaps it should count as much as a sample one-third to one-half its size—14 or 21 or so. I cannot give a formula, but I recommend partially discounting convenience samples, depending on the circumstances.

Cluster sampling refers to sampling in which the unit is a lot or cluster and subunits within the cluster are measured or assessed. All or only some of the subunits can be sampled within each cluster. For example, to ascertain the health status of elementary school students in a certain state, 10 of the state's schools (clusters) might be sampled and all or some of the students (subunits) within each school assessed.

There are two reasons for cluster sampling: (1) convenience and (2) possible contamination among subunits. For example, a program to prevent drug abuse might apply to the whole school with individual students within the school influencing other students in the school. An even clearer example is a vaccine program for a contagious disease. An outbreak of the disease may be avoided in a community (cluster) by immunizing only a fraction of the people (subunits) in the community—this phenomenon is called *herd immunity*.

EXAMPLE 3.13
▷ **Saving energy by junking old refrigerators.** Older refrigerators use energy inefficiently. Power companies encourage customers to junk them so that they will have to produce less power. This may even result in having to build fewer power plants. Some companies buy old refrigerators for $25 to $50 and then scrap them. These programs are costly and their effectiveness is not well understood. One question is how much it reduces a participant's power usage.

Two methods are used to address this question.[12] In one, old refrigerators are monitored in conditions simulating seasonal changes. They use about 1,000 kWh (kilowatt-hours) per year and so this is the estimated savings by eliminating the refrigerator. In the other, the company looks at the total power usage of participants in the 12 months before and in the 12 months after selling their refrigerators. The savings varies considerably, but averages about 400 kWh per customer per year. At 1,000 kWh per year the program is worthwhile, but at 400 kWh it is not. Which is correct?

Neither estimate is likely to be correct, although the second seems better. Both methods are subject to biases. The first estimate does not consider the possibilities that (1) a refrigerator may not have been in use before the customer sold it, (2) the customer may have replaced the refrigerator, perhaps with a new one but possibly with another old one, and (3) replacing a refrigerator may focus one's attention on saving energy and result in more savings than just from the refrigerator. The second estimate does not allow for time trends in attitudes toward saving energy that may be unrelated to refrigerators. One possibility is to adjust the estimate by comparing the reductions of those who turn in refrigerators with those who do not. This would account for seasonal and other types of variation. But there are possible biases resulting from differences between customers who have and do not have old refrigerators, and between customers who select themselves for the program and those who do not.

What method eliminates bias? Randomization is problematic. A possible control is to not buy some customers' refrigerators and compare their energy use. But it is difficult to buy one customer's refrigerator and not buy that of the customer's neighbor. Even if the company pays the neighbor without taking the refrigerator, the neighbor's power usage is influenced by the information that saving energy is possible through not using the refrigerator—thus, contaminating the results.

Randomizing over geographical clusters is an answer (although, apparently, this method has not been used in practice). A set of neighborhoods (or communities) can be identified as candidates for the program. These should be geographically dispersed so as to minimize contamination. Using a randomization device, half of the neighborhoods are selected for the program and the other half are not. Energy use (before and after) would be compared between the two groups. The experimental unit is the cluster, but a more detailed analysis could consider subunits (individual customers within clusters) as well. Differing characteristics of the clusters can be considered. An alternative is a case–control study (see Section 3.2) in which clusters with similar characteristics are matched. ◁

EXERCISES

3.8 To learn about lifestyles, a researcher[13] sent a questionnaire to 1,400 couples. Of these, 339 were returned and 107 were usable. Of the 214 (= 2 × 107) individuals, 198 had been raised as Protestant, Catholic, or Jewish, and 80 of these (about 40%) still claimed the same religion. How accurate is 60% as an estimate of the proportion of individuals in the United States who leave their religions?

3.9 On November 18, 1994, Jesse Helms (who was in line to head the U.S. Senate Foreign Relations Committee) appeared on the CNN program "Evans and Novak." He was asked whether President Clinton was "up to the job" of commander in chief of the U.S. armed forces. Helms said: "You ask an honest question; I'll give you an honest answer. No, I do not. And neither do the people of the armed forces." He went on to say that "just about every military man who writes to me" thinks that Clinton is not fit for the job. Discuss the reliability of the statement, "neither do the people of the armed forces," given the circumstance of Helms's evidence.

3.10 Example 1.4 (page 7) gives data on putting success from 9, 11, and 14 feet for PGA tour players. Investigators selected one relatively flat green at each of 15 tournaments during the

latter half of 1988 and measured every putt of every golfer. Assuming the same putting distances as in Example 1.4, discuss the relevance of this sample to the following populations:
- (a) PGA tour players on other holes in 1988
- (b) PGA tour players on these same holes next year
- (c) Putts of Ben Crenshaw (a professional golfer) next year
- (d) Your next putt

3.11 An epidemic of cholera invaded some parts of South America in the late 1980s and early 1990s. The World Health Organization set up a study to investigate the effectiveness of a cholera vaccine. Recognizing the possibility of "herd immunity," they used cluster sampling and randomization. Describe how such a study might be conducted.

3.12 A company has two versions of a product and has to decide which version to market. It plans to test-market both to see which sells better. Marketing both versions in the same area may be confusing to customers and stores may be unwilling to assign double shelf space to the product. So the company plans to use cluster sampling and randomization. Describe how such a test-marketing program might be conducted.

Multiplicities and Design

Drawing inferences from observations can be a very difficult process. The next example illustrates a problem of making any type of observation.

EXAMPLE 3.14
▷ **Hair color and intelligence.** There were two redheads in my first-grade class. They were from different families and were both very bright. In my naivete, I associated red hair with intelligence. It took several years of meeting redheads of normal intelligence before I learned that I had read too much into the early data. ◁

Consider the two brightest kids in my first-grade class (identified as brightest in any way you like, including random choice). They *have* to be similar in some other way—perhaps, in several other ways. Perhaps they are both girls, both boys, both tall, both short, extreme in height (one may be very tall and the other very short), both of the same nationality or religion, both overweight, both underweight, have similar hair color, have buck teeth, have freckles, can run fast, cannot run fast, are handsome, are not handsome, and so on. So I was doomed. I was bound to learn something that was wrong!

These many possibilities are multiplicities. Multiplicities are present in every statistical application. They are difficult to recognize and even more difficult to deal with once they are recognized. Some people do not understand that their very presence may be a problem, whereas others recognize the problem and overreact. There may be a reasonable compromise, but no one knows what it is. You should understand the problem of multiplicities, and keep it in mind whenever you assess information.

Studies are subject to multiplicities. In Example 3.6, any set of Parkinson's patients is bound to be at least somewhat different from nonpatients. I do not know which characteristics the investigators considered. In addition to rural/urban, they might have considered gender, height, weight, race, educational level, marital status, and others. At least one of these is likely to show a difference as compared with people who do not have Parkinsonism. So, the rural/urban factor may play no special role in this regard.

Both observational and randomized studies are subject to problems of multiplicities. So they have implications for designing experiments. A way to avoid the problem is to consider only one characteristic. For example, the people setting up the study of Parkinson's patients might have made one and only one measurement, say, height. For weight, they would set up a second study, for rural vs urban a third study, and so on. Such a resolution is silly. It is better to face up to the problems of inferring from multiple measurements. But do not ignore them.

An important aspect of the statistical process is combining information from an experiment with that present before the experiment. Assessing information after an experiment that was available before the experiment is difficult at best. This difficulty is exacerbated in the presence of multiplicities. Therefore, assess your available information at the design stage of a study and keep it separate from the experimental results in your mind and in your analysis.

> **Before carrying out an experiment, identify and assess available information.**

Another major multiplicity issue arises from telling only part of the story. This is lying and is deceptive, even if it is not intentional. Being human, researchers focus on aspects of their investigations that make their point most strongly. Any story is subject to this form of bias. To compound the problem, studies tend to be published if their results are startling. Depending on the circumstances—and politics!—studies that confirm previous studies or scientific dogma are likely to be filed away by the investigators or rejected for publication by scientific journals. This is one type of *publication bias*. Another type is *not* publishing results that conflict with current dogma.

The next example is a clear instance of data that are accurate, but partial, and the partial story is misleading and therefore dishonest.

EXAMPLE 3.15

▷ **PGA putting (revisited).** Example 1.4 (page 7) gives the following success percentages for professional golfers putting from three distances:

Length (ft)	Number tried	Number made	Success %
9	217	69	31.8
11	237	75	31.6
14	174	54	31.0

As was indicated in the earlier example, these percentages are amazingly constant for such a range of putting distances. I lied! What I told you was literally true, but I selected these distances from many others published in the article. The complete data[14] are shown in Table 3.4 and in the scatterplot in Figure 3.1. The data points from Example 1.4 are indicated in boldface type in the table and as ✖'s in the scatterplot. The points from Example 1.4 show no relationship with distance. But this is misleading. I selected them to be misleading and I readily admit that to you. Other reporters may not be as willing—and some have misled themselves.

Table 3.4
Data for Example 3.15

Length of putt (ft)	Number tried	Number made	Percentage success
2	1,443	1,346	93.3
3	694	577	83.1
4	455	337	74.1
5	353	208	58.9
6	272	149	54.8
7	256	136	53.1
8	240	111	46.3
9	**217**	**69**	**31.8**
10	200	67	33.5
11	**237**	**75**	**31.6**
12	202	52	25.7
13	192	46	24.0
14	**174**	**54**	**31.0**
15	167	28	16.8
16	201	27	13.4
17	195	31	15.9
18	191	33	17.3
19	147	20	13.6
20	152	24	15.8

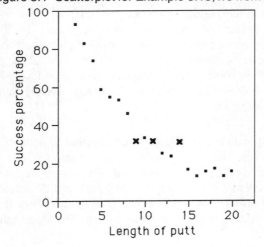

Figure 3.1 Scatterplot for Example 3.15; x's from Example 1.4

The scatterplot in Figure 3.2 shows the areas of the dots as roughly proportional to the corresponding sample sizes. This is in keeping with the principle set forth in Chapter 2 and, especially, in the discussion of area plots in Section 2.4—namely, that data displays should accurately represent numbers of observations at the various points, in addition to showing locations of points. However, in this case the dots are not actual data points but, instead, their locations merely summarize the observations at that particular distance.

Figure 3.2 Scatterplot for Example 3.15 with dot area showing sample size

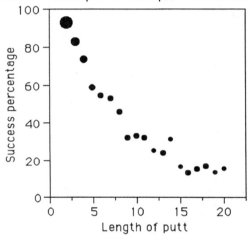

There are other biases in these data. If you are familiar with golf you may find the results surprising. For example, a 55% success rate for 6-foot putts may seem too low. A typical report of golf tournaments on television news programs shows the winner making a 6-foot or longer putt. One comes to think that this is rather typical for golf pros. But the putt shown is chosen precisely because it was made and on any given day the winner makes at least one longish putt.

Professional golfers are presumably less influenced by TV newscasts because while playing they see a representative sample of putts—all their own putts, for example. According to the *Sports Illustrated* article, "Most of the players guessed that at least 70% of their six-footers would drop." Players, too, have biases. Perhaps they pay more attention to successes. Perhaps they think they "should have made that putt," and count it as a near success. ◁

If you keep up with health-related studies published in newspapers and magazines, you are probably frustrated at apparent contradictions from one study to the next. One study says drinking coffee causes cancer; the next says two cups a day is okay; the next says caffeine increases serum cholesterol; the next says decaffeinated coffee is the real culprit; the next says any amount of coffee is okay, decaffeinated or not; the next says coffee increases your sex drive; and on and on. There are other examples: Aspirin prevents heart attacks, aspirin is great, worthless, great; oat bran is worthless, great, worthless; dietary cholesterol is irrelevant, bad, good, both bad and good!

What is going on? To some extent, journalists (and the researchers themselves) overreact to small differences. Sometimes they do not suitably qualify their statements with probabilities—as you are going to learn to do. Sometimes subjects are not the same in the different studies. Some researchers try to adjust for subject differences by comparing those who have similar characteristics: same sex, age, smoking habits, and so on. It may be impossible to adjust. For example, the conclusion that coffee increases the sex drive of elderly men came from an observational study involving some coffee drinkers and some nondrinkers. The result may be spurious for a number of reasons. For example, it may be that a man's propensity to drink coffee is related to a lifestyle more conducive to sexual activity. (Perhaps a man who lives alone is less likely to go through the trouble of making coffee, but if his mate makes some he will join in.) Also, there may be characteristics that are important indicators of response and that are unknown to the researchers. Or, perhaps, there are subject differences related to time or geographical area that are important and that are impossible to make comparable.

Another explanation is the possibility that you are hearing only part of the story. A study might look at several measures—length of life, quality of life, sexual activity, appetite for food, and so on—and conclude that one, say, sexual activity, is increased (or decreased) among coffee drinkers. This is a multiplicity problem. A related problem is the publication bias mentioned earlier: Only certain studies get published and the reasons they are selected for publication are related to their results. Take yourself back to the time coffee was thought to be completely innocuous (which, indeed, it might be!). Some people who investigated the effects of coffee found nothing. Others found interesting things, such as that people who drank coffee died earlier. The latter type of conclusion makes for a more striking story and so is more likely to be published. That opened the gates for publishing contradictory studies—and studies that countered them, and so on.

There is a moral for experimental designers:

> **Before carrying out an experiment, specify the measurements you plan to make. Spurious results are more likely with a greater number of measurements.**

When you are not involved in the design stage, inferences can be difficult. The next example considers not an experiment but an observational study (Section 3.2), in which drawing conclusions is especially difficult.

EXAMPLE 3.16
▷ **Germ warfare testing and birth defects.** In 1953, U.S. Army workers sprayed zinc cadmium sulfide, a mock biological-warfare agent, from a roof near Clinton School in Minneapolis. Apparently, this was one of 239 U.S. sites used by the army to simulate a covert biological attack. But "Of 15 women who were in [the] fourth-

grade class [at Clinton School in 1953] . . . seven are sterile. The other eight have had 25 miscarriages among them. More than a third of the class's offspring are retarded." [15]

These data are startling and tragic. But extremes happen even without a single cause. There are many fourth-grade classes and, therefore, multiple opportunities to observe an extreme—perhaps this is one. Were it not for the army's experiment, they might have been accepted as extreme but random and not having a specific cause. The army claims the agent is harmless and that this is a coincidence, though they admit that there are limited data concerning the agent's effect on childbearing.

A statistical analysis of these data without recognizing the presence of multiplicities would be wrong. But how to incorporate the multiplicities aspect into an analysis is far from clear. ◁

EXERCISES

3.13 An article[16] entitled "Dietary Fat Predicts Breast Cancer's Course" described a study[17] of 220 women who had surgery for breast cancer. A nutritionist asked them about their eating habits. Four years later the researchers compared outcomes (recurrence of disease or not) for women who reported having a diet rich in saturated fat with those who did not. "A statistical analysis revealed that fat intake did influence the outcome of breast cancer treatment, but only in women who had tumors with lots of estrogen receptors." First of all, statistical analyses involve probabilities and do not "reveal truth." But in my reading, the relationship is tenuous at best. There are many other subsets for the investigators to consider and they make claims only for estrogen-receptor-rich tumors. Other possibilities are patients who have tumors with few estrogen receptors, patients with large tumors, small tumors, stage I disease, stage II disease, and so on. Also, the researchers looked at 14 different characteristics of diet: total fat, saturated fat, carbohydrates, protein, alcohol, fiber, and others. Had they found a relationship—however slight but "statistically significant"—between survival and fiber intake for patients with stage I disease, say, an article with a different title might have appeared in *Science News*.

There are two types of multiplicities involved in this report. Describe them and indicate what effect they have on the report's credibility.

3.14 A study[18] considered 13 prostate cancer patients and 14 controls. The report indicates that men were contacted through urology clinics, which, I believe, means these are convenience samples. Thirteen different anthropometric measurements were made on each man: sitting height, standing height, waist circumference, hip circumference, thigh circumference, triceps skinfold, and others. The goal was to identify a way of discriminating between cancer patients and controls. A statistical test showed two of the 13 to be significant discriminators—including waist to thigh ratio, as shown in the accompanying plot.

A report[19] in *Science News* indicates that this might "represent an inexpensive, readily available, and simple means of better defining the one man in 11 who may develop cancer of the prostate." It is not clear that it is such a great predictor for a number of reasons. One is that, if one were presented with these 27 men's waist to thigh ratios and picked the top 13 (those above the dotted line in the plot), only 7 would be correct. This is hardly better than guessing—or than picking the bottom 13! Another reason is because of multiplicities present in this study. Explain this last statement. (An important issue not addressed in this exercise is whether screening for prostate cancer would have any value even if the screening were perfect. The effectiveness of available treatments is controversial,[20] and so the answer is not clear.)

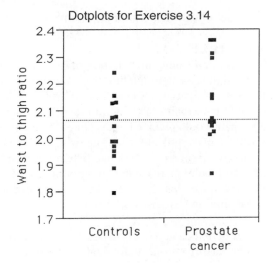
Dotplots for Exercise 3.14

3.5 Paired vs Parallel Designs

Consider comparing a treatment with a control or one treatment with another treatment. There are two types of designs that are standard: paired and parallel. In a **paired design,** experimental units are identified that can receive both treatments, but separately. This allows for a direct treatment comparison within units. Something has to be different or the effects of the treatment could not be assessed separately. One possibility is time; one treatment is assigned at one time and the other at another. A second possibility is space; the treatments could be applied on different parts of the unit. Another is that the unit is really two units that are similar in important ways. An example of the latter is twins.

There may be differences from one experimental unit to the next. A paired design removes these and allows for direct comparison. Consider Example 2.8. Suppose we would like to evaluate a modification of operating conditions of this type of pump. The lot of manufacture was found to be important in that pumps made in the same lot had temperatures and blood factor levels closer to each other than to pumps made in other lots. So we might select two pumps from each lot, randomly choosing one on which to make the modification. In running the experiment, it would be important to record not only which pump was modified, but which lot each pair of pumps represents. In the analysis, pumps from the same lot would be compared with each other (say, by subtracting blood factor level of the modified pump from blood factor level of the unmodified pump). In effect, the experimental unit is the lot of manufacture.

> **In a paired design, two measurements of the same quantity are made on the same experimental unit, one under treatment conditions and the other as control.**

The following is a simple example of a paired design.

EXAMPLE 3.17
▷ **Milk and honey and hemoglobin.** Animal experiments suggested that honey in a diet might raise hemoglobin level. Because hemoglobin levels vary and may increase for reasons apart from honey, a control was essential. A researcher designed a study involving six pairs of twins in a children's home. A coin toss decided which twin in each pair would receive honey. On each day over a period of 6 weeks, all children were given a cup of milk at 9 P.M. The twins assigned to honey always received a tablespoon of honey dissolved in the milk. The researcher then observed which twin in each pair had the greater increase in hemoglobin. So there were six observations: the honey twin had greater or less hemoglobin increase in each pair. (The data and an analysis are considered in Exercises 6.10 and 7.1.) ◁

A paired design tends to decrease variability in comparing responses. Hemoglobin levels vary from one individual to the next. Changes in hemoglobin over time also vary, perhaps from drinking milk even without honey. Because of genetic and environmental similarities one might expect less variability in change in hemoglobin within twinships—the researcher in the example certainly expected this. It would be better yet to consider the *same* individual in both treatment groups at the same time. This would ensure no differences apart from treatment and we could observe the pure treatment effect. But it is not possible. Sometimes, the type of treatment allows for treating the same individual at different times—see Section 3.6. Eliminating or reducing individual response variability would make the effect of honey easier to discern.

EXAMPLE 3.18
▷ **Darwin's comparison of cross- vs self-fertilized plants.** Exercise 2.1 (page 13) gave data from one of many experiments used by Darwin[21] to judge whether cross-fertilized plants are superior to self-fertilized plants. First, the scientific question that he posed:

> It often occurred to me that it would be advisable to try whether seedlings from cross-fertilised flowers were in any way superior to those from self-fertilised flowers. But as no instance was known with animals of any evil appearing in a single generation from the closest possible interbreeding, that is between brothers and sisters, I thought that the same rule would hold good with plants; and that it would be necessary at the sacrifice of too much time to self-fertilise and inter-cross plants during several successive generations, in order to arrive at any result. I ought to have reflected that such elaborate provisions favouring cross-fertilisation, as we see in innumerable plants, would not have been acquired for the sake of gaining a distant and slight advantage, or of avoiding a distant and slight evil. Moreover, the fertilisation of a flower by its own pollen corresponds to a closer form of inter-breeding than is possible with ordinary bi-sexual animals; so that an earlier result might have been expected.
>
> I therefore determined to begin a long series of experiments with various plants, and these were continued for the following eleven years; and we shall see that in a large majority of cases the crossed beat the self-fertilised plants. Several of the exceptional cases, moreover, in which the crossed plants were not victorious, can be explained.

Darwin's design:

> My experiments were tried in the following manner. A single plant, if it produced a sufficiency of flowers, or two or three plants were placed under a net stretched on a frame, and large enough to cover the plant (together with the pot, when one was used) without touching it. This latter point is important, for if the flowers touch the net they may be cross-fertilised by bees, as I have known to happen; and when the net is wet the pollen may be injured. I used at first "white cotton net," with very fine meshes, but afterwards a kind of net with meshes one-tenth of an inch in diameter; and this I found by experience effectually excluded all insects excepting Thrips, which no net will exclude. On the plants thus protected several flowers were marked, and were fertilised with their own pollen; and an equal number on the same plants, marked in a different manner, were at the same time crossed with pollen from a distinct plant. The crossed flowers were never castrated, in order to make the experiments as like as possible to what occurs under nature with plants fertilised by the aid of insects. Therefore, some of the flowers which were crossed may have failed to be thus fertilised, and afterwards have been self-fertilised. But this and some other sources of error will presently be discussed. In some few cases of spontaneously self-fertile species, the flowers were allowed to fertilise themselves under the net; and in still fewer cases uncovered plants were allowed to be freely crossed by the insects which incessantly visited them. There are some great advantages and some disadvantages in my having occasionally varied my method of proceeding; but when there was any difference in the treatment, it is always so stated under the head of each species.
>
> Care was taken that the seeds were thoroughly ripened before being gathered. Afterwards the crossed and self-fertilised seeds were in most cases placed on damp sand on opposite sides of a glass tumbler covered by a glass plate, with a partition between the two lots; and the glass was placed on the chimney-piece in a warm room. I could thus observe the germination of the seeds. Sometimes a few would germinate on one side before any on the other, and these were thrown away. But as often as a pair germinated at the same time, they were planted on opposite sides of a pot, with a superficial partition between the two; and I thus proceeded until from half-a-dozen to a score or more seedlings of exactly the same age were planted on the opposite sides of several pots. If one of the young seedlings became sickly or was in any way injured, it was pulled up and thrown away, as well as its antagonist on the opposite side of the same pot.
>
> The soil in the pots in which the seedlings were planted, or the seeds sown, was well mixed, so as to be uniform in composition. The plants on the two sides were always watered at the same time and as equally as possible; and even if this had not been done, the water would have spread almost equally to both sides, as the pots were not large. The crossed and self-fertilised plants were separated by a superficial partition, which was always kept directed towards the chief source of the light, so that the plants on both sides were equally illuminated. I do not believe it possible that two sets of plants could have been subjected to more closely similar conditions, than were my crossed and self-fertilised seedlings, as grown in the above described manner.

Darwin's paired experiments were meticulously developed and carried out. The point of the "closely similar conditions" is to minimize variability and is typical of well-designed paired experiments. His explanation is reasonably thorough. This was a published version and would be expected to be somewhat abbreviated. In your lab or other report you should give more details about the experimental conditions: temperature, humidity, amount of water, amount of sunlight, and any other information that might affect growth. One treatment (say, crossed) may be better under the conditions of your experiment but not under all conditions. Giving the conditions of your experiment will help others who may try to replicate your results or explore other conditions. ◁

To use a paired design, first identify a pairing mechanism. Sometimes this is impossible or impractical. For example, suppose one wants to know whether machine 1 or machine 2 makes better quality items. For some types of items, it may be possible to use raw material lot or day of manufacture as a pairing mechanism. But the logistics of keeping the pairings together may be impossible. To judge the value of a paired design, decide whether the reduced within-pair variability is sufficient to make up for the logistical problems. In the machine 1 vs machine 2 comparison, say, decide whether the variability across lots or days is sufficiently greater than within lots or days to warrant a more complicated and more expensive design.

An alternative to pairing is to assign some experimental units to treatment (randomly) and other, distinct units to control. This is a **parallel design.** It gives rise to two samples, one for treatment and the other for control. There are various ways of generating parallel designs—the differences depending on the population addressed. For example, suppose we want to compare distances traveled by type A and type B golf balls. We employ a golfer to hit some balls of both types and we record the distances traveled. A problem with this particular parallel design is that its conclusions may be restricted to the golfer chosen. A different parallel design is to have many golfers hitting one ball each; some hit type A and the rest hit type B. There will be more sample variability than in the single-golfer design, but the study's conclusions will apply to more than one golfer. (If we had many golfers each hit one A *and* one B and take the distance on A minus the distance on B as the measurement of interest, there would then be one measurement per golfer and we would have a paired design.)

> **A parallel design has two sets of experimental units, one treatment and the other control.**

The following example uses a simple parallel design. We will address the setting and data in detail in Chapters 8 and 9.

EXAMPLE 3.19
▷ **Malaria and sickle cells.** In the malaria/sickle-cell study[22] of Example 1.3 (page 6), 15 sickle-cell carriers and 15 noncarriers were injected with malaria parasites. A carrier is someone who is *heterozygotic,* that is, has inherited one copy of the sickle-cell gene—from either parent. A noncarrier is someone who does not have a sickle-cell gene (that is, someone who has not inherited the gene from either parent). The quantities of interest were the numbers of cases of malaria in the two groups.

This study was controlled in the sense that it had controls. But it was observational in that the researchers did not make the two groups different—one had the gene and the other did not. All 30 subjects received the injection. So assignment could not have been randomized. The only real control (in the sense of regulate) exercised by the investigators was in selecting subjects. The process of selecting subjects is susceptible to biases, either conscious or subconscious. For example, the investigators might have chosen robust subjects for the noncarrier group and feeble subjects for the carrier group. ◁

The next example is yet another showing the need for controls, as well as that sometimes there are several types of control in a parallel study. It also shows that moderately large sample sizes may be necessary—one observation per group may not be enough. (You will analyze the data from this example in Exercise 12.1.)

EXAMPLE 3.20
▷ **Eyewitness testimony.** An experiment[23] reported by psychologist Elizabeth Loftus addressed the susceptibility of eyewitnesses to false suggestions. She showed 40 students a 3-minute videotape of a classroom lecture being disrupted by 8 demonstrators and immediately administered a questionnaire. One of the questions asked of half the students was whether the leader of the "4 demonstrators" was a male; the other 20 students were asked whether the leader of the "12 demonstrators" was a male. A week later she asked all 40 students how many demonstrators they had seen. Those who had been fed the "4 demonstrators" line reported seeing an average of 6.40 demonstrators, whereas those who had been given the "12 demonstrators" questionnaire reported seeing an average of 8.85. (Deleting the two students who said exactly "4" from the first group and the two who said exactly "12" from the second still gave a substantial difference between the two groups.)

Suppose Loftus had given all 40 students the same misleading question, say, with the suggestion that there were "4 demonstrators" and suppose the average of their responses a week later had been 6.4. Since this is less than 8, one might be tempted to conclude that the misinformation had influenced the eyewitness accounts. But people may be naturally inclined to see a smaller number of objects than actually exist. Such a bias would be present even if she had 40,000 students in the sample.

Students in the second group who were told there were "12 demonstrators" serve as a comparison for the first group; the fact that their average response was substantially greater than the average for the "4 demonstrators" group suggests that the students were influenced by the questionnaire. The investigator might have considered having a third group of students, who would have been given no misinformation about the number of demonstrators. This would help our understanding of how what people see relates to what they think they see. But if the total number of students is 40 and these have to be divided among the groups, a third group would have subtracted from the other two sample sizes. So a three-sample experiment might on balance be less informative than the one actually performed.

With regard to sample size, suppose there were only two students, and that the questionnaire suggested "4" to one and "12" to the other. Then suppose that a week later the first responded "5" and the second "10." You can easily imagine that both responses could have come from the same population, that is, neither had been affected by the misinformation. The actual experiment is convincing only because the number of students involved is moderately large. ◁

In Section 3.1, we described an experiment with one black rock that heats to 95°F and one white rock that heats to 78°F, claiming that this is fairly good evidence of a difference. Why is this conclusion different from the experiment with one "4-demonstrator" student saying "5" and one "12-demonstrator" student saying "10"? The two rocks responded differently, as did the two students. But the heat absorbed by a rock seems much less variable than the observed difference of 17°F, whereas the difference of five demonstrators in the responses of two students seems well within normal variability of responses, with or without being prejudiced by a questionnaire. Saying this in another way, from your experience in similar circumstances you would expect other black rocks to heat to within a few degrees of 95°F and other white rocks to heat to within a few degrees of 78°F, but you would not be too surprised if the next student who was told last week that there were four demonstrators responds that she saw 11. You regard these experiments differently because, in comparison with the observed differences, you regard there to be less variability in the rock experiment than in the eyewitness experiment.

Sometimes a pairing mechanism can be identified in a parallel study. In this case, it is not clear when analyzing the study whether to regard it as parallel or paired. The next example is a case in point.

EXAMPLE 3.21
▷ **6-MP and leukemia.** Example 2.6 (page 16) referred to a study designed to evaluate the effectiveness of 6-mercaptopurine (6-MP) for the treatment of acute leukemia. Patients were randomized to therapy: half to 6-MP and the other half to placebo. Patients were assigned in pairs. The first patient was assigned to receive 6-MP if a coin toss resulted in heads, and to receive placebo if it was tails; the second patient was assigned to the other "therapy." This process was repeated for the third and fourth patients, fifth and sixth patients, and so on. For each pair of patients, the investigators recorded whether the 6-MP patient or the placebo patient stayed in remission longer.

The experimental unit was a pair of patients who happened to arrive next to each other, so the pairing dimension was time of treatment. Members of pairs were not twins, nor did they require similar levels of disease. The investigators relied on the randomization to balance any differences.

If the pairing is artificial, then it would be more appropriate to analyze this as a parallel study and compare the two sets of times of remission, one for 6-MP and the other for placebo. The investigators carried out an analysis appropriate for a parallel design as well as a paired analysis. ◁

Analyzing parallel studies is more difficult than analyzing paired studies. The comparison in the latter is within experimental units and so there is but one sample to consider. In the honey-in-milk study of Example 3.17, there is just one response per set of twins and this response is a comparison of treatment with control. In parallel studies there is no within-unit comparison. Instead, the treatment units are combined and compared with the control units that have been similarly combined—a two-sample comparison.

The analysis used for paired studies depends on the type of response. If there are two possible responses or observations—the honey twin has higher or lower hemoglobin—then Chapters 6 and 7 apply. In the general case, in which any number of responses is possible (such as Darwin's experiment with cross- and self-pollinated seeds in Exercise 2.1 and Example 3.18), then Chapter 9 applies.

Analyzing parallel studies is the subject of Chapter 8 when there are two possible responses or observations—one has malaria or not. The general case, in which any number of responses is possible (such as measurements of force required to remove cannulae in Exercise 2.3), is the subject of Chapter 10.

We will not consider analyses of all possible designs. For example, a design that would be better than either the single-golfer or many-golfer experiments described earlier is to have many golfers, each of them hitting many balls of both types. We will not consider analyses for data from such designs. A two-sample analysis that ignores the possible differences in golfers and combines all distances into two sets of observations—one for type A balls and the other for type B—might lead to wrong conclusions if some of the golfers hit balls substantially farther than others. A paired analysis based on a single indication for each golfer—such as whether the longest ball hit by that golfer was a type A or a type B—overlooks much of the available information. A good general reference for design issues is Snedecor and Cochran.[24]

EXERCISES

3.15 You plan a study to see whether men and women differ in their attitudes to legalized abortion. Describe a parallel design for this study. Describe a paired design, being sure to address the pairing mechanism.

3.16 Consider comparing the durabilities of brand A and brand B bicycle tires. You purchase four tires of each brand from different stores. You hire eight bicyclists to use these tires on their front wheels and ride for 500 miles, at which time you will test the tires' remaining strength. (The rear tires are irrelevant in this study.) How would you assign the tires? Is this a parallel or paired design?

3.17 Exercise 2.1 (page 13) gives data from an experiment carried out by Darwin to learn whether cross- or self-pollination would produce more vigorous seeds. As indicated in Example 3.18, this is a paired design. How would you change the experiment to make its design parallel?

3.18 Exercise 2.2 (page 13) described an experiment for testing a drug's ability to ease the symptoms of congestive heart failure. Was the design paired or parallel? How would you change the experiment to make it the opposite?

Crossover Designs

Generalizing paired designs, more than one treatment can be assigned to each experimental unit. The following is an example.

EXAMPLE 3.22
▷ **Does Mozart make you smarter?** An article[25] purports to show that listening to Mozart improves abstract/spatial reasoning. Thirty-six students participated in an ex-

periment. They took three abstract reasoning tests, one each after three different 10-minute listening conditions. The three listening conditions were (1) a Mozart sonata, (2) relaxation instructions designed to lower blood pressure, and (3) silence. The average IQ scores were 119, 111, and 110. The interpretation of these numbers is not clear. The *Nature* article does not contain the test scores for these 36 individuals so we cannot carry out an appropriate analysis, taking into account the individuals' innate abilities.

More importantly, the article does not indicate the order of administration of the three listening conditions, whether the order was the same for all students, or whether the same test was always administered after the same listening condition. How much faith I put in their conclusion depends on the answers to these questions. If the order was the same for all students or if the same test always followed the same listening condition, then the differences could easily be attributed to training, boredom, or differences in tests and I would give the conclusion essentially no credence. If the design was a *crossover* (defined below)—in which case the authors should have said so—then I am still concerned that the subjects were not blinded to the treatment assignment (nor could they be!) and that the results are therefore biased. But at least then I could place a little credence in the conclusion. ◁

A **crossover design** (also called **changeover design**) with two treatments is a special case of a paired design. The experimental unit is a single subject and the treatments are used in sequence: Half the subjects are chosen randomly to get treatment A followed by B and the other half get treatment B followed by A. In a crossover design with three treatments, A, B, and C, one-sixth of the subjects are assigned to receive them in order ABC, ACB, BAC, BCA, CAB, and CBA. A full analysis of crossover designs is beyond the scope of this book; the interested reader is referred to the book by Senn.[26]

> In a crossover design, subjects are assigned first to one treatment and then to another; the order of treatments is assigned randomly.

The following is an example with two treatments.

EXAMPLE 3.23
▷ **Oral hygiene comparisons.** A study[27] considered improvements in an oral hygiene index among 64 patients with gingivitis, comparing a test compound with placebo. The patients were separated into two groups, randomly. As a result of the randomization, 34 patients were assigned to group 1 and 30 to group 2. Patients in group 1 were assigned to placebo followed by test compound and patients in group 2 were assigned test compound followed by placebo. The results are shown in Table 3.5. (Every entry from the original paper has been multiplied by 6 to make the entries integers, but this change of scale has no effect on any conclusions.) For example, patient

Table 3.5
Data for Example 3.23: Improvements in oral hygiene

	Group 1				Group 2		
	Period Number				Period Number		
Patient Number	1 Placebo	2 Test	t − p	Patient Number	1 Test	2 Placebo	t − p
1	5	11	6	1	2	10	8
2	6	13	7	2	3	15	12
3	4	10	6	3	−1	6	7
4	3	9	6	4	3	10	7
5	3	14	11	5	3	11	8
6	5	11	6	6	2	3	1
7	6	3	−3	7	4	8	4
8	4	2	−2	8	0	8	8
9	4	3	−1	9	1	3	2
10	2	4	2	10	5	13	8
11	0	5	5	11	2	10	8
12	7	8	1	12	0	9	9
13	0	4	4	13	3	8	5
14	3	11	8	14	3	9	6
15	2	9	7	15	0	8	8
16	2	9	7	16	−1	4	5
17	3	7	4	17	3	10	7
18	6	10	4	18	4	15	11
19	0	8	8	19	0	11	11
20	3	9	6	20	4	5	1
21	−3	17	20	21	1	14	13
22	1	14	13	22	3	7	4
23	6	8	2	23	0	8	8
24	6	10	4	24	5	8	3
25	8	4	−4	25	8	2	−6
26	2	5	3	26	7	13	6
27	12	6	−6	27	2	6	4
28	24	1	−23	28	6	2	−4
29	5	10	5	29	1	7	6
30	3	8	5	30	3	3	0
31	3	9	6				
32	3	10	7				
33	13	8	−5				
34	4	7	3				

number 1 in group 1 was assigned to placebo in period 1 and had an improvement of 5 in hygiene score. Then, in the second period, this same patient used the test compound and had an improvement of 11. The improvement of test over placebo (t − p) is also indicated for each patient. For group 1, this is the second period score minus the first period score and, for group 2, this is the first period score minus the second period score.

For the purposes of this book, the analysis of data from a crossover study will be the same as for any other paired study. So the data analyzed in this example will be the 64 patients' improvements, t − p, ignoring sequence of treatment. A defect of this analysis is that giving the test compound first may improve overall hygiene and so the placebo result will look better after the test compound than it does before it—a *residual effect* of the test. Similarly, perhaps giving placebo first improves performance on the test compound. Another defect is that there may be a trend in the patients' hygiene conditions and so period 2 scores will tend to be better (or worse) than period 1 scores, regardless of the treatment used in period 1. These defects can be repaired, but are topics for more advanced courses. You can address them informally, for example, by comparing t − p for group 1 with t − p for group 2 to see whether there is a period effect. ◁

When deciding whether to use a crossover design, decide whether residual or period effects are likely. If they are, then do not use a crossover design. For example, you would not use a crossover design to compare two brands of car wax—the first application provides a foundation for the second and presumably will improve its performance. So if there is no period or residual effect, what is the advantage of a crossover? The answer is that it guards against these possibilities, even without doing a sophisticated analysis. For example, if the results are usually better in the second period than the first period, the crossover allocates the two treatments to the second period equally.

The next example involves three treatments. Its design and analysis are more complicated, of course, but its virtues are the same as for the two-period crossover.

EXAMPLE 3.24
▷ **Three-period crossover trial for asthma.** A double-blind study [28] compared the drugs formoterol and salbutamol in aerosol solutions with a placebo solution for patients suffering from exercise-induced asthma. Patients were randomized to receive these treatments in the six possible arrangements on three different days. Two hours after each treatment, the patients exercised and had their forced expiratory volumes (in ml) measured. There may have been differences in the patients' abilities to exercise, but each patient received all three treatments and so the treatments can be compared *within* patient. The results for the 30 patients at one of the centers in this study are given in Table 3.6. (Patient numbers go up to 31 but patient 29 apparently dropped out of the study.) The sequence indicates the treatment that the patient received in each period. For example, patient 1 (FSP) received F in period 1, S in period 2, and P in period 3. On the other hand, patient 2 (FPS) received F in period 1, P in period 2, and S in period 3.

The analysis of this study may not be obvious to you. The standard analysis considers treatment differences in the presence of possible period and residual effects. This analysis is beyond our scope. But in later chapters we will compare the treatments in this study without worrying about the possibility of period and residual effects.

Table 3.6
Three-period crossover study for Example 3.24:
Forced expiratory volume (ml) after F, S, or P in aerosol

Patient Number	Sequence	Period 1	Period 2	Period 3
1	FSP	35	32	29
10	FSP	34	28	22
17	FSP	23	22	17
21	FSP	23	13	14
23	FSP	30	24	18
2	FPS	31	18	24
11	FPS	28	16	22
14	FPS	31	16	14
19	FPS	23	15	22
25	FPS	30	17	26
28	FPS	31	21	28
3	SFP	21	32	10
12	SFP	16	23	16
18	SFP	16	14	8
24	SFP	31	32	10
27	SFP	28	31	20
4	SPF	22	11	26
8	SPF	28	20	28
16	SPF	24	17	34
6	PFS	22	25	24
9	PFS	22	32	33
13	PFS	8	14	10
20	PFS	9	13	15
26	PFS	17	26	24
31	PFS	14	25	22
5	PSF	9	19	29
7	PSF	15	26	20
15	PSF	12	22	27
22	PSF	24	26	38
30	PSF	19	27	28

EXERCISES

3.19 Exercise 2.2 (page 13) considers a study in which an experimental drug designed to ease the symptoms of congestive heart failure was given to 32 patients. (See also Exercise 3.18, page 91.) A question of interest was whether the patients could walk farther while on the drug than while off. All 32 patients were encouraged to walk as much as possible for two consecutive weeks, during the second week of which they were also given the drug. How would you redesign this study using the same 32 patients, but in a crossover? Tell why such a design could be an improvement. Also, tell why it may be inferior.

3.20 In this section, we considered having golfers hit some balls of type A and some of type B. Suppose we employ 50 golfers and ask each to hit one of each type. How would you use a crossover design for this experiment?

3.21 Design a crossover experiment to evaluate the wear and tear on three different styles (A, B, and C) of tennis shoes after 4 months of use. You enlist 24 volunteers. Each volunteer gets a new pair of shoes for the first 4-month period, trades them in for another new pair for the second 4-month period, and trades in again for a new pair in the third 4-month period. (At the time of analysis, you will compare the wear and tear of the 72 pairs of tennis shoes.)

3.7 Factorial Designs*

This topic is somewhat advanced for an elementary course. But the basic idea is not difficult and it is so important that you should be exposed to it. It is closely related to stratified sampling, but now we are also interested in the variable (or factor) used to stratify.

Suppose you want to understand the effect (on taste, height, or whatever) of increasing yeast and also of increasing flour on baked bread. Consider these experiments:

Experiment 1: Make four loaves at low yeast and four loaves at high yeast levels for a total of eight loaves. Use low flour levels on all eight. We present this schematically, where an entry indicates the number of loaves at that combination:

Experiment 1

		Yeast	
		Low	High
Flour	Low	4	4
	High	0	0

Experiment 2: Make four loaves at low flour and four loaves at high flour levels for a total of eight loaves. Use low yeast levels on all eight. Schematically, these loaves are allocated as follows:

---
*Optional section; not used in the sequel.

Experiment 2

		Yeast	
		Low	High
Flour	Low	4	0
	High	4	0

Running both experiments gives information about both factors and produces 16 loaves. Experiment 1 tells you about the effect of yeast and Experiment 2 tells you about the effect of flour, but both at the low level of the other. In particular, no loaves had both high yeast and high flour.

The following **factorial design** does have this combination, as well as the other three combinations:

Experiment 3: Make two loaves at each of the four possible combinations of levels of yeast and flour for a total of eight loaves. Schematically, these loaves are allocated as follows:

Experiment 3

		Yeast	
		Low	High
Flour	Low	2	2
	High	2	2

This factorial design gives four loaves to assess low yeast, four to assess high yeast, four to assess low flour, and four to assess high flour. This is the same as the combination of Experiments 1 and 2. The rather amazing fact is that Experiment 3 gives estimates that are as reliable about the effects of both yeast and flour as does the combination of the other two experiments. And it does so at half the expense!

There is a bonus to using a factorial design. Suppose a high yeast level is better than low yeast and a high flour level is better than low flour, each at low levels of the other. Experiments 1 and 2 taken together would not tell you the effect of either increasing yeast or increasing flour at high levels of the other. There may be an **interaction.** For example, increasing yeast from low to high levels may not be an improvement when flour is also high.

> **In a factorial design, units are assigned to all possible combinations of factors of interest; usually, a small number of units is assigned to each combination.**

Experiment 4: The preceding design has two factors: yeast and flour. It is called a 2×2 factorial design because both factors appear at two levels, low and high. It is easy to add more factors. For example, including baking time (short or long) gives rise to a three-factor design called a $2 \times 2 \times 2$ factorial:

Experiment 4

			Yeast	
			Low	High
Bake Short	Flour	Low	1	1
		High	1	1
Bake Long	Flour	Low	1	1
		High	1	1

Experiment 4 allows for estimating the effects of yeast and flour as reliably as in Experiment 3, and for the same expense, but it also allows for estimating the effect of baking time—and for assessing various interactions as well.

Experiment 5: The factors can enter at more than two levels. For example, the following is a $3 \times 2 \times 2$ factorial design (using 12 loaves), with yeast at three levels:

Experiment 5

			Yeast		
			Low	Med.	High
Bake Short	Flour	Low	1	1	1
		High	1	1	1
Bake Long	Flour	Low	1	1	1
		High	1	1	1

Experiment 6: If a larger sample size is desired, any of the designs mentioned can be repeated. For example, Experiment 5 can be modified to include 24 loaves:

Experiment 6

			Yeast		
			Low	Med.	High
Bake Short	Flour	Low	2	2	2
		High	2	2	2
Bake Long	Flour	Low	2	2	2
		High	2	2	2

EXAMPLE 3.25
▷ **Blowing away distress.** Children with cancer undergo many venipuncture procedures, usually for drawing blood or administering chemotherapy. Most young children hate needle sticks and many have to be restrained by their parents during venipuncture. Some psychologists theorized that distracting the children might ease their distress during these procedures. Thirty children were admitted to a study [29] designed to address this question. The children ranged in age from 3 to 9 years. All had required restraining by a parent or nurse at their previous venipuncture. Investigators randomized the children into two groups of 15 each. The first group experienced behavioral intervention and the second served as a control.

Children in the intervention group were instructed in attentional distraction, paced breathing, and positive reinforcement. To distract these patients during venipuncture, their parents coached them to blow slowly on a party blower toy and counted out loud. Positive reinforcement consisted of rewards if the child cooperated. Namely, the child was told that he or she could win a sticker of a popular cartoon or television character for holding still during venipuncture, and another sticker for blowing on the toy. Children in the control group received no intervention and their parents were instructed to use whatever techniques they had found helpful in controlling their children's distress at previous venipunctures.

Seven of the 30 children did not complete the study (six withdrew from active therapy and one died). Of the 13 patients in the intervention group who completed the study, seven needed restraining. Of 10 patients in the control group, eight needed restraining.

In Exercise 8.8, you will address the question of whether this intervention is effective. But suppose for now that the intervention is effective. It is unclear which component is responsible. Is it the training? The stickers? The toy? The study's design does not allow for addressing this issue since all 13 of the intervention patients received all three. A more informative design would consider the contribution of each factor. Suppose there were 32 patients in the study. Then four could be assigned to each of the eight combinations in a $2 \times 2 \times 2$ factorial design similar to Experiment 4 above:

Balanced factorial design for Example 3.25

Training?	Stickers?	Toy	No toy
Yes	Yes	4	4
Yes	No	4	4
No	Yes	4	4
No	No	4	**4**

This is balanced in the sense that each combination gets the same number of patients. Balance is not necessary. Because of its importance in comparisons, the investigators might want more than four patients in the no-intervention group—the "4" that is shown in boldface in the table. For example, they could have the same size study (30 patients) and assign as follows (of course, there are many other possibilities):

Unbalanced factorial design for Example 3.25			
Training?	Stickers?	Toy	No toy
Yes	Yes	3	3
Yes	No	3	3
No	Yes	3	3
No	No	3	9

I have a few concerns about some other aspects of the study. These make me wary to conclude that intervention is effective, even if the data were more compelling. First, parents of children in the intervention group may have been more hesitant to restrain their children in order to please the investigators. Second, for obvious reasons the judges could not be blinded as to which treatment was being used, and they may have been biased in favor of the intervention. Third, three different judges were used over the course of the study. While they were trained and tended to agree that a child had or had not been restrained, they might have had slightly different standards in determining whether restraint had occurred. Having a single judge observe all venipuncture sessions would be better; better yet, have the same team of judges at each session. To minimize bias, the judges should be paid for their services, not referenced in any publications, and told in advance that their participation in the study makes them ineligible to participate in any other study run by the investigators. ◁

EXAMPLE 3.26
▷ **Chicken feed.** An experiment[30] was conducted to evaluate the effect on the total weights of sixteen 6-week-old chicks of various components of feed. These components were type of protein [groundnut (GN) and soyabean (SB)], level of protein (0, 1, and 2), and level of fish solubles (0 and 1). In addition, there were two houses (I and II) involved in the experiment and each diet combination was used in each of these houses. The design was $3 \times 2 \times 2 \times 2$ factorial with one chick per combination. The 24 total weights are shown in Table 3.7. The line plot in Figure 3.3 shows the data in pictorial form. (Because showing the vertical axis down to 0 would make all the lines connecting dots look quite flat, I have shown only a section of the vertical axis and not included a horizontal axis at all. The curves may seem to hang in space; I have done this to minimize the chance of misleading you about the comparison of the weights with 0—see Example 2.22 for further discussion of this issue.) The labeling scheme uses the abbreviations already indicated in this order: protein type, level of fish, and house number.

There are several observations one can make from the data. First, in 11 of the 12 diet combinations, the chicks in house I weighed more than those in house II. Second, fish solubles were better in 11 of the 12 combinations for comparing level of fish. Third, increasing soyabean serves to decrease weight (the only exception being increasing from level 1 to level 2 in house II with fish at level 0). On the other hand, increasing the level of groundnut seems to have no benefit. So it is not surprising that the largest chicks were raised in house I, on fish, and soybean at level 0.

Sec. 3.7 / Factorial Designs

Table 3.7
Total weights (in g) of sixteen 6-week-old chicks for Example 3.26

Protein type	Level of fish	Protein level					
		0		1		2	
		House I	II	I	II	I	II
Groundnut	0	6,559	6,292	6,564	6,622	6,738	6,444
	1	7,075	6,779	7,528	6,856	7,333	6,361
Soybean	0	7,094	7,053	6,943	6,249	6,748	6,422
	1	8,005	7,657	7,359	7,292	6,764	6,560

Figure 3.3 Line plot showing weights (in kg) of sixteen 6-week-old chicks depending on diet for Example 3.26

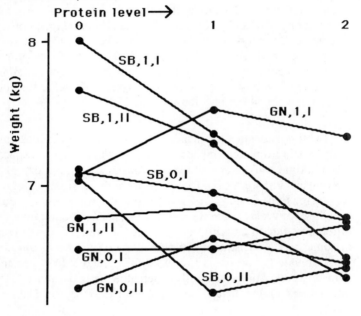

EXAMPLE 3.27
▷ **Cycles to failure of worsted yarn.** A $3 \times 3 \times 3$ factorial experiment[31] related the number of cycles of loading to failure of worsted yarn to length of test specimen (Lng, in mm), amplitude of loading cycle (Amp, in mm), and load (Ld, in g). The schema and results are shown in Table 3.8 (page 102).

Each of the three factors has a clear and consistent effect. Not surprisingly, increasing length increases the number of cycles to failure when the other factors remain constant. Similarly, increasing amplitude or load decreases the number of cycles. The

Table 3.8
Data for Example 3.27: Cycles to failure of worsted yarn

Lng	Amp	Ld	Cycles	Lng	Amp	Ld	Cycles	Lng	Amp	Ld	Cycles
250	8	40	674	300	8	40	1,414	350	8	40	3,636
250	8	45	370	300	8	45	1,198	350	8	45	3,184
250	8	50	292	300	8	50	634	350	8	50	2,000
250	9	40	338	300	9	40	1,022	350	9	40	1,568
250	9	45	266	300	9	45	620	350	9	45	1,070
250	9	50	210	300	9	50	438	350	9	50	566
250	10	40	170	300	10	40	442	350	10	40	1,140
250	10	45	118	300	10	45	332	350	10	45	884
250	10	50	90	300	10	50	220	350	10	50	360

beauty of factorial designs is that there are so many verifications of each of these observations with a total of only 27 trials. For example, consider length. For combinations with Amp = 8 and Ld = 40, number of cycles increases from 674 to 1,414 to 3,636 as Lng increases from 250 to 300 to 350. Think of this as two observations: up and up again. The same is true for the other eight combinations of Amp and Ld for a total of 18 observations of "up." The same is true for both Amp and Ld. So for 27 trials we get 54 observations of "up." ◁

I will not consider the standard analysis of factorial designs in this text. But I will analyze data from factorial designs ignoring the factorial aspect—handling the factors one at a time and collapsing over the other factors. For example, if you are interested in comparing the effect of low yeast to high yeast on bread and you have the above two-factor Experiment 3 or the three-factor Experiment 4, consider the four loaves at each level of yeast as samples from the respective populations. This does not adjust for differences due to any differential effect of flour. A way of controlling for the effects of some factors while analyzing the effect of another is used in Example 13.2. There are various other ad hoc ways of adjusting and I do not want to dissuade you from being creative. But for generally acceptable methods, and also for methods of assessing interactions of the factors in a factorial design, see Chapter 12 of Snedecor and Cochran.[32]

EXERCISES

3.22 In Exercise 3.16 (page 91), you considered comparing the durabilities of brand A and brand B bicycle tires. Again, you have purchased four tires of each brand from different stores. But now you have hired only four bicyclists to ride on the tires for 1,000 miles and you plan to use the tires on both front and back wheels. Which (of the eight) wheels get which tires? Use a 2 × 2 factorial design in which the two factors are brand (A vs B) and wheel (front vs back).

3.23 You want to advise a farmer concerning which of two varieties of corn to plant and how much fertilizer to use. You have one small plot in each of 12 fields to use for an experiment. Set up a

2 × 3 factorial design in which variety of corn has two levels and fertilizer has three levels: low, medium, and high.

3.24 Modifying the design of the previous exercise, you decide you should also address the amount of moisture available. So you add a third factor: irrigation at two levels, low and high. Using these same 12 plots, set up a 2 × 3 × 2 factorial design in which the three factors are variety of corn, level of fertilizer, and amount of moisture.

3.25 Parkinson's disease is a progressive neurological illness affecting hundreds of thousands of people. The largest controlled clinical trial[33] ever conducted for Parkinson's disease involved 800 patients and began in 1987. It was randomized and double-blind. Put yourself at the design stage. You are interested in two factors: treatment with alpha-tocopherol (vitamin E) and treatment with the drug deprenyl. How would you set up a 2 × 2 factorial study? (The results of the actual study will be presented and analyzed in Exercise 9.27.)

3.26 Adriamycin (also called doxorubicin) is among the most important drugs for the treatment of breast cancer. Response to this drug is related to the dose administered, at least up to 60 (in suitable units), but whether it has a benefit beyond 60 units is not clear. A new drug called taxol might also have a benefit. Set up a 3 × 2 factorial design that involves 3,000 patients and considers adriamycin at doses of 60, 75, and 90 units and taxol at two levels: taxol and no taxol. (I helped design such a study. Patients began enrolling in 1994 but the results will not be available for several years.)

Chapter Notes

1. R. T. Clemen, *Making Hard Decisions* (Belmont, Calif.: Duxbury Press, 1991).

2. E. B. Wilson, Jr., *An Introduction to Scientific Research* (New York: Dover Publications, 1990).

3. C. Darwin, *The Effects of Cross and Self Fertilisation in the Vegetable Kingdom,* 2nd ed. (London: John Murray, 1878), Ch. 1. See also G. A. Barnard, "Darwin's data on growth rate of plants," in *Data,* D. F. Andrews and A. M. Herzberg (New York: Springer-Verlag, 1979), 9–12.

4. E. B. Wilson, Jr., *An Introduction to Scientific Research,* (New York: Dover Publications, 1990), 44.

5. J. M. Siegel, *Journal of Personality and Social Psychology* 58 (1990): 1081–1086.

6. D. Freedman, R. Pisani, and R. Purves, *Statistics* (New York: W. W. Norton & Co., 1978), 84 and A-65.

7. *Ibid,* 7–8.

8. J. M. Ruffin, J. E. Grizzle, N. C. Hightower, et al., *New England Journal of Medicine* 281 (1969): 16–19.

9. *Science News* 145 (January 22, 1994): 61.

10. G. A. Green, "Living on borrowed turf," *Natural History* (September 1989): 58–65.

11. Conducted by J. Douglas Bremner and reported in *Science News* 141 (May 16, 1992): 332.

12. I thank Jay Zarnikau of Planery, Inc., for this information.

13. G. A. Thoen, Ph.D. thesis (University of Minnesota, 1977).

14. J. Diaz, "The perils of putting," *Sports Illustrated* 70 (April 13, 1989): 76–79.

15. "Fallout of an invisible war," *Newsweek* (July 25, 1994): 61.

16. *Science News* 143 (January 9, 1993): 22.

17. L.-E. Holm, E. Nordevang, M.-L. Hjalmar, E. Lidbrink, E. Callmer, and B. Nilsson, "Treatment failure and dietary habits in women with breast cancer," *Journal of the National Cancer Institute* 85 (January 6, 1993): 32–36.

18. W. Demark-Wahnefried, D. F. Paulson, C. N. Robertson, et al., "Body dimension differences in men with or without prostate cancer," *Journal of the National Cancer Institute* 84 (September 2, 1992): 1363–1364.

19. "The long and the fat of prostate cancer," *Science News* 142 (September 12, 1992): 171.

20. C. Fleming, J. H. Wasson, P. C. Albertsen, et al., "A decision analysis of alternative treatment strategies for clinically localized prostate cancer," *Journal of the American Medical Association* 269 (May 26, 1993): 2650–2658. Also C. C. Mann, "The prostate-cancer dilemma," *The Atlantic Monthly* (November 1993): 102–118.

21. C. Darwin, *The Effects of Cross and Self Fertilisation in the Vegetable Kingdom,* 2nd ed. (London: John Murray, 1878), 10–13. Quote taken from G. A. Barnard, "Dar-

win's data on growth rate of plants," in *Data,* D. F. Andrews and A. M. Herzberg (New York: Springer-Verlag, 1979), 9–12.

22. Jared Diamond, *Natural History* (February 1989): 8–18.

23. E. F. Loftus, *Cognitive Psychology* 7 (1975): 560–572.

24. G. W. Snedecor and W. G. Cochran, *Statistical Methods,* 6th ed. (Ames, Iowa: Iowa State University Press, 1967).

25. F. H. Rauscher, G. L. Shaw and K. N. Ky, "Music and spatial task performance," *Nature* 365 (1993): 520.

26. S. Senn, *Cross-over Trials in Clinical Research* (Chichester, England: John Wiley and Sons, 1993).

27. D. D. Zinner, L. F. Duany, and N. W. Chilton, "Controlled study of the clinical effectiveness of a new oxygen gel on plaque, oral debris and gingival inflammation," *Pharmacology and Therapeutics in Dentistry* 1 (1970): 7–15.

28. A. N. Tsoy, O. V. Cheltzov, V. Zaseyeva, L. A. Shilinsh, and L. A. Yashina, *European Respiratory Journal* 3 (1990): 235. *Also* S. Senn, *Cross-over Trials in Clinical Research* (Chichester, England: John Wiley and Sons, 1993), 152.

29. S. L. Mann, W. H. Redd, P. B. Jacobsen, K. Gorfinkle, O. Schorr, and B. Rapkin, *Journal of Consulting and Clinical Psychology* (in press, 1990).

30. J. A. John and M. H. Quenouille, *Experiments: Design and Analysis,* 2nd ed. (London: Griffin, 1977). *See also* D. R. Cox and E. J. Snell, *Applied Statistics* (London: Chapman and Hall, 1981): 103–106.

31. E. J. Snell, *Applied Statistics, A Handbook of BMDP Analyses* (London: Chapman and Hall, 1987): 61–66.

32. G. W. Snedecor and W. G. Cochran, *Statistical Methods,* 6th ed. (Ames, Iowa: Iowa State University Press, 1967).

33. The Parkinson Study Group, "Effects of tocopherol and deprenyl on the progression of disability in early Parkinson's disease," *New England Journal of Medicine* 324 (January 21, 1993): 176–183.

4 Probability and Uncertainty

THE EXAMPLES in Chapter 2 deal with samples of experimental results. In particular, a single type of experiment is repeated, each time giving an observation that becomes part of a sample. As the sample size gets larger, an observer learns more and more about the process that is producing the sample. Consider the data in Example 2.6 (B means 6-MP is Better for that pair):

BWBBB   WBBBB   BBBWB   BBBBB   B

Cover these letters with a sheet of paper and move the sheet slowly to the right, uncovering the data one pair at a time. The next result seems less surprising as time unfolds. That is, it becomes clearer that the process producing these letters seems to be giving more B's. The objective of statistical inference is to learn about the process based on the observations produced.

Recall Hawking's definition of a scientific theory from Section 1.1: "... a model of the universe, or a restricted part of it, and a set of rules that relate quantities in the model to observations that we make." A model of this process is a specification of the effectiveness of 6-MP. In particular, a model specifies the proportion of B's we would see if we were to observe the population being sampled. In this case, the population does not exist in advance—it comes into being only as pairs of patients are randomized to treatment in the somewhat artificial setting of the trial. A model is a specification of the proportion of B's in the population. One can make predictions assuming any particular model. Checking a model's predictions is precisely the purpose of the sample. If one model says 6-MP is very effective and another says it is not, then we can get information about which is right by seeing whether B's or W's predominate in the sample.

Exactly what it means to predominate is not always clear. The sampling process is unpredictable. Suppose the study had stopped after six pairs—with 4 B's and 2 W's. But the second six pairs could just as well have been the first six—and all were B's. Understanding this variability in the sampling process is critical to understanding statistics. The language of variability in sampling—as well as more generally—is **probability**, which is the subject of this chapter.

A fundamental concept is an **experiment**. This includes the usual notion of causing something to happen—perhaps making a measurement—and observing the result. For example, mixing two chemicals together and taking the temperature of the combination, administering a drug to a patient and asking whether the pain is relieved, and rolling a die and observing the number on the side facing up when it stops rolling are experiments. But experiments also include simply making observations. Examples are asking someone about their sexual orientation, counting the number of aphids on a plant, and, in the spirit of Example 3.6, finding out whether a Parkinson's patient lives in the city or country. We are interested in experiments or observations that cannot be predicted with certainty. Some of the experiments discussed in this chapter involve gambling devices such as dice, coins, and wheels of fortune. These are useful in developing your skills for dealing with probability in other settings.

> **An experiment is making an observation, usually under known or reproducible circumstances.**

Probability is not limited to experiments, but applies whenever there is uncertainty. For example, no one knows which pass Hannibal used when crossing the Alps, although some historians feel more confident than others about his route. We want to allow questions such as: What is the probability that Hannibal used the Little St. Bernard Pass?

In July 1988, the U.S. Geological Survey said there was a "50 percent probability" that a magnitude 7 or greater earthquake would hit the San Francisco Bay Area in the next 30 years. (The early reports of the Bay Area quake of October 1989 said its magnitude was 7.1, but an earthquake expert tells me that the current consensus among monitoring stations is 6.9.) In 1990, two biophysicists calculated a probability of 1 in 10 billion that a chain of amino acids will fold into a biologically useful form—that is, life. Weather forecasters give a probability of rain. In cases of disputed paternity, blood banks calculate a probability that an alleged father is indeed the father. How should these probabilities be interpreted?

It is not easy to give a clear answer. I have seen representatives of blood banks struggle with the question in court. Few weather forecasters can provide satisfactory answers. Scientists seem little better at answering some questions than are lay people. Most attempts are hopelessly circular: My dictionary defines probability as *chance,* chance as *likelihood,* and likelihood as *probability*! In statistics we need a workable interpretation of probability. I will delay addressing interpretations until later sections. No matter how one *interprets* probabilities, they should behave in accordance with certain rules. The next section considers such rules.

4.1 Properties of Probability

We speak of the probability *of* something or *that* something is true. This "something" is an **event.** We use the notation $P(A)$ for the phrase "probability of event A."

The building blocks of events are **outcomes.** Consider rolling a six-sided die on a flat surface. The die's sides are labeled 1, 2, 3, 4, 5, 6. By convention, the outcome or result of the roll is the number on the side facing up when the die stops rolling. Each outcome is also an event. So, for example, 1 is an outcome and it is also an event. There are six possible outcomes, but there are more than six possible events. Imagine betting on the roll. Each bet defines an event. For example, you might bet that the outcome is odd; call this event {1, 3, 5}. (I will sometimes drop the brackets and, for example, write this event simply as 1, 3, 5.) Or you might bet that the outcome is not a 6. Our notation for the **complement** of 6, read "not 6," is ~6. So ~6 = {1, 2, 3, 4, 5}.

Because we will want to consider probabilities of this *or* that, we will need to add events together. The sum or **union** of two events includes those outcomes that are in either event or both events. For example, {1, 2} ∪ {3} = {1, 2, 3}. So the cup symbol, ∪, means the same thing as a comma. It would be redundant to repeat the labels of outcomes that two events being added have in common: {2, 3, 4} ∪ {3, 4, 5} = {2, 3, 4, 5}.

The biggest possible event is U, for universe. In rolling a six-sided die, U = {1, 2, 3, 4, 5, 6}. The smallest possible event is empty; denote the empty event as ∅.

The basic properties of probability are intuitive. I will convey some of this intuition before listing them.

Since {1, 2, 3} is the result of adding events {1, 2} and {3}, the probability of 1, 2, 3 should be the sum of the probabilities of {1, 2} and {3}: that is, $P(1, 2, 3) = P(1, 2) + P(3)$. And probabilities should not be negative. Taken together, these two statements mean that event 1, 2, 3 cannot have a smaller probability than does event 1, 2—the former can be gotten from the latter by adding 3 to it and if its probability were smaller, $P(3)$ would have to be negative. Also, since 1, 2, 3 = 1 ∪ 2 ∪ 3, we will want $P(1, 2, 3) = P(1) + P(2) + P(3)$. Conventional practice is to set the probability of the biggest possible event, U, equal to 1, or 100%; so in our example, $P(U) = P(\{1, 2, 3, 4, 5, 6\}) = 1$.

> **Basic Properties of Probability**
>
> 1. If A is any event, then $P(A) \geq 0$.
> 2. If U is the largest event, then $P(U) = 1$.
> 3. If events A and B have no outcomes in common, then $P(A \cup B) = P(A) + P(B)$.

Other properties can be derived from these basic properties. I will mention one, by way of example, and give others as the need arises. For any event A, every outcome in U is contained either in A or in its complement ~A. So, $U = A \cup$ ~A. This means that $P(A \cup$ ~$A) = 1$. Also, no outcome can be in both A and ~A, so according to property 3, $P(A \cup$ ~$A) = P(A) + P($~$A)$. Combining these two gives $P(A) + P($~$A) = 1$. Rewriting this:

$$P(A) = 1 - P(\sim A)$$

This derived property comes in handy when calculating probabilities. It says that one way to calculate the probability of an event is to find the probability of its complement and subtract that from 1.

Since $\sim U = \emptyset$ and $P(U) = 1$, it follows from this derived property that $P(\emptyset) = 0$. An interpretation of U is that it is a **certain** event, occurring with probability 1. But certainty is relative. Our probability of rolling either 1, 2, 3, 4, 5, or 6 on a die may be 1. But a mouse could come from nowhere, grab the rolling die, and disappear. We did not include that possibility in U. There are always possibilities we have to ignore or we would never be able to make progress. Specify a U and recognize that your probabilities depend on it. If U changes, your probability of an event may change as well.

Similarly, $\emptyset$ is an **impossible** event, occurring with probability 0. However, there may be events other than $\emptyset$ and U that have probabilities 0 or 1. For example, the number 5 might be listed as an outcome for a die roll and we regard it to be impossible—perhaps the opposite side is rounded and we think the die could not possibly come to rest with that side down. As always, $U = \{1, 2, 3, 4, 5, 6\}$ would have probability 1, but in such a case so would $\{1, 2, 3, 4, 6\}$.

$$P(U) = 1 \quad \text{and} \quad P(\emptyset) = 0$$

All probabilities obey the basic properties. But keep these properties in perspective. They tell you how probabilities behave, but they do not tell you how to calculate or assign probabilities in any particular application. The latter requires an additional definition or interpretation for probability. Various interpretations have been suggested. The two standard ones are based on (1) long-run frequencies and (2) degrees of belief. Either approach is possible and both lead to fruitful methods of data analysis. Which interpretation one uses depends on one's philosophy. Unfortunately, some aspects of any approach to statistics depend greatly on the interpretation given to probability. I will describe both interpretations in the next three sections.

I find the arguments for degrees of belief more compelling and so we will adopt this interpretation. One benefit is that probability based on degrees of belief is more intuitive. Another is that while interpretation (1) can work very well for analyzing data from a given experiment, interpretation (2) is consistent with the overall inferential aspects of the scientific method discussed in Section 1.1.

There is one particular probability assignment that is easy to discuss and manipulate—the assignment for which each possible outcome has the same probability. Such outcomes are called **equally likely.** This assignment plays a role in the development of statistics regardless of the interpretation given to probability. For example, regarding the die roll, to have *equally likely* outcomes means $P(1) = P(2) = P(3) = P(4) = P(5) = P(6) = \frac{1}{6}$; such a die is called *fair*. A related notion is selecting an object from a set in such a way that each possible object has the same probability; this is called **random selection.**

When outcomes are equally likely, calculating probabilities of events simply entails counting the number of outcomes in the event.

EXAMPLE 4.1
▷ **Selecting from a deck of cards.** A card is selected randomly from a deck of 52 cards; that is, all the cards are equally likely to be selected. What is the probability that the card selected is a spade? Of the 52 cards, 13 are spades: ♠A, ♠2, ♠3, ♠4, ♠5, ♠6, ♠7, ♠8, ♠9, ♠10, ♠J, ♠Q, ♠K. Each spade contributes probability $\frac{1}{52}$ to the total. Therefore, the probability of a spade is $\frac{13}{52}$, or $\frac{1}{4}$. ◁

EXERCISES

4.1 Consider a single roll of a six-sided die and events $\{1\}$, $\{5\}$, $\{2, 4\}$, $\{2, 6\}$, $\{5, 6\}$, $\{1, 3, 5\}$, $\{2, 4, 6\}$, $\{3, 4, 5\}$, $\{1, 2, 3, 4\}$, $\{1, 3, 4, 5\}$, $\{1, 3, 5, 6\}$, $\{2, 4, 5, 6\}$, $\{1, 2, 3, 4, 6\}$, $\{2, 3, 4, 5, 6\}$. Pick one of these events to match with each of the following events:
(a) The outcome is odd.
(b) The outcome is larger than 4.
(c) The outcome is larger than 2 and smaller than 6.
(d) The outcome is 5.
(e) The outcome is not a 5.
(f) The union of (a) and (c).
(g) The union of (b) and (d).
(h) The complement of (a).
(i) The complement of (f).

4.2 Suppose $U = \{a, b, c\}$ and $P(a) = \frac{1}{2}$, $P(b) = \frac{1}{3}$, $P(c) = \frac{1}{6}$. Find $P(a, b)$, $P(b, c)$, and $P(\sim b)$.

4.3 Consider the roll of a fair six-sided die. A die roll can be represented using the *tree diagram* shown. Find the following:
(a) $P(1, 2, 3)$
(b) $P(\sim 6)$
(c) $P(\sim\{1, 2\})$
(d) $P(\{1, 2\} \cup \{4, 6\})$
(e) P(outcome is odd)

Tree diagram for Exercise 4.3

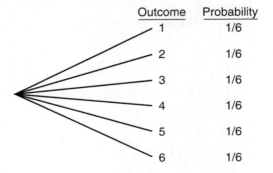

4.4 Suppose you roll two dice, a white one (with black dots) and a black one (with white dots). The 36 possible outcomes are shown in the figure.

The 36 possible outcomes for Exercise 4.4

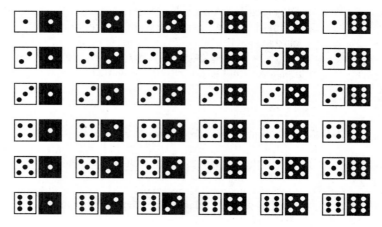

Another way to show these 36 outcomes is using ordered pairs, as follows, with the number on the white die shown first:

```
11  12  13  14  15  16
21  22  23  24  25  26
31  32  33  34  35  36
41  42  43  44  45  46
51  52  53  54  55  56
61  62  63  64  65  66
```

Still another way to view the possible outcomes is by using the tree diagram shown on the facing page. Assume that these 36 outcomes are equally likely—as indicated on the tree diagram. Find the probabilities of these events:

(a) A = black die results in 4.
(b) B = white die result is greater than black die result.
(c) C = sum of results of white and black dice is 2, 3, or 12 (this is called *crapping out* when playing craps).
(d) D = bigger of the two numbers is 5 (including the outcome 55).

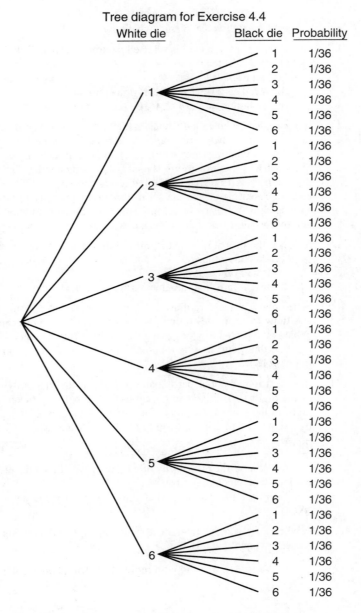

Tree diagram for Exercise 4.4

4.5 Ms. X orders cod, Mr. Y orders trout, and Ms. Z orders tuna. The server forgets who ordered which and, instead of asking, simply places them down randomly on the table. What is the probability that:
  (a) Ms. X gets the cod?
  (b) Mr. Y gets the trout?
  (c) Ms. Z gets the tuna?

(d) All three get what they ordered?
(e) Exactly two people get what they ordered?

4.6 As in Example 4.1, a card is selected randomly from a deck of 52 cards. What is the probability that the card selected is
(a) Not a spade?
(b) A red card (that is, a heart or a diamond)?
(c) An honor card (**10, J, Q, K, A**), but not a spade?

4.7 In a carnival game called Chuck-a-Luck, the player pays \$1, picks a number from 1 to 6, and then rolls three dice. The player wins back \$1 for each die that shows her number. (So her possible net winnings are $-1, 0, 1, 2$.) Assuming each of the 216 outcomes has the same probability, what is the probability she ends up losing her \$1? [*Hint:* A tree diagram is too big to draw, but think of a tree with three levels and six possibilities at each level, or 216 branches, each representing a possible set of three outcomes. The number she picks does not matter—suppose it is 4. Losing on all three dice corresponds to another tree. This tree has three levels and five possibilities at each level, one for each losing number: 1, 2, 3, 5, 6.]

4.8 Suppose the probability that it rains both today and tomorrow is $\frac{1}{5}$, today but not tomorrow is $\frac{1}{10}$, not today but tomorrow is $\frac{1}{10}$, and neither day is $\frac{3}{5}$. What is the probability that:
(a) It will rain today?
(b) It will rain at some time during these next 2 days?

4.9 When playing five-card draw poker, you are dealt ♦5, ♥6, ♠8, ♦9, ♠J. If you toss the ♠J and draw to an inside straight, you get to draw one more card to go with ♦5, ♥6, ♠8, ♦9. What is the probability of getting a straight? (You must get one of the four **7**'s to get a straight. Assume that your draw is random from among the remaining 47 cards—the ♠J now being impossible.)

4.10 In some football pools, participants pay \$1 and are randomly assigned a ticket that contains two of the digits 0, 1, 2, and so on, up to 9. The winning ticket is the one with the same last two digits of the score at the end of the game. For example, the ticket $\boxed{2; 6}$ wins if either team wins by a score of 6 to 2, 12 to 6, 36 to 12, 42 to 16, and so on. There are 55 tickets possible:

$\boxed{0; 0}, \boxed{0; 1}, \boxed{0; 2}, \ldots, \boxed{0; 9}, \boxed{1; 1}, \boxed{1; 2}, \boxed{1; 3}, \ldots, \boxed{9; 9}$

(a) What is the probability of winning if you buy a ticket? (Answer from the perspective of not knowing which digits you receive.)
(b) If you buy a ticket, what is the probability you get a double number ($\boxed{0;0}$, $\boxed{1;1}$, and so on)?

4.11 A wheel of fortune with eight stops is shown. Upon spinning the wheel, assume the eight outcomes are equally likely.

Wheel of fortune with eight stops for Exercise 4.11 (showing outcome 6)

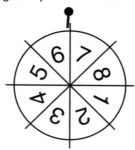

Find the probability that the wheel stops at
- (a) An odd number.
- (b) A number bigger than 5.
- (c) The union (∪) of the events in (a) and (b).
- (d) The complement (∼) of the event in (b).

4.12 Consider the following bowl with five chips, three Reds and two Greens:

Someone mixes up the chips and asks you to select one chip from the bowl without looking. Assuming that the selection is random, what is the probability that you get a green chip?

4.13 Continuing Exercise 4.12, suppose you got a green chip, which you put in your pocket. Now the bowl contains only four chips. You now select a second chip randomly from the four. What is the probability that this second chip is also green?

4.14 There are 75 chips in the game of BINGO, 15 under each letter as indicated in the figure. The chips are drawn one at a time and without replacing chips that were drawn previously. (Drawing goes on until someone wins, but I am not going to get into how the game is played or how someone wins.) Assume that the first chip is drawn randomly from the 75.
- (a) What is the probability that the first chip drawn is under the B?
- (b) What is the probability that the first chip drawn is under the I?
- (c) Calculate the probability that the first chip drawn is under the B or under the I by counting chips.
- (d) Calculate the probability that the first chip drawn is under the B or under the I using basic property 3 of probability and your answers to parts (a) and (b).

Numbering of chips in BINGO for Exercise 4.14

B	I	N	G	O
1	16	31	46	61
2	17	32	47	62
3	18	33	48	63
4	19	34	49	64
5	20	35	50	65
6	21	36	51	66
7	22	37	52	67
8	23	38	53	68
9	24	39	54	69
10	25	40	55	70
11	26	41	56	71
12	27	42	57	72
13	28	43	58	73
14	29	44	59	74
15	30	45	60	75

4.2 Probability as Long-Run Frequency

The long-run frequency interpretation of probability requires a sequence of repeatable experiments. The probability of an event is the long-run proportion of the time it happens compared with the total number of observations—where *long-run* means in the limit as the total number of observations tends to infinity. Infinite sequences are not possible, so interpret a probability to be the proportion of occurrences in a large but finite number of observations.

> **The long-run frequency of an event is the proportion of the time it occurs in a long sequence of trials.**

Suppose you repeatedly roll a lopsided die and keep track of the outcomes. The die has six sides with labels 1, 2, 3, 4, 5, 6. Define $P(4)$ to be the proportion of the time that the 4 comes up. This makes perfectly good sense if we plan to select one of the *previous* rolls. And it seems reasonable for the *next* roll if the number of previous rolls is very large. But if the number of observations is small—as it will be in many real settings—the definition does not work, and it is not supposed to apply to the short run. For example, if after five rolls we have not yet observed a 4, the frequency is $\frac{0}{5} = 0$. From examining the die it seems clear that 4 has some chance of coming up and so taking its probability to be 0 is unreasonable. In particular, we would not be surprised to get a 4 on the next roll.

Similarly, consider a couple that has five children and all five are boys. They are expecting their sixth child. What is the probability that it is a girl? Again, $\frac{0}{5} = 0$ seems to put too much stock in the rather small amount of available information. Many of you will put no stock at all in the sexes of the previous children and say that the probability is $\frac{1}{2}$, since the long-run frequency of girls in the larger population of births is about $\frac{1}{2}$ (actually slightly less than $\frac{1}{2}$). But this is questionable because the larger population is not perfectly appropriate to the couple in question. The propensity to have girls varies from one couple to the next (it may even vary over time within the same couple, but let's ignore this complication). While one should not assign too much credibility to the small number of previous children, the sex distribution of their children contains some information about their propensity to have girls.

Restricting to settings in which the experiment can be repeated limits the applicability of the long-run frequency interpretation of probability. Repetitions may not be possible. For example, no frequency interpretation can be given to the event that Hannibal passed through the Little St. Bernard, or that Al Edged is Suzy Smith's father, or that the New York Jets will win the Super Bowl next year. Hannibal either did or did not and we did not observe which. We have never discovered the identity of the father of a previous Suzy Smith. And we have not observed the Jets play their next season. Moreover, if we had made any of these observations, there would be no need to discuss probabilities because we would know whether the event occurred. We may have relevant information about these experiments even though we have not observed them. The problem is to convert this information into probabilities—and the long-run frequency interpretation is not much help in this regard.

EXERCISES

4.15 I rolled an ordinary six-sided die 10,000 times and it came up 6 on 1,511 rolls. Using the long-run frequency interpretation, what is the probability of rolling a 6 with this die? Since 10,000 is not infinity, give only two significant digits in your answer. (A possible probability assignment is $\frac{1}{6}$ for each side and for 6 in particular. This may apply for an ideal die, but such a die may be impossible to construct. The die I tossed seems far from ideal. The dots indicating the outcomes 1, 2, 3, 4, 5, and 6 were gouged out of the plastic and then painted. This shifts the center of gravity of the die away from the 6—the 6 is opposite the 1 and so six times as much plastic has been removed from the 6-face as from the 1-face.)

4.16 A stock market index has increased from its previous closing level on 2,593 of the last 5,000 business days (about 20 years). From a long-run frequency perspective, what is the probability that it increases on the next business day? Give two-digit accuracy in your answer.

4.17 Toss a coin a total of 13 times. Keep track of the proportion of heads as you toss. Predict the result of every toss—except for the first—using the proportion of heads from all *previous* tosses. Keep track of your cumulative score. For example, if you have observed five heads and three tails on your first eight tosses, predict $\frac{5}{8}$ for the next (ninth) toss.

Here is how the scoring goes. Suppose you get heads on your first toss. Use $\frac{1}{1} = 1$ to predict the next toss. If the next toss is heads, then you score $(1-1)^2 = 0$; if it is tails, then you score $(0-1)^2 = 1$. Low scores are good, in the sense that they indicate accurate predictions, and large scores are bad. Suppose you have gotten two heads in your first three tosses; the current proportion of heads and your prediction for the fourth toss is $\frac{2}{3}$. If you get heads on your fourth toss, then you score $(1-\frac{2}{3})^2 = \frac{1}{9}$ and if you get tails, then you score $(0-\frac{2}{3})^2 = \frac{4}{9}$. Add up your scores as you toss; after the 13 tosses you should have the sum of 12 numbers.

I carried out an example for such a toss sequence, rounding off the sum of scores to the nearest tenth. (For example, $0 + \frac{4}{4} + \frac{4}{9} + \frac{4}{16} = \frac{61}{36}$, which rounds off to 1.7—my cumulative score after four tosses.) The box identifies my final score:

Toss	1	2	3	4	5	6	7	8	9	10	11	12	13
Result	H	H	T	T	T	H	T	T	H	T	H	H	T
Proportion	$\frac{1}{1}$	$\frac{2}{2}$	$\frac{2}{3}$	$\frac{2}{4}$	$\frac{2}{5}$	$\frac{3}{6}$	$\frac{3}{7}$	$\frac{3}{8}$	$\frac{4}{9}$	$\frac{4}{10}$	$\frac{5}{11}$	$\frac{6}{12}$	$\frac{6}{13}$
Score	—	0	$\frac{4}{4}$	$\frac{4}{9}$	$\frac{4}{16}$	$\frac{9}{25}$	$\frac{9}{36}$	$\frac{9}{49}$	$\frac{9}{64}$	$\frac{25}{81}$	$\frac{16}{100}$	$\frac{36}{121}$	$\frac{36}{144}$
Sum	—	0	1	1.4	1.7	2.1	2.3	2.5	2.9	3.1	3.4	3.7	4.0

My score would have been even larger had I ignored the results and predicted 1 (heads for sure) on all 12 tosses (score = 7.0) or 0 (tails for sure) on all 12 (score = 5.0). So I did better using the accumulating data. But I would have done better yet had I simply predicted $\frac{1}{2}$ every time; this conservative procedure always gives the score of $\frac{1}{4}$ for each toss, no matter the result. My total score after 12 predictions would have been 3.0. So I suffered a penalty for taking the current (small) number of previous tosses too seriously. You will find that you too will be taking the results very seriously—it is most unlikely that your total score will be as low as 3 (although even lower scores are possible if you get an abundance of heads or an abundance of tails).

4.3 Betting Odds and Probability*

The next section considers the other standard interpretation of probability: subjective assessment of degrees of belief. Such assessment requires choosing among monetary propositions. These choices are aided by some understanding of betting odds, which is the subject of the current section. The material in this section is rather easy because it is based on keeping track of money and on simple arithmetic. But it is not essential for understanding later material and so is optional.

There are two reasons for discussing gambling in this course. First, betting odds are closely related to probability. More importantly, every decision is a gamble. When you buy a product, you are betting it is in good condition. When you buy life insurance, you are betting you are going to die sooner rather than later. When you take a new job, you are betting that it is going to be fruitful. A company that markets a product is betting it will be profitable. When the U.S. Food and Drug Administration approves a drug for marketing, it is betting that the drug is safe and effective. To study decision making is to study gambling.

EXAMPLE 4.2
▷ **Pari-mutuel betting.** Betting odds are set by racetracks using a very simple principle: They want to make money. A track guarantees this by taking a fixed percentage—called the **track take**—of every dollar that is bet. So the track does not gamble at all. After it takes its percentage, the remaining money is distributed to the winning bettors. This distribution is based on **track odds.** These depend only on the amounts of money wagered on the various horses (or dogs, or whatever); in particular, objective criteria of ability—if there are any—are irrelevant to the track.

Consider a race among five horses, A through E. The same letter stands for both the name of the horse and for the event that the horse wins in the race. The following table shows the amounts bet:

Horse	Amount bet on horse	"Probability"
A	$500,000	.50
B	250,000	.25
C	100,000	.10
D	100,000	.10
E	50,000	.05
Total	$1,000,000	1.00

The total amount bet on a race, $1,000,000 in this case, is the **track handle.** I put "probability" in quotes in this table because the entries are calculated simply by divid-

---
*Optional section; not used in the sequel.

ing the amounts bet by the track handle. These "probabilities" may not represent the abilities of the horses. Instead, they represent those abilities as perceived by the bettors, who vote with their money. So, in a sense, they represent consensus probabilities. It is easy to see that assigning in this way obeys the properties of probability given in Section 4.1.

Before telling you how track odds are calculated, I will show how odds and probability are related, then define *fair odds*. Suppose $P(A)$ is the probability of event A and $O(A)$ is the **odds** *against* event A. Then,

Relating odds and probability:

$$O(A) = \frac{1 - P(A)}{P(A)} \quad \text{or} \quad P(A) = \frac{1}{1 + O(A)}$$

Suppose you bet $1 on event A and you are paid $O(A)$ in dollars if A happens, where $O(A)$ is defined according to this formula. Then this is a **fair bet** and $O(A)$ is the **fair odds.** Both notions depend on the probabilities: What is fair for one probability is not for another.

Defining probabilities using amounts bet means that the fair odds equal the track odds only if the track take is zero. The following table shows the fair odds for the "probabilities" given in the previous table. It also shows the corresponding payoff for each $2 bet (the smallest bet possible at U.S. tracks).

Horse	Amount bet on horse	"Probability"	Fair odds against	Fair payoff for $2 bet
A	$500,000	.50	50:50 = 1	$ 4.00
B	250,000	.25	75:25 = 3	8.00
C	100,000	.10	90:10 = 9	20.00
D	100,000	.10	90:10 = 9	20.00
E	50,000	.05	95:5 = 19	40.00
Total	$1,000,000	1.00		

The odds and payoff on this table are those the track would give if it paid out the entire handle. For example, suppose horse A wins. The $500,000 bet on horse A is equivalent to 250,000 bets of $2 each. Each of these $2 bettors would be paid $4, so the track would pay out the entire $1,000,000. The same is true if any other horse wins.

Racetracks actually subtract the take from the handle before they pay money back to the bettors. Suppose the track's take is 17%, which is typical. If the handle is $1,000,000, the track keeps $170,000 and returns the remaining $830,000. Suppose horse A wins and that 250,000 people had bet $2 each on A. Each $2 is returned to the bettor, along with a pro rata share of the additional $330,000 paid by the track: $330,000/250,000 = $1.32. So the payoff to someone who bet on A is $2 + $1.32 = $3.32. Calculating in this way for the other four horses gives the table at the top of page 118.

Horse	Amount bet on horse	Adjusted "Probability"	Track odds against	Track payoff for $2 bet
A	$500,000	50/83	33:50 = .66	$ 3.32
B	250,000	25/83	58:25 = 2.32	6.64
C	100,000	10/83	73:10 = 7.30	16.60
D	100,000	10/83	73:10 = 7.30	16.60
E	50,000	5/83	78:5 = 15.60	33.20
Total	$1,000,000	100/83		

The adjusted "probabilities" in the third column of this table are calculated by dividing the amount bet on the horse by the amount returned by the track ($830,000). These are not probabilities at all because they sum to $\frac{100}{83}$ and not to 1. You can always find the track take without knowing the amounts bet if you know this sum: Subtract its inverse from one. In this case, inverting $\frac{100}{83}$ to $\frac{83}{100}$ and subtracting $\frac{83}{100}$ from 1 gives 17%. This method of finding the track take requires finding the adjusted "probabilities." These can be calculated from the amounts to be paid per $2 bet. Subtract $2 from each number in the column of payoffs. Divide the result by 2. This gives the "track odds against" column. Convert these to probabilities using the formula $1/(1 + \text{odds})$. (There are methods that are somewhat easier than this for calculating the track take from a listing of track payoffs.) ◁

EXAMPLE 4.3

▷ **Betting on baseball.** Bookmakers (bookies) have various ways to guarantee making money. When they take bets on certain kinds of sports, such as baseball and boxing, they use a technique similar to that used by racetracks. In 1989, the Oakland Athletics played the San Francisco Giants in U.S. baseball's World Series, the "Earthquake Series." The bookies in Las Vegas picked Oakland as the 1:2 favorite. Let A stand for the event: Oakland wins the World Series. Odds of 1:2 *against* this event means that $O(A) = \frac{1}{2}$ and so $P(A) = 1/(1 + \frac{1}{2}) = \frac{2}{3}$. If you had bet $5 on San Francisco and San Francisco had won, then you would have been paid $10 (plus you would get your $5 back), and you would have lost your $5 if Oakland had won. On the other hand, if you had bet $10 on Oakland and Oakland had won, then you would have been paid $5 (plus your $10 back) and nothing if San Francisco had won.

This description of payoffs is not quite right and, in fact, the bookies would not make money if it were. Suppose 100 people had bet $10 each with the bookie on Oakland and another 100 people had bet $5 each on San Francisco. If Oakland had won, then the bookie would have taken in $1,500 and paid out $1,500. If San Francisco had won, then the bookie again would have taken in $1,500 and again paid out $1,500. So there would be nothing in this for him. What the bookmakers actually advertised, as reported in my newspaper on the day before the Series, is that Oakland was the betting favorite, with "odds 9–11." There is no way you can figure out what this means unless you know that there are two 5's understood; in our definition of odds, it means **9**:**5** against San Francisco and **5**:**11** against Oakland. If you had bet $5 on San Francisco and San Francisco had won, then you would have won only $9 (plus your $5 back) and, just as before, you would have lost your $5 if Oakland had won. On the other hand, you would have had to put down $11 in order to win $5 (plus your $11 back), if Oakland had won.

A bookie's task is a good deal more difficult than a racetrack's. The racetrack sets the odds after seeing how much money is bet on the various horses (so the bettor does not know the odds until all betting on the race ends). Bookies do not have this luxury: They have to set the odds before a bet is made. To fix the odds, they have to guess how much money is going to be bet on the various possibilities. Suppose 100 people bet $11 each on Oakland and 200 bet $5 each on San Francisco. The bookie takes in a total of $2,100. If Oakland wins, he pays out $1,600 and so nets $500; but if San Francisco wins, he pays out $2,800 and so *loses* $700. To keep from losing, bookies try to set the odds so that the number of people betting on the two sides will be approximately balanced. In the case at hand, it would be ideal for the bookie if for each seven people who bet $11 on Oakland, eight people would bet $5 on San Francisco. For each set of 15 bettors, he takes in a total of $117. If San Francisco wins, he pays out $112 (= $14 × 8), netting $5. If Oakland wins, he pays out $112 (= $16 × 7), again netting $5. (To build an appropriate balance a bookie may have to refuse bets, or perhaps *lay off* some of his bets with other bookies.) ◁

Discussing gambling does not mean that I condone it. I see no merit whatever in betting at racetracks, casinos, or with bookmakers, or in buying lottery tickets. Even though some people derive pleasure from such endeavors, not all pleasure has merit! People do not seem to understand how foolish gambling is, and if they did, perhaps fewer of them would do it. Here is a fundamental principle of gambling: Never bet with people who make money gambling. They make it from people like you. Las Vegas, Atlantic City, and other gambling resorts are particularly depressing in this regard: Multimillion-dollar gambling palaces are built using gambling proceeds. Lotteries are even worse in that they are operated by governments as a means of taxing the gullible. Life insurance is one of the few exceptions to this principle: It may be wise to bet that you are going to die, even though insurance companies make money making the opposite bet. While many people cannot afford to leave their families without their income, they can afford insurance premiums as a hedge against such an eventuality.

EXERCISES

4.18 Consider the following payoff schedule for a three-horse race:

Horse number:	1	2	3
Payoff per $2 bet:	$3.24	$5.40	$8.10

Figure the odds against each horse and compute the percentage that the track takes.

4.19 The 1989 Belmont Stakes was won by a horse named Easy Goer (horse number 7). The order of finish, together with the track odds, are given here:

Horse number:	7	6	8	9	4	5	2	1	3	10
Odds:	1.3	.7	67.2	*	48.3	31.2	9.0	59.0	60.6	132.1

Calculate the total payoff for a $2 bet on Easy Goer and find the dollar amount the track makes for each $1,000,000 in the handle. [The asterisk indicates that horses 9 and 7 were paired—a bet (at odds 1.3:1) was a bet on both. (Imagine that horse 9 is horse 7's jockey!) So ignore horse 9 in your calculations.]

4.20 In September 1978, Las Vegas bookmakers favored Muhammed Ali to regain the World Boxing Association heavyweight title, putting the odds against him at 2:5. An Associated Press report stated: "A $5 bet would win $2 if Ali wins. The line on Spinks . . . was 2 to 1, meaning a $1 bet would win $2 if Spinks retained the crown." A bookmaker wants to make exactly $100, regardless of which boxer wins. How many $5 bets should he take in on Ali and how many $1 bets should he take in on Spinks?

[*Hint:* Use algebra or trial and error. I will start you out using trial and error. Suppose a total of $3,000 is bet on Ali. The bookie will pay out $1,200 if Ali wins. To make $100 means the bookie must take in a total of $1,300 in bets on Spinks. He will pay out $2,600 if Spinks wins, making $3,000 − 2,600 = $400. Since $400 is more than the required $100, the initial guess of $3,000 was too big. So, let's try $2,000 in bets on Ali. The bookie pays out $800 if Ali wins. To make $100 means $900 must be bet on Spinks. He will pay out $1,800 if Spinks wins, making $2,000 − 1,800 = $200. Again, this is more than $100, so $2,000 is also too big—but we are getting closer.]

4.4 Probability as Degree of Belief

Section 4.2 addresses probabilities based on long-run frequencies. Section 4.3 relates probabilities to betting odds and amounts of money wagered. This section deals with probability based on degree of belief. Is belief appropriate for science or should it be confined to religion? Scientists entertain various theories of the world. No one knows which theory is correct. Some scientists have strong beliefs in particular theories; sometimes two different scientists have strong beliefs in theories that contradict each other—not both of them can be true. In *A Brief History of Time*,[1] physicist Stephen Hawking makes it clear that his view of science is subjective: He discusses the "climate of thought" and, regarding scientific theories, uses language such as: "It was generally accepted," "we now believe," "They believed," and "if you believe."

Degree of belief depends on the person who has the belief (as well as on the event in question). In statistical inferences, this person could be any experimenter or observer. Sometimes I will use a pronoun (such as "you" or "me") to refer to the person whose belief is at issue. I will say things like, "Your probability that the Mets win is greater than my probability." I will reserve *the* probability for circumstances in which it is clear whose probability it is and, also, for when I want you to assume a particular probability in doing a calculation. I will usually be able to avoid simultaneously considering the probability assignments of two different people. You should always be aware that there is a person who is understood whenever I use $P(A)$.

Obviously, some opinions are based on more information than others. Consider a paternity case. Al Edged is named by Suzy Smith's mother to be Suzy's father. You might have different information than I do, yet neither of us has information as good as either Suzy's mother or Al Edged. So it is quite reasonable that our probabilities differ. Moreover, if we discuss the issue—your telling me what you know and my telling you what I know—our probabilities may change. Of course, I will allow for the possibility that you are lying to me or that your information is faulty—I have already incorporated the possibility that my information is faulty. It is likely that the difference between our probabilities will decrease as we talk, but this is not necessary, even if we trust each other.

There are several advantages to a subjective interpretation of probability. First, it applies anytime the person in question has an opinion. Counting ignorance as an opinion, though obviously a very weak one, this includes every setting. Also, subjective probabilities can change as information accrues. It is this aspect of the subjective interpretation that makes it so useful in statistical applications.

> **A probability based on degrees of belief is a subjective assessment concerning whether the event in question will occur (or has occurred).**

To measure one's degree of belief requires a scale, just like any other measurement. For degrees of belief the scale is a **calibration experiment.** The assessor—let's take it to be you—must be able to imagine an experiment with equally likely outcomes. To decide whether outcomes are equally likely, suppose you get to choose any one of the possible outcomes. I promise to pay you $100 should the experiment result in the outcome you choose. Outcomes in the set are *equally likely for you* if you are indifferent among them. In particular, you would strictly prefer any one outcome over all others if I were to increase the reward on that one by an arbitrarily small amount, say one cent. (Statements about preferences are always delicate because other people set the ground rules—me, in this case. Imagine that there is no chicanery afoot and that I am not trying to persuade you to make a choice I believe to be bad for you.)

For example, suppose I offer to pay you $100 if you call the roll of a six-sided die correctly. If you are indifferent as to which one you call, then the six sides are equally likely for you. An experiment with equally likely outcomes is a calibration experiment for you.

> **An experiment is a calibration experiment for you if all its outcomes are equally likely for you.**

There are many candidates for calibration experiments. However, whether a particular experiment serves to calibrate depends on the assessor. One possibility is rolling a die. Another is selecting a chip from a bowl that contains chips of the same size and shape. Another is a wheel of fortune with equally spaced stops, such as the one with eight stops in Exercise 4.11. Another is a random number key on a calculator or a random number command on various types of computer software. Quibbles about limitations of random number generators or the physical impossibility of making stops on a wheel of fortune exactly equally spaced are out of order—the outcomes do not have to *be* equally likely in any objective sense. All that is required is that the assessor *perceive* them as equally likely.

To be specific, I will use *chip-from-bowl experiments* to calibrate. So whenever I say there are chips in a bowl, I am assuming that the person whose probabilities are being assessed regards the chips as equally likely, each having probability 1 divided by the number of chips in the bowl.

Consider a specific setting. I would like to know your probability that average adult male emperor penguins weigh more than 50 lb—call this event A. Since I cannot obtain answers to my questions directly from you, I will guess them. My guesses will be wrong for many readers. Some of you know more about penguins than do others, and some of you may even be penguin experts. You should follow along in any case, but use your actual answers to modify what I say in a way that I hope will be clear to you.

A bowl contains a green chip and a red chip—see the first bowl in Figure 4.1. I offer you the choice of getting $100 if a chip selected from the bowl is green and $100 if A is true (that is, if average adult male emperor penguins do weigh more than 50 lb). If you choose to select from the bowl and the chip is red or if you choose A and it turns out that $\sim A$ is true, then you receive nothing. You say you prefer A. Since you chose A over an event of probability $\frac{1}{2}$, I interpret this as saying that your probability of A is at least $\frac{1}{2}$.

Figure 4.1 A chip-from-bowl experiment

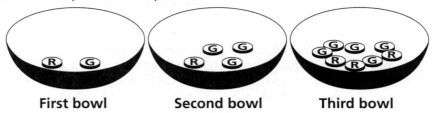

First bowl **Second bowl** **Third bowl**

Consider a new bowl, the second bowl in Figure 4.1: three green chips and one red chip. Again, you may choose between $100 if a chip selected from the bowl is green and $100 if A is true. Now you prefer the chip. Taken together, your two answers mean that $\frac{1}{2} \leq P(A) \leq \frac{3}{4}$. Now I ask you to consider a bowl with five green chips and three red chips—the third bowl of Figure 4.1. Same offer. You prefer A. Taken together, your last two responses mean $\frac{5}{8} \leq P(A) \leq \frac{3}{4}$. We proceed in this way, each time halving the interval by doubling the total number of chips in the bowl, until we know $P(A)$ sufficiently accurately.

There are several problems with this approach. The most important is that I have to be kidding about the money—the problem has little to do with my having limited finances. I could be serious only if we could find out whether A is true. We *can* find out how much emperor penguins weigh (I think A is true—my encyclopedia says that some males reach 75 lb, so I suspect average males weigh more than 50 lb). But I want to be able to assess probabilities of events for experiments that are not observable. Then you have to imagine that there are payoffs. Luckily, in most practical circumstances the assessor is motivated to take the assessment procedure seriously.

Another problem is that the assessor soon faces very difficult decisions. I suspect that by the time the bowl contained 16 chips, you were having a hard time deciding between a green chip and A (unless you "knew" that A was true or that A was false). Again, luckily, a high degree of accuracy in specifying $P(A)$ is seldom required.

Chapter 5 is an extension of this chapter. It describes calculations involving probability that we will apply in the remainder of this text.

EXERCISES

4.21 Consider event *B:* The current population of the state of California is greater than the current population of Canada. Draft someone and use the technique described in this section to help that person assess his or her *P(B)*. Report this probability to the nearest $\frac{1}{16}$, for example, by concluding that $\frac{5}{16} \leq P(B) \leq \frac{6}{16}$. [Report who the person is. Give all your questions and his or her answers. If there are extenuating circumstances that affect your answer, then please explain them. For example, the person might have left the room during the process and—you suspect—peeked at an encyclopedia.]

4.22 In the introduction to this chapter I indicated that in July 1988, the U.S. Geological Survey said there was a "50 percent probability" that a magnitude 7 or greater earthquake would hit the San Francisco Bay Area in the next 30 years. Call this event *C*. Using chip-from-bowl experiments and the comparisons and questions described in this section, indicate the answers given by someone whose probability of *C* is "50 percent."

4.23 You have a serious disease and you want to know your doctor's probability that you will survive. Your doctor is reluctant to use words such as *chance* and *probability*. Describe a process that you would follow to find out your doctor's probability that you will survive the next year. (A complication in this exercise is that the doctor will have to express preferences concerning money and your living. To avoid such comparisons, ask questions using language such as, "Is it more likely that . . . or that . . . ?" Do not mention the monetary rewards.)

Chapter Note

1. S.W. Hawking. *A Brief History of Time: From the Big Bang to Black Holes.* (New York: Bantam Books, 1988).

5 Conditional Probability and Bayes' Rule

THE FUNDAMENTAL problem of statistics is drawing conclusions about a population based on a sample from the population, or about a process based on observations from the process. We need a set of rules—to use Hawking's language: a way to learn about the population based on the sample. We will use probability and its set of rules. Following the scientific method, one formulates models for the population being sampled and updates information about the models based on the sample. The tack we will take is to specify a set of population models that we suppose contains the true model. For example, in observing cards selected from a deck, we may have prior information that only one of these two models is possible: (1) The deck contains 10% spades. (2) The deck of cards contains 20% spades. As we select cards (either with or without replacement) from the deck, we decide how strongly the observations support (1) as opposed to (2). We will never be sure which model is correct, but we ought to be able to discriminate reasonably well if the number of cards selected is large. And we ought to be willing to rethink our set of models should we get something inconsistent with both, such as 40 spades among the first 50 cards observed.

Every set of models is limited by the process of specifying it, so any particular set may fail to contain the true model. When a set of models does not contain the true model, a large sample will be most consistent with those models in the set that are *close* to the true one. A realistic set to specify is one in which the models are sufficiently spread out that at least some will be close to those models in the realm of possibility, and therefore also close to the true model.

Consider Example 2.1 (viscosity of dimethylaniline, page 11). Since the biggest observation is 154 cP, it would be surprising if the next observation were 180 cP. On the other hand, four of the 12 observations were 140 cP and so it would not be too surprising if the next observation were 140 cP. The sample size is small. But it is large enough to suspect that the process producing these observations—the population—produces fewer at 180 cP than at 140 cP.

The relatively large sample size of 100 in Example 2.15 (piglet weight gains, page 33) suggests that the population model is quite close to the sample distribution. For example, since the maximum gain in the sample is 57 lb, it would be very surprising if the next piglet were to gain more than 100 lb.

If we knew the population, we would have complete information when predicting the next observation. We do not know the population, but we do have partial information, since the sample we have is a subset of the population. The basic problem of statistics is to use the information in the sample to make inferences about the population and, in turn, to predict future observations. Using—or conditioning on—information is the subject of this chapter.

We use probability to measure uncertainties such as those just discussed. Using information means updating probabilities. How do we modify probabilities in the light of accumulating evidence? For example, a week after the San Francisco Bay Area earthquake in October 1989, *The New York Times* reported that examination of surface fissures had some geologists "wondering whether [earlier] probability estimates of future quakes are too low." The fundamental problem of statistics is using observations to update probabilities concerning the models that have produced those observations.

In our everyday language, we allow probabilities to change: "Have the chances improved?" I indicated in Chapter 4 that Las Vegas bookmakers set the probability of Oakland winning the 1989 baseball World Series at about 67%—before the series started. Oakland won the first game. Bookmakers then increased this probability to about 78%. [My newspaper said "odds 16–20." We have to read in the "5" that is understood in interpreting these numbers (as described in Example 4.3, page 118). The bettor would receive $16 for each $5 bet on San Francisco should San Francisco win and $5 for each $20 bet on Oakland should Oakland win. So the bookmakers pegged the odds against Oakland as about the average of 5 to 16 and 5 to 20, or about 5 to 18. Odds of 5:18 correspond to a probability of 18/(5 + 18) or about 78%.] They were *conditioning* on the results of the first game, in addition to the information available before the first game.

5.1 Joint Probabilities

The goal of this chapter is to describe how probabilities change as we condition on events. This requires *joint* probabilities. These are probabilities of several events occurring simultaneously. The simultaneous occurrence of two events A and B is their product or **intersection**. The standard notation for the intersection of events A and B is $A \cap B$. The event $A \cap B$ contains those outcomes that are in *both* A and B. For example, $\{1, 2, 3, 4\} \cap \{3, 4, 5\} = \{3, 4\}$. I sometimes use the word *and* for intersection: An outcome is in the intersection of two events if it is in the first *and* it is also in the second.

When two events contain no outcomes in common, then their intersection is empty. For example, $\{1, 2, 3\} \cap \{5, 6\} = \emptyset$. Such events are **mutually exclusive** or **disjoint.**

The use of the word *and* is ambiguous and makes the distinction between union ∪ and intersection ∩ confusing. The union of two events includes those outcomes in the first *and* those in the second. The intersection of two events includes those outcomes in the first *and* in the second. The only difference in these sentences is that one contains an extra *those*. The first *and* refers to adding outcomes; the second *and* refers to the simultaneous occurrence of two conditions.

What about probabilities of intersections? Consider rolling a six-sided die. Let $A = \{1, 3, 5\}$—the number rolled is odd. Let $B = \{4, 5, 6\}$—the number rolled is greater than 3. The event $A \cap B$ happens if the number rolled is *both* odd and greater than 3. So $A \cap B = \{5\}$. Assuming all six outcomes have the same probability, as in

Exercise 4.3, $P(A \cap B) = P(5) = \frac{1}{6}$. Similarly, $P(\sim(A \cap B)) = \frac{5}{6}$ and $P((\sim A) \cap B) = \frac{2}{6}$. Since an event and its complement are always disjoint, $P((\sim A) \cap A) = P(\emptyset) = 0$.

Now consider three events: A, B, and $C = \sim 6$. The simultaneous occurrence of all three is $A \cap B \cap C = \{1, 3, 5\} \cap \{4, 5, 6\} \cap \{1, 2, 3, 4, 5\} = \{5\}$. Another way of describing this event is to say that both $A \cap B$ and C happen; that is, $A \cap B \cap C = (A \cap B) \cap C$. Similarly, $A \cap B \cap C = A \cap (B \cap C)$. In the case at hand, $P(A \cap B \cap C) = P(5) = \frac{1}{6}$. Also, $P((\sim A) \cap B \cap C) = P(4) = \frac{1}{6}$.

Figure 5.1 Tree diagram for rolling two dice

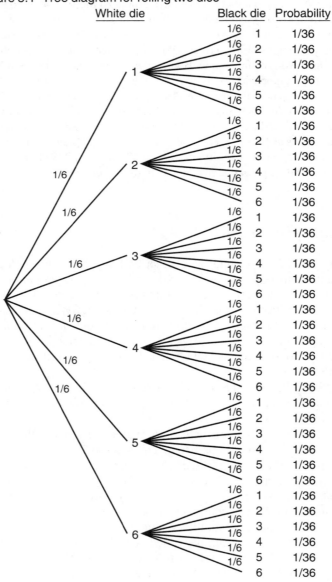

Sec. 5.1 / Joint Probabilities   127

We discussed a special case of probabilities of intersections in Chapter 4. When rolling two dice, each outcome is an event. For example, 53 is the intersection of two events: 5 on the white die and 3 on the black die. If the 36 possibilities are equally likely, then the probability of each pair of digits is $\frac{1}{36}$. This has an important implication. The probability of 5 on the white die is the sum of the probabilities of 51, 52, 53, 54, 55, and 56. Since the probability of each of these events is $\frac{1}{36}$, the probability of 5 on the white die is $\frac{6}{36} = \frac{1}{6}$. Similarly, the probability of 3 on the black die is the sum of the probabilities of 13, 23, 33, 43, 53, and 63—again $\frac{1}{6}$. So multiplying the probability of 5 on the white die by the probability of 3 on the black die gives the probability of the pair 53: $(\frac{1}{6})(\frac{1}{6}) = \frac{1}{36}$. This same calculation applies for every pair (see the tree diagram in Figure 5.1).

In terms of a chip experiment, there are 36 chips in the population, as pictured below. Each chip indicates one of the 36 possibilities.

Figure 5.2 36 possible chips when rolling two dice

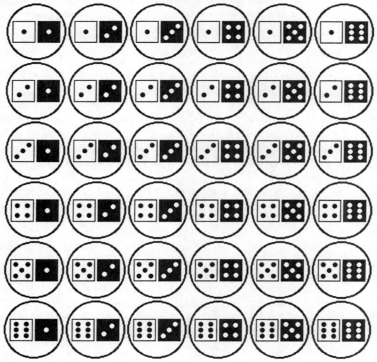

When the probability of the intersection of two (or more) events is the product of individual probabilities, these events are said to be independent (for the person in question). In symbols, events A and B are independent whenever

$$P(A \cap B) = P(A)P(B)$$

More generally,

> $A, B, C, \ldots$ are independent events whenever
> $$P(A \cap B \cap C \cap \cdots) = P(A)P(B)P(C)\cdots$$

Since probability depends on the individual, independence does as well. For example, two events may be independent for you but not for me.

Now consider two (or more) *experiments*. In addition, consider any combination of outcomes, one from each of the experiments. If these outcomes are independent, then so are the experiments:

> Experiments are independent if the outcomes from one are independent of the outcomes from the other, and this holds for any pair of outcomes, one from each experiment.

For example, the two dice rolls are independent when the pairs of outcomes have probability $\frac{1}{36}$, since all outcomes in the individual rolls then have probability $\frac{1}{6}$: $\frac{1}{36} = (\frac{1}{6})(\frac{1}{6})$.

EXERCISES

5.1 Suppose that before the 1989 baseball World Series, you had bet $55 on Oakland to win at 5:11 odds. You will win $25 ($80, including your own $55) if Oakland wins the Series. (The winner of the Series is the first team to win four games.) After Oakland won the first game, my newspaper indicated that the Las Vegas odds on San Francisco were 16:5. This means that each $1 you bet on San Francisco would return $3.20 (in addition to your $1) if they were to win the Series. It happens that you can now bet on San Francisco at these 16:5 odds and win money, regardless of the outcome. You could even place a bet on San Francisco at 16:5 odds, so that your net winnings will be the same no matter who wins the Series. How much would you bet on San Francisco to effect this? And what will you win?

(*Hints:* Rounded to the nearest dollar, the answers are $19 and $6. You are to find them to the nearest cent—though I doubt that you could find a bookie who pays off in pennies! You can proceed using algebra or you can use trial and error. For example, start with a bet of $19. If Oakland wins, you net $25 − 19 = $6. If San Francisco wins, you will net $19 × $\frac{16}{5}$ − 55 = $5.80. So you win in any case. But you have got to increase the $19 bet on San Francisco to decrease the $6 you win in the first case and increase the $5.80 you win in the second case. You are supposed to find the bet that makes them agree.)

5.2 Consider rolling two dice. The following scheme is repeated from Exercise 4.4 (page 110), where the first digit is the white die result and the second is the black die result:

11	12	13	14	15	16
21	22	23	24	25	26
31	32	33	34	35	36
41	42	43	44	45	46
51	52	53	54	55	56
61	62	63	64	65	66

Again, suppose that each of the 36 outcomes has probability $\frac{1}{36}$, and define

A = black die results in 4
B = white die result is greater than black die result
C = sum of results of white and black dice is 2, 3, or 12 (*crapping out when playing craps*)
D = bigger of the two numbers is 5 (including the outcome 55)

Find the probabilities of these intersections:

(a) $A \cap B$ (b) $A \cap C$ (c) $A \cap D$ (d) $B \cap C$ (e) $B \cap D$ (f) $C \cap D$

5.3 You are going to select a colored chip from a bowl. The chip is either blue (B) or yellow (Y) and has a number on it, with the possibilities being 1, 2, 3, and 4. Your probabilities for the various outcomes are given in this table:

		\multicolumn{4}{c}{Number}			
		1	2	3	4
Color	B	3/20	3/20	3/20	3/20
	Y	2/20	2/20	2/20	2/20

(a) What is your probability of selecting a blue chip?
(b) What is your probability of selecting a chip numbered 2?
(c) What is your probability of selecting a blue chip *or* a chip numbered 2?
(d) By calculating and comparing the appropriate probabilities, argue that for you color and number are independent.

5.4 The estrogen-receptor and progesterone-receptor statuses of tumors from 20 patients with locally advanced breast cancer were assessed, with the following proportions:

		Progesterone	
		Positive	Negative
Estrogen	Positive	8/20	4/20
	Negative	1/20	7/20

Think of this as a population of 20 tumors and the proportions as probabilities. Let A be the event that a tumor is estrogen-receptor positive and B be the event that it is progesterone-receptor positive.

(a) What is the probability of A?
(b) What is the probability of B?
(c) What is the probability of $A \cap B$?
(d) Are A and B independent?

5.5 A bowl contains 10 chips, five of which are red and five blue. You select two chips from the bowl in such a way that all possible pairs are equally likely. What is your probability of selecting two red chips if

(a) You return the first chip to the bowl before selecting the second (in which case there are 100 pairs possible)?
(b) You do not return the first chip to the bowl before selecting the second (in which case there are 90 pairs possible)?

5.6 I toss a bent coin and then roll a die. My probabilities of the various pairs are given in the tree diagram shown at the top of page 130.

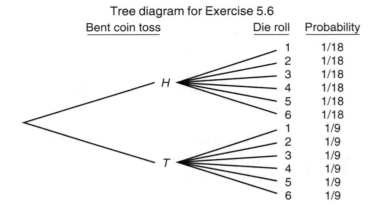

Tree diagram for Exercise 5.6

(a) Find my probability that the coin lands H.
(b) Find my probability that the coin lands H *and* the number on the die is even.
(c) Find my probability that the coin lands H *or* the number on the die is even.
(d) Argue that for me the coin toss and the die roll are independent experiments. (You must calculate probabilities; words are not enough. But you need only calculate enough of the relevant probabilities to show that you understand.)

5.7 Important control systems on aircraft have built-in redundancies. Some systems have as many as three alternatives in case they fail. Suppose four systems function independently and the probability of a failure on any one of them during a flight is 1%. What is the probability that all four fail?

5.2 Conditional Probability

Conditioning on events or circumstances—assuming that they have occurred or that they apply—is standard in our thinking and in our discourse. But it is difficult to keep the logic straight. For example, during the health care debate of 1993–1994, an ad placed by the League of Women Voters and the Kaiser Family Fund said "84 percent of Americans who lack health insurance are in families that work hard and pay taxes, but don't get health insurance on the job." The less than meticulous reader may read "84 percent of Americans . . . don't get health insurance on the job." Clearly wrong—only about 15% of workers did not have health insurance. To further confuse the issue, the ad continues: "That's eight out of ten of us, and that's a fact." The conditional in this latter statement is "of us," apparently referring to "all Americans" rather than "Americans who lack health insurance." The problems with language in this example are easy for the careful reader to sort out. Other examples are more difficult to decipher and you will encounter some of them in this chapter.

Suppose you assess your probability of event B. Then I tell you that event A happened. I ask how this knowledge changes your probability of B. You may tell me that it does not. An obvious example is when A is the universe U, and so I have given you no new information. More generally, A and B are independent if the information contained in A is irrelevant for B.

But your probabilities will sometimes change. For example, suppose $A = \sim B$. Now when I tell you A happened, you know that B did not happen. The new probability of B is 0. Or, suppose $A = B$. Then you know B did happen. Its new probability is 1. So your probability of B can change dramatically—regardless of how big or small it was initially—depending on the new information, A.

EXAMPLE 5.1
▷ **Probability of survival with breast cancer.** A 60-year-old woman has just had a lumpectomy. My probability that she will survive at least 5 more years with optimal care is 90%. (My probabilities are based on a reasonable amount of information, but I am giving them in round numbers.) Now we are told that she had 25 axillary lymph nodes removed and 15 of them tested positive, meaning that her breast cancer has spread to the lymph nodes. My probability of her survival decreases to 50%. Now suppose tests carried out before her lumpectomy are reviewed and it is discovered that her breast cancer has spread to her liver. Now my probability that she survives the next 5 years drops to 10%. ◁

It is easier to define conditional probability if we have a notation for the probability of B given A. I will use $P(B \mid A)$. The bar between B and A is vertical and should not be confused with the slanted bar used for fractions. I will give a definition and then explain why it is reasonable:

> **The conditional probability of B given A is**
> $$P(B \mid A) = \frac{P(B \cap A)}{P(A)}$$

The intersection of B and A is shown in Figure 5.3, where A is shaded. The intersection includes everything in B except that part of B ruled out because it is not in A. That is, the only part of B that is included is the shaded part. Conditioning on A eliminates everything not in A, as shown in Figure 5.4. The universe becomes A. The probability in A is distributed as before the conditioning, but now the total probability is only $P(A)$, which is typically less than 1. The division by $P(A)$ in the definition boosts the total probability back up to 1: $P(A \mid A) = P(A \cap A)/P(A) = P(A)/P(A) = 1$.

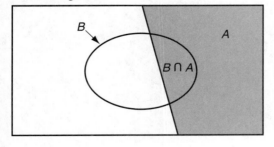

Figure 5.3 Intersection of elliptical region B and shaded region A

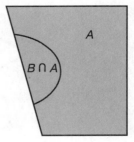

Figure 5.4 Conditioning on event A changes the universe to A

EXAMPLE 5.2
▷ **Rolling a die.** Consider rolling a die. Let $B = \{1, 3, 5\}$ and let $A = \{4, 5, 6\}$. Then $P(A \cap B) = \{5\}$. Assume that each outcome has probability $\frac{1}{6}$. Then $P(B \mid A) = P(5)/P(4, 5, 6) = (\frac{1}{6})/(\frac{3}{6}) = \frac{1}{3}$. The unconditional probability of B is $P(B) = \frac{1}{2}$. So the information that the number is greater than 3 lowers the probability that it is odd. On the other hand, $P(B \mid \sim A) = \frac{2}{3}$, as you can easily check. ◁

According to the definition of conditional probability, the probabilities of outcomes in A have not changed relative to each other. In Example 5.2, $P(4) = \frac{1}{6}$ and $P(5) = \frac{1}{6}$; their ratio is 1. This ratio does not change when given A, since both outcomes 4 and 5 are contained in A: $P(4 \mid A) = \frac{1}{3}$ and $P(5 \mid A) = \frac{1}{3}$.

The roles played by labels in a definition are arbitrary. So exchanging labels A and B in the definition of conditional probability gives an equivalent expression:

$$P(A \mid B) = \frac{P(A \cap B)}{P(B)}$$

The next example shows that you have to be careful when stating a condition to ensure that your calculations are based on the correct assumptions. The idea will be repeated several times in later examples, but under seemingly different guises.

EXAMPLE 5.3
▷ **How many girls?** A family has two children. You are told that at least one is a girl; call this event C. What is your probability that the other is a girl? Call D the event "both children are girls."

Suppose your probabilities are given by the tree diagram in Figure 5.5. The joint probabilities indicate that the sexes of the two children are independent. The branches for outcomes that make up event C are darker. To find your probability of C, add the probabilities of the three outcomes that make up C: $P(C) = \frac{1}{4} + \frac{1}{4} + \frac{1}{4} = \frac{3}{4}$. Now use the definition of conditional probability:

$$P(D \mid C) = \frac{P(C \cap D)}{P(C)} = \frac{P(\text{both girls})}{P(C)} = \frac{1/4}{3/4} = \frac{1}{3}$$

Figure 5.5 Tree diagram for Example 5.3

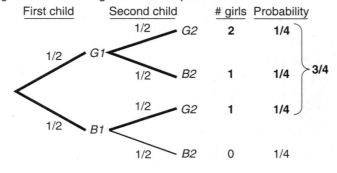

This answer is counterintuitive for many people. In some cases, they are confusing the event C (at least one girl) with a particular child being a girl. For example, conditioning on $G1$ instead of on C gives

$$P(D \mid G1) = \frac{P(G1 \cap D)}{P(G1)} = \frac{P(\text{both girls})}{P(G1)} = \frac{1/4}{1/2} = \frac{1}{2}$$

Saying "at least one is a girl" (or, equivalently, "not both are boys") is easily confused with "the first child is a girl." The difficulty is semantic and not with the calculations. (Exercise 5.12 is similar to this example.) ◁

What happens if A contains no information about B? That would mean $P(B) = P(B \mid A)$. Recall the definition of conditional probability:

$$P(B \mid A) = \frac{P(B \cap A)}{P(A)}$$

So when $P(B) = P(B \mid A)$, it follows that

$$P(B) = \frac{P(B \cap A)}{P(A)}$$

Multiplying both sides by $P(A)$—assuming $P(A)$ is not 0—gives

$$P(B \cap A) = P(A)P(B)$$

This is the defining equation for the independence of events A and B (see Section 5.1). So when A contains no information about B, the two events are independent. This relationship between independence and conditional probability may make the definition of independence seem even more reasonable: If I tell you A happened and this information does not change what you think about B, then it seems appropriate to say that A and B are independent.

There is a neat generalization of the multiplication rule for independent events. For any two events A and B, multiply both sides of the definition of conditional probability by $P(A)$:

$$P(A)P(B \mid A) = P(A)\frac{P(B \cap A)}{P(A)} = P(B \cap A)$$

This is an intuitive and very useful rule:

Multiplication Rule

$$P(A \cap B) = P(A)P(B \mid A)$$

This rule is equivalent to the definition of conditional probability. It generalizes the definition of $P(A \cap B)$, assuming independence: When $P(B \mid A) = P(B)$, then $P(A)P(B \mid A)$ reduces to $P(A)P(B)$ and so $P(A \cap B) = P(A)P(B)$.

EXAMPLE 5.4

▷ **Two aces.** Consider selecting two cards from an ordinary deck of playing cards. Suppose the 52 cards are equally likely initially. Suppose further that the remaining 51 cards are equally likely at the second selection. What is the probability that both cards selected are aces? Define the events:

$$A = \text{first card is an ace}$$
$$B = \text{second card is an ace}$$

Both are aces when and only when both A and B occur—that is, when $A \cap B$ occurs. The multiplication rule means that $P(A \cap B) = P(A)P(B \mid A)$. So we require $P(A)$ and $P(B \mid A)$. Since there are four aces:

$$P(A) = \frac{4}{52}$$

If an ace is selected initially, then there are three aces and 48 non-aces of the 51 cards available for the second selection. So

$$P(B \mid A) = \frac{3}{51}$$

It follows that

$$P(\text{both are aces}) = P(A \cap B) = P(A)P(B \mid A) = \frac{4}{52} \cdot \frac{3}{51} = \frac{1}{221} = .0045$$

It is easy to extend the multiplication rule to three or more events. The point is to condition on everything assumed to the current time and multiply conditional probabilities. Consider selecting three cards from a deck of playing cards. What is the probability that all three are aces? Again assume equally likely selections from the remaining cards. Define A and B as previously and now let C stand for ace on the third card selected. Then

$$P(\text{three aces}) = P(A \cap B \cap C) = P(A)P(B \mid A)P(C \mid A \cap B)$$
$$= \frac{4}{52} \cdot \frac{3}{51} \cdot \frac{2}{50} = .00018 \quad ◁$$

Calculations like those in Example 5.4 seem intuitive to many people. For these people, the definition of conditional probability will seem natural.

Exchangeable Experiments

I defined the notion of independent experiments in the previous section. This definition is quite restrictive. For example, repeated observations to generate a sample from a population are not independent—conditioning on the first observation changes the probabilities for the second. An assumption that is not so restrictive is **exchangeability**: Two experiments are *exchangeable for you* if all the following hold:

(1) The possible outcomes are the same in both.

(2) The probability of each outcome in one is the same as it is in the other.

(3) The conditional probabilities for the second experiment, given the results of the first, are the same as the conditional probabilities for the first, given the results of the second.

So exchangeable experiments are symmetric. In particular, the implications of observations from two exchangeable experiments are the same regardless of which experiment produced which observation. Observing a 12 on the first and a 27 on the second is the same as observing a 27 on the first and a 12 on the second.

EXAMPLE 5.5
▷ **Selecting two chips from a bowl.** Consider randomly selecting two chips one at a time from the following bowl:

Each selection is an experiment. The tree diagram in Figure 5.6 shows the probabilities. The two selections are not independent because the result of the first changes the probabilities for the second—from $\frac{1}{4}$ for each to $\frac{1}{3}$ for the remaining chips.

Figure 5.6 Tree diagram showing probabilities for Example 5.5

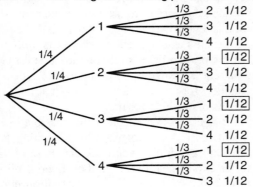

To see that the two selections are exchangeable, refer to the numbering system in the definition: (1) The possible outcomes of the first selection are 1, 2, 3, 4 and for the second selection they are 1, 2, 3, 4. (2) The probabilities for the first selection are $\frac{1}{4}$ for each outcome. The probabilities for the second selection are also $\frac{1}{4}$ for each outcome. To see this, refer to the tree diagram. The boxes show the probabilities for getting a 1 on the second selection: $\frac{1}{12} + \frac{1}{12} + \frac{1}{12} = \frac{1}{4}$; similarly for other outcomes. (3) Suppose the first selection is a 1. Then the probabilities for the second selection are $\frac{1}{3}$ for each of 2, 3, and 4—see the tree diagram. Suppose the second selection is a 1. Then the probability of a 2, say, on the first is

$$\frac{1/12}{1/12 \ + \ 1/12 \ + \ 1/12} = \frac{1}{3}$$

Similar reasoning holds for the other possibilities. ◁

> **Two or more experiments are exchangeable for you if:**
> (1) The sets of outcomes are the same in both.
> (2) The corresponding outcomes have the same probability in both.
> (3) The experiments are symmetric in that the joint probabilities are the same regardless of the order in which the experiments are observed.

We will apply exchangeability to populations and to sampling. Consider a population—patients in a medical setting, plants in a botanical setting, and so on. We plan to select a member of the population and make a measurement—response to a drug, insecticide, or whatever. If your probabilities of the various responses are the same for all members of the population, then the corresponding experiments (making these measurements) are exchangeable, and so we say the members are themselves exchangeable.

For an example of a population whose members are not exchangeable, consider testing a drug designed to lower blood pressures to normal levels—say, to less than 90 mmHg diastolic. Patient 1's blood pressure is 100 and patient 2's is 110. Because patient 2's blood pressure has further to drop to reach normal levels, my probability of a success for patient 2 is less than that for patient 1; so these patients are not exchangeable for me. If, instead, the measurement of interest is *amount of drop* in blood pressure, in mmHg, then my probabilities might be the same for both. For example, a drop of 10 would bring patient 1's down to 90 and patient 2's down to 100, but these may have the same probability for me. In such a case—that is, for such a measurement—the patients would be exchangeable for me.

Sample observations are exchangeable if the possible outcomes are the same, the probability assignment is the same for each, and learning is symmetric, as just described. Since probability assignments are subjective, observations exchangeable for one person may not be exchangeable for another. Even if observations are exchangeable for me, my probabilities for the second observation may change in view of the first observation. Indeed, this is a fundamental aspect of learning.

EXAMPLE 5.6

▷ **Two-headed or two-tailed?** Suppose I have a coin that I believe to be either two-headed or two-tailed. I do not know which, and my probabilities are $\frac{1}{2}$ for each. I plan to toss it twice. The probabilities are shown in the tree diagram of Figure 5.7. (Compare Exercise 5.11.)

Figure 5.7 Tree diagram showing probabilities for Example 5.6

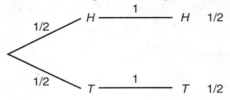

This is a special case of exchangeability. My probability of heads on the first toss is $\frac{1}{2}$. My probability of heads on the second toss is also $\frac{1}{2}$. What I learn from the first toss is the same as what I learn in the second: that the other toss will be the same as the one observed. ◁

The previous example was extreme. Observations tend to be alike in some types of sampling, but they are exactly alike only in extreme cases. (The opposite case—in which observations tend to be different—occurs in Example 5.5. In that example, the population was known to contain four different chips and the chip drawn first could not be drawn second.)

EXERCISES

5.8 Consider rolling two dice, a white one and a black one. The 36 ordered pairs of outcomes are repeated here from Exercise 4.4, again with the number on the white die first and the black die second:

```
11  12  13  14  15  16
21  22  23  24  25  26
31  32  33  34  35  36
41  42  43  44  45  46
51  52  53  54  55  56
61  62  63  64  65  66
```

Reconsider the events defined in Exercises 4.4 (page 110) and 5.2 (page 128):

A = black die results in 4
B = white die result is greater than black die result
C = sum of results of white and black dice is 2, 3, or 12 (*crapping out* when playing craps)
D = bigger of the two numbers is 5 (including the outcome 55)

Find the following:

(a) $P(A \mid B)$ (b) $P(B \mid A)$ (c) $P(B \mid C)$ (d) $P(C \mid B)$ (e) $P(A \mid D)$ (f) $P(D \mid C)$

5.9 Assume the probabilities given in the tree diagram for Example 5.3 (repeated at the top of page 138). Find the probability that both children are boys given that the second child is a boy.

138   Ch. 5 / Conditional Probability and Bayes' Rule

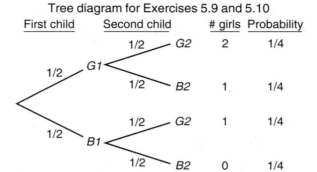

5.10 For the probabilities in the tree diagram for Example 5.3 (repeated above), I claimed that the sexes of the two children are independent. Explain why this is so. (You must use probabilities in your explanation.)

5.11 Consider the probabilities in the tree diagram that follows. (These are modifications of the probabilities in Example 5.3 and the previous two exercises; the tree is essentially the same as the one in Example 5.6.)

Tree diagram for Exercise 5.11

First child	Second child	# girls	Probability
1/2 G1	1 → G2	2	1/2
	0 → B2	1	0
1/2 B1	0 → G2	1	0
	1 → B2	0	1/2

    (a) Explain why the sexes of the two children are not independent. (Conveying your intuition using words is fine, but calculating probabilities is required.)
    (b) What are $P(G1)$, $P(G2)$, $P(B1)$, and $P(B2)$? (They are the same now as they were in Example 5.3.)
    (c) Calculate $P(G1 \mid G2)$.
    (d) As in Example 5.3, define C to be "at least one is a girl" and define D to be "both children are girls." Calculate $P(D \mid C)$.

5.12 A hat contains three cards. Card 1 has a ♦ on both sides. Card 2 has a ♠ on both sides. Card 3 has a ♦ on one side and a ♠ on the other side. You select a card from the hat and feel that each card has the same probability of being selected. You look at one side and see a ♦. What is your probability that the other side is also a ♦?

    [*Hints:* Do not trust your intuition. Make each probability calculation with care. Use reasoning similar to that of Example 5.3.]

5.13 Russian roulette is carried out with a revolver loaded with a single cartridge. The participant spins the cylinder, points the muzzle at his or her head, and pulls the trigger! Suppose the revolver is a six-shooter, with six chambers in the cylinder. Assume that spinning the cylinder makes the six chambers equally likely. So the player loses—and loses big!—on a single play with probability $\frac{1}{6}$. The player plans to play twice (assuming he survives the first). What is the probability of losing in these two plays if he does not plan to spin the cylinder between the plays? [*Hint:* Calculate the probability of survival. Use the multiplication rule, though you may see an easier way to do this.]

5.14 Winning the top prize in most state lotteries requires correctly picking six numbers from {1, 2, 3, ..., N}. The value of N varies from one state to another, but it is always quite large. Suppose $N = 49$, as used in some states. The six winning numbers are selected one at a time and without replacement from 1 to 49; at each selection assume the numbers that remain are equally likely. A ticket that you buy contains six numbers, and it wins if each number agrees with one selected in the lottery. The order of the numbers is irrelevant. Calculate the probability of winning if you buy a single ticket. [*Hint:* Use the multiplication rule. Fix any six numbers. (Most people feel that six numbers such as 4, 17, 19, 31, 37, and 43, are more likely to win than are six special numbers, such as 1, 2, 3, 4, 5, and 6; but these two and all other combinations of six have the same chance.) Your probability that the first number selected matches any one of your six is $\frac{6}{49}$. If the first was a match, the second is a match with probability $\frac{5}{48}$, and so on.]

5.15 In New York City during the summer of 1990, a "zodiac killer" claimed he was going to murder one person for each of the 12 signs of the zodiac. He shot four men to death (on Thursdays, 3 weeks or multiples of 3 weeks apart), correctly announcing each time the victim's zodiac sign. He was never caught. In the summer of 1994, someone claimed to be the "zodiac killer" and the murderer of five men who had been killed in 1994. Four of these latter five murders were known to the police. But there were various aspects of these crimes that led police to question whether the killer was the same person. For one thing, the murderer did not announce the zodiac sign of the victim. However, three of the victims' birthdays were known and, indeed, all three had zodiac signs different from each other and from all four previous victims. If these three victims had been picked randomly (and not because of their signs), what is the probability that their signs would be different and would not duplicate any of the signs of the previous four victims?

5.16 Exercise 4.14 (page 113) deals with drawing from among the 75 chips in the game of BINGO, as shown in the figure at the top of page 140. The chips are drawn one at a time and without replacement. Assume that the chips are drawn randomly from the 75, then from the 74 remaining, then from the 73 remaining, and so on.

(a) Suppose the first chip drawn is under the letter B. What is the probability the second one is also under the B?

(b) Suppose the first chip drawn is not under the B. What is the probability the second one is under the B?

(c) Suppose the first two chips drawn are under the B. What is the probability the third one is also under the B?

(d) Use the multiplication rule to find the probability that both the first and second chips drawn are under the B.

(e) Use the multiplication rule to find the probability that the first three chips drawn are under the B.

Numbering of chips in BINGO for Exercise 5.16

B	I	N	G	O
1	16	31	46	61
2	17	32	47	62
3	18	33	48	63
4	19	34	49	64
5	20	35	50	65
6	21	36	51	66
7	22	37	52	67
8	23	38	53	68
9	24	39	54	69
10	25	40	55	70
11	26	41	56	71
12	27	42	57	72
13	28	43	58	73
14	29	44	59	74
15	30	45	60	75

5.3 Law of Total Probability

There are many circumstances in which you would like to know the probability of an event, but you cannot calculate it directly. You may be able to find it if you know its probability under some conditions. The desired probability is a weighted average of the various conditional probabilities.

EXAMPLE 5.7
▷ **Winning at chess.** You are in a chess tournament and will play your next game against either Joe or Mary, depending on results of some other games. Suppose your probability of beating Joe is $\frac{7}{10}$, but of beating Mary is only $\frac{2}{10}$. You assess your probability of playing Joe as $\frac{1}{4}$. How likely is it that you win your next game?

Let J = play Joe, M = $\sim J$ = play Mary, and W = win. Then, these are the assumptions made previously:

$$P(W \mid J) = \tfrac{7}{10} \quad P(W \mid \sim J) = \tfrac{2}{10}$$
$$P(J) = \tfrac{1}{4} \quad P(\sim J) = \tfrac{3}{4}$$

We want $P(W)$, the *unconditional* probability of winning. For *any* events W and J, we can write

$$W = (W \cap J) \cup (W \cap (\sim J))$$

The diagram in Figure 5.8 shows this, where $M = \sim J$ is the shaded region and W is the elliptical region. Some of W is in the shaded region and some is not. (The diagram is supposed to be typical, with areas of events not meant to be proportional to their probabilities.)

Since events $W \cap J$ and $W \cap M$ are disjoint, property 3 of probability (Section 4.1) means that their probabilities sum to give the probability of W:

$$P(W) = P(W \cap J) + P(W \cap (\sim J))$$

We can rewrite both terms on the right side of this expression, using the multiplication rule twice:

$$P(W \cap J) = P(W \mid J)P(J)$$

and

$$P(W \cap (\sim J)) = P(W \mid \sim J)P(\sim J)$$

Substituting these into the previous formula gives the law of total probability:

Law of Total Probability

$$P(W) = P(W \mid J)P(J) + P(W \mid \sim J)P(\sim J)$$

In the chess example,

$$P(W) = \frac{1}{4} \cdot \frac{7}{10} + \frac{3}{4} \cdot \frac{2}{10} = \frac{7}{40} + \frac{6}{40} = \frac{13}{40} = .325$$

This is a weighted average of the probability of winning if you play Joe and the probability of winning if you play Mary. It is three-quarters of the way from $\frac{7}{10}$ to $\frac{2}{10}$.

This calculation is especially easy to see with the aid of a tree diagram (see Figure 5.9). Remember that $M = \sim J$. The event "win" is the union of the two branches

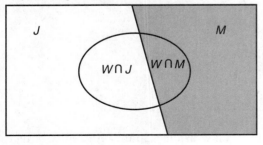

Figure 5.8 Diagram for Example 5.7 showing elliptical region W and shaded region $M = \sim J$

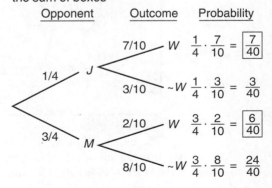

Figure 5.9 Tree diagram for Example 5.7: $P(W)$ is the sum of boxes

whose probabilities are in boxes. These two branches correspond to the two terms in the law of total probability. The numbers in the two boxes are those we added to give the probability of win, $P(W)$. ◁

EXAMPLE 5.8
▷ **Second card is an ace?** Consider selecting two cards from a deck of playing cards. We saw in Example 5.4 that the probability that both are aces is $\frac{1}{221}$. But what is the probability that the second card is an ace? Using the law of total probability, we add the numbers in the boxes in the tree diagram in Figure 5.10. The result is $\frac{17}{221}$, or $\frac{1}{13}$. This is also clear from symmetry: Think of drawing the second card first!

Figure 5.10 Tree diagram for Example 5.8

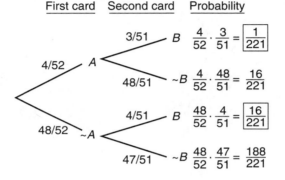

General Law of Total Probability

The law of total probability has a more general form. Suppose you have an arbitrary number of possible opponents: $A, B, C, \ldots$. The event W can be split up as $W = (W \cap A) \cup (W \cap B) \cup (W \cap C) \cup \ldots$. Such a division is shown in Figure 5.11 for the case of five opponents.

Figure 5.11 Diagram showing division of elliptical region W by events A through E

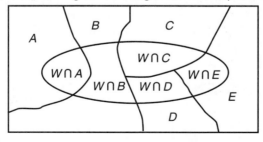

The events $W \cap A$, $W \cap B$, $W \cap C$, ... are disjoint. So the probability of their union is the sum of their individual probabilities. Since $P(W \cap A) = P(W \mid A)P(A)$, $P(W \cap B) = P(W \mid B)P(B)$, and so on, we have a more general law of total probability:

Law of Total Probability

$$P(W) = P(W \mid A)P(A) + P(W \mid B)P(B) + P(W \mid C)P(C) + \cdots$$

where $A, B, C, \ldots$ subdivides the universe U

EXAMPLE 5.9

▷ **"Craps."** Finding the probability of winning in the dice game called craps is somewhat complicated. The law of total probability makes it easier. Here are the rules of the game. The player—say, you—rolls two dice. The total or sum on the two dice is all that matters. You win immediately if you roll a sum of 7 or a sum of 11. You lose immediately if you roll 2, 3, or 12. If you roll any other sum, this sum becomes your *point*. For example, if one die shows a 1 and the other shows a 4, then your point is 5 (= 1 + 4). You proceed to roll repeatedly until getting either your point or a sum of 7. You win in the former case and lose in the latter. In this example, you win if you roll a 5—including 2 + 3 as well as 1 + 4—before you roll a 7.

We are going to need the probability of each possible sum when rolling two dice. I have reproduced the tree diagram from Exercise 4.4, but now I have also included the sum of the numbers along each branch (Figure 5.12 on page 144). The probabilities of branches whose sum equals 5 is shown in boxes. There are four boxes and so $P(5) = \frac{4}{36}$. An easier way to count is to use Figure 5.13. This figure keeps the white die the same in rows and the black die the same in columns. To find the probabilities of sums, we can exploit the fact that the sum is constant along diagonals. By counting along the diagonals in this figure (or counting paths in the tree diagram), you will see that the probabilities for the various possible sums are as follows:

Sum	2	3	4	5	6	7	8	9	10	11	12
P(Sum)	$\frac{1}{36}$	$\frac{2}{36}$	$\frac{3}{36}$	$\frac{4}{36}$	$\frac{5}{36}$	$\frac{6}{36}$	$\frac{5}{36}$	$\frac{4}{36}$	$\frac{3}{36}$	$\frac{2}{36}$	$\frac{1}{36}$

We are also going to need the probability of winning for each possible result of the first roll. We already know that this probability is 0 when the first roll is 2, 3, or 12. We know it is 1 when the first roll is 7 or 11. Suppose your point is 5. You then roll until you get another 5 or a 7. Every intermediate roll is irrelevant. So consider this final roll. It is either a 5 or a 7. The probability that you win is simply the conditional probability that you get a 5 as opposed to a 7 on this final roll:

$$P(5 \mid 5 \text{ or } 7) = \frac{P(5)}{P(5 \text{ or } 7)} = \frac{4/36}{10/36} = \frac{2}{5}$$

This calculation is shown schematically in Figure 5.13 on page 145. We conditioned on the event {5, 7}, shown shaded, and took the ratio of the probability of the points (the darker shaded region) to the total probability (the two shaded regions).

Figure 5.12 Tree diagram for Example 5.9

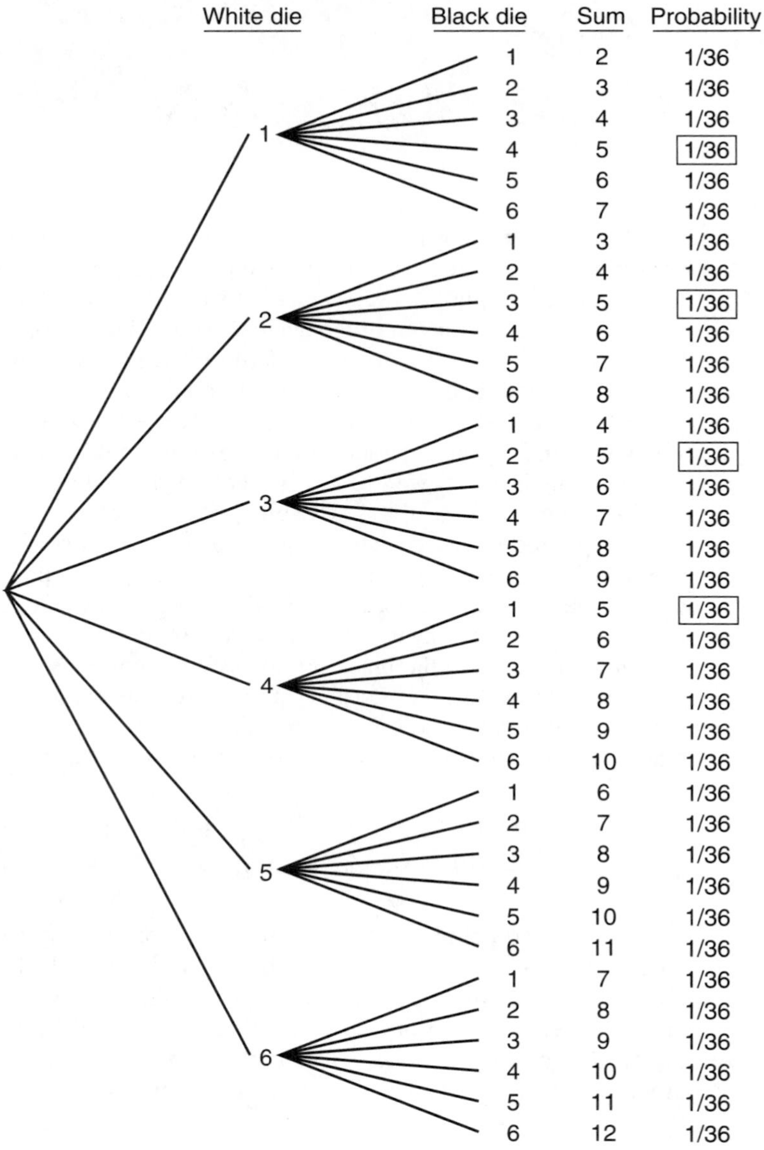

White die	Black die	Sum	Probability
1	1	2	1/36
	2	3	1/36
	3	4	1/36
	4	5	⟦1/36⟧
	5	6	1/36
	6	7	1/36
2	1	3	1/36
	2	4	1/36
	3	5	⟦1/36⟧
	4	6	1/36
	5	7	1/36
	6	8	1/36
3	1	4	1/36
	2	5	⟦1/36⟧
	3	6	1/36
	4	7	1/36
	5	8	1/36
	6	9	1/36
4	1	5	⟦1/36⟧
	2	6	1/36
	3	7	1/36
	4	8	1/36
	5	9	1/36
	6	10	1/36
5	1	6	1/36
	2	7	1/36
	3	8	1/36
	4	9	1/36
	5	10	1/36
	6	11	1/36
6	1	7	1/36
	2	8	1/36
	3	9	1/36
	4	10	1/36
	5	11	1/36
	6	12	1/36

Calculating the probability of winning for other points (sums) is similar; the following table shows them all [along with P(Sum) from the previous table]:

Sum	2	3	4	5	6	7	8	9	10	11	12
P(Sum)	$\frac{1}{36}$	$\frac{2}{36}$	$\frac{3}{36}$	$\frac{4}{36}$	$\frac{5}{36}$	$\frac{6}{36}$	$\frac{5}{36}$	$\frac{4}{36}$	$\frac{3}{36}$	$\frac{2}{36}$	$\frac{1}{36}$
$P(W\mid\text{Sum})$	0	0	$\frac{1}{3}$	$\frac{2}{5}$	$\frac{5}{11}$	1	$\frac{5}{11}$	$\frac{2}{5}$	$\frac{1}{3}$	1	0

Figure 5.13 Example 5.9: Sums of 5 and 7 shaded; these have respective probabilities $\frac{4}{36}$ and $\frac{6}{36}$

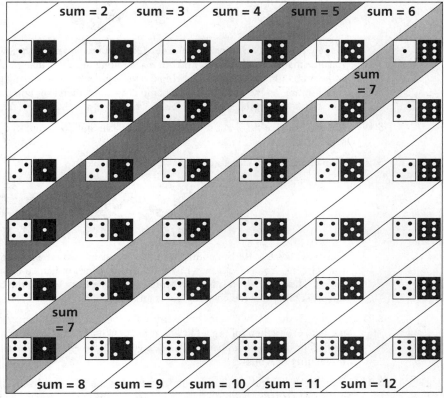

With these preliminaries, we can finally calculate the probability of winning. To use the law of total probability we multiply the last two rows of the preceding table and add the products:

$$P(W) = P(W \mid 2)P(2) + P(W \mid 3)P(3) + P(W \mid 4)P(4)$$
$$+ \cdots + P(W \mid 12)P(12)$$
$$= (0)(\tfrac{1}{36}) + (0)(\tfrac{2}{36}) + (\tfrac{1}{3})(\tfrac{3}{36}) + \cdots + (0)(\tfrac{1}{36})$$
$$= \tfrac{244}{495} = .493$$

This is close to $\frac{1}{2}$. The fact that it is less than $\frac{1}{2}$ means that craps is less than fair to the player. ◁

EXERCISES

5.17 Repeating the assumptions from Exercise 4.8: Suppose the probability that it rains both today and tomorrow is $\frac{1}{5}$, today but not tomorrow is $\frac{1}{10}$, not today but tomorrow is $\frac{1}{10}$, and neither day is $\frac{3}{5}$. Find the probability that:
 (a) It rains tomorrow.
 (b) It rains tomorrow, given that it rains today.

5.18 Suppose a bowl contains five chips; three are **R**ed and two are **G**reen:

You are to select two chips from the bowl, without replacing the first when you select the second. Assume that each selection is random from among the chips that are currently in the bowl. What is the probability that the second chip drawn is green? Answer using the law of total probability, conditioning on the color of the first chip drawn and then averaging. (When you get the answer, you may see an easier way to do this.)

5.19 Repeat Exercise 5.18, but now assume that the first chip is replaced before making the second selection.

5.20 Consider the following three bowls, each with five chips:

One bowl has five **R**eds, another has three **R**eds and two **G**reens, and the third has one **R**ed and four **G**reens. You select a bowl with the probabilities indicated below them in the figure ($\frac{1}{6}$, $\frac{1}{3}$, and $\frac{1}{2}$). Then you select randomly from among the five chips in the bowl you selected. What is the probability that the chip you get is green?

5.21 A game is played by first tossing a **fair coin** (one for which your probabilities of heads and tails are both $\frac{1}{2}$) and then rolling a fair die. If the coin lands heads, you win the game provided the die lands 5 or 6. If the coin lands tails, then you win provided the die lands 1, 2, or 3. What is your probability of winning?

5.22 Exercise 5.16 deals with drawing from among the 75 chips in the game of BINGO. Use your answers to parts (a) and (b) of that exercise and the law of total probability to find the probability that the second chip drawn is under the B.

5.23 Exercise 5.4 gives the following table of proportions of estrogen- and progesterone-receptor status for 20 tumors:

		Progesterone Positive	Negative
Estrogen	Positive	8/20	4/20
	Negative	1/20	7/20

Again, let A be the event that a tumor is estrogen-receptor positive and B be the event that it is progesterone-receptor positive.
    (a) What is $P(B \mid A)$?
    (b) What is $P(B \mid \sim A)$?
    (c) Using your answer from Exercise 5.4 that $P(A) = \frac{12}{20}$, use parts (a) and (b) of this exercise and the law of total probability to verify your earlier answer that $P(B) = \frac{9}{20}$.

5.24 When playing five-card draw poker, you are dealt ♦5, ♥6, ♠8, ♦9, ♠J. You assess your chances of winning if you draw to an inside straight by throwing away the ♠J and drawing one card as follows: Your probability of winning is $\frac{9}{10}$ if you draw a **7**, $\frac{1}{10}$ if you draw a **5, 6, 8,** or **9** (giving you a pair), and 0 if you draw any other card. As in Exercise 4.9, assume that each of the remaining 47 cards has the same probability of selection. What is your probability of winning if you throw away the ♠**J** and draw a card?

5.25 There are two players in a game called Risk. One type of confrontation between players is as follows. Player A rolls two dice and player B rolls one die; all three dice are fair in that the probabilities of all six outcomes are equal, and all three dice rolls are independent. Player A's score is the bigger of the two numbers on his two dice and player B's score is the number on her die. The player with the higher score wins, except that player B also wins if both players' scores are equal. (For example, suppose A rolls a 4 and a 2. A's score is 4. Suppose B also rolls a 4, which is then B's score as well. Player B wins.) What is the probability that player B wins? [*Hint*: Condition on player B's score and use the law of total probability.]

5.4 Bayes' Rule

Bayes' rule* indicates how probabilities change in the light of evidence. So it is the most important tool in statistics, wherein the evidence is usually data. We will use it in every statistics problem we address.

EXAMPLE 5.10
▷ **Who was your opponent?** In Example 5.7, we calculated the probability that you would win your next chess game by averaging over your possible opponents. Now suppose you tell me you won your next chess game—the "evidence" mentioned above. Who was your opponent?

Recall these probabilities:

$$P(W \mid J) = \tfrac{7}{10} \quad P(W \mid \sim J) = \tfrac{2}{10}$$
$$P(J) = \tfrac{1}{4} \quad\quad P(\sim J) = \tfrac{3}{4}$$

So, without conditioning on the evidence, you are three times as likely to have played Mary as Joe. Now I learn that you won and I want to find $P(J \mid W)$. Since you are more likely to win playing Joe than playing Mary, it seems reasonable to expect $P(J \mid W)$ to be bigger than $P(J)$. Bayes' rule gives $P(J \mid W)$—and verifies that it is bigger.

The definition of conditional probability (Section 5.2) says

$$P(J \mid W) = \frac{P(W \cap J)}{P(W)}$$

Using the multiplication rule (Section 5.2) in the numerator gives

$$P(J \mid W) = \frac{P(W \mid J) P(J)}{P(W)}$$

Using the law of total probability (Section 5.3) to expand the denominator gives the following important result:

*Named after Thomas Bayes, who was an 18th century English minister and philosopher.

> **Bayes' Rule**
>
> $$P(J \mid W) = \frac{P(W \mid J)P(J)}{P(W \mid J)P(J) + P(W \mid \sim J)P(\sim J)}$$

This applies for any events W and J, in any setting. In our example,

$$P(J \mid W) = \frac{(7/10)(1/4)}{(7/10)(1/4) + (2/10)(3/4)} = \frac{7/40}{13/40} = \frac{7}{13}$$

This result is pretty intuitive. Think about it in terms of a chip experiment. Suppose a population of 40 chips: seven are labeled JW, three are labeled $J(\sim W)$, six are labeled MW, and the remaining 24 are labeled $M(\sim W)$. Select a chip. Assume you got one with a W. Your probability that it also has a J is $\frac{7}{13}$.

Bayes' rule relates *inverse probabilities,* giving $P(J \mid W)$ in terms of $P(W \mid J)$ [and, also, in terms of $P(W \mid \sim J)$]. In a sense, it says how to "climb back down a tree." Consider the tree diagram from Example 5.7 (Figure 5.9, page 141). To calculate $P(J \mid W)$, we find the proportion of weight on the J branch when restricting consideration to the branches labeled W (those with probabilities in boxes). ◁

The conventional terminology for $P(J \mid W)$ is the **posterior probability of J given W** and for $P(J)$ is the **prior probability of J,** since it applies *before* or not conditionally on the information that W occurred. For reference in later chapters, I will rewrite Bayes' rule in two different ways. Consider the expression

$$P(J \mid W) = \frac{P(W \mid J)P(J)}{P(W \mid J)P(J) + P(W \mid \sim J)P(\sim J)}$$

Dividing the numerator into both numerator and denominator gives a useful expression:

> **Equivalent Version of Bayes' Rule**
>
> $$P(J \mid W) = \frac{1}{1 + \dfrac{P(W \mid \sim J)}{P(W \mid J)} \dfrac{P(\sim J)}{P(J)}}$$

Subtracting this from 1 gives

$$P(\sim J \mid W) = \frac{\dfrac{P(W \mid \sim J)}{P(W \mid J)} \dfrac{P(\sim J)}{P(J)}}{1 + \dfrac{P(W \mid \sim J)}{P(W \mid J)} \dfrac{P(\sim J)}{P(J)}}$$

Dividing this latter expression by the previous expression cancels out the denominator in both and gives still another version of Bayes' rule:

Bayes' Rule in Terms of Odds

$$\frac{P(\sim J \mid W)}{P(J \mid W)} = \frac{P(W \mid \sim J)}{P(W \mid J)} \frac{P(\sim J)}{P(J)}$$

The ratio $P(\sim J \mid W)/P(J \mid W)$ is the **posterior odds against** J. The second factor on the right, $P(\sim J)/P(J)$, is the **prior odds against** J. The other factor on the right, $P(W \mid \sim J)/P(W \mid J)$, is the **Bayes factor against** J.

We can also write the inverse of this expression as follows:

$$\frac{P(J \mid W)}{P(\sim J \mid W)} = \frac{P(W \mid J)}{P(W \mid \sim J)} \frac{P(J)}{P(\sim J)}$$

The ratio $P(J \mid W)/P(\sim J \mid W)$ is the **posterior odds in favor of** J. Now, the second factor on the right, $P(J)/P(\sim J)$, is the **prior odds in favor of** J; $P(W \mid J)/P(W \mid \sim J)$ is the **Bayes factor in favor of** J.

Bayes factors are ratios of probabilities of the information at hand ($W =$ win in the previous example) given particular models ($J =$ played Joe and $\sim J =$ played Mary in the previous example). These probabilities are called **likelihoods**. So $P(W \mid J)$ is the likelihood of J (for observation W) and $P(W \mid \sim J)$ is the likelihood of $\sim J$.

> **The likelihood of a model is the probability of the observations assuming that model.**

The next example is a simplified version of a typical statistics problem. You are uncertain about a mechanism that produces data (a model), you observe some data, and update what you knew before.

EXAMPLE 5.11

▷ **Which die?** I have two dice, one is a four-sider and the other is a 20-sider. The sides of the first are labeled 1, 2, 3, 4 and of the second are labeled 1, 2, ..., 20. Considering the dice separately, you regard the sides to be equally likely. I pick one of the two dice and get either the four-sider, event F, or the 20-sider, event $T = \sim F$. For you, $P(F) = P(\sim F) = \frac{1}{2}$. I roll the die I picked. I tell you the result is 13. You do not need Bayes' rule to know that I picked the 20-sider: $P(F \mid \text{rolled } 13) = 0$.

Suppose I tell you the result was 3. Which die did I pick? The likelihood of F is $P(3 \mid F) = \frac{1}{4}$ and the likelihood of $\sim F$ is $P(3 \mid \sim F) = \frac{1}{20}$. Bayes' rule says

$$P(F \mid 3) = \frac{P(3 \mid F)P(F)}{P(3 \mid F)P(F) + P(3 \mid \sim F)P(\sim F)}$$
$$= \frac{(1/4)(1/2)}{(1/4)(1/2) + (1/20)(1/2)} = \frac{5}{6}$$

While this single observation is far from conclusive, it does have a rather marked effect: It increases your probability that the die has four sides from $\frac{1}{2}$ to $\frac{5}{6}$.

The Bayes factor against F, or in favor of $\sim F$, is

$$\frac{P(3 \mid \sim F)}{P(3 \mid F)} = \frac{1/20}{1/4} = \frac{1}{5}$$

The Bayes factor in favor of F is the inverse of the Bayes factor against F:

$$\frac{P(3 \mid F)}{P(3 \mid \sim F)} = \frac{1/4}{1/20} = 5$$

This says simply that the observation 3 is five times more likely for the four-sider than for the 20-sider. ◁

The next example is a more realistic application of Bayes' rule. It is more involved than the previous example.

EXAMPLE 5.12
▷ **Deciding paternity.** Legal cases of disputed paternity in many countries are resolved using blood tests. Laboratories make genetic determinations concerning the mother, child, and alleged father. (Some cases involve different types of evidence. The mother may not be available. The alleged father may not be available, but his brother is available, and so on.) Most labs apply Bayes' rule in communicating the testing results. They calculate the probability that the alleged father is in fact the child's father given the genetic evidence. (You should ask yourself, Whose probability is it?) For the sake of brevity, I will pare down the genetic evidence usually introduced and deal only with ABO blood type. Knowing some genetics may help you to follow this example, but such knowledge is not required—I will tell you all the probabilities you need.

Suppose you are on a jury considering a paternity suit brought by Suzy Smith's mother against Al Edged. (Many paternity suits in the United States are initiated by welfare departments in the name of the mother.) The following is part of the background information: Suzy's mother has blood type O and Al Edged is type AB. All your probabilities are calculated conditional on this information, although I will not include it explicitly to the right of the vertical bar in probability expressions.

You have other information as well. You hear testimony concerning whether Al Edged and Suzy's mother had sexual intercourse during the time that conception could have occurred, about the timing and frequency of such intercourse, about Al Edged's fertility, about the possibility that someone else is the father, and so on. You put all this information together in assessing $P(F)$, your probability that Al is Suzy's father.

The evidence of interest is Suzy's blood type. If it is O, then Al Edged is excluded from paternity—he is not her father, unless there has been a gene mutation or a laboratory error. Suzy's blood type turns out to be B; call this event B. According to Bayes' rule,

$$P(F \mid B) = \frac{P(B \mid F)P(F)}{P(B \mid F)P(F) + P(B \mid \sim F)P(\sim F)}$$

According to Mendelian genetics—which you accept—$P(B \mid F) = \frac{1}{2}$. You also accept

Sec. 5.4 / Bayes' Rule

the blood bank's $P(B \mid \sim F)$. They calculate this as the proportion of B genes (this is not the same as the proportion of people with type B blood) to the total number of ABO genes in their previous cases. A typical value among Caucasians is 9%. So,

$$P(F \mid B) = \frac{(1/2)P(F)}{(1/2)P(F) + (.09)P(\sim F)} = \frac{50 \cdot P(F)}{41 \cdot P(F) + 9}$$

This is a substantial increase over $P(F)$; for example, it is about 85% when $P(F) = \frac{1}{2}$. The reason such a large increase is possible is that Suzy's paternal gene (B) is relatively rare. Your probability of paternity would increase for any male who has a B gene.

The relationship between our unconditional probability, $P(F)$, and our conditional probability, $P(F \mid B)$, can be shown using a table such as this:

$P(F)$	0	.100	.250	.500	.750	.900	1
$P(F \mid B)$	0	.382	.649	.847	.943	.980	1

Another way to show the same thing is to use a graph, such as the one in Figure 5.14. The diagonal on this graph corresponds to evidence which contains no information about F (in which case B and F would be independent). Comparing this diagonal with the actual curve shows how much the evidence changes one's prior probability of paternity. Tables and graphs are effective ways for juries and others to understand the strength of the evidence.

Blood banks and other laboratories that analyze genetic factors in paternity cases have a name for the Bayes factor in favor of F:

$$\text{Paternity index} = PI = \frac{P(B \mid F)}{P(B \mid \sim F)} = \frac{1/2}{.09} = 5.56$$

Figure 5.14 Posterior vs prior probabilities ($P(F \mid B)$ and $P(F)$) for Example 5.12

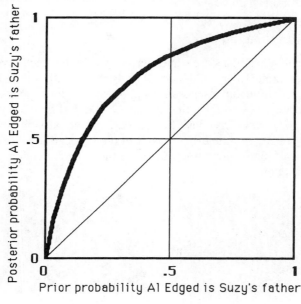

The evidence (child has type B blood) is 5.56 times as likely if Al Edged is the father than if he is not. The posterior probability of paternity (based on the equivalent version of Bayes' rule) is

$$P(F \mid B) = \frac{1}{1 + \dfrac{P(B \mid \sim F)}{P(B \mid F)} \dfrac{P(\sim F)}{P(F)}} = \frac{1}{1 + \dfrac{1}{PI} \dfrac{P(\sim F)}{P(F)}} = \frac{PI}{PI + \dfrac{P(\sim F)}{P(F)}}$$

Laboratories choose $P(F) = \frac{1}{2}$ and report a probability (or likelihood) of paternity as though there is no prior probability involved. This policy is arbitrary and misleading. For example, on February 18, 1994, news wire services carried this report: "Experts who examined blood samples from John Bobbitt, Beatrice Williams and her son, Andrew, determined there is a 99.99 percent likelihood that Bobbitt is the boy's father." This is stated as though it is a posterior probability $P(F \mid$ blood data), but it is such only if the other evidence in the case weighs equally on F as on $\sim F$ from the perspective of the jury. In effect, the experts are preempting the jury in assessing this evidence.

To find the paternity index in the Bobbitt case, carry out the calculations in reverse: the answer is 9,999. Check this, using the formula for probability in terms of PI and assuming $P(F) = .5$ and, therefore, $P(\sim F)/P(F) = 1$:

$$P(F \mid B) = \frac{PI}{1 + PI} = \frac{9{,}999}{1 + 9{,}999} = .9999$$

[There is only one significant digit in calculating the PI to be 9,999—any PI close to 10,000 gives a $P(F \mid B)$ that rounds to .9999.]

A PI as large as 10,000 is possible only by combining evidence from many genetic factors in addition to ABO blood type. These factors are combined by multiplying their individual likelihoods, which is appropriate if the factors are independent—see Exercise 5.40 for an example. ◁

Generalized Bayes' Rule

The preceding examples are somewhat special in that each involves two possibilities: Joe or Mary, four-sider or 20-sider, Al Edged or not. In other examples, there may be several possibilities: many possible opponents, various types of dice, and several alleged fathers. The generalization of Bayes' rule is slightly more involved because the denominator is more involved. For the case of two possibilities, we expanded the denominator of Bayes' rule using the law of total probability. We can generalize this rule to the case of many alternatives by using the general version of that law given in the previous section:

Generalized Bayes' Rule

$$P(A \mid W) = \frac{P(W \mid A)P(A)}{P(W \mid A)P(A) + P(W \mid B)P(B) + P(W \mid C)P(C) + \cdots}$$

where $A, B, C, \ldots$ subdivide the universe, U.

Bayes' rule is fundamental in statistics because it shows how probabilities change when new evidence—data—becomes available. Before seeing the data, you have some information about whether the die is fair, whether Al Edged is the father, whether the drug is effective, and so on. You express this information in terms of probabilities. Then you calculate likelihoods and update your probabilities using Bayes' rule.

Prior probabilities are based on information available separately from an experiment (or other type of observation) and the likelihoods are derived from the experiment itself. In the process of updating, it is important not to use the same information twice:

> **Principle of Separation of Prior Information and Current Data**
>
> **When assessing prior probabilities, use only information not included in the likelihoods.**

If you have seen the data from an experiment, you may have difficulty assessing what you thought before the experiment. You may be forced to make a posterior assessment without using Bayes' rule. Bayes' rule can still serve as a guide. For example, calculate what your posterior probabilities would have been for different assignments of prior probabilities that you might have had.

Since Bayes' rule is critical in making statistical inferences, we will apply the ideas developed in this section throughout this course. Calculations such as those in the next two examples will be especially relevant.

EXAMPLE 5.13

▷ **How many greens?** A bowl contains five chips. I would like to know how many are green—the remaining chips can be different colors, but for convenience suppose all nongreens are red. Let 0, 1, 2, 3, 4, and 5, respectively, stand for the event that there are the corresponding number of greens. For me, $P(0) = P(1) = \cdots = P(5) = \frac{1}{6}$. The six alternative **chip models** (labeled by the number of greens) are given in the drawing along with my prior probabilities:

$P(G\mid 0)$	$P(G\mid 1)$	$P(G\mid 2)$	$P(G\mid 3)$	$P(G\mid 4)$	$P(G\mid 5)$
0	$\frac{1}{5}$	$\frac{2}{5}$	$\frac{3}{5}$	$\frac{4}{5}$	1

To get more information about the correct model, I select a chip randomly. It is green. Call this event G. The likelihoods of the various models are shown in the following table:

The updated probability of any model follows from Bayes' rule:

$$P(\text{model} \mid G) = \frac{P(G \mid \text{model})P(\text{model})}{P(G)}$$

where $P(G)$ follows from the law of total probability:

$P(G)$
$= P(G \mid 0)P(0) + P(G \mid 1)P(1) + P(G \mid 2)P(2) + P(G \mid 3)P(3)$
$\quad + P(G \mid 4)P(4) + P(G \mid 5)P(5)$
$= (0)(1/6) + (1/5)(1/6) + (2/5)(1/6) + (3/5)(1/6) + (4/5)(1/6) + (1)(1/6)$
$= \frac{1}{2}$

(In view of the answer you may guess that there is an easier way to calculate this. My prior probabilities of $0, 1, \ldots, 5$ are symmetric with regard to green and red. Therefore, my probability of G must equal my probability of $\sim G = R$. Since they sum to 1, both must equal $\frac{1}{2}$.) For example, my posterior probability of model 4 is the following:

$$P(4 \mid G) = \frac{P(G \mid 4)P(4)}{P(G)} = \frac{(4/5)(1/6)}{1/2} = \frac{4}{15}$$

To find the probability for other models, put the appropriate term from the preceding expansion of $P(G)$ into the numerator. The results are given in both Table 5.1 and the following figure.

Table 5.1
Posterior probabilities for Example 5.13

Number of greens	0	1	2	3	4	5
P(number of greens $\mid$ G)	0	$\frac{1}{15}$	$\frac{2}{15}$	$\frac{3}{15}$	$\frac{4}{15}$	$\frac{5}{15}$

These calculations can also be accomplished using the tree diagram in Figure 5.15.

Figure 5.15 Tree diagram for Example 5.13

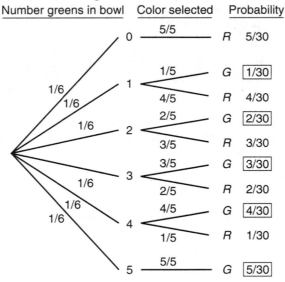

Given event G, my probabilities of the various possible numbers of green chips are proportional to the probabilities in the boxes on the right-hand side of the tree. Dividing these probabilities by $\frac{1}{2}$ gives the results in Table 5.1. ◁

In a practical setting, where a green chip might be a successful medical treatment, knowing your probabilities of the number of greens is not enough. You also need to know the probability that the next chip selected is green—or that the next treatment is successful.

EXAMPLE 5.14
▷ **Is the next chip green?** Continuing in the vein of Example 5.13, suppose I select a second chip from the bowl. Using what I learned on my first draw, I want to know whether the second will also be green. Now I need to modify the notation because G could refer to the color of the first chip or of the second chip. So let $G1$ be the event First chip is green (the old event G). Let $G2$ be the event Second chip is green. I want to calculate my $P(G2 \mid G1)$.

There are at least two cases: with replacement (replace the first chip before selecting the second) and without replacement. (In both cases, the two selections are exchangeable.) In the first case, the bowl contains an extra green chip, so it seems clear that $P(G2 \mid G1)$ will be greater in that case.

First consider the case with replacement. The following table is the same as Table 5.1 in Example 5.13, with $G1$ in place of G.

With Replacement

Number of greens	0	1	2	3	4	5
$P(\text{number of greens} \mid G1)$	0	$\frac{1}{15}$	$\frac{2}{15}$	$\frac{3}{15}$	$\frac{4}{15}$	$\frac{5}{15}$

To calculate $P(G2 \mid G1)$ using the law of total probability requires conditioning on $G1$ in every probability. For example, I will need $P(4 \mid G1)$. This is given in the table as $\frac{4}{15}$. I will also need $P(G2 \mid 4, G1)$. Both events $G1$ and 4 are assumed to be given in evaluating this probability. But if you tell me there are four green chips, I do not get any additional information from your telling me $G1$. Therefore, $P(G2 \mid 4, G1) = P(G2 \mid 4) = \frac{4}{5}$. The law of total probability gives

$$
\begin{aligned}
P(G2 \mid G1) &= P(G2 \mid 0, G1)P(0 \mid G1) + P(G2 \mid 1, G1)P(1 \mid G1) \\
&\quad + P(G2 \mid 2, G1)P(2 \mid G1) + P(G2 \mid 3, G1)P(3 \mid G1) \\
&\quad + P(G2 \mid 4, G1)P(4 \mid G1) + P(G2 \mid 5, G1)P(5 \mid G1) \\
&= (0)(0) + (1/5)(1/15) + (2/5)(2/15) \\
&\quad + (3/5)(3/15) + (4/5)(4/15) + (1)(5/15) \\
&= 11/15
\end{aligned}
$$

This is an increase from my *unconditional* probability of green on the second chip: $P(G2) = \frac{1}{2}$. [This equals $P(G1)$ by symmetry.]

Clearly, $P(G2 \mid G1) \neq P(G2)$, and so $G1$ and $G2$ are dependent. The fact that $P(G2 \mid G1) > P(G2)$ means they are positively related. This seems reasonable since, for one thing, I might draw the same chip both times.

Although I will not do much with sampling without replacement in later chapters, I will consider it here—mainly to contrast it to sampling with replacement. I observe that the first chip is green. I keep the chip and select the second one, necessarily different from the first. You might think that now $P(G2 \mid G1) = P(G2)$. But this is not so. The event $G1$ is more likely if there are many greens in the bowl than if there are few. So observing $G1$ means that I think there are more greens, and my probability of green is increased.

The appropriate probabilities of the colors of the four remaining chips are easy to find from the previous table by simply subtracting a green chip.

Without Replacement

Number of greens left	0	1	2	3	4
$P(\text{number of greens} \mid G1)$	$\frac{1}{15}$	$\frac{2}{15}$	$\frac{3}{15}$	$\frac{4}{15}$	$\frac{5}{15}$

The law of total probability now gives

$$\begin{aligned} P(G2 \mid G1) &= P(G2 \mid 0, G1)P(0 \mid G1) + P(G2 \mid 1, G1)P(1 \mid G1) \\ &\quad + P(G2 \mid 2, G1)P(2 \mid G1) + P(G2 \mid 3, G1)P(3 \mid G1) \\ &\quad + P(G2 \mid 4, G1)P(4 \mid G1) \\ &= (0)(1/15) + (1/4)(2/15) + (2/4)(3/15) + (3/4)(4/15) \\ &\quad + (4/4)(5/15) \\ &= 2/3 \end{aligned}$$

As advertised, this is an increase from $P(G2) = \frac{1}{2}$, but it is not as large as $\frac{11}{15}$, the corresponding probability with replacement. ◁

Conditional probabilities are sometimes difficult to sort out. Example 5.3 is a case in point. The next example is similar. It uses Bayes' rule. You will understand this example and Example 5.3 better if you refer to the latter example after reading this one.

EXAMPLE 5.15

▷ **"Let's Make a Deal."** A contestant on the TV show "Let's Make a Deal" was given the choice of three boxes; one contained the key to a Lincoln Continental and the other two were empty. She chooses box 2 and decides that her probability of winning the Lincoln is $\frac{1}{3}$. The emcee of the show, Monte Hall, then opened box 1 and showed her that it was empty. Since either box 2 or box 3 must contain the key, he claimed that her probability of winning had increased to $\frac{1}{2}$.

Whether or not this is correct depends on the emcee's knowledge and how he acts in light of that knowledge. There are at least four possibilities. Only the last one is consistent with the emcee's actual knowledge. But some of you will assume things I do not want you to assume, and so I will consider the other two possibilities first.

Suppose that he opens one of the three boxes at random (as far as the contestant is concerned). Let **1** be the event that he opens box 1 and it is empty. Also, let $\boxed{1}$ stand for the event that the key is in box 1, $\boxed{2}$ stand for the event that the key is in box 2, and $\boxed{3}$ stand for the event that the key is in box 3. The relevant probabilities for the contestant are these:

$$P(\boxed{1}) = P(\boxed{2}) = P(\boxed{3}) = \tfrac{1}{3},$$

$$P(\mathbf{1} \mid \boxed{1}) = 0 \qquad P(\mathbf{1} \mid \boxed{2}) = \tfrac{1}{3} \qquad P(\mathbf{1} \mid \boxed{3}) = \tfrac{1}{3}.$$

Using Bayes' rule,

$$P(\boxed{2} \mid \mathbf{1}) = \frac{P(\mathbf{1} \mid \boxed{2})P(\boxed{2})}{P(\mathbf{1} \mid \boxed{1})P(\boxed{1}) + P(\mathbf{1} \mid \boxed{2})P(\boxed{2}) + P(\mathbf{1} \mid \boxed{3})P(\boxed{3})}$$

$$= \frac{(1)(1/3)}{(0)(1/3) + (1)(1/3) + (1)(1/3)} = \frac{1}{2}$$

So the emcee's statement is correct under this assumption.

Now, suppose the emcee would not open the box containing the key, but he might open box 2 (the box chosen by the contestant)—if it did not contain the key. Event **1** is as before. Now her conditional probabilities change:

$$P(\mathbf{1} \mid \boxed{1}) = 0 \qquad P(\mathbf{1} \mid \boxed{2}) = \tfrac{1}{2} \qquad P(\mathbf{1} \mid \boxed{3}) = \tfrac{1}{2}$$

Using Bayes' rule as before,

$$P(\boxed{2} \mid \mathbf{1}) = \frac{(1/2)(1/3)}{(0)(1/3) + (1/2)(1/3) + (1/2)(1/3)} = \frac{1}{2}$$

the same as in the first possibility.

Here is the third possibility: The emcee would not open box 2, the one chosen by the contestant, but he might open the box containing the key. When conditioning on either $\boxed{1}$ or $\boxed{2}$, the contestant's probability that the emcee opens box 1 is the same as in the previous case, since she had assumed there that he would not open the box containing the key. But it does change when conditioning on $\boxed{3}$. These are her conditional probabilities under the current assumptions:

$$P(\mathbf{1} \mid \boxed{1}) = 0 \qquad P(\mathbf{1} \mid \boxed{2}) = \tfrac{1}{2} \qquad P(\mathbf{1} \mid \boxed{3}) = 1$$

Now Bayes' rule gives a different answer:

$$P(\boxed{2} \mid \mathbf{1}) = \frac{(1/2)(1/3)}{(0)(1/3) + (1/2)(1/3) + (1)(1/3)} = \frac{1}{3}$$

Finally, the fourth possibility: Not only would the emcee not open the box containing the key, he would not open box 2, the box chosen by the contestant. The calculations are identical with the third possibility and so, again, $P(\boxed{2} \mid \mathbf{1}) = \tfrac{1}{3}$. I believe this possibility corresponds with the emcee's actual practice in the show. Since $P(\boxed{2} \mid \mathbf{1}) = P(\boxed{2})$, the emcee has successfully avoided giving any information at all to the contestant about the event $\boxed{2}$. However, since $P(\boxed{3} \mid \mathbf{1}) = \tfrac{2}{3}$, telling her $\sim\boxed{1}$ (under the assumptions made in this possibility) serves to move all the probability she once associated with box 1 to box 3. ◁

If the calculations in the previous example do not convince you, actually play the game with a friend. You should see what is going on after a few plays.

This example caused a stir in 1991 after a national columnist, who calls herself Marilyn Vos Savant, used it in her column. She gave the correct answer. A surprising number of PhD mathematicians wrote to her saying that she was wrong.

EXERCISES

5.26 In Exercise 5.20, you considered the following three bowls:

I have now labeled them for easy reference: Bowl A has five **R**eds, bowl B has three **R**eds and two **G**reens, and bowl C has one **R**ed and four **G**reens. You select a bowl with the probabilities indicated below them in the figure ($\frac{1}{6}$, $\frac{1}{3}$, and $\frac{1}{2}$). Then you select randomly from among the five chips in the bowl you selected. The chip turns out to be green. What are the updated probabilities that the bowl you selected is A? B? C?

5.27 This is a continuation of Exercise 5.26. You replace the first chip you selected (the green one) and again select randomly from the five chips in the bowl. You get a green chip the second time. On the basis of the two green chips selected, what are the updated probabilities that the bowl you selected is A? B? C?

5.28 This is a variation of Exercise 5.27. This time you do *not* replace the first chip you selected (the green one). Again, you select a second chip, this time randomly from the *four* chips in the bowl. You get a green chip for the second time. On the basis of the two green chips selected, what are the updated probabilities that the bowl you selected is A? B? C?

5.29 Exercises 5.4 and 5.23 give the following table of proportions of estrogen- and progesterone-receptor status for 20 tumors:

		Progesterone	
		Positive	Negative
Estrogen	Positive	$\frac{8}{20}$	$\frac{4}{20}$
	Negative	$\frac{1}{20}$	$\frac{7}{20}$

Again, let A be the event that a tumor is estrogen-receptor positive and B be the event that it is progesterone-receptor positive. Using Bayes' rule and your calculations of $P(B \mid A)$ and $P(B)$ from Exercise 5.23, calculate $P(A \mid B)$. [In this instance, you can check your answer by calculating $P(A \mid B)$ directly from the table, and you should do so.]

5.30 Someone gives you a fax number over the telephone and you write it down. You send a fax message and wonder whether you got the correct number. There is always a chance that the person gave you the wrong number, that you wrote it down wrong, or that you will dial it wrong. Your probability that the number is correct (and dialed correctly) is 90%. It turns out that you do connect with a fax machine and your message goes through. That is a good sign. To find how good, calculate your updated probability that your fax message got to the person's machine. Use your assessment that only 2% of telephone numbers are fax numbers. Assume that numbers are randomly assigned to fax machines and telephones, and that there are a very large number of both.

5.31 A firm has 50 employees: five are executives (E), 20 are white-collar workers (W), and 25 are blue-collar workers (B). Some have annual salaries in excess of $75,000 (S) and some do not: Four of the five executives do; six of the 20 white-collar workers do; none of the 25 blue-collar workers do. Interpreting probabilities as proportions of the total (consistent with sampling randomly):

(a) Calculate $P(S)$ directly based on the fact that $4 + 6 = 10$ of the 50 employees earn more than \$75,000.

(b) Calculate $P(S)$ using the law of total probability:
$$P(S) = P(S \mid E)P(E) + P(S \mid W)P(W) + P(S \mid B)P(B)$$

(c) Calculate $P(E \mid S)$ directly based on the fact that four of the employees who earn more than \$75,000 are executives.

(d) Calculate $P(E \mid S)$ using Bayes' rule.

5.32 A bowl contains four chips; some are red and some are green. You do not know how many there are of each color, but you believe that there are either 1, 2, or 3 reds, and your probability is evenly divided among these three possibilities. These possibilities and the corresponding probabilities are depicted in the drawing:

(a) You select a chip from the bowl and find that it is red. Calculate your new probabilities of the three possibilities.

(b) After replacing the red chip you selected in part (a), you select a second chip from the bowl and find that it also is red. Find your updated probabilities from those you calculated in part (a) using this new information.

5.33 A young friend of mine was diagnosed as having a type of cancer that occurs extremely rarely in young people. Naturally, he was very upset. I told him it was probably a mistake. I reasoned as follows. No medical test is perfect: There are always incidences of false positives and false negatives. Let C stand for the event that he has cancer and let + stand for the event that an individual responds positively to the test. Assume $P(C) = 1/1,000,000$, $P(+ \mid C) = .99$, and $P(+ \mid \sim C) = .01$. (So only one per million people his age have the disease and the test is extremely good relative to most medical tests—giving only 1% false positives and 1% false negatives.) Find the probability that he has cancer given that he has a positive response. (After you make this calculation you will not be surprised to learn that he did not have cancer.)

5.34 In Exercise 5.15, you calculated the probability that three victims—picked at random and not because of their zodiac sign—would not duplicate any of the signs of the previous four victims or of each other. Assume, on the basis of the other evidence, that the probability the killer is the same as the 1990 "zodiac killer" is 10%. Assume further that, if he is the "zodiac killer," he would not duplicate any signs in choosing new victims and, if he is not the "zodiac killer," he would select victims irrespective of their signs (that is, randomly). Since none of the victims' signs were duplicated, what is the updated probability he is the "zodiac killer"?

5.35 You and two friends, Carol and Luke, are being held in a prison in a faraway land. The authorities have decided to hang two of you tomorrow, but you do not know which two. You decide that you have a $\frac{2}{3}$ chance of being hanged. The suspense is killing you! So you ask a guard for more information. He will not tell you whether you are one of the two, but he does agree to tell you which one of the other two people is to be hanged. When he says that Carol is going to be hanged, you rejoice because your chances of being hanged seem to have decreased to $\frac{1}{2}$. But this is wrong. Find the correct probability of your being hanged, given the information you got from the guard (making an assumption about what he would answer if both Carol and Luke are to be hanged). Tell why you would like to switch places with Luke. [*Hint:* "Let's Make a Deal!"]

5.36 Reconsider the bowl containing five chips in Example 5.13. Again, I would like to know how many are green. Now take the prior probabilities to be as follows:

(This is what I ended up with in Example 5.13; so you are starting where I ended.) You select a chip from the bowl and it turns out to be green; call this event G. Find $P(1 \mid G), P(2 \mid G), \ldots, P(5 \mid G)$. [*Hint:* Your $P(G)$ is the same as my $P(G2 \mid G1)$, calculated assuming with replacement in Example 5.14.]

5.37 Consider your posterior probabilities found in Exercise 5.36. Suppose you replace the green chip and select another chip. What is your probability that the next chip is green?

5.38 I have two dice. As in Example 5.11, suppose that one is a four-sider and the other is a 20-sider. I pick one of the dice, getting either the four-sider, event F, or the 20-sider, event $T = {\sim}F$. For you, $P(F) = P({\sim}F) = \frac{1}{2}$. I roll the die I picked and get a 3. I am planning to roll the same die a second time. What is the probability that the second roll is a 2? [*Hint:* The answer is neither $\frac{1}{4}$ nor $\frac{1}{20}$, but something in between.]

5.39 A burglar cut himself on a broken window in the process of committing a crime. A DNA analysis reveals that only 1 person in 100,000 would match the culprit's blood. A man is charged with committing a crime and a DNA analysis of his blood finds that it indeed matches that on the glass.
(a) Find the Bayes factor in favor of guilt.
(b) Find the posterior probability of the man's guilt, assuming, in turn, each of the following six prior probabilities of guilt: .00001, .0001, .001, .2, .5, .99.

5.40 In the 1994 preliminary hearing in the O. J. Simpson murder case and before the DNA evidence was assessed, prosecutors presented genetic information for Simpson and also for a drop of blood found at the crime scene (prosecution item #49). The following table shows phenotypes (observed characteristics) for three genetic systems, as reported by the Los Angeles Police Department laboratory, including frequencies of the phenotypes of item #49 in the general population. These phenotypes did not match those of either victim. Let S be the event that item #49 is Simpson's blood—which is distinct from "Simpson is guilty." Since Simpson's phenotypes matched item #49, the likelihood of S is 1. Assuming independence of these systems, the likelihood of ${\sim}S$ is the product of the probabilities shown in the table.
(a) Find the Bayes factor in favor of S.
(b) Find the posterior probability of S assuming, in turn, each of the following six prior probabilities of S: .00001, .0001, .001, .2, .5, .99.

Exercise 5.40: Phenotypes for three systems in the Simpson murder case

System	Simpson	Item #49	Probability
ABO	A	A	.347
EsD	1	1	.79
PGM	2+2−	2+2−	.016
Product			.0043

5.41 About 30% of human twins are identical and the rest are fraternal. Identical twins are necessarily the same sex—half are males and the other half are females. One-quarter of fraternal twins are both male, one-quarter are both female, and one-half are mixed: one male, one female. You have just become a parent of twins and are told they are both girls. Given this information, what is the probability they are identical?

5.42 This is a continuation of Exercise 5.41. In addition to the information already given, you find that both children have the same type of blood. Based on the blood types of you and your spouse, you calculate the probability that identical twins would have this particular blood type as $\frac{6}{10}$, and the probability that fraternal twins would both have this particular blood type as $\frac{1}{10}$. Given this information, and *also the information* from Exercise 5.41, what is your probability that they are identical?

Appendix: Using Minitab for Conditional Probability and Bayes' Rule

The Minitab macros '**bayes_se**' and '**bayes**' are designed to implement Bayes' rule for a finite collection of models and an independent sequence of observations. To illustrate the use of these programs, consider Example 5.13, where you wish to learn about the number of green chips in a bowl of five chips. To set up the problem, you run the macro 'bayes_se' by typing the command

```
exec 'bayes_se'
```

The program first asks for the number of models and the names of each model. In this example, a model corresponds to the number of greens. There are six possible models (corresponding to 0, 1, 2, 3, 4, 5 greens) and we will call the models '0 green,' '1 green,' etc. Next, you input the (prior) probabilities for the six models. In the example, we give each model the same probability $\frac{1}{6} = .17$. The program next asks for the number of outcomes. You will be choosing one chip from the bowl—the two possible outcomes are green (which we will refer to as 'G') and red (denote by 'R'). You input the names of the two outcomes on the next two 'DATA' lines. Last, you give the likelihoods or the probabilities of the two outcomes for each model. For model 1 (0 green), the probabilities of choosing a green and red are 0 and 1, respectively; for model 1 (0 green), the probabilities of choosing a green and red are 0 and 1, respectively; for model 1 (1 green), the probabilities are .2, .8, and so on. After you specify the likelihoods for all models, the program displays the models, prior probabilities, and likelihoods in table form.

```
MTB > exec 'bayes_se'

INPUT NUMBER OF MODELS:
DATA> 6

INPUT NAMES OF MODELS (ONE NAME ON EACH LINE):
DATA> 0 green
DATA> 1 green
```

(continued)

```
DATA> 2 green
DATA> 3 green
DATA> 4 green
DATA> 5 green

INPUT PRIOR PROBABILITIES OF MODELS:
DATA> .17 .17 .17 .17 .17 .17

INPUT THE NUMBER OF POSSIBLE OUTCOMES:
DATA> 2

INPUT THE NAME OF EACH OBSERVATION:
(ONE OBSERVATION ON A LINE)
DATA> G
DATA> R

INPUT LIKELIHOODS OF EACH MODEL:

MODEL
    1

DATA> 0 1

MODEL
    2

DATA> .2 .8

MODEL
    3

DATA> .4 .6

MODEL
    4

DATA> .6 .4

MODEL
    5

DATA> .8 .2

MODEL
    6

DATA> 1 0
```

Appendix / Using Minitab for Conditional Probability and Bayes' Rule

```
TABLE OF PROBABILITIES OF MODELS AND OUTCOMES:

ROW        MODEL        NAME          PRIOR        OUT_1        OUT_2
  1          1        0 green       0.166667        0.0          1.0
  2          2        1 green       0.166667        0.2          0.8
  3          3        2 green       0.166667        0.4          0.6
  4          4        3 green       0.166667        0.6          0.4
  5          5        4 green       0.166667        0.8          0.2
  6          6        5 green       0.166667        1.0          0.0
```

Suppose now that you choose a chip at random and observe a green. To update your probabilities, the macro 'bayes' is run using the **exec 'bayes'** command. In this setting, you have one observation, G, which corresponds to a green. You input the number of observations, 1, and the particular observation, G. The program summarizes the calculations for Bayes' rule in a table. Your current probabilities about the number of greens are contained in the column 'POST.' Note that the model '0 green' has probability 0 and the most probable model is '5 green,' which has probability .33.

```
MTB > exec 'bayes'

INPUT NUMBER OF OBSERVATIONS;
DATA> 1

INPUT OBSERVATIONS:
 (ONE OBSERVATION NAME ON A LINE:)
DATA> G

OUTCOME
  G

ROW     MODEL      NAME         PRIOR       LIKE      PRODUCT       POST
  1       1      0 green      0.166667      0.0      0.000000    0.000000
  2       2      1 green      0.166667      0.2      0.033333    0.066667
  3       3      2 green      0.166667      0.4      0.066667    0.133333
  4       4      3 green      0.166667      0.6      0.100000    0.200000
  5       5      4 green      0.166667      0.8      0.133333    0.266667
  6       6      5 green      0.166667      1.0      0.166667    0.333333
```

The program 'bayes' can also be used to update model probabilities for a sequence of independent observations. In our example, suppose that you select four chips with replacement from the bowl, and observe a green, a red, a red, and a green. You run the program 'bayes,' say that four observations are observed, and input the letters G, R, R, G. The program will update your probabilities after each observation. One can see from the output of this program how your probabilities get modified as you acquire more and more information about the composition of the bowl by sampling. The final table (after all four observations) is shown below. The column 'PRIOR' gives the model probabilities after observing the first three observations: G, R, R; the 'POST' column gives the model probabilities after observing the final G.

```
MTB > exec 'bayes'

INPUT NUMBER OF OBSERVATIONS;
DATA> 4

INPUT OBSERVATIONS:
 (ONE OBSERVATION NAME ON A LINE:)
DATA> G
DATA> R
DATA> R
DATA> G
```

ROW	MODEL	NAME	PRIOR	LIKE	PRODUCT	POST
1	1	0 green	0.00	0.0	0.000	0.000000
2	2	1 green	0.32	0.2	0.064	0.153846
3	3	2 green	0.36	0.4	0.144	0.346154
4	4	3 green	0.24	0.6	0.144	0.346154
5	5	4 green	0.08	0.8	0.064	0.153846
6	6	5 green	0.00	1.0	0.000	0.000000

6 Models for Proportions

THIS CHAPTER begins a formal development of statistical inference. As I have said several times, the goal is learning. We update probabilities about the various possible models based on observations. The process is easier if we have a clear way of dealing with models and understanding what it means to make an observation from a model. We will think of a model as a bowl with chips:

> **A model for a population is a bowl with chips labeled according to the possible outcomes of the experiment. The numbers of the various types of chips characterize the model.**

Updated probabilities are conditional probabilities. The approach in this chapter relies heavily on manipulations of probability that you learned in the previous two chapters. An example in which the population and inference are easy to describe follows.

EXAMPLE 6.1
▷ **Does Felix have ESP?** In a segment of the TV show "The Odd Couple," Felix claimed to have extrasensory perception (ESP). Oscar was skeptical and suggested testing this claim with the following experiment. Oscar selected one card from four large cards, each having a different geometric figure on it. Oscar concentrated on the card selected and Felix tried to identify it. They repeated this 10 times. The population is made up of corrects and wrongs. The proportion of corrects indicates Felix's level of ESP, which is unknown. We gain information about the population by sampling from it. A model for the population is a bowl with chips; each chip is labeled C (for correct) or W (for wrong).

The proportion of corrects depends on the model. Oscar was pretty sure that Felix had no extrasensory ability. A theory of no ability or no effect is called a **null hypothesis.** So the null hypothesis is that Felix was only guessing and that they were, in effect, making observations from a population composed of 25% corrects and 75% wrongs, such as from the following bowl containing two C's and six W's:

165

The null hypothesis may or may not be the right one. This is but one possible model of the population.

The results of Felix and Oscar's experiment are shown in the bar chart in Figure 6.1: Felix was correct on six and wrong on four. We will test the null hypothesis by finding its probability, given these data. To find posterior probabilities, we will have to assign prior probabilities to this and also to competing models (see Exercise 6.8).

Figure 6.1 Bar chart for Example 6.1

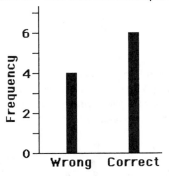

This example is a typical statistics problem. We have a sample and we want to make inferences about the population producing it. In this case, the population indicates Felix's ability. We know how to calculate probabilities assuming a particular population model. We will turn these probabilities around using Bayes' rule to associate probabilities with the populations. We summarize this process, as follows:

Steps of Statistical Inference

1. Specify a set of models.
2. Assign a (prior) probability to each model.
3. Collect data.
4. Calculate the likelihood of each model: $P(\text{data} \mid \text{model})$.
5. Use Bayes' rule to calculate the posterior probabilities of the models, $P(\text{model} \mid \text{data})$.
6. Draw inferences, finding the probabilities of various models and predicting the next observation.

Sometimes, we are interested in a subset of models. A particular model or a subset of models is a **hypothesis.** The (prior or posterior) probability of a hypothesis is the sum of the probabilities of the individual models in the hypothesis. Any model or particular model of interest is the **null hypothesis**—usually because its description contains the word *no*, such as in *no effect*.

> **A null hypothesis is a model of particular interest—usually, one with no treatment effect, no ability, or no difference.**

> **Testing a null hypothesis means finding its posterior probability.**

6.1 Sampling from Populations Having Two Types of Members

Chapters 6–9 deal with analyzing proportions. They apply when there are two possible outcomes. Applications are abundant: success or failure, hit or miss, yes or no, vote Democrat or vote Republican, female or male, and so on. So it is important to understand and be able to analyze proportions of successes, of misses, of yeses, of Democratic voters, of females, among others.

Moreover, proportions can play a role in essentially *any* statistical application, regardless of the number of outcomes. Consider the effect of a drug on blood pressure. You run an experiment and observe blood pressure changes. You could summarize the results by indicating the proportions of patients who experienced various blood pressure changes. The proportion who experienced some decrease is a good example. But you might also give the proportion who dropped at least 5 mmHg, the proportion who dropped at least 10 mmHg, and so on. (An alternative is to calculate a single measure: the *average* drop. This may provide a less complete description, but it is simpler because it avoids considering many proportions. Averaging is the subject of Chapters 9 and 10.) Section 13.5 addresses a particular set of statistical methods in which proportions of experiencing particular changes play a central role.

The study of proportions is important for us for another reason. The *ideas* involved in handling this relatively simple kind of data are the same as those for analyzing more complicated measures. A good part of this text is dedicated to proportions. In a sense, the later chapters will merely amplify and extend the basic ideas that we develop in this chapter. While there still will be more things to learn after you have mastered proportions, the ideas tend to be technical rather than novel.

In this chapter, an observation can take one of two values. I will usually call them success and failure, or S and F for short. A sample is a collection of S's and F's. A sample of $n = 25$ observations might look like this:

$$\text{SFFSS} \quad \text{FFFSF} \quad \text{SFFFS} \quad \text{SSSFS} \quad \text{SFSFS}$$

When the observations are exchangeable, the data can be summarized using just two pieces of information: s = number of S's and f = number of F's, where $n = s + f$. In the sample above, $s = 13$ and $f = 12$. Observations are *exchangeable* if the probability of S is the same for every observation—see Section 5.2. However, exchangeability does not preclude the probability of S changing during the course of an experiment—the accumulating data may change what we think about the population being sampled.

Think of the above as a sample from a population of S's and F's. We want to conclude something about the population in view of the sample. We know that the population contains only S's and F's, but we do not know their proportions. It seems reasonable to expect that the sample proportion of S's, s/n, is not very different from the population proportion, especially when n is large. This notion is formalized in the *law of large numbers*, the subject of the next section.

6.2 Law of Large Numbers

I tossed a coin 500 times. The results are shown here, where H stand for heads and T for tails. This is a sample from the population of all such tosses. For a fair coin, the population proportion of heads is by definition $\frac{1}{2}$. But this does not mean that the sample proportion of heads will be $\frac{1}{2}$. It will likely be close to $\frac{1}{2}$, but unlikely to be exactly $\frac{1}{2}$.

Results of 500 coin tosses

TTHHHTTHTHTTHHTHHTHTHTHHHTTHTHTHHTHHTHHHTHTHTHTTHH
TTHTTHHHHTHHTHTTHTTHHHHTHTTHTTTTTTHTHTHHHHTTTTHTTTT
HHTTHHHTTHTHHHTTHTTHTHHTHHHHTTHTTHHTTTTHTHTHHTHHHH
TTHHHHTTHTTHHHTTTTHHHHTTHTTTTHTTTHHHHTHTTTHHTHTTTT
THTHTHHTHHHHHTHTHTHTTTTHHHTHTTHHHTTTTTHTHTHTHTTHTT
TTHHTHHHHHTTTHTHTHTTTHTHTHHHTTHTHTTHHTTTHTHHHHHTTH
HTTHHHTTHTTTTHTTHTTTTHTHTHTHTHTTTTTTTHHHTHHTTHHTTHHHH
HHTHHTHTHTHHHHHTTTTTHTHTHHTTTTHHTTTHHHHHTHTHTTHHH
THHTHHHHTTHTHTTHHHTHHTTTHHHHHTTHHTTTTTTHHTHHHHTHHT
THHTHHTTTHHHHTHTHHHTHTTHHTTHTHTTTTTHTHTHTHHTHTHHTHHHH

An important question in statistics is how well a sample approximates the population. The sample size, n, has a big effect on the answer. In the case of my $n = 500$ tosses, there were $s = 253$ heads and $f = 247$ tails. The sample proportion of heads was $s/n = 253/500 = 50.6\%$. This proportion changed throughout the experiment. For example, after five tosses, there were three heads, and the proportion was $\frac{3}{5} = 60\%$. In the second five, the proportion was $\frac{2}{5}$ and these combined to give $\frac{5}{10} = 50\%$ after 10 tosses. Figure 6.2 shows the proportion of heads as it changed during the course of the experiment.

Figure 6.2 Plot demonstrating law of large numbers for coin-toss experiment

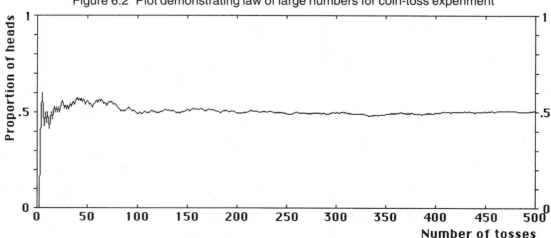

The curve in this figure bounces up and down initially, but levels out as the number of tosses increases. This is because single observations affect the proportion of heads less and less as the denominator increases. All coin-toss experiments have this characteristic. The tendency to settle down is called the **law of large numbers.** Another way of stating this law is that a curve such as the one in the figure tends to a limit, and that oscillations are eventually eliminated. The limit of a sample proportion of successes is the corresponding proportion in the population being sampled. The limit for the curve in Figure 6.2 seems to be around .5, but the *exact* limit cannot be gleaned from this or from any other particular set of tosses.

> **Law of Large Numbers**
>
> **The proportion of successes in a sample from a population tends to the proportion of successes in the population as the sample size gets larger.**

EXERCISE

6.1 You will need a telephone book for this exercise. Locate the page on which the S's begin. Find the first 50 names starting with S. Consider whether the last digits of their telephone numbers are even (0, 2, 4, 6, 8) or odd (1, 3, 5, 7, 9). Starting with the first name, calculate the proportion of evens. For example, if the first entry is Saab, Aaron . . . 636-1350, then the first observation is 0 (even); the proportion of evens is $\frac{1}{1} = 1$. If the next is Saab, Abigail . . . 684-3063, then the second observation is 3 (odd); the proportion of evens is $\frac{1}{2}$. Proceeding through the 50 entries, calculate the cumulative proportion of evens just as I did for the proportion of heads in the coin-toss example of this section. Make a plot of the cumulative proportion versus the number of names considered. (This plot will settle down but its limit may not be clear.)

6.3 Assessing Information about Models and Finding Means

We saw in Section 1.1 that assessing available information is an important step in applying the scientific method. Since this information is usually available *before* running an experiment, it is *prior information* and, as indicated in Section 5.4, the probabilities are called **prior probabilities.** When we update probabilities using data collected during the experiment, they are **posterior probabilities.** But one's probabilities depend on what is known at the time. So my *current* probabilities are *posterior* to the study just completed but *prior* to the study that is being planned.

The following example shows how this process works.

EXAMPLE 6.2
▷ **Who's better?** In many sports, one player (or team) plays another on several occasions during some relatively short period of time. As examples, two tennis players or two horses may meet several times in 1 year. Consider the first time they meet during that period—who's going to win?

If we knew the answer, they would not be playing; or, at least, no one would watch them play! We might ask for the probability that one player, say player A, beats the other, player B. You have learned by now that there is no universal probability. So let's interpret this as asking for *my* probability that A wins.

Asking "Who's going to win?" completes the first step of the scientific method. The second step is to assemble the available information. I have two particular tennis players in mind. A few years ago, player B was the undisputed queen of tennis. Player A is younger than B and has won only a few games in their meetings before this year. But A has a better record against comparable competition during the past year. Based on this information, my suspicion is that player A is better. However, it is only a suspicion, and even if I knew that A was better, it is far from clear to me how much better. Moreover, even if she is better, she would not win *every* match against B.

Let A stand for the event that player A wins their next match. I could try to assess my prior probability $P(A)$ directly. But for reasons that will become clear shortly, I will proceed along the lines of Examples 5.11 and 5.12. In particular, I will think of their matches in terms of chips in bowls. I imagine that a match between A and B is determined by selecting a chip from a population. The chips are labeled either A or B. The proportion of A's is the proportion of the time that A will win. If there are more A's, then player A is better than B, and if there are more B's, then player B is better. I do not know this proportion. But I entertain various possible population models.

The total number of chips in the various bowls does not really matter, and different bowls can contain different numbers of chips. But I will keep the total number of chips constant at 10. Since I do not know the proportion of A's in the population, I will entertain one model for each possible number of A's, 11 in all, as in Figure 6.3. Each model is characterized by the number of A's that it contains. It is convenient, although redundant, to use this number also as a model label, as the figure shows. It can as well be characterized by the proportion of A's—a device that is useful because it is the probability of selecting an A if the corresponding model is correct.

Figure 6.3 Population models entertained in Example 6.2, labeled by number of *A*'s and also by proportion of *A*'s

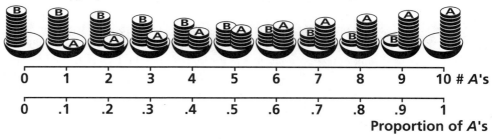

The proportions of A's and B's in a model indicate the relative abilities of players A and B. For example, when they are both .5, then A and B are equally likely to win. But if the proportion of A's is .8 (that is, there are eight A's and two B's) then A is 4 times as likely to win as B (the long-run frequency that A wins is .8). I will assume that the players' relative abilities do not change over the time period in question (so the matches in this period are exchangeable—as defined in Section 5.2). But my *percep-*

Sec. 6.3 / Assessing Information about Models and Finding Means   171

tion of their abilities—and therefore of the various models—will change as I get more information.

Assuming that relative abilities do not change over time is not appropriate in all sports. Obviously, if the time period is long enough, then they can change, but I specified in this example that the period be short. In some sports they change from day to day; for example, a baseball team's chances may depend on who is pitching. Also, as we saw in Example 2.10, the home team may have an advantage.

Using the methods described in Section 4.4, I have assessed my **prior probabilities** concerning the various models and these are shown in Figure 6.4. I will usually give such probability distributions in table form (as in Table 6.1) without showing pictures of the population models. Sometimes I will show them graphically in a bar chart. Each point on the horizontal axis in the bar chart in Figure 6.5 corresponds to

Figure 6.4 Population models of Example 6.2, with my prior probabilities indicated

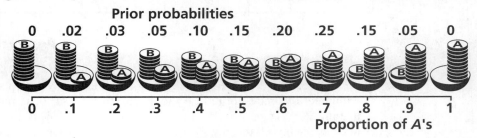

Table 6.1
My prior probabilities for Example 6.2

Proportion of *A*'s in model	0	.1	.2	.3	.4	.5	.6	.7	.8	.9	1
My probability	0	.02	.03	.05	.10	.15	.20	.25	.15	.05	0

Figure 6.5 Bar chart showing my prior probabilities for Example 6.2

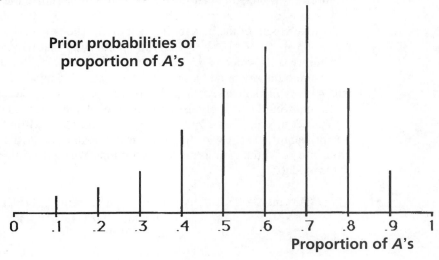

one of the 11 population models; the heights of the bars are proportional to my probabilities. I have not labeled the heights of the bars in the bar chart, but the total of their heights is 1. We are interested mostly in *relative* heights. So, for example, the alternative bar chart in Figure 6.6 is equivalent to the preceding chart because every bar is reduced (i.e., multiplied) by the same factor.

Figure 6.6 Equivalent version of bar chart in Figure 6.5

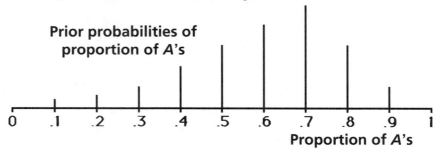

For reasons discussed in Section 4.4, I did not have to specify my probabilities with great accuracy, nor did I try to do so. If I were to spend more time, I might find that my $P(6)$ is closer to .17 or to .23 than it is to .20, say. But I did work hard enough to convince myself that my $P(6)$ is not as small as .10. One type of assessment in which I exercised extra care was specifying very small probabilities. In particular, when I indicate that, for me, a particular number of A's has probability 0, I mean I really think that number is impossible. For example, I say $P(10) = 0$; even though I think player A is better, I am convinced that there is *some* chance B would beat A in a particular match. (Even if A were otherwise invincible, she might get sick.)

I can calculate probabilities for various events using the given distribution. For example, my probability that A is better than B is the sum of the probabilities for model proportions .6, .7, .8, .9, and 1: $.20 + .25 + .15 + .05 + 0 = .65$. In the bar chart, this is the total height of the bars to the right of .5. Also, my probability that B is better is $.02 + .03 + .05 + .10 = .20$, and my probability that they are equally good is my probability at .5, which is .15. (*Check*: The sum of .65, .20, and .15 is 1.)

A more important and more interesting calculation is my (predictive) probability that A *wins their next match*. If I knew the correct population model (or thought I did!), this would be simply the proportion of A's in that model. Since I do not know the correct model, I can use the law of total probability: Condition and then average. The probability that A wins their next match is the weighted average or **mean** of that probability over the various model proportions, where the weights are the probabilities of the models:

$$P(A \text{ wins}) = P(A \text{ wins} \mid 0)P(0) + P(A \text{ wins} \mid .1)P(.1)$$
$$+ \ldots + P(A \text{ wins} \mid 1)P(1)$$
$$= 0(0) + .1(.02) + .2(.03) + .3(.05) + \ldots + 1(0)$$
$$= 0 + .002 + .006 + .015 + \cdots + 0 = .598, \quad \text{or about } 60\%$$

The mean of a bar chart can be thought of as the mean or average model, although it may not itself be a model that has positive probability. The mean is related to a bar chart in the same way that the sample average is related to a histogram of sample values.

The mean proportion of *A*'s is shown on the horizontal axis of the bar chart in Figure 6.7. The mean is the center of gravity of the distribution. Imagine that the bars are beams standing upright on a long (very thin) plank; the mean is the balance point of the plank.

Figure 6.7 Mean of the bar chart for Example 6.2

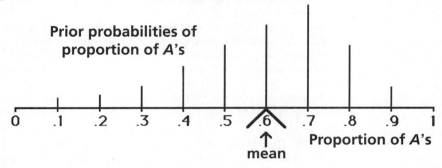

An easy way to calculate a mean is to use prior probabilities (Table 6.1). Table 6.2 includes the products of these probabilities and their corresponding proportions. The mean is the sum of these products: 60%.

Table 6.2
Prior probabilities for Example 6.2 showing calculation of mean

												sum
Proportion of *A*'s	0	.1	.2	.3	.4	.5	.6	.7	.8	.9	1	—
Probability	0	.02	.03	.05	.10	.15	.20	.25	.15	.05	0	**1.00**
Product	0	.002	.006	.015	.040	.075	.120	.175	.120	.045	0	**.60**

To check whether I was too hasty in writing down my probability distribution of the various models, I considered whether betting odds of 60:40 or 3:2 in favor of A were reasonable for me; I decided they were. Had these odds not seemed reasonable I would have reevaluated my probabilities. ◁

In the previous example, I described the probability that A wins the next match as the weighted average over the various model proportions. As such, it balances the bar chart showing *P*(proportion of *A*'s). I called this quantity the mean of the bar chart. Some people use *average value* and others use *expected value*. Here is the definition:

> **The mean of a bar chart is its balance point. The mean is calculated by adding the possible values multiplied by their probabilities.**

When the values on the horizontal axis of the bar chart are proportions of A's in population models, then the calculation of the mean—weighting by probabilities—is the same as the law of total probability. Hence, it is the probability that A wins a particular match.

EXERCISES

6.2 Pick a setting similar to that of Example 6.2. It might involve two players (or teams), A and B, that will play each other in the coming season. These could be football teams, horses, golfers, College Bowl teams, bridge teams, or whatever. (You will be more interested if you choose an area you know something about.)

(a) Describe the setting.

(b) Construct a table of your probabilities that A wins a particular encounter along the lines of Table 6.1. Your table should specify your prior probabilities concerning population models that are analogous to encounters between A and B.

(c) Use your table in (b) and the law of total probability to calculate $P(A)$, the mean of your bar chart.

(d) Adjust the entries in your table (if necessary) until you are comfortable with it and also with the resulting $P(A)$ that is implied by the table and calculated as in (c).

6.3 Bridge is a card game with one partnership of two players pitted against another partnership of two players. On any deal of the cards, only one partnership scores points and so wins that particular deal. My partner and I are reasonably good players. But the randomness in a deal means that we are not assured of winning. Even very strong players against very weak opponents could not win more than 70% of the deals. Consider nine models, one for each proportion of winning from 30% to 70% in steps of 5%. Against a particular partnership of my acquaintance, these are my probabilities that we would win the indicated proportion of deals, if we were to play a very large number of times:

Proportion of W's	.30	.35	.40	.45	.50	.55	.60	.65	.70
Probability	.00	.01	.02	.03	.09	.40	.25	.15	.05

What is my probability that we win the next deal?

Likelihoods, Posterior Probabilities, and Predictions

If I were interested in $P(A$ wins next match$)$, why did I go through the trouble of specifying population models analogous to A playing B in Example 6.2? Why not simply assess this probability directly? Suppose A wins the next match and I want to update my probability that A wins the match after that—what then? Our minds are not very good at keeping all the probabilities straight. Try this for yourself in the setting you chose for Exercise 6.2: Assume that A wins the upcoming encounter with B and assess your probability that A wins their *next* encounter. That is exactly what you will be doing in part (a) of Exercise 6.4. Then in part (b), you will calculate the same quantity, but using your probabilities in Exercise 6.2 and Bayes' rule.

EXAMPLE 6.3

▷ **Updating "Who's better?"** I used population models in Example 6.2 to assess my probability that A beats B in their next match. Now I have more information: A and B have since played each other three times; A won the first match and B won the next two (remember that I thought A was better than B). Call this information D (for data). Now they are going to play a fourth time. Let A now stand for the event that A wins this upcoming match. My probabilities of the various models change when conditioning on D. I want to calculate my conditional probability of A winning given D; this is my **posterior probability** of A based on D.

To keep things simple, I have ignored some possibly important considerations. For example, I have not told you how well A and B played in their three games. Nor have I given you information about the results of matches that A and B have had against other players.

Recall that my probability of A winning was about 60%. Since B has won two out of the three matches, it seems reasonable to expect that $P(A \mid D)$ is less than 60%. But how much less is by no means clear. I can find out by updating my probabilities of the various population models using Bayes' rule (see Section 5.4). For example, the new probability of model .6 (in which there are six A's and four B's) is

$$P(.6 \mid D) = \frac{P(D \mid .6)P(.6)}{P(D)}$$

Evaluate the denominator by expanding it using the law of total probability, just as in the generalized Bayes' rule of Section 5.4:

$$P(D) = P(D \mid 0)P(0) + P(D \mid .1)P(.1) + \cdots + P(D \mid 1)P(1)$$

The factor $P(.6)$ in the numerator is just my prior probability of proportion .6 as given in the table in the previous section: $P(.6) = .20$. What remains is to find $P(D \mid .6)$ and other probabilities of D, given the various candidate population models. These are called **likelihoods**.

> **The likelihood of a model is the probability of the data calculated assuming that model.**

The tree diagram in Figure 6.8 (page 176) shows the calculation of the likelihood of model .6. There are eight possible ways for the three matches to end. We are interested only in the one that is highlighted: $D = ABB$. The probabilities indicated on the branches assume model .6 for all three matches (and that the sampling is without replacement). So the likelihood of model .6 is $P(D \mid 6 A\text{'s}, 4 B\text{'s}) = \frac{96}{1,000}$ or .096. We have to make this calculation for the other models as well.

Multiplying the probabilities of individual observations assumes that they are independent when given any particular model. In terms of sampling from a bowl with chips, this assumes the sampling is with replacement.

> **To calculate the likelihood of a model, multiply the probabilities of the individual observations *assuming that model*.**

Figure 6.8 Tree diagram showing likelihood calculation for model .6 in Example 6.3

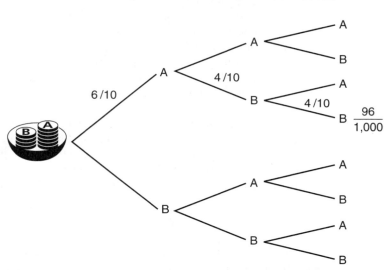

It is convenient to display the calculations in table form. To make certain that you see the progression of the calculations, I will give them as fractions—using a level of accuracy that you should not worry about imitating. The first column in Table 6.3 identifies the various candidate models, given in terms of their proportions of A's—that is, #A's/10. The second column repeats my prior probabilities (Table 6.1, page 171). The third column shows the likelihoods of the various models (for data D). I calculated the entry for model .6 in the tree diagram (Figure 6.8) (boxed in the table). I will calculate one other entry (also boxed) in this column: the likelihood for model .3 is $.3 \times .7 \times .7 = .147$.

Table 6.3
Calculations for Example 6.3; D = "A wins one game and B wins two"

Model	Prior	Likelihood	Prior × Likelihood	Posterior	Model × Posterior
0	0	0	0	0	0
.1	.02	.081	.00162	162/8,616 = .02	162/86,160
.2	.03	.128	.00384	384/8,616 = .04	768/86,160
.3	.05	.147	.00735	735/8,616 = .09	2,205/86,160
.4	.10	.144	.01440	1,440/8,616 = .17	5,760/86,160
.5	.15	.125	.01875	1,875/8,616 = .22	9,375/86,160
.6	.20	.096	.01920	1,920/8,616 = .22	11,520/86,160
.7	.25	.063	.01575	1,575/8,616 = .18	11,025/86,160
.8	.15	.032	.00480	480/8,616 = .06	3,840/86,160
.9	.05	.009	.00045	45/8,616 = .01	405/86,160
1	0	0	0	0	0
Sum	1.00		.08616	1.00	45,060/86,160

The likelihoods represent the data in calculating the posterior probabilities. The bar chart in Figure 6.9 shows the likelihoods. In this bar chart, the total height of the bars (the total of the likelihoods) is not 1, and the total height is irrelevant. The total probability of models is always 1, but likelihoods are probabilities of data given models and not of models themselves. (Adding up likelihoods over the possible data does indeed give 1, but we want to condition on the data actually observed and so this fact is not helpful.) While it is difficult to keep them straight, there is a big difference, conceptually, between the probability of a model and the probability of the observed data given the model.

Figure 6.9 Likelihoods of various models for $D = $ "A wins one game and B wins two"

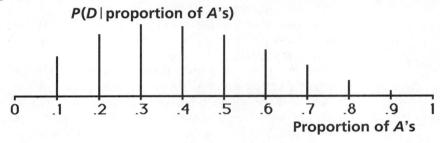

According to the law of total probability we can calculate $P(D)$, the denominator in Bayes' rule, by multiplying the prior probabilities and likelihoods (second and third columns in the table) and adding:

$$P(D) = P(D \mid 0)P(0) + P(D \mid .1)P(.1) + \cdots + P(D \mid 1)P(1)$$

The individual terms in this sum are shown in the fourth column of the table. The sum of the entries in this column is $P(D) = .08616$.

The fifth column of the table gives the posterior probabilities of the models:

$$\text{Posterior probability of model} = \frac{\text{Prior} \times \text{Likelihood}}{P(D)}$$

These probabilities are also shown above the various population models in Figure 6.10, a version of the one in Figure 6.4 (the models are the same, but the probabilities have been updated). They are also shown in the bar chart in Figure 6.11 (page 178).

Figure 6.10 My posterior probabilities for the bowl models in Example 6.3

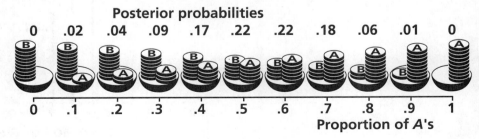

178 Ch. 6 / Models for Proportions

Figure 6.11 Bar chart showing my posterior probabilities (given *D*) in Example 6.3

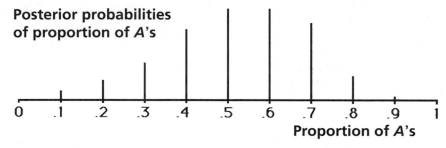

The posterior bar chart is the result of multiplying the prior bar chart by the likelihoods. This is shown schematically in Figure 6.12. My convention of drawing bar charts without the vertical scale has the advantage of simplifying this representation: $P(D)$ does not appear in this schema.

Comparing the posteriors with the priors shows the effect of the data. As is true in general, the likelihoods move the probabilities in the direction of the data. In this case, the shift is to the left, as is shown in Figure 6.13—both bar charts are on the same graph. The total height of the bars representing prior probabilities equals the total height of the bars representing posterior probabilities equals 1. However, the latter distribution is shifted to the left because the data suggest that player A is not as good as originally thought. The shift is not dramatic because there is not very much data.

Figure 6.12 Schematic representation of Bayes' rule in Example 6.3

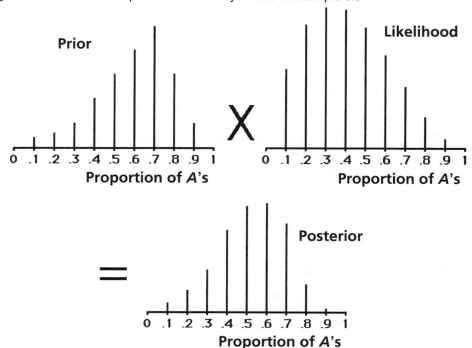

Figure 6.13 Prior and posterior bar charts for Example 6.3 on same graph

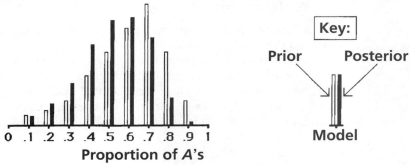

Now that I have updated my probabilities of the various model proportions, I can calculate my posterior probability of various events. For example, my posterior probability, given D, that player A is better than B is $(1{,}920 + 1{,}575 + 480 + 45)/8{,}616 = 47\%$, down from my prior probability of 65%. My updated probability that B is better is $(162 + 384 + 735 + 1{,}440)/8{,}616 = 32\%$, up from 20%. And my probability that they are equally good is $1{,}875/8{,}616 = 22\%$, up from 15%. (The sum of the given numbers is 101% due to rounding off.)

The mean of this new bar chart is my predictive probability that A wins the next (fourth) match, given D. I proceed in the same way that I calculated my unconditional probability that A wins the first match in Example 6.2. The last column of Table 6.3 shows this calculation: $P(A \mid D) = 45{,}060/86{,}160 = 52\%$. The mean has changed from $P(A) = 60\%$ that we calculated in Example 6.2. Again, this change is not great because the data are not very conclusive. ◁

In a typical statistics problem we want to find probabilities of models given the data—that is, posterior probabilities. According to Bayes' rule, posterior probabilities depend on prior probabilities and on the experimental (or observational) results. The experimental results contribute to the posterior probabilities only through the likelihoods. Likelihoods depend only on the data at hand—priors depend only on evidence available beforehand.

Here is a restatement of Bayes' rule:

> **Calculate the posterior probability of a model by multiplying its prior probability and its likelihood. Then divide by the sum over the models to make the total probability equal to 1.**

Summarizing, the process of updating goes as follows:

1. Write down the prior probabilities for the various models.
2. Calculate the likelihoods of the models.
3. Obtain the posterior probabilities of the models:
   (a) Multiply the prior probability and likelihood for each model

(b) Sum the products in (a) over all the models.
(c) Divide each product in (a) by the sum from (b).

The first two steps require the most care. The various parts of the third step are trivial, but they can be tedious.

The next example demonstrates the process. It also serves as an example of testing a null hypothesis. And it has an additional purpose: It shows one way that the single-proportion approach of this chapter can be used to address a two-proportion question. This approach applies whenever the samples are large and the numbers of successes (or failures) are small. (Other approaches will be considered in Chapters 8 and 9.)

EXAMPLE 6.4

▷ **Abortion drug as morning-after contraceptive?** A study [1] addressed the question of whether the controversial abortion drug RU 486 could be an effective "morning-after" contraceptive. The 800 study participants were women who came to a health clinic in Edinburgh, Scotland, asking for emergency contraception after having engaged in sexual intercourse within the previous 72 hours. Investigators randomly assigned the women to receive either RU 486 or a standard therapy consisting of high doses of the sex hormones, estrogen and a synthetic version of progesterone. Of the 402 women assigned to RU 486 (R), none became pregnant. Of the 398 women who received standard therapy (C, for control), four became pregnant. How strongly does this information indicate that R is more effective than C?

Consider only the four pregnancies, turning a two-proportion problem into a one-proportion problem. How likely is it that a pregnancy occurs in the R group? If R and C are equally effective (null hypothesis), then a pregnancy has a 402/800 = 50.25% (round this off to 50%) chance of being in the R group. Once a pregnancy occurs in the C group, then this probability would change, but the change is negligible and so let's ignore this sampling-without-replacement adjustment. There were no R's and four C's in the sample of $n = 4$. So $s = 0$ and $f = 4$.

To test the null hypothesis, consider the nine bowl models shown in Figure 6.14. (As compared with earlier examples, I have dropped models with proportions of 0 and 1, since I am convinced that no treatment can be perfect.) The middle model (proportion of R's of .5) is the null hypothesis of no difference between R and C. The models on the left indicate benefits for RU 486 of $1:9$, $2:8$, $3:7$, and $4:6$. Models on the right indicate that C is better. The model on the far left means that a pregnancy is 9 times as likely to occur in the C group as in the R group. This is a lot, but it is conditional on a pregnancy occurring. Both probabilities of pregnancy may be very small, say, .001 in R and .009 in C.

Figure 6.14 Models for proportions of pregnancies in R and C groups for Example 6.4

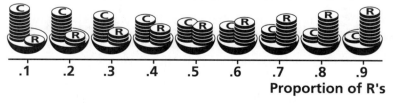

Step 1 of the updating process is to assess my prior probabilities; they are indicated in the second column of Table 6.4. My prior information about the benefit of RU 486 is symmetric—it is as likely to be better as worse—and all models, except the null, have equal probabilities.

Step 2 is to calculate likelihoods of the model proportions. These are the probabilities of data D: $s = 0$ and $f = 4$, assuming the model. Letting r stand for the model proportion of R's, the probability of data D, given r, is given by

$$\text{Likelihood} = (1 - r)^4$$

For example, the probability all four pregnancies occur in the C group assuming the leftmost model in Figure 6.14 is $(1 - .1)^4 = .6561$. This is shown (rounded off) along with the other likelihoods in the third column of the table.

Table 6.4
Calculations for Example 6.4; D = All four observations are R's

Model	Prior	Likelihood	Prior × Likelihood	Posterior	Model × Posterior
.1	.06	.656	.0394	.326	.033
.2	.06	.410	.0246	.204	.041
.3	.06	.240	.0144	.119	.036
.4	.06	.130	.0078	.064	.026
.5	.52	.063	.0325	.269	.135
.6	.06	.026	.0015	.013	.008
.7	.06	.008	.0005	.004	.003
.8	.06	.002	.0001	.001	.001
.9	.06	.000	.0000	.000	.000
Sum	1.00		.1208	1.000	.281

Step 3 is to multiply the prior probabilities and their corresponding likelihoods (fourth column), sum the products (bottom of fourth column), and divide the products by the products' sum (fifth column). For example, for model $r = .1$, .0394 divided by .1208 is the posterior probability, .326.

Because the sample size is so small, the study is far from conclusive. But the probability of the null hypothesis has dropped from 52% to about 27%. Also, the probability that C is better than R has dropped considerably (from 24% to 1.8%).

The sixth column in the table is extra; its sum (.281) is the predictive probability of an R. This has the following interpretation. Suppose the study were to continue. This is the probability that the next pregnancy would occur in the R group. ◁

EXERCISES

6.4 Refer to the setting you described in Exercise 6.2. Suppose that since you assessed your probabilities in that exercise, A and B have had one encounter, which A won. Call this information D.

(a) Using only your intuition, report your probability that A wins their *next* encounter (given D). (This is the posterior mean of your bar chart.)

(b) Now calculate the probability in part (a) using Bayes' rule.

[Your two numbers may disagree. If they disagree, either your first estimate is wrong or the

6.5 Repeat parts (a) and (b) of Exercise 6.4, but let D stand for the information that A won three of four matches and lost the other one. Find your probability that A wins their *fifth* encounter (given D).

6.6 Consider the bridge setting of Exercise 6.3 (page 174). My prior probabilities of the various models are repeated here:

Proportion of W's	.30	.35	.40	.45	.50	.55	.60	.65	.70
Probability	.00	.01	.02	.03	.09	.40	.25	.15	.05

After I assessed these probabilities, my partner and I played five deals against the other partnership, winning all five. Update my probabilities based on this information. Then calculate the mean model for the updated (i.e., posterior) probabilities—this is my probability that we will win the sixth deal.

6.7 A study[2] addressed the question of whether cancer runs in families. The investigator identified 200 women with breast cancer and another 200 women without breast cancer and asked them whether their mothers had had breast cancer. Of the 400 women in the two groups combined, 10 of the mothers had had breast cancer. If there is no genetic connection, then about half of these 10 would come from each group—the null hypothesis is that the proportion of daughters with breast cancer is .5. It turned out that seven of the daughters had cancer and three did not. Is this evidence in favor of the null hypothesis or not?

Consider the models in the following table. The table also gives my prior probabilities. Take $s = 7$ and $f = 3$. W stands for a mother with breast cancer having a daughter **W**ith breast cancer. (It is difficult to imagine that the population proportion of W's is less than .5—mother with breast cancer makes it less likely that daughter will have breast cancer—so these models have little probability for me.)

Proportion of W's	.1	.2	.3	.4	**.5**	.6	.7	.8	.9
Prior probabilities	.01	.01	.01	.01	.31	.15	.25	.20	.05

The model proportion .5 (in boldface type) is the null hypothesis and has prior probability 31%. The model on the far right means that a woman with breast cancer is 9 times as likely to have a daughter with breast cancer as without breast cancer. On the basis of $s = 7$ and $f = 3$, find the following:

(a) The posterior probability of the null hypothesis.

(b) The posterior probability that a woman with breast cancer is 9 times as likely to have a daughter with breast cancer.

(c) The posterior probability that there is some genetic linkage in the sense that the proportion of W's is greater than .5.

(d) Suppose the study were to continue, with equal numbers of daughters added to the two groups. Find the predictive probability that the next mother with breast cancer occurs in the group of daughters **W**ith breast cancer.

6.5 Calculations in an Example

Example 2.6 deals with a study designed to evaluate the effectiveness of 6-MP for the treatment of acute leukemia. This study will be the focus of the current section and serves as an example of the calculations described in the previous section.

Sec. 6.5 / Calculations in an Example

There were $n = 21$ pairs of patients in the study. The results were as follows, where B means that the 6-MP patient fared **B**etter than the placebo patient in that pair and W means that the 6-MP patient fared **W**orse:

$$BWBBB \quad WBBBB \quad BBBWB \quad BBBBB \quad B$$

So 6-MP was better in $s = 18$ of the 21 pairs of patients. The proportion of pairs in which it was better is $s/n = \frac{18}{21} = 86\%$.

Proportions are numbers between 0 and 1. So proportions can be thought of as probabilities. Suppose I label chips with the symbols from the list of B's and W's—18 chips are labeled B and three are labeled W—and place them in a bowl. I do this in such a way that the chips are exchangeable for me. So when I select a chip from the bowl, my probability of a B is $\frac{18}{21}$. When confined to the 21 pairs in the experiment, my probability that 6-MP is better than placebo is $\frac{18}{21}$.

But the study was wasted if the strongest conclusion one can make applies only to the patients in the study. The important question is whether we can say anything about patients who are candidates for treatment with 6-MP but are not in the study. Consider someone who has acute leukemia and whom we regard as exchangeable with the patients in the trial before they were treated. Suppose that the only possible treatments are 6-MP and placebo (the latter being essentially no treatment). Should this patient be given 6-MP?

This is not an easy question to answer. It involves weighing any possible beneficial effects with possible adverse effects and with other costs (including monetary costs) of 6-MP therapy. Two quite rational patients might select differently. To help with the decision, let's simplify things. Restrict consideration to finding the probability that this next patient will stay in remission longer if treated with 6-MP than if not treated. (Assuming that the patients are exchangeable, this is the same as saying that for the next pair of patients treated, one with 6-MP and the other with placebo, the 6-MP patient will stay in remission longer.)

What is the posterior probability that 6-MP is effective in prolonging remission? We proceed as in Examples 6.3 and 6.4, but the current setting is more complicated (as well as being more realistic). We need to talk about population models that are analogous to treating leukemia patients. Suppose the population models contain B's and W's. Selecting a B is equivalent to the 6-MP patient being in remission longer. The prior probabilities of the various distributions of B's and W's depend on the person making the assessment. An important person in this regard is a clinician who was involved in the study.

I do not know what the clinicians thought about the effectiveness of 6-MP before the study. They must have had some reason to think that it was effective or they would not have organized the study. They also must have given some prior probability to the possibility that 6-MP was not effective or they could not have condoned giving a placebo to some of the patients. I am going to specify *two* different prior probability distributions, one for Dr. X and the other for Dr. Y. While it is possible that one of these corresponded to an actual clinician's opinions before the study, I do not know whether this is so. My main purpose is to show how someone's prior opinion is updated using data in a study. A secondary purpose is to show you how much more closely the two physicians agree after they observe the data than they did before.

Suppose the total number of chips in the models is 10, with either 0, 1, 2, up to 10 B's. When things get crowded, I will let p stand for the proportion of B's, which then is either 0, .1, .2, up to 1. If $p > .5$, then 6-MP is effective, and if $p < .5$, then it is detrimental. The null hypothesis of no effect is $p = .5$. Dr. X is quite open-minded about the effectiveness of 6-MP, whereas Dr. Y is somewhat more optimistic about its effectiveness. Both feel that it is possible that 6-MP is detrimental, but Dr. X rates this as more likely than Dr. Y. Their probabilities are given in Figure 6.15, tabulated in Table 6.5 and, finally, presented in bar chart form in Figure 6.16 (see next page).

Figure 6.15 Prior probabilities of Drs. X and Y under various models

p:	0	.1	.2	.3	.4	.5	.6	.7	.8	.9	1
Dr. X:	1/11	1/11	1/11	1/11	1/11	1/11	1/11	1/11	1/11	1/11	1/11
Dr. Y:	0	1/55	2/55	3/55	4/55	5/55	6/55	7/55	8/55	9/55	10/55

Table 6.5
Prior probabilities of Drs. X and Y

Proportion of B's	0	.1	.2	.3	.4	.5	.6	.7	.8	.9	1
Dr. X's prior	1/11	1/11	1/11	1/11	1/11	1/11	1/11	1/11	1/11	1/11	1/11
Dr. Y's prior	0	1/55	2/55	3/55	4/55	5/55	6/55	7/55	8/55	9/55	10/55

Consider the first pair of patients, one assigned to 6-MP and the other to placebo. By the law of total probability, Dr. X's predictive probability of event B—the 6-MP patient will stay in remission longer—is

$$P(B \text{ for Dr. X}) = 0 \times \frac{1}{11} + .1 \times \frac{1}{11} + .2 \times \frac{1}{11} + .3 \times \frac{1}{11}$$
$$+ \cdots + 1 \times \frac{1}{11}$$
$$= \frac{1}{2}$$

For Dr. Y,

$$P(B \text{ for Dr. Y}) = 0 \times 0 + .1 \times \frac{1}{55} + .2 \times \frac{2}{55} + .3 \times \frac{3}{55}$$
$$+ \cdots + 1 \times \frac{10}{55}$$
$$= \frac{7}{10}$$

These are the means of the corresponding bar charts (Figure 6.16). The mean for Dr. X is .5, whereas for Dr. Y it is .7. Obviously, they had different prior information, or at least they assessed the available information differently.

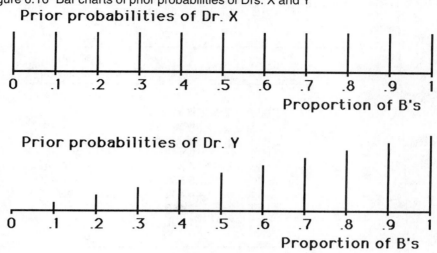

Figure 6.16 Bar charts of prior probabilities of Drs. X and Y

What happens when they learn the results of the study? In particular, do they come closer together in their opinions? We want to find the likelihoods. Suppose that Drs. X and Y both regard the pairs in the study to be exchangeable. (This means, for example, that it is as likely that 6-MP will be better in the first pair as in any other pair. However, if the first pair results in B, then the probability that the second results in B may well increase.) The data are as follows:

$$B\ W\ B\ B\quad W\ B\ B\ B\quad B\ B\ B\ W\ B\quad B\ B\ B\ B\ B\ B$$

The likelihood is the probability of the data. The probability of the data depends on the model, that is, on the proportion of B's in the model, which we have called p. To be specific, consider $p = .8$. The likelihood of $p = .8$ is the probability of the data assuming the model has eight B's and two W's. This is the product of the 21 separate probabilities:

$$(.8)(.2)(.8)(.8)(.8)(.2)(.8)(.8)(.8)(.8)(.8)(.8)(.8)(.2)(.8)(.8)(.8)(.8)(.8)(.8)(.8)$$

This equals $(.8)^{18}(.2)^3 = .000144 = 1.44\text{E}-4$. (Exponential notation $\text{E}-4$ means to move the decimal point four places to the left. Similarly, $\text{E}-13$ would mean to move the decimal point 13 places to the left and $\text{E}+13$ would mean to move the decimal point 13 places to the right.)

Likelihoods for all 11 models are shown in the third column of Table 6.6 (page 186). The table applies for Dr. X and shows the calculations that lead to posterior probabilities—the next to last column in the table. Numbers too small to matter in the end are called "tiny." My practice is to carry a high degree of accuracy except when writing down a final answer. This guarantees that when I write down a three-digit final answer, it is accurate to three digits. If you check my calculations using only the accuracy shown in the table, you will get slightly different answers. For example, in the row for $b = .9$, $1.36E-5$ divided by $3.14E-5$ gives .433 instead of .435. That is because $1.36E-5$ is rounded off from $1.3645E-5$ and $3.14E-5$ is rounded off from $3.1379E-5$.

Table 6.6
Calculations for Dr. X

Model	Prior	Likelihood	Prior × Likelihood	Posterior	Model × Posterior
0	1/11	0	0	0	0
.1	1/11	tiny	tiny	tiny	.000
.2	1/11	tiny	tiny	tiny	.000
.3	1/11	tiny	tiny	tiny	.000
.4	1/11	tiny	tiny	tiny	.000
.5	1/11	4.77E−7	4.33E−8	.001	.001
.6	1/11	6.50E−6	5.91E−7	.019	.011
.7	1/11	4.40E−5	4.00E−6	.127	.089
.8	1/11	1.44E−4	1.31E−5	.418	.334
.9	1/11	1.50E−4	1.36E−5	.435	.391
1	1/11	0	0	0	0
Sum	1		3.14E−5	1.000	.826

This table is in the same format as the tables in Examples 6.3 and 6.4. The sum of the fourth column (3.14E-5) is $P_X(D)$. The fifth column contains Dr. X's posterior probabilities, calculated by dividing the entries in the fourth column by the sum of the entries in the fourth column. Because the prior probabilities are equal, the third, fourth, and fifth columns in this table are proportional to each other. So a picture of the likelihoods is the same as the bar chart of Dr. X's posterior probabilities—see Figure 6.17. The last column of the table serves to find the mean of this bar chart, which is .826.

Figure 6.17 Bar chart of Dr. X's posterior probabilities (proportional to likelihoods)

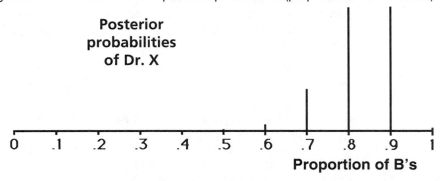

Table 6.7 gives the analogous quantities for Dr. Y. The likelihoods are the same as for Dr. X, so the third column of Dr. Y's table is the same. Dr. Y's posterior probabilities (fifth column in the table) are shown in the bar chart in Figure 6.18.

Sec. 6.5 / Calculations in an Example

Table 6.7
Calculations for Dr. Y

Model	Prior	Likelihood	Prior × Likelihood	Posterior	Model × Posterior
0	0	0	0	0	0
.1	1/55	tiny	tiny	tiny	.000
.2	2/55	tiny	tiny	tiny	.000
.3	3/55	tiny	tiny	tiny	.000
.4	4/55	tiny	tiny	tiny	.000
.5	5/55	4.77E−7	4.33E−8	.001	.001
.6	6/55	6.50E−6	7.09E−7	.014	.008
.7	7/55	4.40E−5	5.60E−6	.108	.076
.8	8/55	1.44E−4	2.10E−5	.405	.324
.9	9/55	1.50E−4	2.45E−5	.472	.425
1	10/55	0	0	0	0
Sum	1		5.19E−5	1.000	.834

Figure 6.18 Bar chart of posterior probabilities of Dr. Y

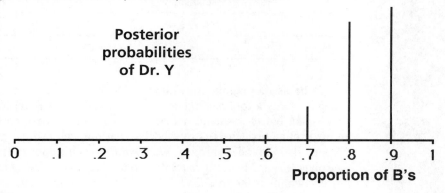

The posterior bar charts of Drs. X and Y are quite similar, even though their prior bar charts are different. For example, both now associate a posterior probability of .999 with 6-MP having a beneficial effect (this is the sum of the probabilities of proportions of B's greater than .5). Also, the two means (the predictive probabilities that the next chip drawn is a B) are the same to two decimal places:

For Dr. X: $P(B \mid D) = .826$
For Dr. Y: $P(B \mid D) = .834$

In terms of the clinical study, these are their probabilities that the next patient will stay in remission longer if treated with 6-MP. The small difference in these two values is due to the difference in the two prior probability distributions.

So even though Drs. X and Y started out far apart (with means of .5 and .7), their posterior probabilities are quite similar. They changed their views as they observed the data. This is possible because both were reasonably open-minded about the effectiveness of the treatment. In terms of population models, they allowed for various possible proportions, b. Consider a know-it-all—call him Dr. Z—who is singularly convinced of the null hypothesis that 6-MP has no effect. Dr. Z "knows" that the better response within each pair of patients is as likely to be on placebo as on 6-MP. Dr. Z's prior probabilities would not change for *any* data—which is what I mean by know-it-all. So, Dr. Z has a single type of population model in mind, one with proportion b equal to .5. One such model has five chips of each type:

The probability table for Dr. Z (Table 6.8) is analogous to those for Drs. X and Y. Calculations using Bayes' rule are now trivial.

Table 6.8
Calculations for Dr. Z

Model	Prior	Likelihood	Prior × Likelihood	Posterior	Model × Posterior
.5	1	4.77E−7	4.77E−7	1.000	.500
Sum	1		4.77E−7	1.000	.500

To be as thoroughly convinced as Dr. Z would require impossibly strong prior evidence. Only a fool is certain. One's probabilities may reasonably be very large or very small, but they should not be 1 or 0. I seem to be ignoring my own advice by giving probability 1 to sets of models. Learning can take place only in the restricted framework of a model universe. We *condition* on the correct model being in a particular universe—and learning is relative to this universe. But we can never *know* that our universe contains the true model. Considering a larger universe may give greater confidence that the model is contained therein, but there is no all-encompassing universe containing all possible models.

Being called a fool convinced Dr. Z to back off from his prior certainty of the null hypothesis. He still feels strongly, but he has backed off from being 100% sure to 99% sure. He spreads his remaining 1% evenly among the other 10 possible b's. The initial table of probabilities changes accordingly—see Table 6.9. While 100% certainty was not changed by the data, the 99% probability of the null hypothesis drops to about 58%. Dr. Z's probability that the next patient will stay in remission longer if treated with 6-MP has increased from 50% (no benefit) to about 64%.

While Drs. X and Y were somewhat different in their prior views, they were quite far from the new Dr. Z. All three came closer together in their views after seeing the data, but Dr. Z is still rather different from the other two. In particular, he still gives a substantial probability (58%) to the null hypothesis. They might resolve their differences by discussing their prior information and the bases for their disagreements. But there is no reason that they should agree.

Table 6.9
Calculations for new Dr. Z

Model	Prior	Likelihood	Prior × Likelihood	Posterior	Model × Posterior
0	.001	0	0	0	0
.1	.001	tiny	tiny	tiny	.000
.2	.001	tiny	tiny	tiny	.000
.3	.001	tiny	tiny	tiny	.000
.4	.001	tiny	tiny	tiny	.000
.5	.990	4.77E−7	4.72E−7	.578	.289
.6	.001	6.50E−6	6.50E−9	.008	.005
.7	.001	4.40E−5	4.40E−8	.054	.038
.8	.001	1.44E−4	1.44E−7	.176	.141
.9	.001	1.50E−4	1.50E−7	.184	.166
1	.001	0	0	0	0
Sum	1		8.17E−7	1.000	.639

What about people other than Drs. X, Y, and Z? How should they decide whom to believe? Consider a regulatory agency (such as the U.S. Food and Drug Administration) deciding whether to approve 6-MP for standard use based on this study. I cannot do justice to this important issue here, but I will mention some considerations. If the drug is not approved, then more studies would have to be conducted. This would delay delivering possibly efficacious therapy to patients not in the studies. If it is approved, then the opportunity for further experimentation may be lost. The benefits suggested by this study may not be real and an ineffective therapy might be given to thousands of patients who could perhaps be treated better with other therapies. A consideration is that Dr. Z's attitudes may be typical among oncologists and they would not use the drug even if it were approved. Perhaps another study would convince them. Finally, I have considered only effectiveness. Decisions to use the drug or not and to approve the drug or not depend on its side effects. A drug that prolongs life might make that life of sufficiently low quality that the benefits are more than offset by the drawbacks.

The calculations in this chapter are easy to describe, but some of them are tedious. In the next chapter we will continue dealing with models of proportions, but we will greatly reduce the calculations.

EXERCISES

6.8 Example 6.1 (page 165) describes an experiment from a segment of the TV show "The Odd Couple." Put yourself in Oscar's place. He is pretty sure, say, 95% sure—of the null hypothesis that Felix has no ESP ability whatever. That is, Oscar's probability is .95 that Felix's answers are those that would result by drawing chips from a model with one-quarter of the chips giving the correct answer, such as this one:

He distributes his remaining 5% on "some ability"—Felix did not claim that he was perfect.

Consider other population models having eight chips, with the proportion of C's being the population proportion of corrects. Oscar's prior probabilities on these models are as follows (where the null model is in boldface type):

Proportion of C's	0	1/8	**2/8**	3/8	4/8	5/8	6/8	7/8	1
Prior probabilities	.001	.001	.950	.008	.008	.008	.008	.008	.008

(You may think that any probability on the two left-hand models is misplaced; Oscar feels that Felix may have some ability, but that he may have it backwards!) As indicated in Example 6.1, out of 10 tries, Felix was correct on $s = 6$ and wrong on $f = 4$. (I do not know the order of the resulting corrects and wrongs, but when assuming exchangeability the order does not matter: Assume the first six were correct and the last four were wrong.) Find Oscar's posterior probability of the null hypothesis. (That is, find the updated probability of the model pictured in the chip-from-bowl drawing.)

6.9 I have a friend who claims that she can tell the sex of an unborn child from the way it "rides"—high means boy and low means girl. She claims to be good but not perfect. I do not believe her and associate the probability .7 that she is only guessing—the null hypothesis of no ability. Consider three other population models. View each of the four models as containing eight chips. The null model has four C's (**C**orrect) and four N's (**N**ot correct). The other models contain C's in the proportions $\frac{5}{8}, \frac{6}{8}$ and $\frac{7}{8}$—I give these latter models a prior probability of .1 each. The models are given in picture form here.

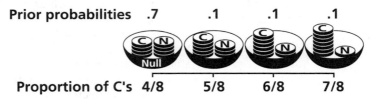

(a) Consider an experiment using $n = 10$ expectant mothers to test my friend's claim. Suppose she is correct on $s = 8$ of the 10 and not correct on the other two. [The probability that she is correct when she says "boy" might be different from when she says "girl," but I am assuming the two are equal.] Find my updated probabilities.

(b) After the experiment described in part (a), when she predicts the sex of another unborn child (the 11th overall), what is my probability that she will be correct?

6.10 Example 3.17 (page 86) describes a study in which a researcher gave to one member of a set of twins honey with milk every night for 6 weeks, while the other member got milk without honey. Before the experiment, the researcher formulated a probability distribution of the effect on hemoglobin levels of honey administered in this fashion. The probability of the null hypothesis of no effect was $\frac{1}{2}$—see drawing of models that follows. The probability that the honey twin would always have a higher increase (regardless of the number of twins studied) was $\frac{1}{4}$—the "All H" model shown. The researcher associated the remaining $\frac{1}{4}$ probability with $\frac{3}{4}$ of the hemoglobin levels of those given honey increasing more than that of their twin—model "$\frac{3}{4}$".

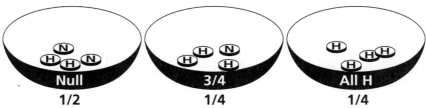

The researcher recorded the following increases in hemoglobin (these are from an actual study):

Pair	1	2	3	4	5	6
Honey	19	12	9	17	24	24
No honey	14	8	4	4	11	11

Assume that the six pairs of twins are exchangeable. Use only the information that the hemoglobin level of each of the six children who received honey increased more than that of their twin. (So $s = 6$ and $f = 0$.)

(a) What is the researcher's posterior probability of the null hypothesis that honey has no effect on hemoglobin level?

(b) Consider a seventh set of twins—one receives honey in milk and the other receives only milk. Find the researcher's predictive probability that the honey twin's hemoglobin will have a greater increase.

6.11 Burrowing owls sometimes build their nests in holes that were dug by prairie dogs, coyotes, or badgers and have been since abandoned. They sometimes line the nests with cattle or horse dung. Why? Perhaps it insulates the nest from temperature extremes. A possibility suggested by biologist Dennis Martin is that the owls use the dung to keep predators away. To test this theory, biologist Gregory Green observed lined and unlined owl nests in the Columbia River basin.[3] We will deal with both samples in Exercise 9.14, but for now consider only the lined nests. When a nest is **R**aided, that is analogous to selecting an R chip from a bowl model; **N**ot raided is selecting an N chip. Suppose his prior probabilities are as indicated on the population models shown:

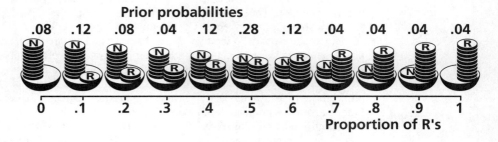

Green identified $n = 25$ nests lined with dung and observed that only $s = 2$ were raided. Assume exchangeability of the 25 nests.

(a) Find his posterior probabilities of the 11 population models.

(b) Find his predictive probability that the next lined nest he observes (that is, the 26th) is raided.

6.12 A friend of yours claims to be able to toss a coin and get heads every time. You suspect he may have a two-headed coin and convince him to let you toss it 10 times (without examining it). Your probability of two-headed is .01 and your remaining .99 probability is associated with the coin being *fair* (in that it gives heads half the time). The two population models are shown at the top of page 192, with your prior probabilities indicated below the bowls. In 10 tosses, you get heads every time. Now what is your probability of two-headed? [The increase from .01 is dramatic and may seem surprising. These 10 tosses are very informative.]

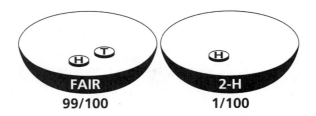

6.13 Researchers[4] followed 127 adults under the age of 70 who had undergone an angioplasty procedure to widen at least one constricted heart artery. After 18 months, 28 of them had experienced serious reactions (R): severe chest pains, heart attacks, or sudden death. Consider the 11 proportion models with the prior probabilities of Dr. X as follows:

Proportion of R's	0	.1	.2	.3	.4	.5	.6	.7	.8	.9	1
Prior probabilities	$\frac{1}{11}$	$\frac{1}{11}$	$\frac{1}{11}$	$\frac{1}{11}$	$\frac{1}{11}$	$\frac{1}{11}$	$\frac{1}{11}$	$\frac{1}{11}$	$\frac{1}{11}$	$\frac{1}{11}$	$\frac{1}{11}$

(a) Calculate the posterior probabilities of these models.

(b) Find the predictive probability of a serious reaction within 18 months for the next person in this population who receives angioplasty.

6.14 Have you ever noticed in which arm a mother carries her baby? Apparently, 80% of women carry their babies with their left arms, and this habit is unrelated to whether they are left-handed. The question is, Why? Suggesting that the behavior has deep evolutionary roots, scientists John Manning and Andrew Chamberlain studied 23 female chimpanzees, gorillas, and orangutans in zoos.[5] Five carried their infants behind them or on their backs or necks. Of the $n = 18$ that carried their infants in their arms, $s = 14$ used their left arms. Since $\frac{14}{18}$ is about 80%, they took this as evidence that apes resemble humans in this regard.

How strong is the evidence? Consider two possible population models. Both have 10 chips. Model (1) corresponds to the hypothesis that ape mothers are the same as human mothers in that 80% of them carry their infants with their left arm: It has eight chips labeled "left" and two labeled "right." Model (2) corresponds to the hypothesis that female apes have no preference as to the arms they use to carry their infants: It has five of each type of chip. Assign equal prior probabilities to the two models, corresponding to equal degrees of belief in the two possibilities. Find the posterior probability of model (1).

[Both models in this exercise are candidate null models. For the $\frac{8}{10}$ model, there is *no difference* between apes and humans and for the $\frac{5}{10}$ model there is *no preference* between right and left.]

6.15 A study[6] was designed to see whether there would be fewer injuries in baseball with breakaway bases. In a large number of games with breakaway (B) bases, there were two sliding injuries. In about the same number of games with stationary (S) bases, there were 10 sliding injuries. To find out how strongly this information indicates that B bases are safer, consider only the 12 injuries. If there is no difference (null hypothesis) between B and S, then each injury has a 50% chance of being a B. There were two B's and 10 S's. (So $s = 2$ and $n = 12$.) Test the null hypothesis by considering the five bowl models shown. The model on the right is the null hypothesis and it has prior probability .6. Each of the other models has an advantage for breakaway bases; each of these four models has prior probability .1. Find the posterior probability of the null hypothesis.

Prior probabilities: .1 .1 .1 .1 .6

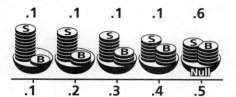

Proportion of B's: .1 .2 .3 .4 .5

Appendix: Using Minitab with Models for Proportions

The Minitab programs '**p_disc**' and '**p_disc_p**' can be used to perform Bayesian calculations for a proportion p in the case where a finite collection of models is of interest. To illustrate these programs, consider Example 6.3, in which you are interested in the probability p that player A defeats player B in a single match. The program 'p_disc' focuses on learning about the proportion p. To use this program, create two columns in Minitab. The first column, named 'p,' contains the possible values of p and the column 'prior' contains the corresponding prior probabilities. In the output below, these columns are defined (with Minitab 'name' and 'set' commands) using the prior probabilities in Example 6.3. Next, the two players play three matches—player A wins 1 and B wins 2 (one success and two failures). To compute the posterior probabilities after observing this data, type the command

```
exec 'p_disc'
```

You input the numbers of successes and failures in the experiment. The output for Example 6.3 is given below. The calculations of the posterior probabilities for p are illustrated in table form. The first two columns of the table list the values of p and the prior probabilities. The 'P_×_PRIO' column contains the products of the elements of 'p' and 'prior'; the sum of this column (.598), listed at the bottom, is the mean of the prior distribution. The last four columns of the table contain the posterior calculations. The 'like' column contains the likelihood values, the 'product' column contains the product of the 'prior' and 'like' columns, and the 'post' column contains the values of the posterior distribution. The final column, 'P_×_POST,' contains the products of the 'P' and 'POST' columns; the sum of this column (.523) is the mean of the posterior distribution. The program graphs the posterior probabilities with a spike chart.

```
MTB > name c1 'p' c2 'prior'
MTB > set 'p'
DATA> 0 .1 .2 .3 .4 .5 .6 .7 .8 .9 1
DATA> end
MTB > set 'prior'
DATA> 0 .02 .03 .05 .10 .15 .20 .25 .15 .05 0
DATA> end

MTB > exec 'p_disc'

INPUT OBSERVED NUMBER OF SUCCESSES AND FAILURES;
DATA> 1 2
```

Row	p	prior	P_X_PRIO	LIKE	PRODUCT	POST	P_X_POST
1	0.0	0.00	0.000	0	0	0.000000	0.000000
2	0.1	0.02	0.002	551020	11020	0.018802	0.001880
3	0.2	0.03	0.006	870748	26122	0.044568	0.008914
4	0.3	0.05	0.015	1000000	50000	0.085306	0.025592
5	0.4	0.10	0.040	979592	97959	0.167131	0.066852
6	0.5	0.15	0.075	850340	127551	0.217618	0.108809
7	0.6	0.20	0.120	653061	130612	0.222841	0.133705
8	0.7	0.25	0.175	428571	107143	0.182799	0.127960
9	0.8	0.15	0.120	217687	32653	0.055710	0.044568
10	0.9	0.05	0.045	61225	3061	0.005223	0.004701
11	1.0	0.00	0.000	0	0	0.000000	0.000000
12			0.598				0.522981

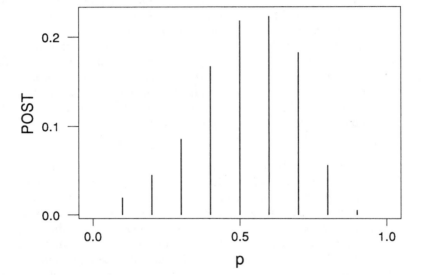

A second type of inference is predicting the results of a future experiment. Suppose that, after observing the results of these three matches, you wish to predict the results of five future matches between A and B. A table of predictive probabilities is obtained by typing the commands

exec 'p_disc_p'

The program first asks if the current *p* probabilities are contained in the column 'prior' or the column 'post.' It asks for the number of trials for the future experiment and the range of values for which we wish to compute predictive probabilities. In this example, we wish to predict results based on the posterior probabilities for *p*, the number of trials in the experiment is 5, and we are interested in predictive probabilities for all possible values of successes from 0 to 5. The output of this problem is a table of the predictive probabilities and a spike chart of these probabilities. In this example, the most likely numbers of wins for player A in these five matches are 3 and 2; the chance that one or the other of these two results will happen is approximately .5.

```
MTB > exec 'p_disc_p'

INPUT 1 IF PROBABILITIES ARE IN 'PRIOR' OR
      2 IF PROBABILITIES ARE IN 'POST':
DATA> 2

INPUT NUMBER OF TRIALS:
DATA> 5

INPUT RANGE (LOW AND HIGH VALUES) FOR NUMBER OF SUCCESSES:
DATA> 0 5

PREDICTIVE PROBABILITIES OF NUMBER OF SUCCESSES:

ROW    SUCC      PRED
  1       0    0.062585
  2       1    0.155125
  3       2    0.241020
  4       3    0.265467
  5       4    0.197679
  6       5    0.078124
```

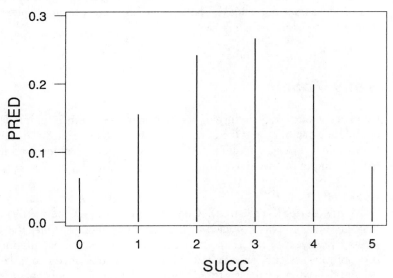

Chapter Notes

1. *Science News* 142 (October 10, 1992): 228
2. D. P. Murphy and H. Abbey, *Cancer in Families* (Cambridge, Mass.: Harvard University Press, 1959).
3. G. A. Green, *Natural History* 97 (September 1988): 58–65.
4. *Science News* 146 (August 6, 1994): 87.
5. *Newsweek* (August 13, 1990): 65.
6. D. H. Janda, R. Maguire, D. Mackesy, R. J. Hawkins, P. Fowler, and J. Boyd, "Sliding injuries in college and professional baseball: A prospective study comparing standard and breakaway bases," *Clinical Journal of Sports Medicine* 3 (April 1, 1993): 8.

7 Densities for Proportions

TO STREAMLINE the types of calculations we made in Chapter 6, we will first make things harder by increasing the number of chips in the population models. This will serve to smooth things out, and we will eventually take advantage of this smoothness. The ideas in Section 7.1 are the same as those of Chapter 6. We will even use some of the same examples and exercises. By the time we finish Section 7.2 and get into Section 7.3, the calculations will be easier. But there is always a danger when calculations get easy that you will lose track of the ideas. The roots of this chapter lie in Chapter 6. In the coming examples and exercises, think back to those roots.

7.1 Many Models

Chapter 6 makes frequent use of population models with 10 chips. The number 10 was completely arbitrary and the conclusions would have been similar for any other number of chips. The next example doubles the number of chips used in the "Who's better?" example that we dealt with extensively in Chapter 6.

EXAMPLE 7.1
▷ **Who's better? (revisited).** Consider Example 6.2 (page 169) and now suppose that there are 20 chips in each of the candidate models. Only the *proportion* of A's (wins by player A) in the model matters. So, the same calculations apply as in the probabilities assigned in Example 6.2 to proportions of A's equal to 0, .1, .2, .3, . . . , .9, and 1—whether there are 20 chips or 10 chips. I now want to cut down those probabilities and assign some to the proportions .05, .15, .25, . . . , .95. Player A's probability of beating B might actually be 55%. In Example 6.2, the proportion 55% would have to be carried by the population models with 50% and 60% A's, a role they may not play very well. Now I am entertaining this possibility explicitly by including a model with 11 A's and nine B's. Having a richer set of population models is good. But there is a cost: The number of calculations is now doubled.

Assume population models with 20 chips, an unknown number of them A's. To modify my prior probability distribution in Example 6.2, I can halve my prior probabilities given there for 0, .1, .2, and so on, up to 1 and assign the probability left over to models with proportions of A's of .05, .15, .25, up to .95. The result is shown in the bar chart in Figure 7.1.

Figure 7.1 Bar chart showing prior probabilities of population models for Example 7.1

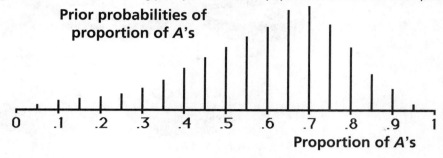

In Example 6.2, we conditioned on data D: Player A won the first of three matches and B won the next two. Given D, my posterior probabilities can be calculated in the same way as in Example 6.3. These probabilities are shown in the bar chart in Figure 7.2. This bar chart is identical to the corresponding bar chart in Example 6.3, except for the extra 10 bars interspersed.

Figure 7.2 Posterior probability bar chart of population models for Example 7.1: Given D = "A wins one game and B wins two"

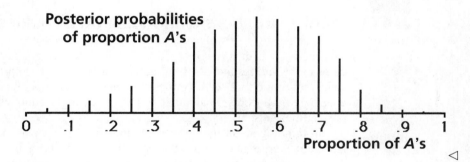

◁

EXERCISE

7.1 In Exercise 6.10 (page 190) we considered an experiment in which one twin was given honey in milk and the other was not. In all six pairs of twins, the honey twin's hemoglobin level increased more than the nonhoney twin's. So there were six successes and no failures. The population models "Null," "3/4," and "All H" shown in the figure at the top of page 198 are equivalent to those in the earlier exercise: The numbers of H's and N's are doubled, but their proportions are the same. Now there are two more models in the population universe. The model "5/8" has $P(H) = \frac{5}{8}$, interpolating between $P(H) = \frac{1}{2}$ for model 2 and $P(H) = \frac{3}{4}$ for model 3. The model "7/8" has $P(H) = \frac{7}{8}$, interpolating between $P(H) = \frac{3}{4}$ for model 3 and $P(H) = 1$ for model 4. I have spread out the prior probabilities somewhat, keeping the probability of population 2 the same because it corresponds to no effect of honey on hemoglobin.

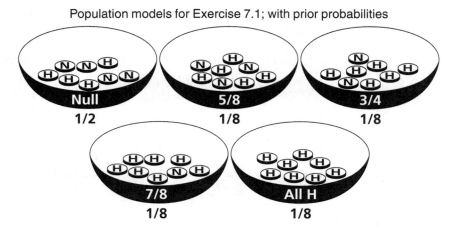

Population models for Exercise 7.1; with prior probabilities

Redo Exercise 6.10; that is, after observing six *H*'s out of six tries:
(a) What is the researcher's posterior probability of the null hypothesis?
(b) Consider a seventh set of twins, one of whom receives honey (in milk, as in the study) and the other receives only milk. What is the researcher's predictive probability (posterior to the data from the study) that the honey twin's hemoglobin will have a greater increase?
[Your answers should be similar to those of Exercise 6.10.]

7.2 Infinite Numbers of Models and Densities

Continuing as in the previous section, I could go on to consider 50, 100, 1,000, or more chips in the various models. Letting the number of chips approach infinity is realistic, since any proportion of *A*'s is possible, and so I should allow for this in the models considered. Of course, calculations get increasingly cumbersome as the number of possibilities gets larger.

In thinking about the number of chips increasing, and the number of bars in the bar chart increasing, assume that the shape of the prior bar chart does not change. To effect this, we will simply connect the tops of the bars and fill in new bars as needed. Adding new bars increases the total height of the bars, but we can simply relabel the vertical scale so the total height is 1. Each model is now a point on the interval from 0 to 1, with the points corresponding to population success proportions.

In carrying out this limiting process, the probabilities spread out into a smoothed bar chart called a **density.** (A density specifies how the probability is distributed and so I will sometimes call a density a *distribution.*) Smoothing out the probabilities means that the probability of any particular model vanishes. But it still helps to think of the "probability" of a model as proportional to the height of the density at the point defining the model. Now, only collections of models (that is, points on the horizontal axis) can have probability greater than 0. In particular, we are interested in probabilities of *intervals,* which are special kinds of collections of points.

Sec. 7.2 / Infinite Numbers of Models and Densities

> **A density is a smoothed bar chart that shows how probability is distributed.**

EXAMPLE 7.2
▷ **Smoothing "Who's better?"** I have connected the tops of the bars and relabeled the horizontal axis for the bar chart in Example 7.1, giving the density shown in Figure 7.3 with bars and without bars.

Regardless of the number of models (or bars), to calculate a probability such as that the proportion of A's is greater than .5—that is, that player A is better than B—I would add the probabilities of the models included in that event—that is, the heights of the corresponding bars. But in the limit as the number of models grows to account for more and more possibilities, there are infinitely many bars. As I have suggested, we are interested only in probabilities of intervals. For example, the interval from .5 to 1 corresponds to player A being better than player B in the sense that player A wins more than half the time. For any number of models, the total height of the bars to the right of .5 is approximately the *area* under the right half of the density (shaded in Figure 7.4 on page 200) divided by the total area under the density. In the limiting case, as the number of models tends to infinity and fills up the interval from 0 to 1, this approximation is exact.

Figure 7.3 (a) Prior bar chart for Example 7.2 with bars connected showing density (b) Same density as in (a), but with bars removed

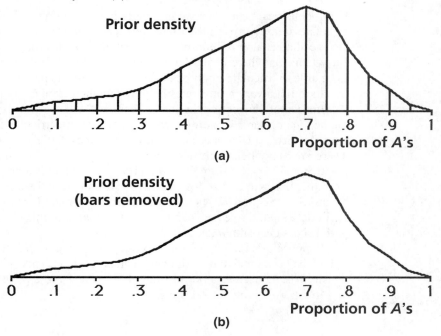

Figure 7.4 Showing probability of models with proportion > .5 for Example 7.2

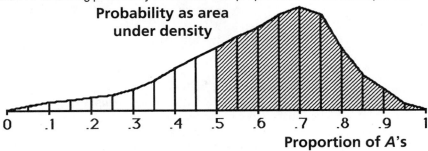

The shaded portion of this density appears to be about 75% of the total. (The heights of the bars over .55, .60, and on up to .95 divided by the total heights of all the bars is a little less: 70%.) The unshaded portion—corresponding to B being the better player—is the remaining 25%. Suppose I allow for a still larger number of models, filling in the spaces between bars with additional bars. If the probabilities of the models are smooth in the sense that the heights of the additional bars are nearly the same as the corresponding heights of the density, then the portion of the area under the density and to the right of .5 will not change much. ◁

Beta Densities

Using densities means that we no longer have to add probabilities, but we do have to calculate areas. (Apparently, I have replaced one difficult problem with another.) We also need an analog of multiplying probabilities and adding the results, as in applying Bayes' rule and the law of total probability. In the "Who's Better?" example, we want to be able to calculate my probability that A wins the next match. Connecting heights of bars is not a very promising smoothing method from either point of view.

Instead of connecting the tops of bars, we will draw in a curve that approximates the heights of bars. To do this, we will consider a particular family of curves, called **beta densities.** (They are named after the Greek letter β, for no particular reason.) There are many beta densities, one for each pair of values a and b, where these are positive numbers. I will include a and b in the name of a particular beta density. For example, if $a = 6$ and $b = 9$, then I will use the term beta(6, 9) density. For reasons discussed in Section 7.3, beta densities are convenient for representing prior opinion. I will develop the roles of a and b over the course of this chapter and, in particular, I will discuss choosing a and b in Section 7.4.

Suppose p is the population success proportion. So p is a number between 0 and 1; it is the horizontal axis in showing a graph of the density. The form of the beta(a, b) density has p raised to a power times $1 - p$ raised to a power. The powers of p and $1 - p$ are $a - 1$ and $b - 1$; thus,

$$p^{a-1}(1 - p)^{b-1}$$

Sec. 7.2 / Infinite Numbers of Models and Densities

The curves in Figure 7.5 show several beta(a, b) densities. The horizontal axis is p, the population proportion. As in Chapter 6, I will sometimes call p the model. For larger a and b, the density becomes narrower. For a larger than b, more of the probability is concentrated on models to the right of .5; for b larger than a, the converse holds.

Figure 7.5 Examples of beta(a, b) densities

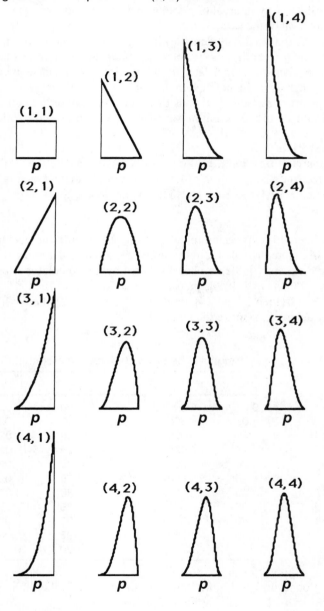

> Beta densities are particular types of prior (and posterior) distributions. There is a different density for each pair of a and b.

An important special case is beta(1, 1). When $a = b = 1$, the beta density is constant over all p-values between 0 and 1. It represents a type of indifference over these values: Any interval of p-values between 0 and 1 has the same probability as any other interval of the same width.

For all the beta(a, b) densities in Figure 7.5, a and b are integers. More generally, either a or b or both may be fractions, but they must be positive. More examples of beta(a, b) densities occur later in this section and in subsequent sections. In some of these—such as in the next example—a and b are not integers.

Constructing beta(a, b) densities is not very important in this course. But it is very easy and it should not seem mysterious to you. An example follows.

EXAMPLE 7.3

▷ **Constructing a beta(a, b) density.** Consider a beta(4.5, 3) density. By this, I mean a beta(a, b) density with $a = 4.5$ and $b = 3$. So the exponents are $a - 1 = 4.5 - 1 = 3.5$ and $b - 1 = 3 - 1 = 2$. The mathematical form is

$$p^{3.5}(1 - p)^2$$

I calculated this expression for $p = 0$, .05, .10, up to 1 and give the results in Table 7.1. I will illustrate the calculation required for a particular entry—$p = .6$. I will show in square brackets, [ ], the number displayed by my calculator after each step. Proceed as follows. Punch in p [0.6], then hit the x^y key [0.6], then 3.5 [3.5], then × (times) [0.1673129], then $1 - p$ [0.4], then x^y [0.4], then 2 [2.], then = [0.0267701]. I rounded to four decimals to make the table.

Table 7.1
Heights of beta(4.5, 3) density for Example 7.3

p	$p^{3.5}(1-p)^2$	p	$p^{3.5}(1-p)^2$
0	0	.50	.0221
.05	.0000	.55	.0250
.10	.0003	.60	.0268
.15	.0009	.65	.0271
.20	.0023	.70	.0258
.25	.0044	.75	.0228
.30	.0072	.80	.0183
.35	.0107	.85	.0127
.40	.0146	.90	.0069
.45	.0185	.95	.0021
		1.00	0

I plotted dots in Figure 7.6(a) for each entry in the table and connected them in Figure 7.6(b). The result is the beta(4.5, 3) density.

Figure 7.6 (a) Plot of points from Table 7.1 in Example 7.3; (b) Connecting the dots in (a) to form beta(4.5, 3) density

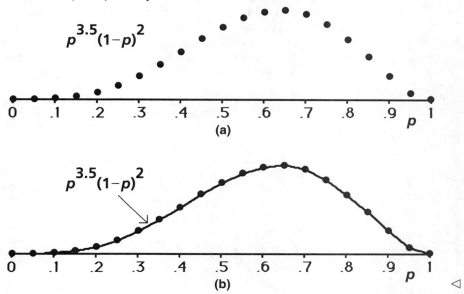

An important feature of beta densities is the ease of calculating probabilities of future observations—these are called predictive probabilities. The next example leads to a simple formula (given after the example) for calculating probabilities for the next observation.

EXAMPLE 7.4
▷ **Predictive probability.** We used the law of total probability in Example 6.2 (page 169) to calculate the mean of a bar chart. As discussed in Chapter 6, this mean is also the (predictive) probability that player A wins her next match with player B. To approximate the mean of a beta density, we begin by drawing bars with the same heights as the density. Figure 7.7 shows this for the beta(4.5, 3) density used in the previous example.

Figure 7.7 Density for Example 7.4

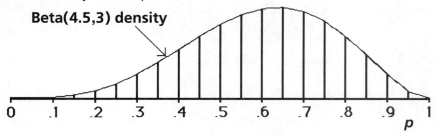

To have a legitimate probability distribution, the total of the heights of the bars has to be 1. To attain this, add the values of $p^{3.5}(1 - p)^2$ in Table 7.1 of the previous

example (the result is .2486) and divide each entry by this sum. Table 7.2 gives the results. For example, the value for $p = .6$ is .0268 (from Table 7.1), and .0268 divided by .2486 is .108, as indicated in Table 7.2. This table also contains a column of products of p and bar heights. The sum of this last column (.6) is the predictive probability of the event in question, and also the mean of the bar density approximating the beta(4.5, 3) density. Even though the bars only approximate the beta(4.5, 3) density, .6 also turns out to be the mean of the beta(4.5, 3) density.

Table 7.2
Predictive probability for Example 7.4

p	Bar height	Product	p	Bar height	Product
0	0	0	.50	.089	.044
.05	.000	.000	.55	.101	.055
.10	.001	.000	.60	.108	.065
.15	.004	.001	.65	.109	.071
.20	.009	.002	.70	.104	.073
.25	.018	.004	.75	.092	.069
.30	.029	.009	.80	.074	.059
.35	.043	.015	.85	.051	.044
.40	.059	.023	.90	.028	.025
.45	.074	.033	.95	.008	.008
			1.00	0	0
			Sum	1.001	.600

It is not an accident that the predictive probability in the previous example is such a tidy fraction. This is a pleasant characteristic of beta densities generally—namely, predictive probabilities are given by a simple formula:

$$\text{Predictive probability of success}[1] = \frac{a}{a + b}$$

I have emphasized that the predictive probability of success is also the mean of the density of population models. At the risk of saying this too often:

$$\text{Mean of beta}(a, b) \text{ density} = \frac{a}{a + b}$$

In Example 7.4, $a = 4.5$ and $b = 3$; so

$$\frac{a}{a + b} = \frac{4.5}{7.5} = .6$$

which is exactly what we found using the bars in the example.

The fact that there is such a neat formula for the mean makes using beta densities appealing. The problem is to find a beta density that fits my prior probabilities. The following example begins to address this problem.

EXAMPLE 7.5

▷ **Fitting a beta(a, b) density.** Let's find a beta density that approximates my prior probabilities in Examples 6.2 and 7.1. In Example 6.2, I calculated the mean to be .598. This is very close to the mean of .6 just calculated for the beta(4.5, 3) density. So the latter is a candidate for approximating my density of prior probabilities. Figure 7.8 compares this density with the bar chart (Figure 7.1) of Example 7.1.

Figure 7.8 Comparing a bar chart with a beta(4.5, 3) density for Example 7.5

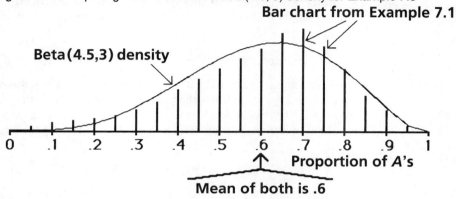

The fit is not great. In particular, the height of the beta(4.5, 3) curve for very small proportions of A's does not approximate the heights of the corresponding bars very well. I tried several other beta densities and found that I could not do much better than the beta(4.5, 3). Not every bar chart is well approximated by a beta density. Even though the fit is not great in this instance, calculations in the next section will show that there are important ways in which the fit provided by the beta(4.5, 3) density is quite satisfactory. ◁

7.3 Updating Rule for Beta Densities

In the general study of proportions, only two types of responses are possible: success or failure, hit or miss, Republican or Democrat, and so on. I will use success/failure to be specific. You have some data and, among other things, you are interested in the probability that the next observation is a success. Suppose the observations are exchangeable for you. One implication of exchangeability is that you regard a success on the first observation and a failure on the second as having the same probability as a failure on the first and a success on the second. The same statement holds if *first* and *second* are replaced with any other two observations.

Suppose you observe s successes and f failures. What is your probability that the next trial results in a success? You know by now that answering this question requires assessing your prior probabilities of population models and using Bayes' rule to find posterior probabilities. Bayes' rule says to multiply prior probabilities (or densities) by likelihoods. When your prior density is the beta(a, b), there is a very convenient updating formula. I will suggest how it can be derived and then I will convince you that it is correct by using examples.

Recall from Section 7.2 the mathematical form of the beta(a, b) density:

$$p^{a-1}(1 - p)^{b-1}$$

The likelihood is the probability of observing s successes and f failures, assuming model p. In calculating the likelihood, whenever a success appears, multiply by p, and whenever a failure occurs, multiply by $1 - p$. As described in Section 6.4, this assumes that the individual observations are independent, assuming model p. So the likelihood has s factors of p and f factors of $1 - p$:

$$(p)(p) \cdots (p)(1 - p)(1 - p) \cdots (1 - p) = p^s(1 - p)^f$$

[See Section 6.5 (page 185) for a calculation using $s = 18$ and $f = 3$ and $p = .8$.] Bayes' rule says to multiply the prior probabilities [in this case, multiply the density $p^{a-1}(1 - p)^{b-1}$ by the likelihood $p^s(1 - p)^f$] to obtain the posterior probabilities for p. When multiplying factors with exponents, add the exponents. So the posterior density is

$$p^{a-1}(1 - p)^{b-1} p^s(1 - p)^f = p^{a+s-1}(1 - p)^{b+f-1}$$

The right-hand side is the mathematical expression for yet another beta density. To see this, let a take the place of $a + s$ and let b take the place of $b + f$; then $p^{a-1}(1 - p)^{b-1}$ takes the place of $p^{a+s-1}(1 - p)^{b+f-1}$. That is, the updated density is the beta($a + s, b + f$). So when your prior is a beta density, you can use this updating property to find your posterior:

> **Updating Rule for Beta Densities**
> When the prior density is beta(a, b) and you observe s successes and f failures, the posterior density is beta($a + s, b + f$).

Consider Figure 7.5 (page 201), and suppose your current density is one of those pictured. A failure moves you one density to the right and a success moves you one density down. For example, when starting at beta(2, 3), a failure moves you to beta(2, 4) and a success moves you to beta(3, 3).

It is difficult to exaggerate the importance of this updating rule. You will like it because it allows you to calculate posterior probabilities in a flash—*provided your prior is a beta density.* If your prior probabilities are not well approximated by a beta(a, b) density, then you will have to go through the tedious calculations demonstrated in the previous chapter.

Which density to use depends on the circumstances. Before an experiment, the information available is given by the prior density, say, beta(a, b). After an experiment in which there are s successes and f failures, the current density is beta($a+s, b+f$). So, using the predictive probability of success from the previous section, the probability that the first observation in the experiment is a success is $a/(a+b)$. Then, the probability of success in the next observation after the experiment (assuming s and f

are known) is the same formula, but with $a+s$ playing the role of a and $b+f$ playing the role of b; that is, $(a+s)/(a+s+b+f)$.

Again consider Figure 7.5 (page 201). Suppose your current density is beta(1, 1), the flat density at the top left. Suppose you observe, in sequence: failure, success, success, success, failure. In the figure, move right, down, down, down, right, so you move through these beta densities: (1, 2), (2, 2), (3, 2), (4, 2), (4, 3). The corresponding probabilities of these five observations (given where you are at the time of each observation) are $1/(1+1) = \frac{1}{2}$, $1/(1+2) = \frac{1}{3}$, $2/(2+2) = \frac{2}{4}$, $3/(3+2) = \frac{3}{5}$, $2/(4+2) = \frac{2}{6}$.

EXAMPLE 7.6
▷ **Checking a beta density as a posterior.** Once again, consider the "Who's better?" example. Assume the prior probability is the beta(4.5, 3) density of Example 7.5. I observed $s = 1$ success and $f = 2$ failures. Since $a + s = 4.5 + 1 = 5.5$ and $b + f = 3 + 2 = 5$, the updating rule gives my new density as the beta(5.5, 5). The beta(5.5, 5) density is shown in Figure 7.9, superimposed on the posterior probabilities calculated in Examples 6.3 and 7.1. Like the fit provided by the beta(4.5, 3) density to my prior probabilities in Example 7.5, this fit is not perfect—but it is a little better than the earlier fit.

Figure 7.9 The beta(5.5, 5) density and posterior bar chart for Example 7.6

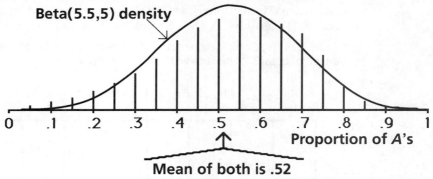

In Example 6.3, I calculated the mean of my posterior bar chart to be .523. (This is also my predictive probability that A wins her next match with B.) To find the analogous probability in the current setting, I can use the formula for the mean of a beta(a, b) density: $a/(a + b)$. Both a and b have changed; a has become $a + s = 5.5$ and b has become $b + f = 5$. So the updated mean is

$$\frac{a + s}{a + b + s + f} = \frac{5.5}{10.5} = .52$$

To two-digit accuracy, this is the same as we found before—indicating at least one sense in which the beta(5.5, 5) density fits the posterior bar chart well. But beta updating is much easier. ◁

EXAMPLE 7.7
▷ **6-MP for leukemia (revisited).** Section 6.5 showed how the opinions of Drs. X and Y change in view of the data concerning 6-MP. We could let p refer to the model

proportion of B's (6-MP is better) or of W's (6-MP is worse); let's do the former. In the study we considered, 6-MP was better for $s = 18$ pairs and worse for $f = 3$ pairs. The likelihood is shown in Figure 7.10.

Figure 7.10 Likelihoods of p for Example 7.7 with 18 successes and 3 failures

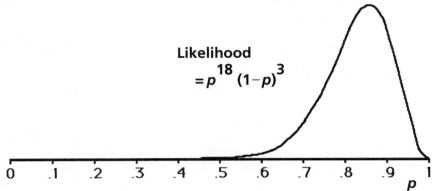

Figure 7.11 shows that the prior bar charts of Drs. X and Y can be fit perfectly by beta densities: (a) beta(1, 1) fits Dr. X's and (b) beta(2, 1) fits Dr. Y's. Using the updating rule for beta densities, Dr. X's posterior would be the beta(1+18, 1+3) = beta(19, 4) density, while Dr. Y's would be the beta(2+18, 1+3) = beta(20, 4) density.

Figure 7.11 (a) Comparison of beta(1, 1) density with Dr. X's prior bar chart; (b) Comparison of beta(2, 1) density with Dr. Y's prior bar chart

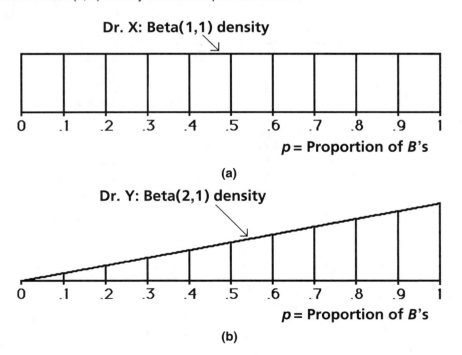

These are shown in Figure 7.12, superimposed on the posterior probability bar charts calculated in Section 6.5. The updating rule for beta(a, b) densities means that these should fit perfectly, and they do.

Figure 7.12 (a) Comparison of beta(19, 4) density with Dr. X's posterior bar chart from Section 6.5; (b) Comparison of beta(20, 4) density with Dr. Y's posterior bar chart from Section 6.5

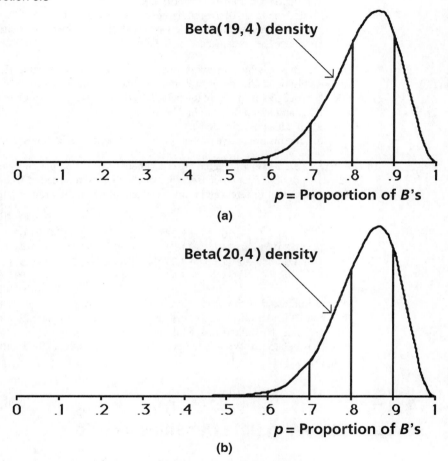

Let's check the predictive probability formula. For Dr. X,

$$\frac{a + s}{a + b + s + f} = \frac{19}{23} = .826$$

This is the same (to three decimal places) as that calculated in Example 6.5. For Dr. Y,

$$\frac{a + s}{a + b + s + f} = \frac{20}{24} = .833$$

as compared with .834 calculated earlier. ◁

EXERCISES

7.2 In Exercise 6.10 (page 190) and again in Exercise 7.1 (page 197), we considered an experiment in which one twin was given honey in milk and the other was not. In all six pairs of twins, the honey twin's hemoglobin level increased more than the nonhoney twin's. So there were six successes ($s = 6$) and no failures ($f = 0$). Suppose the researcher's prior opinion concerning the proportion of H's is represented by a beta(1, 1) density.
  (a) Find the researcher's posterior density.
  (b) Find the mean of the researcher's posterior density.
  (c) Find the researcher's predictive probability that the honey twin in a seventh pair of twins will experience the greater increase in hemoglobin.

7.3 In the matter of why burrowing owls line their nests with dung (Exercise 6.11, page 191), suppose biologist Green's prior opinion concerning the proportion of dung-lined nests that are raided by badgers is given by the beta(2, 2) density. (Recall that Green identified 25 nests lined with dung, and found that two had been raided.)
  (a) Find his posterior density.
  (b) Find his predictive probability that the next lined owl nest he observes will be raided.

7.4 In Exercise 2.2, we considered an experiment in which 32 patients were encouraged to walk as much as possible for two consecutive weeks; during the second week they were given a drug. These were their differences in miles walked while on the drug as compared with the previous week:

.00	+.56	+3.27	−2.55	+8.42	+1.07	−1.31	+3.19
−.59	+10.75	+11.73	−.05	+1.65	−3.42	+1.73	−1.44
+6.04	+12.21	+4.97	+1.68	+2.28	−6.57	−2.11	+.75
−.96	+1.68	+8.85	+7.45	−.59	+2.91	.00	+5.40

We are interested in the proportion of all patients who would walk farther while on the drug than not. Consider only the signs of these differences: 20 patients walked farther while on the drug (+) and 10 walked less while on the drug (−); ignore the two 0's. Suppose the prior density is beta(5, 5). What is the probability that the next patient will walk farther if given the drug than if not?

7.4 Choosing Beta Densities as Priors

Section 4.4 described assessing one's probabilities based on available information. In Section 6.1, we used this assessment process for choosing a prior distribution of probabilities among 11 different models. This section uses the same ideas in selecting a beta(a, b) density as a prior. There are many benefits associated with assuming beta prior densities. But *beta densities are not always appropriate.* For example, suppose your distribution for the proportion of S's looks like the bar chart in Figure 7.13. No beta density has two humps, and so fitting this with a beta density would be difficult.

Selecting a beta density means selecting a and b. The assessor has to make two decisions—more if the assessor is willing to check the implications of these decisions.

Sec. 7.4 / Choosing Beta Densities as Priors

Figure 7.13 Example of a bar chart *not* well approximated by a beta density

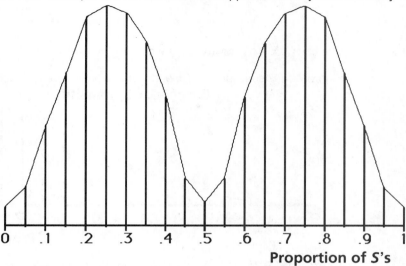

Proportion of S's

The first requirement is the assessor's probability of success on the first trial. This is the mean of the beta density—call it r. Suppose that using the methods of Section 4.4, the assessor indicates that $r = .7$. We know that the mean of the beta(a, b) density is $a/(a+b)$. So

$$r = \frac{a}{a+b}$$

The second requirement is a bit harder to describe, and also a bit harder for assessors to provide. The assessor imagines that the first trial is a success; what is the probability of success on the second trial given this information? This number has to be at least as big as r. [If it is not, then the observations from the correct model (whichever it is) are not independent.] Suppose the assessment process gives .75 as this probability. The updating rule for beta(a, b) densities (Section 7.3) says that the assessor's updated density is beta($a+1, b$). Therefore, the probability of success on the second trial is $(a+1)/(a+1+b)$. Call this

$$r^+ = \frac{a+1}{a+b+1}$$

Now we know $r^+ = .75$. Solving the above two conditions simultaneously gives

$$a = \frac{r(1-r^+)}{r^+ - r} \quad \text{and} \quad b = \frac{(1-r)(1-r^+)}{r^+ - r}$$

In the example with $r = .7$ and $r^+ = .75$,

$$a = \frac{.7(1 - .75)}{.75 - .7} = 3.5 \quad \text{and} \quad b = \frac{(1 - .7)(1 - .75)}{.75 - .7} = 1.5$$

The beta(3.5, 1.5) density is shown in Figure 7.14.

Figure 7.14 Assessed beta(a, b) density in "Who's better?" example

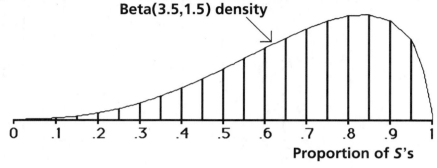

Figure 7.15 Beta($a+s$, $b+f$) densities in "Who's better?" example: (a) $s = 1$ and $f = 0$; (b) $s = 0$ and $f = 1$

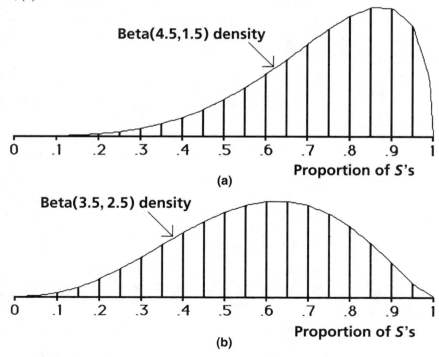

The updating rule for beta(a, b) densities means that the beta(3.5, 1.5) prior density will become beta(4.5, 1.5) density after a success and beta(3.5, 2.5) density after a failure. These two densities are pictured in Figure 7.15(a)–(b).

Sec. 7.4 / Choosing Beta Densities as Priors

The size of r^+ is important, especially in relation to r. This is because the difference $r^+ - r$ appears in the denominator of both a and b. Suppose the assessor were to specify .72 instead of .75. Then

$$a = \frac{.7(1 - .72)}{.72 - .7} = 9.8 \quad \text{and} \quad b = \frac{(1 - .7)(1 - .72)}{.72 - .7} = 4.2$$

The beta(9.8, 4.2) density shown in Figure 7.16 is very different from the beta(3.5, 1.5) density. [To better fit the beta(9.8, 4.2) density on the page, the total height of its bars is less than that for the previous three densities.]

Compare the beta(3.5, 1.5) and beta(9.8, 4.2) densities. When $a + b$ is larger, the predictive probability of success is less affected by additional data. For the beta(3.5, 1.5) density—with $a + b = 5$—the predictive probability of success increases from .7 to .75 after observing a success. For the beta(9.8, 4.2) density—with $a + b = 14$—it increases from .7 to only .72.

Figure 7.16 Alternative beta(a, b) density in "Who's better?" example

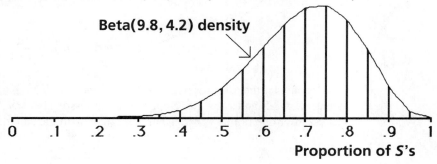

Consistency Checks

Predictive probabilities—for the next observation and also for observations well into the future—are sensitive to the choice of (a, b). An assessor should not be satisfied with specifying r and r^+ and calculating a and b as suggested above. Rather, the assessments should be checked. I will list three types of consistency checks.

The **first consistency check** of the values of a and b is to assess the probability of success on the second trial assuming that the first trial results in a *failure*. The updating rule for beta densities says that the posterior density is beta(a, $b + 1$). So this predictive probability is

$$r^- = \frac{a}{a + b + 1}$$

Once a and b have been calculated using the formulas involving r and r^+, r^- can be calculated using this formula. If the specified and calculated values of r^- disagree, then a and b should be reassessed until they do agree.

The tree diagram in Figure 7.17 may help you visualize the predictive process; S stands for success and F stands for failure.

Figure 7.17 Tree diagram showing first two observations

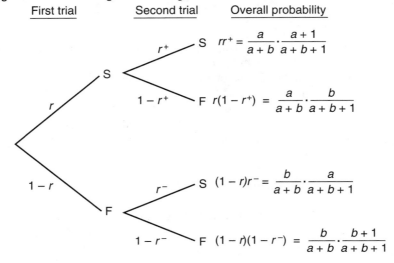

Reconsider the assessor who said $r = .7$ and $r^+ = .75$. As we saw previously, this means $a = 3.5$ and $b = 1.5$. Suppose the assessor also says that the probability of success after one failure is 68%. This disagrees with the calculated value of r^-:

$$r^- = \frac{a}{a + b + 1} = \frac{3.5}{3.5 + 1.5 + 1} = \frac{3.5}{6} = 58\%$$

This discrepancy should be pointed out to the assessor, who then has three options: (1) change a and b, (2) agree that the conditional probability of success is 58% after all, and (3) no longer assume a beta density as the prior.

Suppose the assessor feels that, in retrospect, $r^+ = .75$ was rather cavalier, relying too heavily on a single observation to change the probability of success. The assessor is willing to drop this back toward $r = .7$. On the other hand, the assessor feels that $r^- = .68$ was too conservative and that $r^- = .65$ is more reasonable. The following numbers are consistent with the beta(9.8, 4.2) density pictured in Figure 7.16:

$$r = \frac{9.8}{9.8 + 4.2} = .7; \quad r^+ = \frac{9.8 + 1}{9.8 + 4.2 + 1} = .72; \quad r^- = \frac{9.8}{9.8 + 4.2 + 1} = .65$$

The **second consistency check** also involves $a + b$. I indicated earlier that predictive probabilities change less radically when $a + b$ is large. The updating rule for beta densities says that when s successes and f failures are observed, the new density is beta($a+s$, $b+f$). So the new sum is the old sum plus $s + f$, the total number of observations.

This gives a neat interpretation for $a + b$. Suppose that before formulating a prior distribution, the assessor had experienced a number of observations deemed to be exchangeable with the current observations. Then the assessor might use the results of these trials in choosing a and b: namely, a is the number of prior successes and b is the number of prior failures; $a + b$ is the total number of prior observations. So $a + b$ is a measure of reliability in the prior distribution that compares directly with the number of observations in the experiment. The assessor could be asked whether the assessed value of $a + b$ indeed corresponds with the "prior sample size."

This might provide a good primary means of assessing a and b, or at least of $a + b$. However, the inherent dangers in this method persuade me to suggest it as secondary, providing only a check for the recommended method. One danger is that people cannot remember their prior observations very well. But a more important problem is that prior observations should seldom if ever be regarded as exchangeable with current observations. In particular, there are usually differences related with time. A possible solution is to discount prior observations as compared with current observations. I think this is a great idea, but I am not confident enough in any such discounting procedures to recommend them for practical use.

A **third consistency check** is provided by asking the assessor about the probabilities of various sets of models. One might ask for the assessor's probability of models with the proportion of success less than .5, less than .2, less than .7, and so on. Figure 7.18 (page 216) shows the first of these as shaded areas in the assessor's tentative density [Figure 7.18(a)] and also in the alternative beta(9.8, 4.2) density [Figure 7.18(b)] considered previously. Using an approximation described in the next section, I estimate these areas as 16% and 6% of the total, respectively. If the assessor feels that a probability of 16% does not adequately reflect his or her uncertainty that the proportion in question is less than .5, then some further adjustment in a and b may be appropriate.

None of these checks should dominate the final assessment. They all have flaws that reflect people's inability to accurately quantify their knowledge (and to understand what it is that is being asked!). Rather, as each check is applied, the assessor should weigh the consequences of various assumptions and make adjustments and compromises.

A final suggestion regarding assessment: When in doubt, lean in the direction of being *open-minded*. This means favoring smaller values of $a + b$. The result is to give the current data more influence. In this regard, the beta(1, 1) density (Dr. X's density in Example 7.7) counts for only two prior observations: $a + b = 2$. So it represents a high degree of open-mindedness: A small number of observations will have more influence on the posterior density than will the prior. [For this reason, the beta(1, 1) density is sometimes called *noninformative*. Whether any density can be truly noninformative is open to question, but it is clear that the beta(1, 1) density is not based on much information.]

> **Principle of open-minded priors:** Smaller values of a and b correspond to more open-minded assessors.

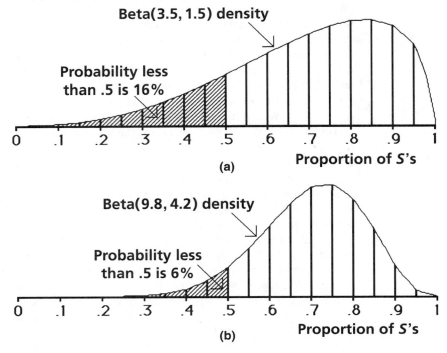

Figure 7.18 Probability of proportion of S's less than .5 for (a) beta(3.5, 1.5) density and for (b) beta(9.8, 4.2) density

In many scientific, industrial, medical, and other settings, experiments follow one after the other. Yesterday's posterior is today's prior. But observations from the next experiment may not be exchangeable with those of the previous experiment. For example, the laboratory may be different or the laboratory conditions may be different. Even a slight change in ancillary factors, such as ambient temperature, may have an effect. So it is *never* really clear that observations are exchangeable. When using yesterday's posterior, lower the sizes of a and b. Choosing an amount by which to lower them is not easy, except to say that the decreases should be greater if exchangeability is less clearly satisfied. It may be wise to reassess one's prior probabilities whenever entering into a new experimental setting, but keep the earlier results and the earlier posterior density in mind.

> **If the circumstances of an experiment make the observations not clearly exchangeable with previous evidence, then use smaller values of a and b.**

EXERCISES

7.5 You plan to survey 10 people's opinions concerning legalized abortion. Before the survey, you have a beta(a, b) prior density concerning the proportion of the population who favor legalization. To find a and b, you assess your probability that the first person is in favor and decide that it is .40. Also, if the first person you ask happens to be in favor, your probability that the second is also in favor is .45. Of the 10 people you ask, seven say they favor and three say they do not. Based on this information, what is your predictive probability that the next (11th) person is in favor? [First find your prior and then your posterior density.]

7.6 I want to consider a setting in which you can actually collect some data, but I do not want to make things too difficult. So I have to restrict to commonly available experiments. Consider an ordinary thumbtack (an English drawing pin). The experiment consists of dropping the tack on a hard, flat surface and observing whether it comes to rest with point up—call point up a success. You should hold the tack 2–3 feet above the surface and drop it vertically—do not throw it.

(a) Assuming a beta-type density, assess your prior density of the proportion of successes using r and r^+ as indicated in this section.

(b) Assess your value of r^- and then compare it with the value calculated using your a and b from part (a). Adjust your a and b until the resulting values of r, r^+, and r^- are consistent with your prior information.

(c) Drop the tack a total of 10 times, keeping track of the number of successes. Find your posterior density. Using your updated a and b, calculate your revised r, r^+, and r^-. [These will be closer together than your r, r^+, and r^- in part (b).]

7.7 Repeat Exercise 7.6 [parts (a), (b), and (c)], but now *roll* the thumb tack. To roll the tack, hold it near a hard, smooth surface and release it while moving your hand parallel to the surface. You should throw it with enough force that it travels at least a few feet. (In assessing your prior distribution, you can use the information gained from your experiment in Exercise 7.6 in any way you like, but do not regard the results from the two experiments as exchangeable.)

Finding Areas and Probabilities of Intervals

Many statistical problems require calculating probabilities of sets of models. For bar charts, this means adding heights of bars. For densities, this means calculating areas. So we need to calculate areas for beta densities.

One way to calculate areas is to use calculus (which is not a prerequisite for this course). Another way applies when a and b are large. In that case, the beta(a, b) density looks like a bell curve; in particular, it is close to being symmetric [in the sense that flipping it about the point $r = a/(a+b)$ leaves it nearly unchanged]. We will use **bell curves** (also called **normal densities**) extensively in this text. The Standard Normal Table inside the front cover of the text gives areas under normal densities. I will explain how to use this table for approximating areas under beta densities.

To find the area to the left of .5 under the beta(9.8, 4.2) density, first convert the number .5 into a **z-score** or **standardized score** and then use the Standard Normal Table. Recall from Section 7.4 that

$$r = \frac{a}{a+b} \quad \text{and} \quad r^+ = \frac{a+1}{a+b+1}$$

The quantity

$$t = \sqrt{r(r^+ - r)}$$

is the **standard deviation** of the beta(a, b) density. In the example,

$$r = \frac{9.8}{9.8 + 4.2} = .7 \qquad r^+ = \frac{9.8 + 1}{9.8 + 4.2 + 1} = .72$$

$$t = \sqrt{.7(.72 - .7)} = .118$$

To calculate the z-score corresponding to proportion .5, first subtract the mean, r, and then divide by the standard deviation, t:

$$\text{z-score:} \quad z = \frac{.5 - r}{t}$$

In the example, the z-score is

$$z = \frac{.5 - .7}{.118} = -1.69$$

This z-score is highlighted with a box in Figure 7.19. The figure also shows the z-scores corresponding to proportions of .1, .3, .7, and .9. As is always true, the z-score at the mean, $r = .7$, is 0:

$$z = \frac{r - r}{t} = 0$$

Also, as always, z is negative for models to the left of r and positive for models to the right. The magnitude of the z-score is the number of standard deviations from the mean r.

Figure 7.19 z-score and area under density to left of 50%

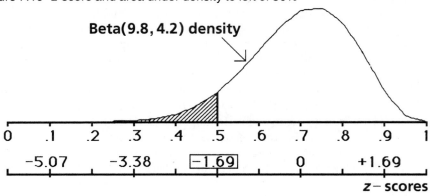

The Standard Normal Table gives the *areas to the left* of the z-scores under a standardized normal curve from $z = -3.9$ to $z = +3.9$. The leftmost column gives values of z to one decimal: −2.9, −2.8, −2.7, ..., +2.9. The 10 main columns of the table are used for interpolating; they give values of z to two decimals: −2.80, −2.81, −2.82, and so on. For example, the area to the left of z = −2.80 is .0026, to the left of −2.81 is .0025, to the left of −2.82 is .0024, and so on. The first and last rows are different; they interpolate in tenths rather than hundredths: −3.0, −3.1, −3.2, ..., −3.9 and 3.0, 3.1, 3.2, ..., 3.9. For example, the area to the left of z = −3.4 is .0003. Finally, for z outside the range of the table the last entry in the table applies: For z < −3.9 the area to the left is .0000 and for z > 3.9 the area to the left is 1.0000.

> **Use the Standard Normal Table to find probabilities to the left of a z-score.**

In our example, z = −1.69. The entry in the Standard Normal Table under z = −1.69 is .0455, or about 4.6%. This is an estimate of the area to the left of .5 under the beta(9.8, 4.2) density. An exact method gives 5.8%. The answer using the normal curve is not very close because the beta(9.8, 4.2) density is not symmetric. To see this, estimate the area to the *right* of .9—the shaded area in Figure 7.20. As shown in this figure, the z-score for .9 is 1.69:

$$z = \frac{.9 - r}{t} = \frac{.9 - .7}{.1183} = +1.69$$

To find an *area to the right* of a model proportion using the Standard Normal Table, find the area to the left and subtract it from 1. According to the Standard Normal Table, the area to the left of 1.69 is .9545. So 1 − .9545 = .0455, which is the same as the area to the left of z = −1.69; this relationship always holds because the normal curve is symmetric. But if you compare the two areas in Figures 7.19 and 7.20, the one to the right of .9 looks much smaller than the one to the left of .5. An exact method gives 2.5%, only about half that calculated using a normal curve, and only about 40% as big as 5.8%, the actual area to the left of .5. As I indicated before, the normal curve method is not very accurate when a and b are moderate or small.

Figure 7.20 z-score and area to right of 90% S's

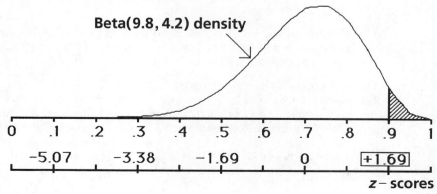

Another way to see that the beta(9.8, 4.2) density is not symmetric about $r = .7$ is to calculate the area to the left of this—its mean. A symmetric density has an area of .5 to the left of its mean (as well as to the right of its mean). For example, according to the Standard Normal Table, the area to the left of $z = 0$ is .5. The exact probability less than .7 is 46.9% (calculated using Minitab), less than 50%.

Accuracy of numerical answers is an important concern when using calculators especially when subtracting and then dividing, as when calculating z-scores. You should carry many digits in your intermediate calculations but advertise less accuracy in your final answer. Carrying too few digits in the intermediate calculations can make your answer nonsensical. Consider $3.31828/(12.3456 - 12.3345)$. Because each of the three numbers involved is accurate to six digits, you may think you can advertise six-digit accuracy: $3.31828/(12.3456 - 12.3345) = 298.944$. You cannot. The problem lies in the subtraction in the denominator: $12.3456 - 12.3345 = .0111$. So the denominator is accurate to only three digits. Therefore, the result is accurate to only three digits and you should round off 298.944 to 299. It is alright to carry six digits in intermediate calculations, but in the end you should give, at most, three. Look what happens if you carry only four digits: $3.318/(12.35 - 12.33) = 165.9$, which is nonsense. Carrying five digits is an improvement: $3.3183/(12.346 - 12.335) = 301.66$, but it is accurate to only two digits. So carry as many digits as you can in intermediate calculations and give a smaller number (such as two) in the final answer.

EXERCISES

7.8 Find the area to the left of .6 for the beta(6, 10) density using areas under the normal curve (see the Standard Normal Table).

7.9 Find the area to the right of .5 under the beta(19, 4) density. (In the context of Example 7.7, this is Dr. X's posterior probability that 6-MP is effective.)

7.10 In Exercise 7.2(a) on the effect of honey on hemoglobin levels (page 210), you found a researcher's posterior density. Find the area under the researcher's density to the right of .5. Use the normal curve approximation.

7.11 In Exercise 7.3 (page 210), assume that half of all owl nests not lined with cattle or horse dung are raided by badgers. So the researcher's probability that lining nests wards off badgers is the area under his density to the left of .5. Use your answer to Exercise 7.3(a) and the method of normal curves to find this probability.

7.12 Consider the posterior distribution you found in Exercise 7.6(c) (page 217) for the population proportion of point up when dropping a thumbtack. Find your probability that point up is more likely than point down. (That is, find your posterior probability to the right of .5.)

7.13 A Riverside High School (Durham, N.C.) student surveyed 19 fellow students concerning their favorite music.[2] Possibilities included rap/rhythm and blues (R&B), classic rock, classical, country, blues, heavy metal, folk, alternative, rock/pop, jazz, and oldies. Suppose these 19 students form a random sample from this high school. Eleven of the 19 said they preferred rap/R&B. Is this sufficient evidence to conclude that more than half of the students at this high school prefer rap/R&B? Assume a beta(1, 1) prior and calculate the posterior probability that the population proportion is greater than .5.

7.6 Percentiles and Probability Intervals

Sample results give information about populations, but the information is incomplete. The way we indicate residual uncertainty is to give the posterior probabilities of the various models. In the setting of this chapter, we give the posterior density of p. A way to do this is to specify a and b. When a and b are large, then the normal density is an approximation and we can calculate probabilities using mean $r = a/(a+b)$ and standard deviation $t = \sqrt{r(r^+ - r)}$. But the consumer may not understand these numbers. A convenient and standard way of expressing uncertainty is to give an interval of values of p that contain a specified amount of probability—such as 95%. Showing how to find such an interval is the purpose of this section. It depends on finding **percentiles** of normal densities. A 60th percentile, for example, is a value such that 60% of the probability (area under the density) is to the left of that value.

> **A $100x$ percentile of a distribution of proportion p is the value of p having probability x to the left of it.**

Recall that in Section 2.6 we defined the median of a sample as the middle number in the ordered sample. The median of a density has the same interpretation—namely, half to the left and half to the right.

> **The median of a density for p is its 50th percentile.**

EXAMPLE 7.8
▷ **HIV positivity and thoughts of suicide.** A study[3] of 301 people who had just been told that they had tested positive for the AIDS-associated virus (HIV) indicated that 91 of them (30.2%) had thoughts of suicide. (This rate dropped 1 week later and it continued to drop.) This is a sample from some population (people living in New York City who signed up for free HIV tests, have just been told they have HIV, and are willing to submit to a depression inventory). Assuming a beta(1, 1) prior density for the population proportion who have thoughts of suicide, the posterior density is beta(1+91, 1+210) = beta(92, 211). So

$$r = \frac{a}{a+b} = \frac{92}{303} = .30363$$

$$r^+ = \frac{a+1}{a+b+1} = \frac{93}{304} = .30592$$

$$t = \sqrt{r(r^+ - r)} = \sqrt{\frac{92}{303}\left(\frac{93}{304} - \frac{92}{303}\right)} = .0264$$

Consider the interval of values of p from .2519 to .3554 (I picked these numbers for a reason that will be clear shortly). The z-scores are

$$z = \frac{.2519 - .3036}{.0264} = -1.96 \quad \text{and} \quad z = \frac{.3554 - .3036}{.0264} = 1.96$$

The probability to the left of -1.96 is .025 (from the Standard Normal Table) and to the left of 1.96 is .975. The probability between these two z-scores is $.975 - .025 = .950$. The probability that p is between .2519 and .3554 is also 95%. So the set of proportions from .2519 to .3554 is a **95% probability interval**. ◁

In the previous example, we found a 95% probability interval from $z = -1.96$ to $z = 1.96$. According to the Standard Normal Table, the fifth percentile of the normal density is -1.65 and the 95th percentile is 1.65. Since $95\% - 5\% = 90\%$, going from $z = -1.65$ to $z = 1.65$ gives a 90% probability interval of p-values. The following short table was constructed in this way from the Standard Normal Table. It gives the z-score, z_{perc}, used for constructing probability intervals containing **perc%** probability.

68% probability interval: $z_{68} = 1.00$
80% probability interval: $z_{80} = 1.28$
90% probability interval: $z_{90} = 1.65$
95% probability interval: $z_{95} = 1.96$
98% probability interval: $z_{98} = 2.33$
99% probability interval: $z_{99} = 2.58$

Constructing a Probability Interval for a Proportion p Using Normal Curves

Applies for large a and b. Calculate:

$$r = \frac{a}{a+b} \qquad r^+ = \frac{a+1}{a+b+1}$$

$$t = \sqrt{r(r^+ - r)}$$

A perc% probability interval* for p is

$$r \pm z_{perc}\, t$$

where z_{perc} is given in the preceding table.

How large is large for a and b? The normal density can be used to approximate beta(a, b) for any a and b, but it is better for larger a and b. As a rule of thumb, both a and b should be at least 5, and 10 or more is better yet. For any given $a + b$, the approximation will be better for r near $\frac{1}{2}$—that is, for a and b nearly equal.

*A perc% probability interval is closely related to a *perc% confidence interval* in a classical statistical approach.

EXAMPLE 7.9

▷ **Opinion poll: Gays in the military.** Opinion and political polls taken by the Gallup, Harris, and other organizations use random samples from particular populations. Typical sample sizes for a national poll are 600 to 1,500. For example, on January 21–22, 1993, the *Newsweek* poll asked several questions of 663 adults. The questions ranged from support for Zoë Baird to gays in the military. To the question, "Can gays serve effectively in the military if they keep their sexual orientation private?" 72% said "yes."[4] *Newsweek* does not say exactly how many responded "yes," but 72% of 663 is about 477. The remainder is 186. How good is 72% as an estimate of the population proportion who would say "yes?"

Suppose a prior beta(1, 1) density. Because 663 is so large in comparison with the strengths of most people's prior opinions, the actual values of a and b used in the prior density do not matter much. The posterior density is beta(478, 187). Then

$$r = \frac{a}{a+b} = \frac{478}{665} = .71880, \quad r^+ = \frac{a+1}{a+b+1} = \frac{479}{666} = .71922$$

$$t = \sqrt{r(r^+ - r)} = \sqrt{\frac{478}{665}\left(\frac{479}{666} - \frac{478}{665}\right)} = .0174$$

Since $1.65 \times .0174 = .0287$, a **90% probability interval** is $.7188 \pm .0287$. The lower limit of this interval is $.7188 - .0287 = .690$ and the upper limit is $.7188 + .0287 = .748$. Another way of specifying this interval is to say that it extends from 69.0% to 74.8%.

Since $1.96 \times .0174 = .0341$, a **95% probability interval** is $.7188 \pm .0341$ or from 68.5% to 75.3%.

If the sample size were smaller then these intervals would be larger, and if the sample size were larger then these intervals would be smaller. For example, if there were 36 out of 50 yes responses (72%), then the 95% probability interval would be $72\% \pm 12\%$, whereas if there were 3,600 out of 5,000 yes responses, then the 95% probability interval would be $72\% \pm 1.2\%$. In general, increasing the sample size by a factor of k (in the example, 50 to 5,000, $k = 100$) decreases the width of the probability interval by a factor of $\sqrt{k}$ (in the example, 12% to 1.2% is a factor of $\sqrt{100} = 10$). ◁

In Chapter 6, we tested a null hypothesis by finding its probability. When using beta densities, as in this chapter, such calculations require a computer. Calculations are carried out in the Bayesian statistics software package, *Bayesian Computation Using Minitab,* by James Albert (Duxbury Press, 1996). An alternative way to test null hypotheses using posterior probability intervals is suggested by the following example.

EXAMPLE 7.10

▷ **6-MP for leukemia (revisited).** Reconsider the example of Section 6.5. In Example 7.7 (page 207), we found that Dr. X's posterior density is beta(19, 4). So $r = 19/23 = .826$, $r^+ = 20/24$, and $t = .07737$. The 95% posterior probability interval for the proportion of pairs for which 6-MP is better is $.826 \pm 1.96(.07737) = .826 \pm .152$, or from about 67% to 98%. The null hypothesis proportion is 50%. That the null proportion is not contained in the 95% probability interval means the evidence does not support the null hypothesis. ◁

An alternative way to test a null hypothesis proportion is to decide whether it is in the 95% posterior probability interval. If the null hypothesis proportion is not in the 95% posterior probability interval then it is not supported, and if it is then the null hypothesis is supported. However, regarding the latter conclusion, if the interval is very wide, then the posterior density is quite spread out (that is, it has a large standard deviation) and so many other hypotheses are supported as well. So support is weak if the interval is wide.

> **Testing a null hypothesis means finding its posterior probability.**

> **Alternative test of null hypothesis that the success proportion is $\frac{1}{2}$:** Decide whether the 95% posterior probability interval contains $\frac{1}{2}$; if yes, then the null hypothesis is supported—otherwise, it is not.

> **Equivalently, find $z = (\frac{1}{2} - r)/t$, where r and t are calculated from the posterior beta density; if $-1.96 < z < 1.96$, then the null hypothesis is supported—otherwise it is not.**

I suggest using the following language when you test a hypothesis. When $\frac{1}{2}$ is not in the interval: "The null hypothesis is not supported since the null proportion is not in my 95% posterior probability interval." Or, getting rid of both negatives when the null hypothesis corresponds to no difference in effectiveness between two groups (or treatments), "A difference between the two groups is supported." When $\frac{1}{2}$ is in the interval: "The null hypothesis is supported since the null proportion is in my 95% posterior probability interval." The phrase "is supported" refers to posterior probabilities, which makes hypothesis testing subjective—a null hypothesis may be supported for one person but not for another. "Testing a hypothesis" is a standard procedure in a frequentist approach to statistics; an approximate connection is the following: A null hypothesis is "rejected" in the classical sense if it is not supported for someone who has a beta(1, 1) prior density.

Using a 95% probability interval (or ± 1.96 as bounds for the z-score) is conventional, but any such convention is crude. For example, $z = 1.95$ and $z = 1.97$ provide essentially the same evidence regarding the null hypothesis. Saying that one supports it and the other does not is arbitrary and artificial.

A 99% probability interval contains more values than does a 95% interval. Should a 99% interval not contain the null proportion, then this would be stronger evidence against the null hypothesis than if it is not contained in a 95% interval. If the null proportion is not in a 95% interval, then this is stronger evidence against the null hypothesis than if it is not in a 90% interval. And so on. To fine tune a test of a hypothesis, you might indicate whether a null proportion that is in a 95% interval is also in a 90% interval. If it is, then this would be stronger support. Also, you might indicate whether a null proportion that is not in a 95% interval is in a 99% interval. If it is not, then this implies even stronger conviction against the null hypothesis.

When giving intervals, use the same accuracy for the upper and lower limits. For example, if you are using two-digit accuracy, from .0023418 to .0033129 becomes from .0023 to .0033 (or from 2.3E−3 to 3.3E−3, using exponential notation), but from .0023418 to .33129 becomes from 0.00 to .33.

EXERCISES

7.14 Do opposites attract? Three Duke undergraduates[5] conducted a study of this question in 1993. They asked questions of 28 female undergraduates concerning their own personalities. These included the following five questions: Are you (1) introverted, (2) easy-going, (3) studious, (4) serious, (5) gregarious? They asked the same five questions about the respondent's ideal mates. Take "opposites attract" to mean that the respondent answers differently for their ideal mates than they do for themselves in at least two of the five questions. In this sample of 28, only eight were attracted to opposites. Deeming this to be a random sample of Duke undergraduates, give a 90% posterior probability interval for the population proportion of those attracted to opposites. Use a beta(1, 1) prior density.

7.15 A study[6] by Charles J. Graham and others at Arkansas Children's Hospital in Little Rock addressed the question of whether left-handed children are more accident prone than right-handed children. Of 267 children between the ages of 6 and 18 who were admitted to a pediatric emergency room for trauma, 44 of them (or 16.5%) were left-handed. The investigators claimed that about 10% of all children are left-handed. (Indeed, this was about the proportion of children who were admitted to the same emergency room for nontrauma reasons.) Consider the population of all pediatric trauma patients. Assume a beta(2, 18) prior density for the proportion of left-handers in this population. (For this density, the predictive probability of left-handed is 10%.)

(a) Find the 95% posterior probability interval for the population proportion of left-handers among all pediatric trauma patients.

(b) Test the null hypothesis that in the population of children brought to the emergency room, the proportion of left-handed children is 10%.

(c) Find the posterior probability that the population proportion is greater than 10%.

7.16 A company commissioned an audit of its accounts receivable. The auditor selected 25 accounts randomly and found that eight were in error. Assume a beta(1, 1) prior density for the proportion of accounts that are in error in the company's total accounts receivable. Find the 90% posterior probability interval for the proportion of accounts in error.

7.17 In the same *Newsweek* poll mentioned in Example 7.9, the 663 respondents were asked, "Should Zoë Baird have been confirmed as attorney general after she admitted that she hired illegal aliens, mainly for child care, and that she failed to pay the federal taxes required on their salaries until recently?" A total of 106 respondents (16%) answered "yes." Find the 95% posterior probability interval for the population proportion who would have answered "yes," assuming a beta(1, 1) prior density.

7.18 In a *Newsweek* poll taken on February 18–19, 1993,[7] of 753 adults who were asked, "Would you favor additional taxes to pay for reforming and expanding health care in the United States?" 65% (or 489) said "yes." Assuming a beta(5, 5) prior density, find the 90% posterior probability interval for the population proportion who would have answered "yes."

7.19 In a *Newsweek* poll taken on February 3–4, 1994,[8] of 750 adults who were asked, "Do you have a gay friend or acquaintance?" 43% (or 322) said "yes." Assume a beta(1, 1) prior density.

(a) Find the 95% posterior probability interval for the population proportion who would have answered "yes."

(b) Suppose 4 times as many people (3,000) had been asked the same question and again 43% said "yes." The width of the interval in your answer to part (a) would be divided by what factor?

7.20 Are college women too thin? Two Duke undergraduates[9] conducted a survey of heights and weights of freshmen women students at Georgetown University in 1993. They surveyed 75 students and deemed 36 of them to be underweight for their height (as judged by a chart published in 1989 by *The Complete Home Medical Guide*). It is not clear whether this is a random sample of the population of freshmen women at Georgetown University or any other set of freshmen women. (Indeed, the surveyors were not interested in estimating the proportion of underweight women, but in determining whether underweight women were more likely to subscribe to fashion or beauty magazines—the subject of Exercise 9.15.) Assume it is a random sample of some population and assume a beta(1, 1) prior density for the proportion of underweights in this population. Find the 95% posterior probability interval for this proportion.

7.21 In the documentary film *A Private Universe,* only two of 23 gowned graduates at a Harvard University commencement were able to correctly answer the question: "Why is it hotter in summer than in winter?" In view of the results of the survey, it is difficult for me to believe that these 23 are a random sample of all Harvard graduates, but assume that they are for the sake of calculation. Also assume a beta(1, 1) density for the proportion of Harvard graduates who would answer this question correctly and find a 90% posterior probability interval for this proportion. (You will see that this interval comes close to 0. This is an instance where the beta density is not very close to being symmetric and so less than 5% of the area is to the left of the interval you calculated and more than 5% is to the right. If you have access to Minitab you can find the exact answers.)

7.22 In November 1994, a *USA Today*/CNN/Gallup poll[10] indicated that 54% of a sample of 1,020 adults surveyed felt that the Republican party would do a better job handling the economy than would the Democrats (33% said the Democrats would do a better job and 13% were undecided). Find the 95% posterior probability interval for the proportion of the population who feel the Republicans would do a better job. Assume a beta(1, 1) prior density and use $s = 551$ and $f = 469$.

7.23 In 1994, a Roper Starch Worldwide poll surveyed 503 U.S. students in grades 9–12. Half (252) were males. Thirty-six percent (about 181) said they had engaged in sexual intercourse. Assuming this is a random sample of all such students, find a 95% probability interval for the proportion of all such students who would answer similarly. Use a beta(1, 1) prior density. (The way I posed this problem sidesteps the question of whether the students in the sample answered truthfully.)

7.24 On September 26, 1994, CNN announced a poll of 500 people aged 18–34. Only 45/500 (or 9%) believe Social Security will still be available for them when they retire. Assuming a beta(5, 5) prior density for the population proportion who believe it will be available, find a 90% posterior probability interval for this proportion.

7.25 In the same poll of Exercise 7.24, an incredible 46% (or 230/500) say they believe in unidentified flying objects (UFOs). Assuming a beta(1, 9) prior density for the population proportion who believe in UFOs, find a 99% posterior probability interval for this proportion.

7.26 In Exercise 6.15 (page 192), you tested the null hypothesis that the population proportion of injuries with breakaway bases was $\frac{1}{2}$. Among roughly equal numbers of games with the two types of bases, two injuries occurred with breakaway bases and 10 occurred with stationary bases. Now test this null hypothesis using the approach of this section. Assume a beta(1, 1) prior density and use $s = 2$ and $f = 10$.

Prediction

We have seen that for a beta prior density and s successes and f failures, the posterior density is beta($a+s$, $b+f$). The posterior mean is the same as the **predictive probability** of success on the *next* observation:

$$\frac{a + s}{a + b + s + f}$$

Suppose instead of the next observation, one is interested in the next two or more observations. How many successes will there be?

EXAMPLE 7.11
▷ **Cause of glass breakage.** Glass panels in high-rise buildings sometimes break and fall to the ground. One case involved 39 panels. The owners replaced all remaining panels (numbering in the thousands) and sued the manufacturer and builder for the replacement costs. The legal case centered on the cause of the breakage. If flaws in the glass had caused the breakage, then the manufacturer was at fault; otherwise, the builder was at fault.

The origins of the cracks were, for the most part, destroyed and so the cause of the breakage was known for only three of the panels. Nickel sulfide (NiS) stones were found at all three origins. So the manufacturer was at fault for these three, but what about the other 36?

A glass expert testified that among panels that break, only 5% are caused by NiS stones. However, while NiS is rare, any NiS in a production lot tends to be pervasive and will likely be responsible for most of the breakages. In particular, if the expert is told that two panels from the same lot break, and that one breakage was caused by the presence of NiS stones, his probability that the other was also caused by NiS stones increases to 95%. In the notation of Section 7.4,

$$r = .05 \quad \text{and} \quad r^+ = .95$$

Solving for a and b:

$$a = \frac{.05(1 - .95)}{.95 - .05} = \frac{1}{360} \quad \text{and} \quad b = \frac{(1 - .05)(1 - .95)}{.95 - .05} = \frac{19}{360}$$

Let's check:

$$r = \frac{a}{a+b} = \frac{\frac{1}{360}}{\frac{1}{360} + \frac{19}{360}} = \frac{1}{1+19} = .05$$

$$r^+ = \frac{a+1}{a+b+1} = \frac{\frac{1}{360} + 1}{\frac{1}{360} + \frac{19}{360} + 1} = \frac{1 + 360}{1 + 19 + 360} = .95$$

The beta(1/360, 19/360) density has the U-shape shown in Figure 7.21. The vertical lines at 0 and 1 are not part of the density and serve only to show that the density explodes at these two endpoints. The density is not as close to being symmetric as the limited view presented in the figure suggests: There is much more probability near 0 than near 1 and, indeed, the mean r is only .05.

Figure 7.21 Prior and posterior densities for Example 7.11

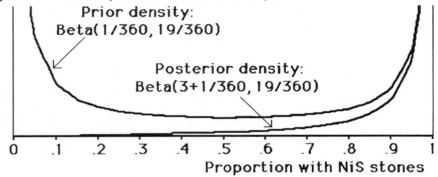

Using the updating rule for beta densities, the expert's posterior density (given $s = 3$ "successes" out of three trials) of the proportion with NiS stones in the lot in question is beta(3+1/360, 19/360). The posterior density is shown in the figure along with the prior density. The two curves do not look as though they can both have area 1, but they do. The reason for this illusion is that most of the area under the posterior density occurs near model proportion 1 and the figure is truncated and so does not show this. Since most of the posterior probability is close to 1, the mean is also:

$$\frac{a+s}{a+b+s+f} = \frac{\frac{1}{360} + 3}{\frac{1}{360} + \frac{19}{360} + 3 + 0} = \frac{1,081}{1,100} = .983$$

This mean is the predictive probability that any *one* of the other 36 panels broke because of NiS stones. This is interesting information, but the court wants to know the expert's probability distribution of the *number* of the remaining 36 panels that broke because of NiS stones. For example, it wants to know the expert's probability that all 39 of the panels broke because of NiS stones. (This example is to be continued.) ◁

Prompted by the previous example with its 39 panels, let's address a similar problem, but one that is easier to assess. Consider a beta density (either prior or posterior) for a proportion of successes. Suppose we will make just *two* more observations. How many of the two will be successes? The possibilities are 0, 1, and 2. What are the corresponding probabilities? The answer can be calculated easily using a tree diagram from Section 7.4. I have reproduced it in Figure 7.22 and added a column that indicates the number of successes along each path.

Figure 7.22 Tree diagram showing probabilities for number of successes in two tries

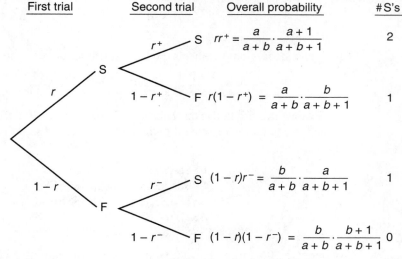

As the tree diagram indicates, the **predictive probabilities of the number of successes** are given by the following table:

Number of successes	2	1	0
Predictive probability	$\frac{a}{a+b} \cdot \frac{a+1}{a+b+1}$	$2 \cdot \frac{a}{a+b} \cdot \frac{b}{a+b+1}$	$\frac{b}{a+b} \cdot \frac{b+1}{a+b+1}$

In case a and b are both 1, all three of these probabilities equal $\frac{1}{3}$.

Now let's make things a little more interesting. Suppose there are 10 future observations. How many will be successes? Trees with 10 levels of branches are too complicated to draw, but you may be able to picture one in your head. The reasoning is the same as for two branches. To calculate the probability that all 10 are successes, think of the path along the tree that always follows the *success* branch—just as there is only one such path in the case of two observations, there is only one path for 10 observations.

Just as in the case of two observations, the probability along the first S branch is the (predictive) probability of success on the initial trial: $a/(a+b)$. After one success, the new density is beta$(a+1, b)$; so the probability along the second S branch, given a success at the first trial, is $(a+1)/(a+b+1)$. If we get two successes, then the new density is beta$(a+2, b)$; so the probability along the third S branch, given that the first two trials resulted in successes, is $(a+2)/(a+b+2)$. Proceeding in this way, we find that the **predictive probability of 10 consecutive successes** is the product of 10 factors, one for each level of branches:

$$\frac{a}{a+b} \frac{a+1}{a+b+1} \frac{a+2}{a+b+2} \cdots \frac{a+9}{a+b+9}$$

This formula generalizes the predictive probability formula given in Section 7.2. As an example, suppose a and b are both 1. Then this expression becomes

$$\frac{1}{2} \frac{2}{3} \frac{3}{4} \cdots \frac{10}{11}$$

Since each denominator cancels the subsequent numerator, this product becomes

$$\frac{1}{\cancel{2}} \frac{\cancel{2}}{\cancel{3}} \frac{\cancel{3}}{\cancel{4}} \frac{\cancel{4}}{\cancel{5}} \frac{\cancel{5}}{\cancel{6}} \frac{\cancel{6}}{\cancel{7}} \frac{\cancel{7}}{\cancel{8}} \frac{\cancel{8}}{\cancel{9}} \frac{\cancel{9}}{\cancel{10}} \frac{\cancel{10}}{11} = \frac{1}{11}$$

Finding the predictive probability of nine successes in 10 trials is more complicated, since there are several paths that have nine S's and one F. Here is the complete list:

```
FSSSSSSSSS    SSSSSFSSSS
SFSSSSSSSS    SSSSSSFSSS
SSFSSSSSSS    SSSSSSSFSS
SSSFSSSSSS    SSSSSSSSFS
SSSSFSSSSS    SSSSSSSSSF
```

So there are 10 paths with exactly nine S's. The probability of the first is

$$\frac{b}{a+b} \frac{a}{a+b+1} \frac{a+1}{a+b+2} \frac{a+2}{a+b+3} \cdots \frac{a+8}{a+b+9}$$

The probability of the second is

$$\frac{a}{a+b} \frac{b}{a+b+1} \frac{a+1}{a+b+2} \frac{a+2}{a+b+3} \cdots \frac{a+8}{a+b+9}$$

and of the third,

$$\frac{a}{a+b} \frac{a+1}{a+b+1} \frac{b}{a+b+2} \frac{a+2}{a+b+3} \cdots \frac{a+8}{a+b+9}$$

All three of these and the probabilities of the other seven sequences are identical. So *the probability of nine successes out of 10 trials* is 10 times any one of them:

$$10 \frac{b}{a+b} \frac{a}{a+b+1} \frac{a+1}{a+b+2} \frac{a+2}{a+b+3} \cdots \frac{a+8}{a+b+9}$$

Sec. 7.7 / Prediction

When $a = b = 1$, this becomes

$$10 \, \frac{1}{2} \frac{1}{3} \frac{2}{4} \frac{3}{5} \frac{4}{6} \frac{5}{7} \frac{6}{8} \frac{7}{9} \frac{8}{10} \frac{9}{11}$$

Canceling gives

$$\cancel{10} \, \frac{1}{\cancel{2}} \frac{1}{\cancel{3}} \frac{\cancel{2}}{\cancel{4}} \frac{\cancel{3}}{\cancel{5}} \frac{\cancel{4}}{\cancel{6}} \frac{\cancel{5}}{\cancel{7}} \frac{\cancel{6}}{\cancel{8}} \frac{\cancel{7}}{\cancel{9}} \frac{8}{\cancel{10}} \frac{\cancel{9}}{11} = \frac{1}{11}$$

This answer is the same as for 10 successes out of 10 trials.

Finding the predictive probability of eight successes in 10 trials is more complicated yet. Here is the list of possibilities for eight S's and two F's:

```
FFSSSSSSSS    SFFSSSSSSS    SSFSFSSSSS    SSSFSSSFSS    SSSSSFSFSS
FSFSSSSSSS    SFSFSSSSSS    SSFSSFSSSS    SSSFSSSSFS    SSSSSFSSFS
FSSFSSSSSS    SFSSFSSSSS    SSFSSSFSSS    SSSFSSSSSF    SSSSSFSSSF
FSSSFSSSSS    SFSSSFSSSS    SSFSSSSFSS    SSSSFFSSSS    SSSSSSFFSS
FSSSSFSSSS    SFSSSSFSSS    SSFSSSSSFS    SSSSFSFSSS    SSSSSSFSFS
FSSSSSFSSS    SFSSSSSFSS    SSFSSSSSSF    SSSSFSSFSS    SSSSSSFSSF
FSSSSSSFSS    SFSSSSSSFS    SSSFFSSSSS    SSSSFSSSFS    SSSSSSSFFS
FSSSSSSSFS    SFSSSSSSSF    SSSFSFSSSS    SSSSFSSSSF    SSSSSSSFSF
FSSSSSSSSF    SSFFSSSSSS    SSSFSSFSSS    SSSSSFFSSS    SSSSSSSSFF
```

So there are 45 paths with exactly eight S's. The probability of each is the same; multiplying by 45 gives *the probability of eight successes out of 10 trials:*

$$45 \, \frac{b}{a+b} \frac{b+1}{a+b+1} \frac{a}{a+b+2} \frac{a+1}{a+b+3} \cdots \frac{a+7}{a+b+9}$$

When $a = b = 1$, this becomes

$$45 \, \frac{1}{2} \frac{2}{3} \frac{1}{4} \frac{2}{5} \frac{3}{6} \frac{4}{7} \frac{5}{8} \frac{6}{9} \frac{7}{10} \frac{8}{11}$$

Canceling gives

$$\cancel{45} \, \frac{1}{\cancel{2}} \frac{\cancel{2}}{\cancel{3}} \frac{1}{\cancel{4}} \frac{\cancel{2}}{\cancel{5}} \frac{\cancel{3}}{\cancel{6}} \frac{\cancel{4}}{\cancel{7}} \frac{\cancel{5}}{\cancel{8}} \frac{\cancel{6}}{\cancel{9}} \frac{\cancel{7}}{\cancel{10}} \frac{8}{11} = \frac{1}{11}$$

Again—the answer is the same as for both nine and 10 successes out of 10 trials.

Figure 7.23 Predictive probabilities for successes in next 10 tries

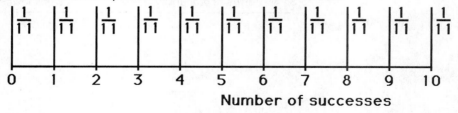

I am not going to take you through the work of counting the paths for other numbers of successes. The number of possibilities increases to 252 for five successes out

of the 10 trials, and then decreases to 45 (again) for two successes, 10 for one success, and one for 0 successes. The answer is complicated in general. But it is simple when $a = b = 1$. In that case, all of the possibilities for the number of successes have the same probability: 1/11—see bar chart in Figure 7.23. (This bar chart happens to be the same as Dr. X's in Section 6.5, but it applies to the number of successes, instead of to the various model proportions.)

Now we return to the issue raised in the glass breakage example.

EXAMPLE 7.12

▷ **Cause of glass breakage (continued).** In Example 7.11, we found the expert's posterior density of the population proportion of broken panels with NiS stones to be beta$(a+s, b+f)$ = beta$(1/360+3, 19/360)$. The court wants to know the expert's probabilities for the number of the 36 panels that had NiS stones.

First, consider the probability that all 36 panels had NiS stones. Applying the previous formula for 10 observations to 36 observations gives

$$\frac{a+3}{a+3+b} \cdot \frac{a+3+1}{a+3+b+1} \cdot \frac{a+3+2}{a+3+b+2} \cdots \frac{a+3+35}{a+3+b+35}$$

$$= \frac{3+\frac{1}{360}}{3+\frac{1}{360}+\frac{19}{360}} \cdot \frac{4+\frac{1}{360}}{4+\frac{1}{360}+\frac{19}{360}} \cdot \frac{5+\frac{1}{360}}{5+\frac{1}{360}+\frac{19}{360}} \cdots \frac{38+\frac{1}{360}}{38+\frac{1}{360}+\frac{19}{360}}$$

$$= (.9827)(.9870)(.9896) \cdots (.9986) = .863$$

This is the critical quantity for the court case and the final answer to our problem. It says that the expert's probability that all the breakages were caused by NiS stones is about 86%.

To be complete, I will give the expert's complete predictive probability distribution for the number of NiS breakages in the other 36 panels in Table 7.3. Except for the extremes—36 and 0—the counting problems are rather involved. (The exercises do not contain such difficult counting calculations.)

Table 7.3
Predictive probabilities of number of broken panels caused by NiS stones

Number of NiS breaks among 36 unknowns	Predictive probability	Number of NiS breaks among 36 unknowns	Predictive probability
36	.863	27	.003
35	.043	26	.003
34	.022	25	.002
33	.014	24	.002
32	.010	23	.001
31	.008	22	.001
30	.006	21	.001
29	.005	20	.001
28	.004	0–19	.005

◁

This chapter and the previous chapter dealt with inferences about a single proportion. The next two chapters treat the important problem of comparing two proportions. Chapter 8 generalizes Chapter 6 and Chapter 9 generalizes the present chapter to samples from two populations.

EXERCISES

7.27 You have a beta(2, 1) density for a proportion of successes and can make observations from the population in question.

(a) What is your probability that the next four observations are all successes?

(b) What is your probability that three of your next four observations are successes and the other is a failure?

(c) Assume that three of your next four observations are successes and the other is a failure—as in part (b). What is your probability that your next (fifth) observation is a success?

7.28 A study[11] reported on the long-term effects of exposure to low levels of lead in childhood. Researchers analyzed children's shed primary teeth for lead content. Of the children whose teeth had a lead content of more than 22.22 parts per million (ppm) (which is rather high), 22 eventually graduated from high school and seven did not. Suppose your prior density for the proportion of all such children who will graduate from high school is beta(1, 1), and so your posterior density is beta(23, 8). Based on this information, of 10 more children who are found to have a lead content of more than 22.22 ppm, what is your (predictive) probability that nine or 10 of them will graduate from high school? [*Hint:* Find the probability of nine and also of 10 and add them together.]

7.29 The study mentioned in Exercise 7.28 also reported on children who had experienced relatively low exposure to lead. Of the children whose teeth had a lead content of less than 5.95 ppm, 25 eventually graduated from high school and two did not. Now suppose that your prior density for the proportion of such children who will graduate from high school is beta(3, 1). For the next 10 children who are found to have a lead content of less than 5.95 ppm, what is your (predictive) probability that eight or more of them will graduate from high school? [*Hint:* Add the probabilities of 8, 9, and 10.]

7.30 The most important prognostic factor in early breast cancer is the number of axillary lymph nodes that test positive for breast cancer, indicating the extent to which the cancer has spread. There is no standard number of lymph nodes to sample. Sometimes surgeons remove three and sometimes they remove 30. The proportion of lymph nodes that are positive has approximately a beta(.1, 5) density. So the probability that the first lymph node sampled is positive is .1/5.1 or about 2%. But if the first one sampled tests positive, the updated density is beta(1.1, 5) and so the conditional probability that the second is positive is 1.1/6.1 or about 18%.

A woman with breast cancer had a mastectomy and, subsequently, the surgeon removed three lymph nodes. None tested positive. Another doctor questions the surgeon, suggesting that had more been removed, perhaps some would have been positive. Should the surgeon have removed more? To address this, suppose the surgeon had removed an additional 20 lymph nodes (for a total of 23). What is the probability that none of the 23 would have tested positive? (Assume that the 23 are exchangeable initially.)

Appendix: Using Minitab with Densities for Proportions

The Minitab program **'p_beta'** summarizes the posterior density for a proportion *p* when one chooses a beta prior distribution. This program will graph the beta distribution and summarize the distribution by the computation of cumulative probabilities and percentiles. In Example 7.6 (page 207), one is interested in the probability, p, that player A wins a single game. Your beliefs about p are represented by a beta(4.5, 3) curve and you observe three games with player A winning one and player B winning two. The posterior density for p is a beta(5.5, 5) curve. To summarize this curve, we type the command

exec 'p_beta'

The program asks for a and b of the beta(a, b) distribution; in this example, we input the numbers 5.5 and 5. Suppose that we wish to view this beta curve. In addition, we are interested in the probability that p is smaller than .5, and wish to find the 5th, 50th, and 95th percentiles of the curve. We type 'y' at the first question in the program to indicate that we wish to graph the curve. We type 'y' to indicate that we wish to compute cumulative probabilities and type .5 on the 'DATA' line to indicate that we are computing the cumulative probability of .5. Last, we type 'y' to compute percentiles of the curve and place the values .05. .5, and .95 on the 'DATA' line. The program outputs the cumulative probabilities and percentiles of interest.

In the following output, we see that the most likely value for p is approximately .5 and the beta curve is roughly symmetric about this value. In addition, the probability that p is smaller than .5 is approximately 44%, and there is 90% probability that p is between .28 and .76.

```
MTB > exec 'p_beta'

INPUT VALUES OF BETA PARAMETERS A AND B:
DATA> 5.5 5

TYPE 'y' TO SEE A PLOT OF THE BETA DENSITY:
y
```

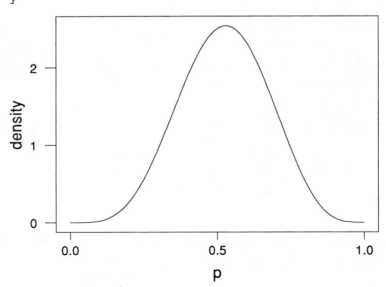

```
TYPE 'y' TO COMPUTE CUMULATIVE PROBABILITIES:
-----------------------------------------------------------------
  Input values of P of interest.  The output is the column of
  values P and the column of cumulative probabilities PROB_LT.
-----------------------------------------------------------------

y
DATA> .5
DATA> end

ROW      P      PROB_LT


 1      0.5    0.436191

TYPE 'y' TO COMPUTE PERCENTILES:
-----------------------------------------------------------------
  Input probabilities for which you wish to compute percentiles.
  The output is the probabilities in the column PROB and the
  corresponding quantiles in the column PERCNTLE.
-----------------------------------------------------------------

y
DATA> .05 .5 .95
DATA> end

ROW    PROB     PERCNTLE


 1     0.05     0.278226
 2     0.50     0.525377
 3     0.95     0.763998
```

The program 'p_beta_p' computes predictive probabilities for the number of successes for a future experiment when the probabilities for p are given by a beta curve. In this example, suppose that players A and B will play 10 future matches and you are interested in predicting the number of matches that A will win. Type

exec 'p_beta_p'

One inputs the parameters of the beta curve, the number of trials in the future experiment, and the range for the number of future successes for which you wish to compute probabilities. In this example, we input the beta values 5.5 and 5 corresponding to the posterior distribution, input 10 future trials, and wish to compute predictive probabilities for numbers of success from 0 to 10. The output is a table of the predictive probabilities and a graph of these numbers. In this example, the most likely number of wins for player A is 5, and A will win either four, five, or six games with probability .5.

```
MTB > exec 'p_beta_p'

INPUT VALUES OF BETA PARAMETERS A AND B:
DATA> 5.5 5
```

```
INPUT NUMBER OF TRIALS:
DATA> 10

INPUT RANGE (LOW AND HIGH VALUES) FOR NUMBER OF SUCCESSES:
DATA> 0 10

PREDICTIVE PROBABILITIES OF NUMBER OF SUCCESSES:
```

ROW	SUCC	PRED
1	0	0.007614
2	1	0.029914
3	2	0.067306
4	3	0.112177
5	4	0.151694
6	5	0.172931
7	6	0.168127
8	7	0.138105
9	8	0.092481
10	9	0.046240
11	10	0.013410

Other Minitab programs for learning about a proportion are contained in *Bayesian Computation Using Minitab* by James Albert (Duxbury Press, 1996). This book contains the macro 'beta_sel', which finds the particular beta density that matches prior beliefs about future observations. The program 'p_cont_t' performs a Bayesian test of the hypothesis that the proportion takes a particular value. Also, in some situations, it may be difficult to specify a beta density, but one can specify probabilities that the proportion p will fall in particular intervals. The program 'p_hist_p' will compute the posterior probabilities of the intervals when the prior has this histogram form.

Chapter Notes

1. This was put forward by the Marquis de Laplace in the early 1800s and is called Laplace's rule of succession.
2. *The Herald-Sun* (Durham, N.C., December 11, 1993): A6.
3. *Science News* 137 (February 3, 1990): 70.
4. *Newsweek* (February 1, 1993): 59.
5. I am indebted to Kristen Fondren, Marjorie Manley, and Barry Persh for these data.
6. *Science News* 141 (May 23, 1992): 351.
7. *Newsweek* (March 1, 1993): 28.
8. *Newsweek* (February 14, 1994): 42.
9. I am indebted to Lisa Hill and Julie McCalden for these data.
10. R. Benedetto, "Poll: Most view GOP favorably," *USA Today* 13 (December 2, 1994): 7A.
11. H. L. Needleman, A. Schell, D. Bellinger, et al., "The long-term effect of exposure to low doses of lead in childhood," *New England Journal of Medicine* 322 (1990): 83–88.

8 Comparing Two Proportions

MOST quantitative questions are relative. "How big is it?" Compared with what? "How effective is it?" Compared with what? "How fast does it travel?" Compared with what? The previous two chapters dealt with a single proportion. In the observational study of Exercises 6.11 and 7.11, of 25 burrowing owl nests lined with dung, two were raided by badgers. Is 2 big, small, or typical? Whether lining with dung wards off badger raids cannot be addressed without a comparison group. Perhaps the same number of nests would have been raided had these 25 nests not been lined. Perhaps more. Perhaps fewer.

To have a comparison proportion in Exercise 7.11, we assumed that half of all unlined nests are raided. The probability that lining helps is then the probability that the population proportion of raids among lined nests is less than $\frac{1}{2}$. But assuming half of unlined nests are raided is arbitrary and artificial. No one knows whether this is the appropriate comparison, not even biologist Green. Recognizing the need for a **control,** he also observed 24 unlined owl nests and found that 13 of the 24 unlined nests were raided by badgers. In view of these data, what is the probability that lining is effective? Now there are two unknown proportions to be modeled: raids among lined nests and raids among unlined nests. Problems with two proportions are the focus of this and the next chapter. The development and calculations in this chapter are similar to those of Chapter 6, but now for two proportions. In Chapter 9, we will consider densities, extending Chapter 7 to the case of two proportions.

Addressing questions regarding two proportions is more difficult than testing the hypothesis that a population proportion is a particular value, such as $\frac{1}{2}$. Now we need two models, one for the population of interest and the other for the control. Both populations are unknown, but we have partial evidence from both: two of 25 lined nests raided and 13 of 24 unlined nests raided. One population is the treatment group (say, lined nests) and the other is the comparison group or control group (unlined nests). In the burrowing owl example, as in other two-sample problems, we will compare treatment and control populations on the basis of samples from both.

8.1 Two Population Models

Bayes' rule works for two populations the same as for one: Multiply likelihood by prior probabilities to give posterior probabilities. Now models are *pairs* of proportions. Some of the ideas in this chapter are developed using the malaria/sickle-cell study of Example 1.4. The first example finds likelihoods. Subsequent examples deal with prior probabilities and multiplying them by likelihoods.

EXAMPLE 8.1
▷ **Malaria and sickle cells: Likelihoods.** In the malaria/sickle-cell study[1] (discussed in Examples 1.4 and 3.19), 15 sickle-cell carriers and 15 noncarriers were injected with malaria parasites. A carrier is someone who is heterozygotic, that is, someone who has inherited one copy of the sickle-cell gene—from either parent, but not from both parents. A noncarrier is someone who does not have a sickle-cell gene (that is, someone who has not inherited the gene from either parent). The data D_C (for Carrier) are: 2 of the 15 were cases—that is, developed malaria. The data D_N (for Noncarrier) are: 14 of the 15 were cases. Are sickle-cell carriers partially protected from malaria? What is the probability that sickle-cell carriers are protected? (For the purposes of this example, ignore problems with observational studies as described in Chapter 2. However, the possibility of selection bias—choosing weaker-looking subjects for the noncarrier group than for the carrier group, say—and other biases would lead me to somewhat discount the strength of conclusions from this study.)

As usual, the likelihood of a population model is the probability of the observed data assuming that model. For exchangeable observations containing s successes and f failures, the likelihood for a population proportion, p, is

$$P(\text{Data} \mid p) = p^s(1-p)^f$$

Now there are two p's, one for each population: p_C is the population proportion of malaria among carriers and p_N is the population proportion among noncarriers. The data D_C are $s = 2$ and $f = 13$, and data D_N are $s = 14$ and $f = 1$. The likelihoods for p_C and p_N are

$$P(D_C \mid p_C) = p_C^2(1 - p_C)^{13} \quad \text{and} \quad P(D_N \mid p_N) = p_N^{14}(1 - p_N)$$

As in Chapter 6, consider 11 possibilities for p_C: 0, .1, .2, .3, .4, .5, .6, .7, .8, .9, and 1. Consider the same 11 possibilities for p_N. Table 8.1 and the bar charts in Figure 8.1 (page 240) give the likelihoods for the 11 models. In the tables, think of $p = p_C$ for $P(D_C \mid p)$ and $p = p_N$ for $P(D_N \mid p)$. The left-hand portion of the table shows the results of using a calculator and the right-hand portion drops the numbers that are too small to matter in the sequel—and that disappear from the bar charts.

Table 8.1

Calculator results for Likelihoods for Carrier and Noncarrier models			Simpler version of Likelihoods for Carrier and Noncarrier models		
p	$P(D_C \mid p)$	$P(D_N \mid p)$	p	$P(D_C \mid p)$	$P(D_N \mid p)$
0	0	0	0	0	0
.1	2.54E−3	9.00E−15	.1	2.54E−3	0
.2	2.20E−3	1.31E−10	.2	2.20E−3	0
.3	8.72E−4	3.35E−8	.3	8.72E−4	0
.4	2.09E−4	1.61E−6	.4	2.09E−4	0
.5	3.05E−5	3.05E−5	.5	3.05E−5	3.05E−5
.6	2.42E−6	3.14E−4	.6	2.42E−6	3.14E−4
.7	7.81E−7	2.04E−3	.7	0	2.04E−3
.8	5.24E−10	8.80E−3	.8	0	8.80E−3
.9	8.10E−14	2.29E−2	.9	0	2.29E−2
1	0	0	1	0	0

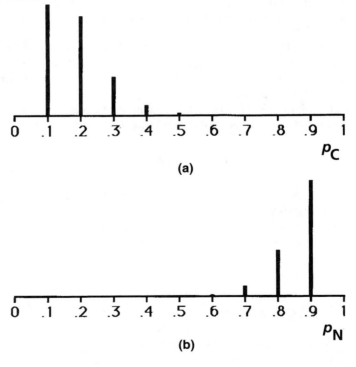

Figure 8.1 Likelihoods for models in Example 8.1 from Table 8.1: (a) Carrier; (b) Noncarrier

There are 11 models for each proportion, or $11 \times 11 = 121$ combinations for the two proportions. Each of these 121 pairs of models is itself a model, one specifying the characteristics of both populations. These 121 possibilities are shown as dots in Figure 8.2. Three regions are highlighted by using boxes in the figure. The 11 dots in the vertical box correspond to $p_C = .3$. The 11 dots enclosed in the horizontal box have $p_N = .9$. These two boxes intersect at the (single) model $p_C = .3$, $p_N = .9$.

The diagonal on this figure (the dots connected by a narrow, diagonal box) has special significance. The proportions are equal for each of the 11 dots along this diagonal: $p_C = p_N$. For such models, a carrier is as likely to get the disease as is a noncarrier. So, if any one of these models is correct, sickle-cell carriers are not protected from malaria. The 55 dots above this diagonal correspond to the sickle-cell gene protecting against malaria. (In my discussion of this study, protecting against malaria is restricted to the setting of the study—that is, when malaria is injected into a person's bloodstream using a needle. Extrapolating to a mosquito instead of a needle seems reasonable, but it is an extrapolation.)

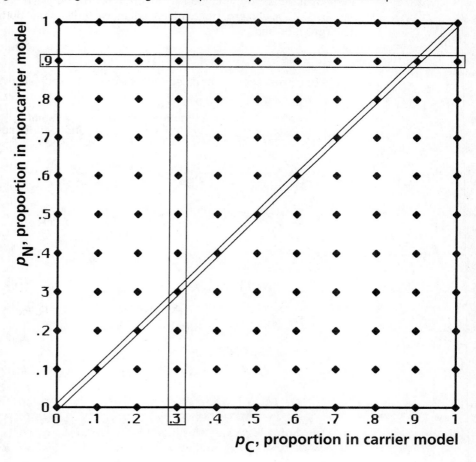

Figure 8.2 Diagram showing the 121 possible pairs of models for Example 8.1

We know the likelihoods for proportions p_C and p_N separately—namely, $p_C^2(1 - p_C)^{13}$ and $p_N^{14}(1 - p_N)$. We need likelihoods for pairs of models, that is, joint likelihoods for p_C and p_N. As usual, the likelihood of a model is the probability of observing the data assuming that model. So we need to calculate the probabilities of observing *both* D_C and D_N for each possible pair of p_C and p_N.

Consider just the first two subjects in the experiment. If they were in the same group—both carriers or both noncarriers—then we would assume independence (for any particular values of p_C or p_N) and multiply probabilities. For example, if they were noncarriers and both got cases of malaria, we would calculate the likelihood as p_N^2. Or if both were carriers and the first was a case but the second was not, then we would calculate the likelihood as $p_C(1 - p_C)$. We will also assume independence across samples. (This assumption would not be appropriate if there were connections between the two samples, as, for example, in paired experiments.)

Suppose the first subject is a carrier and the second is not, and suppose the first is a case and the second is not. Then the likelihood is $p_C(1 - p_N)$. All four possibilities are shown on the tree diagram in Figure 8.3. Not only will we assume that the first members of the two groups are independent, we will assume both samples are independent (given p_C and p_N).

Figure 8.3 Tree diagram for independence of first two subjects in Example 8.1

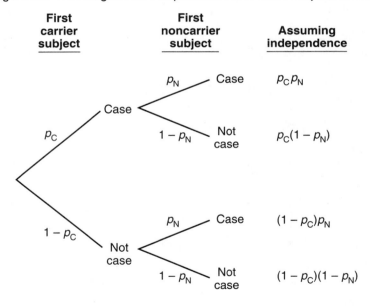

Multiply likelihoods for independent samples.

Sec. 8.1 / Two Population Models

Assuming independence of the two samples given p_C and p_N means that to calculate the likelihoods at the various pairs (p_C, p_N), multiply the individual likelihoods shown in Table 8.1. The likelihood of data D_C evaluated at $p_C = .3$ is $8.72E-4$ (using three-decimal accuracy). The likelihood of data D_N evaluated at $p_N = .9$ is $2.29E-2$. The product is $8.72E-4 \times 2.29E-2 = 2.00E-5 = 200E-7$, or .0000200. This is shown (in boldface type) in Table 8.2 along with the other 120 joint likelihoods. To make things less cumbersome, I moved the decimal place over 7 places in the body of the table and rounded off. So an entry less than $5E-8$ becomes 0. The bottom row and the last column repeat the likelihoods for the individual samples, carrier and noncarrier, respectively. So the entries in the body of the table are simply the product of the bottom row and rightmost column. (Check this out for $p_C = .3$ and $p_N = .9$ and for a few other combinations.)

Table 8.2
Likelihoods for Example 8.1: Noncarrier model p_N vs Carrier model p_C
(factor of $E-7$ dropped from entries in body of table)

p_N	0	.1	.2	.3	.4	.5	.6	.7	.8	.9	1	Sum
1	0	0	0	0	0	0	0	0	0	0	0	0
.9	0	582	503	**200**	48	7	1	0	0	0	0	$2.29E-2$
.8	0	224	193	77	18	3	0	0	0	0	0	$8.80E-3$
.7	0	52	45	18	4	1	0	0	0	0	0	$2.04E-3$
.6	0	8	7	3	1	0	0	0	0	0	0	$3.14E-4$
.5	0	1	1	0	0	0	0	0	0	0	0	$3.05E-5$
.4	0	0	0	0	0	0	0	0	0	0	0	0
.3	0	0	0	0	0	0	0	0	0	0	0	0
.2	0	0	0	0	0	0	0	0	0	0	0	0
.1	0	0	0	0	0	0	0	0	0	0	0	0
0	0	0	0	0	0	0	0	0	0	0	0	0
Sum	0	$2.54E-3$	$2.20E-3$	$8.72E-4$	$2.09E-4$	$3.05E-5$	$2.42E-6$	0	0	0	0	

The table entries are shown in the bar chart in Figure 8.4 (page 244). This figure shows three dimensions: one dimension each for the carrier and noncarrier models (p_C and p_N) and the other for the bar heights. The square in Figure 8.2 has only the first two dimensions; think of knocking it over into a third dimension and erecting bars over the dots.

The evidence in the experiment contributes to the posterior probabilities through the likelihoods. Subsequent examples will address this process. The substance of these later examples is the assessment of prior probabilities and multiplying them by the likelihoods. The likelihoods used will be the ones we calculated in this example.

Figure 8.4 Likelihoods for Example 8.1 (bar chart of Table 8.2)

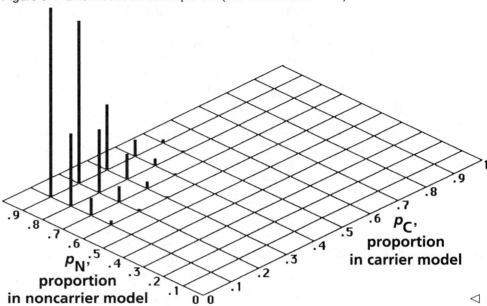

Throughout this chapter we will assume that the two samples are independent (given the two population proportions). Assuming independence is not always appropriate. For example, it is not appropriate for studies with paired designs, but it may be appropriate for parallel designs—see Section 3.5.

You should worry about the assumption of independence, even for parallel designs. But checking for deviations from independence is difficult. It is not enough that the two groups seem unrelated. A critical concern is how the experiment was conducted and how the results were obtained and assessed. If there is bias in assigning treatment, then independence may be violated. For example, suppose the first carrier happened to come down with malaria and *for that reason* the investigator selected someone sickly as the first noncarrier. Also, the matter of deciding whether someone has a case of malaria may not be clear. An investigator might be persuaded to decide that the first noncarrier was a case in part because the first carrier was a case. In general, you should judge whether a study was conducted fairly and honestly. Look for the possibility of biases. If the possibility of biases exists, you may decide to discount the study. For this reason and for other reasons, whenever possible the individual who assesses the results of an experiment should be blinded as to treatment and as to the other responses.

We have assumed independence only when we calculate probabilities of observations for fixed population proportions—p_C and p_N in the example. In general, information about one proportion may give information about the other. So even though the samples are independent for fixed population proportions, they may be dependent when not given the population proportions.

The next two examples consider different prior probability assignments for the

Sec. 8.1 / Two Population Models   245

malaria/sickle-cell question. The assessor in the first example maintains an open-minded position regarding any relationship between susceptibility to malaria and being a sickle-cell carrier; the second uses information from sources other than the experiment in question.

EXAMPLE 8.2
▷ **Malaria and sickle cells—open-minded assessor.** The open-minded assessor—Opey, for short—is similar to the Dr. X considered in Chapter 6 in regarding all possible models as equally likely. As indicated in the previous example, now there are 121 models—one for each of the cells in Table 8.2 or one for each of the dots in Figure 8.2. Opey assigns probability $\frac{1}{121}$ to each. This is shown in Table 8.3 and Figure 8.5 (page 246). The table and figure have the same schemes as those showing likelihoods in the previous example.

Opey's prior probability that sickle-cell carriers are protected from malaria (when malaria parasites are injected into the bloodstream, as in the experiment) is the sum of the probabilities above the diagonal. There are 55 cells with this characteristic, and so this probability is $\frac{55}{121} = \frac{5}{11}$, or about 45%. Similarly, Opey's probability that sickle-cell carriers are at greater risk than noncarriers is $\frac{55}{121}$. The remaining probability of $\frac{11}{121} = \frac{1}{11}$ is on the diagonal and corresponds to equality of p_C and p_N; that is, sickle-cell heterozygosity and malaria are unrelated. (If you think these implications are unrealistic, Infy's opinions in the next example may be more to your liking.)

Table 8.3
Opey's prior probabilities in Example 8.2:
Noncarrier model p_N vs Carrier model p_C

p_N	\multicolumn{11}{c}{p_C}	Sum
	0	.1	.2	.3	.4	.5	.6	.7	.8	.9	1	Sum
1	1/121	1/121	1/121	1/121	1/121	1/121	1/121	1/121	1/121	1/121	1/121	1/11
.9	1/121	1/121	1/121	1/121	1/121	1/121	1/121	1/121	1/121	1/121	1/121	1/11
.8	1/121	1/121	1/121	1/121	1/121	1/121	1/121	1/121	1/121	1/121	1/121	1/11
.7	1/121	1/121	1/121	1/121	1/121	1/121	1/121	1/121	1/121	1/121	1/121	1/11
.6	1/121	1/121	1/121	1/121	1/121	1/121	1/121	1/121	1/121	1/121	1/121	1/11
.5	1/121	1/121	1/121	1/121	1/121	1/121	1/121	1/121	1/121	1/121	1/121	1/11
.4	1/121	1/121	1/121	1/121	1/121	1/121	1/121	1/121	1/121	1/121	1/121	1/11
.3	1/121	1/121	1/121	1/121	1/121	1/121	1/121	1/121	1/121	1/121	1/121	1/11
.2	1/121	1/121	1/121	1/121	1/121	1/121	1/121	1/121	1/121	1/121	1/121	1/11
.1	1/121	1/121	1/121	1/121	1/121	1/121	1/121	1/121	1/121	1/121	1/121	1/11
0	1/121	1/121	1/121	1/121	1/121	1/121	1/121	1/121	1/121	1/121	1/121	1/11
Sum	1/11	1/11	1/11	1/11	1/11	1/11	1/11	1/11	1/11	1/11	1/11	1

Adding the probabilities in Table 8.3 in either direction (across the rows or down the columns) always gives $\frac{11}{121} = \frac{1}{11}$, the entries in the last column and in the bottom row. Therefore, the probabilities of the models within the groups separately are equal for Opey. In addition, each entry in the body of the table is the product of the corresponding value in the bottom row and the last column: $(\frac{1}{11})(\frac{1}{11}) = \frac{1}{121}$. So the two models are independent for Opey.

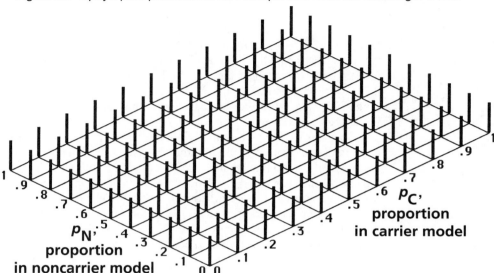

Figure 8.5 Opey's prior probabilities for Example 8.2: Each bar has height 1/121.

Now calculate Opey's posterior probabilities. We calculated likelihoods in the previous example. All that remains is to multiply the likelihoods by Opey's prior probabilities. It helps that all Opey's prior probabilities are the same: The products are proportional to the likelihoods. Figure 8.6 shows the posterior probabilities. This figure is identical with the likelihood figure from Example 8.1, except that the heading is changed. The units for the bar heights would be different, but this dimension has not been labeled in either picture—units matter only for the posterior probability version in which the sum of the bar heights must be 1.

Since probabilities add to 1, finding Opey's posterior probabilities means adding the entries in the likelihood table and dividing each entry by the sum. This gives the probability table, Table 8.4. The entry in the lower right corner of the table reminds you that the overall sum is 1.

From this table we can calculate probabilities of various events—for example, the probability that sickle-cell carriers are protected from malaria. Protection from malaria corresponds with cells above the diagonal: $p_N > p_C$. Summing the probabilities above the diagonal gives 1.000. So there is probability 0.000 on and below the diagonal. These calculations use three-decimal accuracy. Using six-decimal accuracy (calculations not shown), the total probability on or below the diagonal is .000012. Opey's probability that sickle-cell carriers are protected has increased to very nearly 1—up from $\frac{5}{11}$—before seeing the results of the study.

Sec. 8.1 / Two Population Models

Figure 8.6 Opey's posterior probabilities in Example 8.2 (picture same as for likelihoods)

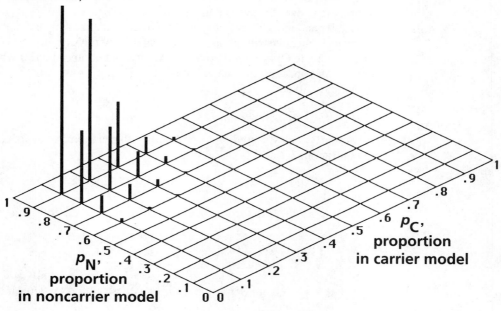

Table 8.4
Opey's posterior probabilities in Example 8.2 (proportional to likelihoods):
Noncarrier model p_N vs Carrier model p_C

p_N	___				p_C						___	Sum
	0	.1	.2	.3	.4	.5	.6	.7	.8	.9	1	
1	0	0	0	0	0	0	0	0	0	0	0	0
.9	0	.292	.252	.100	.024	.004	0	0	0	0	0	.672
.8	0	.112	.097	.039	.009	.001	0	0	0	0	0	.258
.7	0	.026	.022	.009	.002	0	0	0	0	0	0	.060
.6	0	.004	.004	.001	0	0	0	0	0	0	0	.009
.5	0	0	0	0	0	0	0	0	0	0	0	.001
.4	0	0	0	0	0	0	0	0	0	0	0	0
.3	0	0	0	0	0	0	0	0	0	0	0	0
.2	0	0	0	0	0	0	0	0	0	0	0	0
.1	0	0	0	0	0	0	0	0	0	0	0	0
0	0	0	0	0	0	0	0	0	0	0	0	0
Sum	0	.434	.376	.149	.036	.005	0	0	0	0	0	1

Pick an entry in the table—say, .100 for $p_C = .3$ and $p_N = .9$. (There is nothing special about this pair; we simply need an example.) The probability of $p_C = .3$ is .149, the entry in the bottom row for $p_C = .3$. The probability of $p_N = .9$ is .672, which is the entry in the last column for $p_N = .9$. Multiply these together. You will get the entry you started with (except for round-off error)—in our case .100. The same multiplicative property holds for every entry in the table. That means Opey's posterior distribution is independent. This is true generally: Independence a priori implies independence a posteriori whenever likelihoods are calculated by multiplying.

To calculate Opey's predictive probability that another carrier who is injected will contract malaria, multiply the column totals (bottom row) by the corresponding proportion, p_C, and add:

$$(0 \times 0) + (.434 \times .1) + (.376 \times .2) + (.149 \times .3) + \cdots = .180$$

Similarly, Opey's predictive probability that another noncarrier who is injected will contract malaria is

$$(0 \times 0) + (.672 \times .9) + (.258 \times .8) + (.060 \times .7) + \cdots = .859 \triangleleft$$

An important lesson from the previous example is the following:

> **Rule of Independent Models**
>
> **For independent samples—one from each of two populations—if the population models are independent a priori, then they are independent a posteriori.**

This conclusion does not apply in the next example because the models are not independent according to the prior probabilities. You will see other ways in which the informed assessor of the next example differs from Opey.

EXAMPLE 8.3
▷ **Malaria and sickle cells—informed assessor.** In the *Natural History* article[1] referred to previously, Jared Diamond gives several reasons for thinking that sickle-cell carriers are protected from malaria. All involve observational studies. I will summarize them here.

The first clue to a possible association between the sickle-cell gene and malaria occurred in 1946 when a physician in Rhodesia noticed that the blood of children who were sickle-cell heterozygotes contained fewer malaria parasites. This evidence was presumably the motivation for the study we are analyzing; it gave the designers of the study a reasonably high prior probability that sickle-cell carriers were partially protected.

Sickle-cell genes occur naturally in the world only where malaria is common. Diamond says that, in some parts of Africa, as much as 40% of the population carry the gene. (The slave trade brought the gene to the United States, where it still occurs in about 10% of blacks.) This association might suggest that sickle-cell carriers are *more* susceptible to malaria. Here is why the opposite conclusion is more reasonable. Sickle-cell carriers have an evolutionary disadvantage. If two sickle-cell heterozygotes mate, one-quarter of their offspring are sickle-cell homozygotes and die from sickle-cell anemia. This is enough of a disadvantage that sickle-cell genes would eventually disappear. Indeed, it has disappeared from most populations. In those populations where it persists, there must be a compensating evolutionary advantage. So a perfect geographic correlation with malaria suggests that resistance to malaria may be its advantage. This makes sense to our well-informed assessor named Infy, so the possibility that sickle-cell genes provide protection against malaria becomes more believable to Infy.

Diamond says that in malarial regions of Africa, children under 5 years old are "far more likely to die" of malaria if they lack the sickle-cell gene. If true, this settles the issue once and for all. But Diamond does not say what the evidence is for this statement. Is it anecdotal? Is it a comprehensive study? Infy is well informed about malaria and sickle-cell genetics, but Infy is also well informed about commonly held notions in science generally: Sometimes they are right and sometimes they are not. So Infy hears the statement, but, without seeing the evidence, it has little effect on Infy's opinion.

Diamond says that in Nigeria (one of the malarial regions), the fraction of the population carrying the sickle-cell gene increases from 24% at birth to 29% at age 5. The implication is that the noncarriers die off at a faster rate in these early years. This is obviously based on a study. Infy would like to see the actual data and learn how they were collected, but gives some credence to the implication on the basis of Diamond's report.

All the evidence cited above is circumstantial. Diamond gives one piece of direct evidence: a possible protective mechanism. To infect their host with the disease, malaria parasites have to go through a stage in which they grow inside their host's red blood cells. But malaria parasites trigger cell sickling when they infect the red blood cells of sickle-cell heterozygotes. Since sickled cells are usually destroyed by the patient's body, this could have the effect of ridding the body of some of the malaria parasites trying to grow within the cells. Furthermore, he says, the "infected cells that survive provide a poor environment for the parasite's further growth."

Infy now has a lot of information about the relationship between sickle-cell genes and malaria, but not much about how frequently someone injected with malaria will contract the disease—whether carrier or not. So while Infy's prior probability distribution will not be as diffuse as Opey's, it will still be reasonably spread out.

Infy thinks it likely that sickle-cell genes provide protection against malaria. But Infy has some experience in the world and has often heard reasonable-sounding explanations for false hypotheses! Even if the protection hypothesis is true, it is far from clear that the experimental circumstances of injection into the bloodstream imitate being bitten by a malaria-carrying mosquito. So Infy associates 25% probability with there being no relationship between sickle-cell genes and malaria; this is the total probability on the diagonal of Table 8.5 (page 250). On the other hand, it seems unlikely that sickle-cell carriers have an *increased* chance of contracting malaria. So Infy associates only about 5% probability below the diagonal. Infy's prior probabilities are shown in the table. These are Infy's probabilities, but there is nothing natural or compelling about them. You may well have very different prior probabilities. As is true generally, the row and column totals give probabilities for the individual models—the row sums for the noncarrier model and the column sums for the carrier model.

Table 8.5
Infy's prior probabilities for Example 8.3:
Noncarrier model p_N vs Carrier model p_C

p_N	p_C										Sum
	0	.1	.2	.3	.4	.5	.6	.7	.8	.9	1	
1	.040	.040	.040	.030	.021	.010	.008	.006	.004	.001	.040	.240
.9	.040	.034	.034	.032	.016	.010	.008	.006	.004	.030	.001	.215
.8	.040	.034	.034	.032	.020	.010	.008	.006	.024	.001	.001	.210
.7	.010	.020	.020	.012	.005	.005	.005	.020	.001	.001	.001	.100
.6	.004	.006	.006	.003	.002	.005	.020	.001	.001	.001	.001	.050
.5	.002	.002	.002	.002	.002	.020	.001	.001	.001	.001	.001	.035
.4	.001	.001	.001	.001	.020	.001	.001	.001	.001	.001	.001	.030
.3	.001	.001	.001	.020	.001	.001	.001	.001	.001	.001	.001	.030
.2	.001	.001	.020	.001	.001	.001	.001	.001	.001	.001	.001	.030
.1	.001	.020	.001	.001	.001	.001	.001	.001	.001	.001	.001	.030
0	.020	.001	.001	.001	.001	.001	.001	.001	.001	.001	.001	.030
Sum	.160	.160	.160	.135	.090	.065	.065	.045	.040	.040	.050	1

Using Bayes' rule, we update Infy's prior distribution, multiplying it by the likelihoods of Example 8.1. For your convenience, these likelihoods are reproduced in Table 8.6. Remember that likelihoods are not probabilities of models and so their sum is irrelevant. The row and column sums have now been deleted from the table.

Table 8.6
Likelihoods for Examples 8.1 and 8.3:
Noncarrier model p_N vs Carrier model p_C

p_N	p_C										
	0	.1	.2	.3	.4	.5	.6	.7	.8	.9	1
1	0	0	0	0	0	0	0	0	0	0	0
.9	0	582	503	200	48	7	1	0	0	0	0
.8	0	224	193	77	18	3	0	0	0	0	0
.7	0	52	45	18	4	1	0	0	0	0	0
.6	0	8	7	3	1	0	0	0	0	0	0
.5	0	1	1	0	0	0	0	0	0	0	0
.4	0	0	0	0	0	0	0	0	0	0	0
.3	0	0	0	0	0	0	0	0	0	0	0
.2	0	0	0	0	0	0	0	0	0	0	0
.1	0	0	0	0	0	0	0	0	0	0	0
0	0	0	0	0	0	0	0	0	0	0	0

The product of the two tables (prior and likelihood) is shown in Table 8.7, with the decimal points dropped (in effect, every entry is multiplied by 10). For example, consider $p_C = .3$ and $p_N = .9$. The product of the entries from the two tables is $.032 \times 200 = 6.4$ (to two-decimal accuracy); the entry in the table is 64. This table of products is Infy's posterior distribution, except that the sum is not right. To make the sum equal 1, add the entries—either all 121 entries in the table or the row totals or the column totals. (Remember that I carry more digits than I advertise. If you add the numbers in the table, you will get 634—but 633.3 is more accurate.) Now divide each entry by the sum. In the example of $p_C = .3$ and $p_N = .9$, $\frac{638}{6,333} = .101$. The final table of Infy's posterior probability is shown in Table 8.8. Just as for Opey in Example 8.2, Infy's probability above the diagonal is 1.000.

Table 8.7
Products of previous two tables (applying Bayes' rule) in Example 8.3:
Noncarrier model p_N vs Carrier model p_C

p_N					p_C							Sum
	0	.1	.2	.3	.4	.5	.6	.7	.8	.9	1	
1	0	0	0	0	0	0	0	0	0	0	0	0
.9	0	198	171	64	8	1	0	0	0	0	0	441
.8	0	76	66	25	4	0	0	0	0	0	0	170
.7	0	10	9	2	0	0	0	0	0	0	0	22
.6	0	1	0	0	0	0	0	0	0	0	0	1
.5	0	0	0	0	0	0	0	0	0	0	0	0
.4	0	0	0	0	0	0	0	0	0	0	0	0
.3	0	0	0	0	0	0	0	0	0	0	0	0
.2	0	0	0	0	0	0	0	0	0	0	0	0
.1	0	0	0	0	0	0	0	0	0	0	0	0
0	0	0	0	0	0	0	0	0	0	0	0	0
Sum	0	285	246	90	12	1	0	0	0	0	0	633

Table 8.8
Infy's posterior probabilities in Example 8.3 (Entries of Table 8.7 divided by 633):
Noncarrier model p_N vs Carrier model p_C

p_N					p_C							Sum
	0	.1	.2	.3	.4	.5	.6	.7	.8	.9	1	
1	0	0	0	0	0	0	0	0	0	0	0	0
.9	0	.312	.270	.101	.012	.001	0	0	0	0	0	.696
.8	0	.120	.104	.039	.006	0	0	0	0	0	0	.269
.7	0	.016	.014	.003	0	0	0	0	0	0	0	.034
.6	0	.001	.001	0	0	0	0	0	0	0	0	.002
.5	0	0	0	0	0	0	0	0	0	0	0	0
.4	0	0	0	0	0	0	0	0	0	0	0	0
.3	0	0	0	0	0	0	0	0	0	0	0	0
.2	0	0	0	0	0	0	0	0	0	0	0	0
.1	0	0	0	0	0	0	0	0	0	0	0	0
0	0	0	0	0	0	0	0	0	0	0	0	0
Sum	0	.449	.388	.143	.018	.002	0	0	0	0	0	1

To calculate Infy's predictive probability that another carrier would contract malaria if injected, multiply the column totals (bottom row) by the corresponding proportion, p_C, and add:

$$(0 \times 0) + (.449 \times .1) + (.388 \times .2) + (.143 \times .3) + \cdots = .174$$

Similarly, Infy's predictive probability that another noncarrier would contract malaria if injected is

$$(0 \times 0) + (.696 \times .9) + (.269 \times .8) + (.034 \times .7) + \cdots = .866$$

These are not very different from Opey's predictive probabilities of .180 and .859. ◁

Opey's posterior probabilities in Example 8.2 and Infy's in Example 8.3 are quite similar. Both are based on the same data. For both, the data weigh more heavily than does their prior information. Anyone who is reasonably open-minded will come to about the same conclusions as Opey and Infy.

We have seen that the study results were very convincing, even to an open-minded assessor. As indicated in Example 1.3, Jared Diamond thinks the study that we have been analyzing was too small to be conclusive. (He does make it clear though that he is not suggesting that the study should have been bigger, since he too says it was unethical.) Taken together, the reasons cited in Example 8.3 are sufficient to convince Diamond that sickle-cell carriers are at least partially protected from malaria, even without the results of the study. Diamond's research and arguments are a wonderful example of science. But the evidence provided by the study is in fact stronger than he appreciates.

Examples 8.2 and 8.3 address the question of whether sickle-cell carriers are partially protected from malaria. Generally, in comparing a treatment with a control, it is important to know whether the treatment is better. But it may also be important to know how much better it is. As a measure of effectiveness (or protectiveness, in the case of sickle cells), we will consider the difference—call it d—between the treatment (T) and control (C) success proportions: $d = p_T - p_C$. If d is positive, then the treatment is beneficial; if d is negative then the treatment is detrimental. The value of d measures the level of benefit. For example, $d = .3$ means that the treatment is 30 percentage points better than the control.

Sometimes, it will not be clear which group is the "treatment." When calculating differences, it is important to say which is being subtracted from which. Also, it may not be clear that a "success" is beneficial. In the malaria example, we are interested in the proportions of cases of malaria in the two groups. In that example, we want d to represent the benefit of being a carrier. The smaller p_C is in comparison with p_N, the better it is to be a carrier in the sense of protecting from malaria. So we will have $d = p_N - p_C$. If $d > 0$, then $p_N > p_C$, which means that being a carrier is protective.

We will calculate **P**robabilities that **d** is **A**t **L**east as big as .1, .2, and so on, calling these PdAL.1, PdAL.2, and so on.

> The difference between the treatment and control success proportions is
>
> $$d = p_T - p_C$$
>
> The probability that d is at least x is PdALx.

The next example is an illustration using Opey's and Infy's posterior probabilities.

EXAMPLE 8.4
▷ **Difference in malaria rate due to sickle cells.** How effectively does the sickle-cell gene protect its carriers from malaria? Consider Opey's posterior probability distribution—reproduced in Figure 8.7 from Figureshading added to the region in which p_N is larger than p_C by at least .6.

Opey's probability that the difference equals .6 is the sum of the probabilities for the six cells on the diagonal for which $d = p_N - p_C = .6$. This includes the point $p_C = .3$ and $p_N = .9$ that we have considered previously, since $p_N - p_C = .9 - .3 = .6$. The probability of d equal to .6 is $0 + .026 + .097 + .100 + 0 = .223$.

Opey's probability that d is *at least* .6 is the sum of the probabilities for the 15 cells on or above the $d = p_N - p_C = .6$ diagonal. These are shown in boldface type in Table 8.9. The region with $d \geq .6$ is shaded in Figure 8.7. Adding up the bars in the shaded region, I find that Opey's posterior probability that d is at least .6, or PdAL.6, is $.292 + .252 + .100 + .112 + .097 + .026 = .879$.

The difference d is negative when p_C is bigger than p_N—the lower right half of the table. This half contains 0 probability (to three-decimal accuracy). Since all the probability is in the upper left half of the table, PdAL0 = 1.

Table 8.9
Opey's posterior probabilities in Example 8.2, with region of difference having d at least as large as .6 shown in boldface: Noncarrier model p_N vs Carrier model p_C

p_N					p_C							Sum
	0	.1	.2	.3	.4	.5	.6	.7	.8	.9	1	
1	0	0	0	0	0	0	0	0	0	0	0	0
.9	0	**.292**	**.252**	**.100**	.024	.004	0	0	0	0	0	.672
.8	0	**.112**	**.097**	.039	.009	.001	0	0	0	0	0	.258
.7	0	**.026**	.022	.009	.002	0	0	0	0	0	0	.060
.6	0	.004	.004	.001	0	0	0	0	0	0	0	.009
.5	0	0	0	0	0	0	0	0	0	0	0	.001
.4	0	0	0	0	0	0	0	0	0	0	0	0
.3	0	0	0	0	0	0	0	0	0	0	0	0
.2	0	0	0	0	0	0	0	0	0	0	0	0
.1	0	0	0	0	0	0	0	0	0	0	0	0
0	0	0	0	0	0	0	0	0	0	0	0	0
Sum	0	.434	.376	.149	.036	.005	0	0	0	0	0	1

Figure 8.7 Opey's posterior probabilities showing region where *d* is at least .6

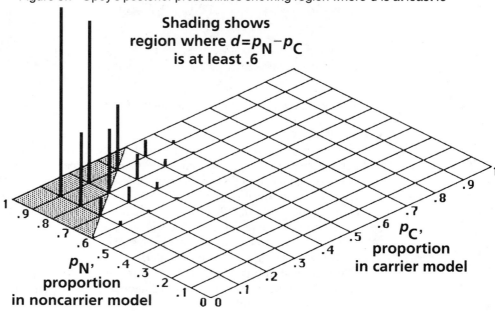

Table 8.10 shows Opey's PdALx for various values of x. It also shows Infy's PdALx, where we used Infy's posterior probabilities from Example 8.3 and proceeded just as we did for Opey's. We have already calculated the PdAL.1 for both Opey (Example 8.2) and Infy (Example 8.3), since PdAL.1 is the probability that sickle-cell carriers are protected from malaria. The table indicates that the level of protection afforded by sickle-cell heterozygosity is likely to be at least 60% (for both Opey and Infy), but that it is not likely to be greater than 80%.

Table 8.10
Posterior PdAL for Opey and Infy for various differences

x	0	.5	.6	.7	.8	.9
Opey's PdALx	1.000	.968	.879	.656	.292	0.000
Infy's PdALx	1.000	.988	.922	.702	.312	0.000

Figure 8.8 shows each PdALx as a smooth curve. Neither the graph nor the table shows PdALx for x less than 0; since PdAL0 = 1 for both Opey and Infy, PdALx also equals 1 when x is negative. The two curves in this figure are quite close to each other. This reflects the similarity between Opey's and Infy's posterior probability distributions, a similarity I have mentioned before.

Figure 8.8 Comparison of PdAL for Opey and Infy, where $d = p_N - p_C$

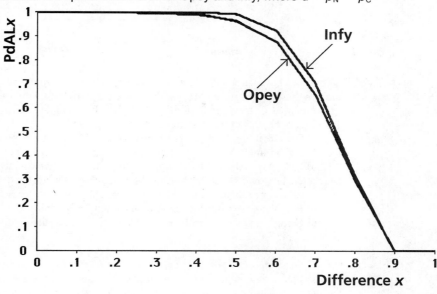

The following steps summarize finding posterior probabilities when the samples are independent:

1. Decide on the set of possible models for each population. (What are the possible proportions in each population?)
2. Make a table of prior probabilities for the various pairs of models. (In the open-minded case, make them all equal.)
3. Calculate likelihoods. Make a table similar to the one in step 2. Then
   (a) Calculate the likelihoods for one population; write them in the appropriate margin (last row or last column).
   (b) Calculate the likelihoods for the other population; write them in the other margin.
   (c) Multiply the margins together to obtain the body of the table. Move all the decimal places, if that is convenient (but move all the same number of places).
4. Multiply (entry by entry) the table of priors from step 2 by the table of likelihoods found in step 3c.
5. Add up the entries in the table of step 4. Divide each entry by the total. This is the table of posterior probabilities.

EXERCISES

8.1 Example 3.12 considered samples of men and women and their attitudes toward violence on television. The data are given at the top of page 256.

	Too much TV violence?	
	Yes	No
Men	8	12
Women	17	5

Suppose we have agreed that there is some population for which these can be viewed as random samples. Let p_M and p_W be the population proportions who would respond "yes" among men and women. Use Opey's prior probabilities given in Example 8.2: Assign prior probability $\frac{1}{121}$ to all 121 pairs of p_M and p_W. Use three-decimal accuracy.

(a) Find likelihoods for all 121 pairs. [*Hint:* To save time, notice that most of these likelihoods will be small. The biggest likelihood is for $p_M = .4$, $p_W = .8$. Set a likelihood to 0 when it is 1,000 times smaller than the biggest one—that is, when divided by the bigger likelihood, the result is less than .001. This hint works generally, but when the point with greatest likelihood has a small prior probability, then the figure .001 should be decreased.]

(b) Find the posterior probability distribution for p_M and p_W.

(c) Using the posterior probabilities from part (b), calculate PdALx (where $d = p_W - p_M$) for $x = 0$, .3, and .6.

The next two exercises consider the same experiment. The second one asks you to analyze the results of the experiment as an open-minded assessor might (see Example 8.2) and the first has you act as an informed assessor (see Example 8.3). In the "well-informed" case, you are to formulate your own prior opinions and do so without knowing the results of the experiment. [As indicated by the principle of separation of prior information and current data of Chapter 5 (page 153), it would be incorrect to use the results first in your prior and a second time in the likelihood.] So the order given in Examples 8.2 and 8.3 has been reversed. Do not look at the results of the experiment until you formulate your prior probabilities in Exercise 8.2, part (a). However, after you have answered part (a) of Exercise 8.2, you may want to go to Exercise 8.3 before returning to parts (b) and (c) of Exercise 8.2.

8.2 An experiment[2] was carried out in 1983 at the University of Michigan to evaluate a radical experimental treatment for newborn babies suffering from severe respiratory failure. A high percentage of such infants die when given conventional therapy (CVT), usually consisting of hyperventilation and drugs. The experimental therapy is extracorporeal membrane oxygenation (ECMO). This is a device that is adapted from a heart–lung machine. It takes over the lungs' function of removing carbon dioxide from the blood and adding oxygen. Its use is far from casual. A vein or an artery (or both) near the heart are cut. Then tubes are inserted to bypass the lungs until they are capable of functioning effectively, usually a matter of days.

Over a period of 9 years ending in 1983, the Michigan investigators achieved results they regarded as better than possible with CVT. They had used ECMO on 40 newborns (who weighed at least 2 kg), and 28 of them had survived. According to their report, "some patients survived who were not expected to recover," and "the risks of the treatment itself were less than the risks of the disease." With this favorable experience behind them, they felt justified in carrying out a randomized experiment to compare ECMO with CVT.

Babies were admitted to the Michigan study if they had "an 80% or greater chance of mortality despite optimal [standard] therapy." The investigators used a randomization procedure that was more likely to assign a therapy if it was performing better than the other therapy. Of 11 infants assigned to ECMO, all 11 survived. The only infant assigned to CVT died. (Most of the infants were assigned to ECMO because it performed so well.) After the study, 10 additional babies who were born at their hospital (or were referred to them) met the criteria for inclusion into the study; eight were treated with ECMO and survived, and two were treated with CVT (apparently the ECMO device was in use when they needed therapy) and both died.

Researchers at Harvard[3] thought that this evidence was not sufficient to conclude that ECMO was safe and effective. They did not think that the historical results were reliable enough to provide a comparison group, and they were concerned that only one infant in the Michigan study had been assigned to CVT. They decided to carry out another randomized study, this time making sure enough infants would be assigned to CVT.[4] The infants admitted to their study had diagnoses similar to those in the Michigan study, but the infants in the two studies may not be exchangeable.

(a) Before you read the following results of the Harvard study, you are to formulate your prior probability distribution (as Infy did in Example 8.3) for p_E and p_C, the success (or survival) proportions for ECMO and CVT. Consider these values for both proportions: 0, .1, .2, ..., 1. So you are to indicate your prior probabilities for 121 pairs of models. In assessing this prior distribution, you should use the historical information given above, including the data from the Michigan study, but do not make any formal calculations; for example, do not apply Bayes' rule using these data.

(b) The Harvard study had two phases. In the first phase, six of the 10 infants assigned to CVT survived, while all nine of the ECMO babies survived. Because it was the winner in the first phase, only ECMO was used in treating the 20 babies admitted during the second phase, and 19 of these survived. Using the prior probabilities you assessed in part (a), calculate your posterior probability distribution for p_E and p_C. Use these data: 28 ECMO infants survived and 1 died; 6 CVT infants survived and 4 died.

(c) Using your answer to part (b), calculate your PdALx where $d = p_E - p_C$ and $x = 0$, .2, .4, and .6.

8.3 Repeat Exercise 8.2 using Opey's prior probabilities given in Example 8.2: Assign prior probability $\frac{1}{121}$ to all 121 pairs of p_E and p_C. Use likelihoods (as calculated in Exercise 8.2) for these data: 28 ECMO infants survived and 1 died; 6 CVT infants survived and 4 died.

(a) Find your posterior probability distribution for p_E and p_C.

(b) Using your answer to part (a), calculate your PdALx where $d = p_E - p_C$ and $x = 0$, .2, .4, and .6.

8.2 Testing a Null Hypothesis*

This section is a continuation of Section 8.1. But the focus is on a particular aspect of the posterior distribution and on a particular type of prior distribution. It would not exist as a separate section but for a common type of question that occurs in step 1 of the scientific method. While this section is not used in the sequel, it helps to motivate a more practical and neater treatment of the issue that is addressed further by Albert.[5]

The question of interest is whether two population models have the same success proportions: $d = 0$. This is the **null hypothesis** of no difference in proportions. For example, the null hypothesis means that susceptibility to malaria is the same for sickle-cell carriers as for noncarriers. Other applications include the following: no difference between treatment and control groups, no ESP, method A has the same effect as method B. Recall from Chapter 6 that to test a null hypothesis means to find its posterior probability.

---

*Optional section; not used in the sequel.

> **Testing a null hypothesis means finding its posterior probability.**

I will show how to calculate the posterior probability of a null hypothesis using an example from an article[6] by Tversky and Gilovich in *Chance* magazine. They were interested in whether there is a "hot hand" in basketball: Do basketball players tend to shoot in streaks? One way they addressed this question was to consider free-throw shooting by the Boston Celtics players during the 1980–1981 and 1981–1982 seasons. They reasoned that if players shoot in streaks, then they would be more likely to hit the second free throw if they had hit the first than had they missed the first. But if a player's attempts are independent, then the proportions of successes should not be too different in these two circumstances. The null hypothesis is that the probability of success after a success is the same as the probability of success after a failure.

The example concerns a Celtics player named Chris Ford. I chose him because he had the fewest free-throw chances in the Tversky/Gilovich table, and so the calculations would be relatively easy. Two of the exercises consider other players: Larry Bird and M. L. Carr. While our conclusions are properly restricted to these three players, at the end of the section I will suggest that the phenomenon is quite general.

EXAMPLE 8.5

▷ **Does Chris Ford have a "hot hand"?** I will formulate my prior probabilities concerning a success proportion p_S after a successful first free throw—the "after-S model"—and another success proportion p_F after a failure—the "after-F" model. As in the examples of the previous section, I will consider models with 10 chips, so the possible proportions are one-digit decimals. However, I will cut down the number of models by ruling out the possibility that either p is 1 (perfection) or 0 (always miss). So there are nine possible values of each p and $9 \times 9 = 81$ models overall. These are shown as dots or ×'s in Figure 8.9. The ×'s indicate diagonal points, those for which the two p's are equal ($p_S = p_F = .1$, $p_S = p_F = .2$, and so on); that is, $d = p_S - p_F = 0$. This is the null hypothesis. Assigning prior probabilities means I have to replace these dots and ×'s with probabilities.

Figure 8.9 81 models for Chris Ford's free-throw shooting in Example 8.5

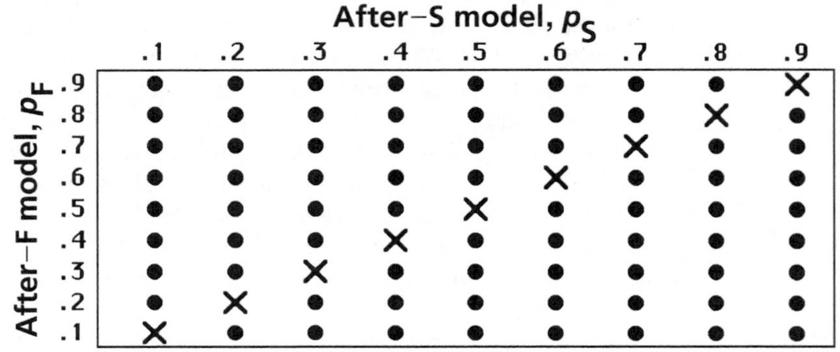

Sec. 8.2 / Testing a Null Hypothesis

What do I think about p_S and p_F for Chris Ford? A survey taken by Tversky and Gilovich among avid basketball fans may help me formulate my prior probabilities. Out of 100 fans, 68 thought it more likely that basketball players are successful after a success than after a failure. Apparently, none thought the opposite. But it seems plausible to me that a player might get overconfident after a success, or that he would try harder after a failure. Fans are just as likely to be wrong as not regarding the direction of an effect, if one is present, and so I will give equal probabilities to $p_S > p_F$ and $p_S < p_F$. Also, I have a rather higher probability of the null hypothesis of no relationship between the two throws: $\frac{5}{12}$ (about 42%) as opposed to the 32% suggested by the survey.

How should I spread out my probability $\frac{5}{12}$ on the null hypothesis points? I am quite ignorant about basketball. I give each of the nine points the *same* probability: $\frac{5}{12}$ divided by 9 is $\frac{5}{108} = .046$. (Those of you who know basketball will appreciate that I do not. After seeing the data in the Tversky/Gilovich article, I realized that knowledgeable people would assign no probability to small success rates: Professional basketball players are more likely than not to hit free throws. However, it happens that any probability assigned to such points is annihilated by the likelihood and has no effect on the posterior. So this particular kind of ignorance is indeed bliss.)

Table 8.11
My prior probabilities for Chris Ford shooting in Example 8.5:
After-failure model p_F vs After-success model p_S

p_F	p_S									Sum
	.1	.2	.3	.4	.5	.6	.7	.8	.9	
.9	.008	.008	.008	.008	.008	.008	.008	.008	.046	.11
.8	.008	.008	.008	.008	.008	.008	.008	.046	.008	.11
.7	.008	.008	.008	.008	.008	.008	.046	.008	.008	.11
.6	.008	.008	.008	.008	.008	.046	.008	.008	.008	.11
.5	.008	.008	.008	.008	.046	.008	.008	.008	.008	.11
.4	.008	.008	.008	.046	.008	.008	.008	.008	.008	.11
.3	.008	.008	.046	.008	.008	.008	.008	.008	.008	.11
.2	.008	.046	.008	.008	.008	.008	.008	.008	.008	.11
.1	.046	.008	.008	.008	.008	.008	.008	.008	.008	.11
Sum	.11	.11	.11	.11	.11	.11	.11	.11	.11	1

Now let's address the remaining 72 points, those off the diagonal. I indicated that if the null hypothesis is not true, then either effect (more likely or less likely after a success) is possible. So I will assign the remaining probability equally to these 72 points: $\frac{7}{12}$ divided by 72 is $\frac{7}{864} = .008$.

Summarizing, my prior probabilities are shown in Table 8.11 and Figure 8.10 (page 260). (Rounding off in the 81 table entries leads to a loss of accuracy in the sum of probabilities, which because of round-off error is apparently .99 and not 1.) There are two points of interest in the table. First, the column and row totals indicate that the models are equally likely for me when viewed separately. Second, the column totals and row totals are the same. This happens whenever probabilities on the diagonal are equal and probabilities off the diagonal are equal.

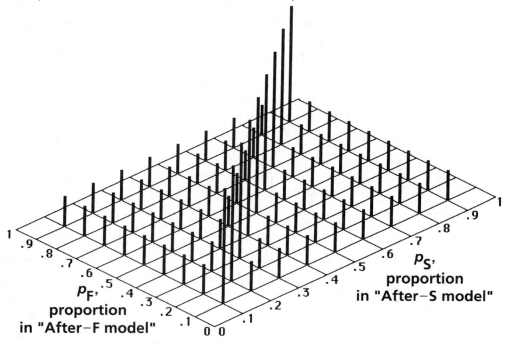

Figure 8.10 Histogram form of Table 8.11: Three-dimensional picture of my prior probabilities for "hot hand in basketball" example

Here are the data: Ford was successful on the first throw 51 times. Of these, he was successful on the second throw 36 times. So for the after-S model, D_S is $s = 36$ and $f = 15$. His first throw failed 22 times, and he was successful on the second throw 17 of these times. So for the after-F model, D_F is $s = 17$ and $f = 5$. [Both models are for the second throw, with the *model* being determined by the result on the first throw. The fact that Ford's second shot was slightly more successful than the first ($36 + 17 = 53 > 51$) is not related to the question we are addressing.]

The likelihood calculations are as usual, and the results are shown in Table 8.12. For D_S and D_F:

$$P(D_S \mid p_S) = p_S^{36}(1 - p_S)^{15}$$
$$P(D_F \mid p_F) = p_F^{17}(1 - p_F)^{5}$$

The entries in the table are the products of these two probabilities, where I have multiplied every number by the same constant, namely E+21. This has the effect of moving the decimal points for these entries to the right 21 places and allows me to drop decimal points and the E−XX notation.

Sec. 8.2 / Testing a Null Hypothesis

Table 8.12
Likelihoods for Chris Ford data in Example 8.5:
After-failure model p_F vs After-success model p_S

p_F	p_S								
	.1	.2	.3	.4	.5	.6	.7	.8	.9
.9	0	0	0	0	1	19	64	18	0
.8	0	0	0	0	3	80	274	77	0
.7	0	0	0	0	3	63	215	60	0
.6	0	0	0	0	1	19	66	18	0
.5	0	0	0	0	0	3	9	3	0
.4	0	0	0	0	0	0	1	0	0
.3	0	0	0	0	0	0	0	0	0
.2	0	0	0	0	0	0	0	0	0
.1	0	0	0	0	0	0	0	0	0

I will calculate one entry in the body of the table—that for $p_S = p_F = .7$. Evaluating the likelihood for D_S at $p_S = .7$:

$$P(D_S \mid p_S) = p_S^{36}(1 - p_S)^{15} = (.7^{36})(.3^{15}) = 3.8049\text{E}-14$$

Evaluating the likelihood for D_F at $p_F = .7$:

$$P(D_F \mid p_F) = p_F^{17}(1 - p_F)^5 = (.7^{17})(.3^5) = 5.6529\text{E}-6$$

Multiplying these two numbers gives $2.1509\text{E}-19$. Moving the decimal place 21 places and rounding gives 215, the entry in the table.

As indicated many times in this text, calculating a table of posterior probabilities from tables of priors and likelihoods is straightforward. First, multiply the two tables together. Table 8.13 is the result. (Actually, this multiplication gives the entries in the

Table 8.13
Products of prior probabilities and likelihoods in Example 8.5:
After-failure model p_F vs After-success model p_S

p_F	p_S									Sum
	.1	.2	.3	.4	.5	.6	.7	.8	.9	
.9	0	0	0	0	0	2	5	1	0	8
.8	0	0	0	0	0	6	22	35	0	64
.7	0	0	0	0	0	5	99	5	0	109
.6	0	0	0	0	0	9	5	2	0	16
.5	0	0	0	0	0	0	0	1	0	1
.4	0	0	0	0	0	0	0	0	0	0
.3	0	0	0	0	0	0	0	0	0	0
.2	0	0	0	0	0	0	0	0	0	0
.1	0	0	0	0	0	0	0	0	0	0
Sum	0	0	0	0	1	22	132	43	0	198

table divided by 10, but they have been multiplied by 10 to get rid of the decimal point.) For example, consider the point $p_S = .7$ and $p_F = .7$. The prior is .046 and the likelihood is 215. The product is 9.9 and the entry is 99.

Add up the entries in the table—add either over the columns first or over the rows first—to get 198. Divide each entry in the table by 198 to give the table of posterior probabilities shown in Table 8.14. For example, for $p_F = p_S = .7$, $\frac{99}{198}$ (to three-digit accuracy: 98.9/197.6) = .501. (Some of the 0's in the table of products are large enough to give probabilities bigger than .0005 and so show up as positive in Table 8.14.) Figure 8.11 shows the posterior distribution using a bar chart.

Table 8.14
Posterior probabilities in Example 8.5:
After-failure model p_F vs After-success model p_S

p_F					p_S					
	.1	.2	.3	.4	.5	.6	.7	.8	.9	Sum
.9	0	0	0	0	.001	.008	.026	.007	0	.041
.8	0	0	0	0	.002	.032	.111	**.178**	0	.323
.7	0	0	0	0	.001	.025	**.501**	.024	0	.551
.6	0	0	0	0	.001	**.045**	.027	.008	0	.079
.5	0	0	0	0	**0**	.001	.004	.001	0	.006
.4	0	0	0	**0**	0	0	0	0	0	0
.3	0	0	**0**	0	0	0	0	0	0	0
.2	0	**0**	0	0	0	0	0	0	0	0
.1	**0**	0	0	0	0	0	0	0	0	0
Sum	0	0	0	0	.004	.111	.668	.213	0	1

As always, there are many things we can calculate from a table of probabilities. One is the mean of the individual model histograms. This is the predictive probability of success. For the after-S model it is

$$(.5 \times .004) + (.6 \times .111) + (.7 \times .668) + (.8 \times .213) = .710$$

For the after-F model it is

$$(.5 \times .006) + (.6 \times .079) + (.7 \times .551) + (.8 \times .323) + (.9 \times .041) = .731$$

The calculation of most interest in this section is the posterior probability of the null hypothesis of no difference between shooting after a success and after a failure. "No difference" refers to the diagonal where $p_S = p_F$, or $d = p_S - p_F = 0$. The diagonal entries in the table are shown in boldface type. The total of these is .723. This is an increase from the prior probability of $\frac{5}{12} = .417$ that $d = 0$. So, while far from conclusive, the data support the null hypothesis to the extent that they increase its probability from about 42% to about 72%.

Figure 8.11 Bar chart of posterior probabilities for Example 8.5: Three-dimensional picture of my posterior probabilities for "hot hand in basketball" example

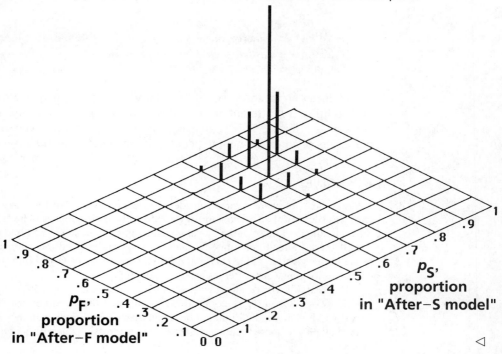

Tversky and Gilovich give data for nine Celtics players. Four of them did better after a success than they did after a failure and five did worse. None did surprisingly better or surprisingly worse. Tversky and Gilovich address the "hot hand" question in basketball in various ways, and conclude that there is none!

Many people think that there are hot streaks in sports. By a "hot streak" I do not mean simply that a team (or player) wins many games in a row. If you toss a coin 100 times, say, you will probably get six or more heads or tails in a row at some time during the 100. I would not call the coin "hot" during such a period. By "hot streak," I mean a period during which a team is more likely to win than it is at other times. "Hot" means *relatively* hot. A football team that wins 30 games in a row is not hot—it is good! I have looked for evidence of hot streaks in various sports and have found none. If there is none, a team on a 10-game winning streak is not "hot"; it is either good or lucky. Similarly, I do not see evidence for the existence of slumps—when a player is not playing as well as usual and not simply when the player has lost a number of games in a row or, in baseball, failed to get a hit a number of at-bats in a row.

An important message from this section is that a way of checking a null hypothesis that two population proportions are equal is to assign prior probability to pairs of models with the same number of success chips (that is, along the diagonal). Then apply the likelihoods in the usual way and add up the posterior probabilities of those pairs of models.

> To test the null hypothesis $d = p_T - p_C = 0$, add the posterior probabilities on the diagonal having values $p_T = p_C$.

If, in an application, you do not have the time or patience to write down prior probabilities, try different priors and see how the posterior probability of the null hypothesis is affected. This is an especially good tactic when you are trying to convey the results of an experiment to people who may not know their prior probabilities.

EXERCISES

8.4 Address the "hot hand" issues considered in Example 8.5, but for Larry Bird in 1980–1982. Bird was successful on the first throw on 285 of his 338 attempts and he failed on the first throw the remaining 53 times. Of the 285 after-S attempts, he made $s = 251$ and missed the remaining 34. Of the 53 after-F attempts, he made $s = 48$ and missed the remaining 5. Use my prior probabilities given in Table 8.11 and calculate my posterior probability that Bird has the same chance of making a shot after a success as after a failure.

[Incidentally, Bird's success rate on his first throw was $\frac{285}{338} = 84\%$ as opposed to his success rate of $\frac{299}{338} = 88\%$ on his second throw. I compared the two rates for all nine Celtics in the Tversky/Gilovich article: Eight did better on their second shots than they did on their first. Are basketball players more accurate with practice? Addressing this would be an interesting research project.]

8.5 Address the "hot hand" issues considered in Example 8.5, but for M. L. Carr in 1980–1982. He made 39 of 57 second shots after a success and 21 of 26 second shots after a failure. Use my prior probabilities given in Table 8.11 and calculate my posterior probability that Carr has the same chance of making a shot after a success as after a failure.

8.6 Collect some data addressing the issue of hot and cold streaks. Select any team or player or players from a sport or game. Record numbers of wins and losses after wins and numbers of wins and losses after losses. (In place of wins and losses, you may use hits and misses, scored goal in game and did not score, or anything appropriate.) Either assess your prior probabilities concerning the after-win and after-loss models, or use the prior probabilities in Table 8.11. Calculate your posterior probability that chances of winning are not affected by the result of the previous game.

8.7 Reconsider the violence-on-television data from Exercise 8.1:

	Too much TV violence?	
	Yes	No
Men	8	12
Women	17	5

Now assume the prior probabilities from Example 8.5, but for p_M and p_W—see following table. Using the likelihoods you calculated in Exercise 8.1,

(a) Find the posterior probability of the null hypothesis, $p_M = p_W$.

(b) Find PdAL0 and PdAL.1 (the posterior probabilities of differences that are at least 0 and at least .1, respectively) and subtract the second from the first. Your answer should agree with that of part (a), since this is the posterior probability that $d = p_M - p_W = 0$.

Prior probabilities for Exercises 8.7 and 8.8: Model for women p_W vs Model for men p_M
(For Exercise 8.8, replace p_M with p_I and p_W with p_C.)

p_W					p_M					Sum
	.1	.2	.3	.4	.5	.6	.7	.8	.9	
.9	.008	.008	.008	.008	.008	.008	.008	.008	.046	.11
.8	.008	.008	.008	.008	.008	.008	.008	.046	.008	.11
.7	.008	.008	.008	.008	.008	.008	.046	.008	.008	.11
.6	.008	.008	.008	.008	.008	.046	.008	.008	.008	.11
.5	.008	.008	.008	.008	.046	.008	.008	.008	.008	.11
.4	.008	.008	.008	.046	.008	.008	.008	.008	.008	.11
.3	.008	.008	.046	.008	.008	.008	.008	.008	.008	.11
.2	.008	.046	.008	.008	.008	.008	.008	.008	.008	.11
.1	.046	.008	.008	.008	.008	.008	.008	.008	.008	.11
Sum	.11	.11	.11	.11	.11	.11	.11	.11	.11	1

8.8 Example 3.25 (page 99) describes a study in which 30 children with cancer were divided into two groups to compare an intervention regimen designed to distract them and possibly alleviate their distress during venipuncture procedures. Half of the 30 were assigned to experience behavioral intervention and the other half served as control. Seven of the 30 children did not complete the study. Of the 13 patients in the **I**ntervention group who completed the study, seven needed restraining and six did not. Of 10 patients in the **C**ontrol group, eight needed restraining and two did not. Given this evidence and assuming the prior distribution of Example 8.5, compute the probability that intervention as defined here is effective.

8.9 A study[7] on women with node-positive breast cancer considered three chemotherapy regimes: low dose, standard dose, and high dose of combinations of drugs. (High dose is actually the same total dose as standard, but compressed into a shorter period of time. Low dose is exactly half that of high dose.) Among patients with a particular tumor marker (called c-*erb*B-2), these were the results for the first 3 years after therapy, where "relapsed" includes women who died:

	High	Standard	Low
Okay	32 (86%)	25 (61%)	13 (36%)
Relapsed	5	16	23
Total	37	41	36

Consider the combination of models and prior probabilities for p_H and p_S—the population proportions of alive and not relapsed (okay) for **H**igh dose and **S**tandard dose, as indicated in the table at the top of page 266. Find the posterior probability of $p_H > p_S$.

Prior probabilities for Exercise 8.9:
Model for High dose p_H vs Model for Standard dose p_S

	p_S						
p_H	.4	.5	.6	.7	.8	.9	Sum
.9	.01	.02	.02	.03	.03	.04	.15
.8	.02	.02	.03	.03	.04	.03	.17
.7	.02	.03	.03	.04	.03	.03	.18
.6	.03	.03	.04	.03	.03	.02	.18
.5	.03	.04	.03	.03	.02	.02	.17
.4	.04	.03	.03	.02	.02	.01	.15
Sum	.15	.17	.18	.18	.17	.15	1

8.10 In view of existing evidence that exposure to phenoxy acid and chlorophenols may increase the risk of certain types of cancer, researchers[8] conducted a case–control study (compare to Example 3.8 on page 70) with 27 agricultural workers who had cancer and 53 who did not. They found that 20 had experienced short-term exposure to these agents and 60 had not been exposed:

	Exposed	Unexposed	Total
Cancer	11	16	27
No cancer	9	44	53
Total	20	60	80

To assess my prior probabilities, I first considered p_U, the population proportion of cases of cancer among those **Unexposed**, as follows:

p_U	.1	.2	.3	.4	.5	.6	.7	.8	.9
Probability	.20	.25	.25	.15	.07	.04	.02	.01	.01

Further, I decided that my probability of $p_U = p_E$ (for **Exposed**) was about $\frac{3}{4}$, irrespective of the value of p_U. And I believed that these agents might be protective and so wanted some probability with p_U greater than p_E. The full table follows. The last row is labeled "Sum" and the entries are supposed to be similar to my probabilities for p_U given previously. It differs because I used three-decimal accuracy and wanted no entry smaller than .001.

According to the table, the prior probability of $p_U = p_E$ is .743 (approximately the $\frac{3}{4}$ indicated before). Find the posterior probability of $p_U = p_E$. (To do this, you will have to find all posterior probabilities, but most of the 81 entries in the posterior probability table will be less than .001.)

Prior probabilities for Exercise 8.10:
Model for Exposed p_E vs Model for Unexposed p_U

p_E	p_U									Sum
	.1	.2	.3	.4	.5	.6	.7	.8	.9	
.9	.006	.008	.008	.004	.002	.001	.001	.001	.001	.032
.8	.006	.008	.008	.004	.002	.001	.001	.003	.001	.034
.7	.006	.008	.008	.004	.002	.001	.010	.001	.001	.041
.6	.006	.008	.008	.004	.002	.030	.001	.001	.001	.061
.5	.006	.008	.008	.004	.050	.001	.001	.001	.001	.080
.4	.006	.008	.008	.100	.002	.001	.001	.001	.001	.128
.3	.006	.008	.200	.004	.002	.001	.001	.001	.001	.224
.2	.006	.200	.008	.004	.002	.001	.001	.001	.001	.224
.1	.150	.008	.008	.004	.002	.001	.001	.001	.001	.176
Sum	.198	.264	.264	.132	.066	.038	.018	.011	.009	1

8.11 In the early 1990s, silicone gel breast implants were implicated in a whole host of health problems, some related to the immune system. (This implication was based on very weak evidence, but that is another story.) In a surprising turn of events, a researcher[9] claimed that silicone leaking from the implants might actually have a health benefit: protecting women from breast cancer. (Silicone had been shown to kill prostate cancer cells in test tubes, so this was not a silly proposition.) The researcher took samples of blood plasma from five women who had had silicone gel breast implants (S) and three who had not (C). He put the plasma into test tubes containing breast cancer cells. Three of the five women with implants had had them for more than 10 years and the other two had had them for less than 6 years, with no sign of leakage of the implant. He theorized that the plasma from women with implants might contain tiny silicone fragments, especially if they had been in place for a long time. He found that the plasma of the three women who had had implants for more than 10 years killed the cancer cells within 10 days, while the plasma from the other five women (two with implants less than 6 years and three with no implants) did not.

So the data are as follows:

	Implant > 10 yr	No implant or < 6 yr	Total
Killed in 10 days	3	0	3
Not killed in 10 days	0	5	5
Total	3	5	8

The existence of multiplicities makes these data suspect (see Section 3.4). If the plasma from the two women with the shorter-term implants had killed the cancer cells, then we would be comparing implant vs no implant and, again, it would be 5 to 0 vs 0 to 3. Also, if the cancer cells had not been killed in 10 days, perhaps the investigator would have reported a different period with perfect kills. Further, while I have no additional information in this particular instance, investigators tend to offer their results in the most favorable light. There may be similar experiments that failed—perhaps on other cancers—but that are not reported.

Analyze the data as in the table. But partly to account for the concerns about multi-

plicities and partly to account for my skepticism, use the conservative prior probabilities indicated in the next table. The total probability along the diagonal with equal population kill rates p_I and p_N [for **I**mplant (> 10 yr) and **N**o implant (or < 6 yr)] is .67. Where $d = p_I - p_N$, find the posterior probability that
(a) $d = 0$ (no difference in kill rates)
(b) $d > 0$ (greater kill rate with long-term silicon gel implants)

Prior probabilities in Exercise 8.11: Implant model p_I vs No implant model p_N

p_I	\multicolumn{11}{c}{p_N}	Sum
	0	.1	.2	.3	.4	.5	.6	.7	.8	.9	1	
1	.003	.003	.003	.003	.003	.003	.003	.003	.003	.003	.110	.14
.9	.003	.003	.003	.003	.003	.003	.003	.003	.003	.050	.003	.08
.8	.003	.003	.003	.003	.003	.003	.003	.003	.050	.003	.003	.08
.7	.003	.003	.003	.003	.003	.003	.003	.050	.003	.003	.003	.08
.6	.003	.003	.003	.003	.003	.003	.050	.003	.003	.003	.003	.08
.5	.003	.003	.003	.003	.003	.050	.003	.003	.003	.003	.003	.08
.4	.003	.003	.003	.003	.050	.003	.003	.003	.003	.003	.003	.08
.3	.003	.003	.003	.050	.003	.003	.003	.003	.003	.003	.003	.08
.2	.003	.003	.050	.003	.003	.003	.003	.003	.003	.003	.003	.08
.1	.003	.050	.003	.003	.003	.003	.003	.003	.003	.003	.003	.08
0	.110	.003	.003	.003	.003	.003	.003	.003	.003	.003	.003	.14
Sum	.14	.08	.08	.08	.08	.08	.08	.08	.08	.08	.14	1

8.12 A researcher[10] collected aphids after spraying them with two concentrations of sodium oleate. Consider models for proportions of population kill, p_L (**L**ow concentration) and p_H (**H**igh). In assigning prior probabilities, you think that it is not possible that the low concentration has a higher kill rate than does the high. So (not worrying about being called a "know-it-all") you give 0 probability to $p_L > p_H$. Where difference $d = p_H - p_L$, your probability of "no difference" is .22. Your full table is given on page 269. The results follow:

Exercise 8.12: Effect of Sodium Oleate

	Concentration		Total
	.65%	1.10%	
Alive	13	3	16
Dead	55	62	117
Total	68	65	133

(a) Find the posterior probabilities of the various pairs of models.
(b) Find the probability that high concentration has a greater kill rate than low (this is PdAL.1).
(c) Find PdAL.2 and PdAL.3.
[Your PdAL.1 should be quite large. Does this mean that high concentration is probably better than low? Perhaps, but what are the appropriate populations? I do not know the experimental

conditions. If there was one spraying per concentration (say, one plant for each) then the populations are defined by those particular sprayings—hardly interesting. But if there were 68 sprayings at low concentration and 65 at high—with one aphid per plant, say, and the spray used was mixed separately each time—then extrapolating to future sprayings seems reasonable. The experimental unit is the spraying; if there was one spraying per concentration, then high bettered low in a sample with one observation per population and not 65 vs 68.]

Prior probabilities for Exercise 8.12:
Model for Low concentration p_L vs Model for High p_H

p_L	p_H										Sum
	0	.1	.2	.3	.4	.5	.6	.7	.8	.9	1	
1	0	0	0	0	0	0	0	0	0	0	.02	.02
.9	0	0	0	0	0	0	0	0	0	.02	.02	.04
.8	0	0	0	0	0	0	0	0	.02	.02	.02	.06
.7	0	0	0	0	0	0	0	.02	.02	.02	.02	.08
.6	0	0	0	0	0	0	.02	.02	.02	.02	.02	.10
.5	0	0	0	0	0	.02	.02	.02	.02	.02	.01	.11
.4	0	0	0	0	.02	.02	.02	.02	.01	.01	.01	.11
.3	0	0	0	.02	.02	.02	.02	.01	.01	.01	.01	.12
.2	0	0	.02	.02	.02	.01	.01	.01	.01	.01	.01	.12
.1	0	.02	.02	.01	.01	.01	.01	.01	.01	.01	.01	.12
0	.02	.01	.01	.01	.01	.01	.01	.01	.01	.01	.01	.12
Sum	.02	.03	.05	.06	.08	.09	.11	.12	.13	.15	.16	1

8.13 How effective is marital therapy? A study[11] compared insight-oriented therapy with behavioral therapy. In the **I**nsight group, psychotherapists helped the partners resolve unconscious emotional conflicts brought into the marriage from relationships outside the marriage. In the **B**ehavioral group, therapists focused on improving specific interpersonal skills of the partners. Married couples seeking counseling were randomized to these two groups and followed for at least 4 years. The question of interest is whether the partners divorced in that period.

Before giving the results of the study, I will relay my prior probabilities for p_I and p_B, the respective 4-year divorce rates. I am skeptical about the effectiveness of any such therapy. (I am glad to see a randomized study of this nature; randomizing some couples to no therapy would be still more convincing.) The fact that the couples are seeking help is both a negative and a positive indication. It is negative because they are having problems. It is positive because they have admitted their problems to someone outside the relationship and they are willing to work to resolve them. For a particular couple, my probability that they divorce in 4 years is around 35–40%. I assign probabilities to models as indicated in the table at the top of page 270. My prior probability of the null hypothesis $p_I = p_B$ (the total probability on the diagonal) is 90%. My suspicion that therapy does not matter means that I think these two therapies are equally ineffective and therefore the same as each other. (I expect that you disagree, but I want you to use my prior probabilities in this exercise.) I have assigned token positive probabilities (.004 or less) on pairs of possible models off the diagonal. It will take convincing data to get my total off-diagonal probability to be very high.

Prior probabilities for Exercise 8.13: Model for Insight-oriented divorce rate p_I vs Model for behavioral therapy divorce rate, p_B

p_I	\	p_B								Sum
	.1	.2	.3	.4	.5	.6	.7	.8	.9	
.9	.001	.001	.001	.001	.001	.001	.001	.001	.005	.013
.8	.001	.001	.001	.001	.001	.001	.001	.015	.001	.023
.7	.001	.001	.001	.001	.001	.002	.027	.001	.001	.036
.6	.001	.001	.001	.002	.002	.057	.002	.001	.001	.068
.5	.001	.001	.002	.003	.116	.002	.001	.001	.001	.128
.4	.001	.002	.004	.200	.003	.002	.001	.001	.001	.215
.3	.001	.004	.250	.004	.002	.001	.001	.001	.001	.265
.2	.002	.150	.004	.002	.001	.001	.001	.001	.001	.163
.1	.080	.002	.001	.001	.001	.001	.001	.001	.001	.089
Sum	.013	.023	.036	.068	.128	.215	.265	.163	.089	1

The results of the experiment are shown in the table that follows. Based on these data,
(a) Calculate my posterior probabilities.
(b) Find my posterior probability of the null hypothesis $p_I = p_B$.
(c) Find my PdAL.1 and PdAL.2, where $d = p_B - p_I$.

Exercise 8.13: Results of marital therapy

	Insight	Behavior	Total
Divorced	1	10	11
Married	28	16	44
Total	29	26	55

Appendix: Using Minitab for Comparing Two Proportions

The Minitab programs **'pp_disc'** and **'pp_discm'** are used to construct priors and summarize inference about two proportions, p_1 and p_2, when sets of models are used for the proportions. In Example 8.1, we are interested in comparing the proportions, p_C and p_N, that carriers and noncarriers, respectively, obtain malaria. As in Example 8.2, we will assume that each proportion can take 11 equally spaced values from 0 to 1. The open-minded assessor (Opey) believes that all pairs of possible values (p_C, p_N) are equally likely. The program 'pp_disc' is used to construct this particular uniform prior and summarize the posterior probabilities for the proportions. To run this program, type

```
exec 'pp_disc'
```

One constructs a uniform prior by inputting the smallest and largest values and the total number of models for each proportion. In the following output of the program, we enter 0 and 1 for the extreme values and 11 for the total number of models. Next, one inputs the number of successes and failures for each sample. In this example, the observed numbers of successes and failures in the two samples are $(s_1, f_1) = (2, 13)$ and $(s_2, f_2) = (14, 1)$.

This program displays the posterior probabilities of the two proportions in table form. The values of the proportions are expressed in percentage format; the values of p_C are listed along the left side and the values of p_N are listed across the top of the table. For example, the probability that $p_C = .4$ and $p_N = .7$ is equal to .002133. This table is displayed using a scatterplot. Each point corresponds to a (p_C, p_N) model and the size of the dot corresponds to the size of the associated probability. The darkest dot corresponds to a probability whose value is larger than one-half of the maximum probability. The $\times$ dots correspond to values between 10% and 50% of the maximum probability and the small dots correspond to probabilities smaller than 10% of the largest probability. It is easy to see from this graph that most of the posterior probability is concentrated on six of the 121 proportion models.

The program also summarizes probabilities for the difference in proportions, $p_N - p_C$. In a table, all possible values of the difference are given together with the associated probabilities. We see from this table that $p_N - p_C$ is equal to .6, .7, or .8 with approximate probability .9.

```
MTB > exec 'pp_disc'

 FOR EACH P DISTRIBUTION:
 ------------------------
 INPUT LO AND HI VALUES:
DATA> 0 1

 INPUT NUMBER OF MODELS:
DATA> 11

INPUT OBSERVED NUMBER OF SUCCESSES AND FAILURES IN FIRST SAMPLE:
DATA> 2 13

INPUT OBSERVED NUMBER OF SUCCESSES AND FAILURES IN SECOND SAMPLE:
DATA> 14 1

Posterior distribution of P1 and P2:

(Rows and columns are expressed in percentage format.)

    ROWS: PER_1     COLUMNS: PER_2

            0        10        20        30        40        50        60        70
  0  0.000000  0.000000  0.000000  0.000000  0.000000  0.000000  0.000000  0.000000
 10  0.000000  0.000000  0.000000  0.000000  0.000021  0.000389  0.003996  0.025940
 20  0.000000  0.000000  0.000000  0.000000  0.000018  0.000337  0.003457  0.022441
 30  0.000000  0.000000  0.000000  0.000000  0.000007  0.000133  0.001371  0.008899
 40  0.000000  0.000000  0.000000  0.000000  0.000002  0.000032  0.000329  0.002133
 50  0.000000  0.000000  0.000000  0.000000  0.000000  0.000005  0.000048  0.000311
 60  0.000000  0.000000  0.000000  0.000000  0.000000  0.000000  0.000004  0.000025
```

(*continued*)

```
 70 0.000000 0.000000 0.000000 0.000000 0.000000 0.000000 0.000000 0.000001
 80 0.000000 0.000000 0.000000 0.000000 0.000000 0.000000 0.000000 0.000000
 90 0.000000 0.000000 0.000000 0.000000 0.000000 0.000000 0.000000 0.000000
100 0.000000 0.000000 0.000000 0.000000 0.000000 0.000000 0.000000 0.000000

             80       90      100

  0 0.000000 0.000000 0.000000
 10 0.112142 0.291657 0.000000
 20 0.097016 0.252319 0.000000
 30 0.038471 0.100055 0.000000
 40 0.009219 0.023978 0.000000
 50 0.001346 0.003502 0.000000
 60 0.000107 0.000277 0.000000
 70 0.000003 0.000009 0.000000
 80 0.000000 0.000000 0.000000
 90 0.000000 0.000000 0.000000
100 0.000000 0.000000 0.000000

          CELL CONTENTS --
                POST:DATA


TYPE 'Y' AND RETURN TO SEE A GRAPH OF THE POSTERIOR DISTRIBUTION:
Y
```

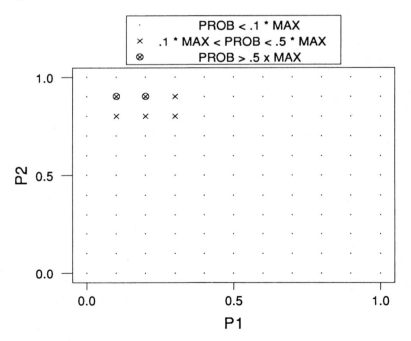

```
TYPE 'Y' AND RETURN TO SEE A TABLE OF THE POSTERIOR DISTRIBUTION
OF THE DIFFERENCE IN PROBABILITIES P2-P1:
Y
```

(The values of the difference are expressed in percentage form.)
```
ROWS: PCT_DIFF
         POST
          SUM

-100  0.000000
 -90  0.000000
 -80  0.000000
 -70  0.000000
 -60  0.000000
 -50  0.000000
 -40  0.000000
 -30  0.000000
 -20  0.000000
 -10  0.000001
   0  0.000011
  10  0.000116
  20  0.000907
  30  0.005484
  40  0.025466
  50  0.088886
  60  0.223011
  70  0.364461
  80  0.291657
  90  0.000000
 100  0.000000
 ALL  1.000000
```

To calculate Opey's PdALx as in Table 8.10 (page 254), add the appropriate probabilities from this table. For example, PdAL.7 = .3645 + .2917 = .656.

The program **'pp_discm'** can be used to summarize posterior probabilities for two proportions when an informative prior distribution is used. To illustrate this program, consider the estimation of proportions in the sickle-cell example using Infy's prior distribution in Example 8.3. Place the values of p_C and p_N in columns C1 and C2 and the matrix of probabilities in columns C3 through C13. The first row of the matrix

↓	C1 pc	C2 pn	C3	C4	C5	C6	C7	C8	C9	C10	C11	C12	C13
1	1.0	0.0	40	40	40	30	21	10	8	6	4	1	40
2	0.9	0.1	40	34	34	32	16	10	8	6	4	30	1
3	0.8	0.2	40	34	34	32	20	10	8	6	24	1	1
4	0.7	0.3	10	20	20	12	5	5	5	20	1	1	1
5	0.6	0.4	4	6	6	3	2	5	20	1	1	1	1
6	0.5	0.5	2	2	2	2	2	20	1	1	1	1	1
7	0.4	0.6	1	1	1	1	20	1	1	1	1	1	1
8	0.3	0.7	1	1	1	20	1	1	1	1	1	1	1
9	0.2	0.8	1	1	20	1	1	1	1	1	1	1	1
10	0.1	0.9	1	20	1	1	1	1	1	1	1	1	1
11	0.0	1.0	20	1	1	1	1	1	1	1	1	1	1

gives the probabilities for all values of p_N for the first value in p_C, and so on. For ease of typing, 1,000 times the probability values are entered; the program will ensure that the probabilities used sum to 1. When the program is run, input the numbers of the columns that contain the proportion values and the number of the first column which contains the probability matrix. Input the data from the two samples. The output of the program is identical to that of 'pp_disc.' The probabilities of the two proportions are displayed in table and graphical form, and the probabilities for the difference in population proportions are tabulated. One can summarize Infy's prior information by inputting 0 successes and 0 failures for each sample. In this case, the prior probability that $p_N - p_C = 0$ (the two proportions are equal) is .254.

```
MTB > exec 'pp_discm'

  INPUT THE NUMBER OF THE COLUMN WHICH CONTAINS THE P1 VALUES:
DATA> 1

  INPUT THE NUMBER OF THE COLUMN WHICH CONTAINS THE P2 VALUES:
DATA> 2

  INPUT THE NUMBER OF THE FIRST COLUMN WHICH CONTAINS THE
  PROBABILITIES:
DATA> 3

  INPUT OBSERVED NUMBER OF SUCCESSES AND FAILURES IN FIRST SAMPLE:
DATA> 0 0

  INPUT OBSERVED NUMBER OF SUCCESSES AND FAILURES IN SECOND SAMPLE:
DATA> 0 0

Posterior distribution of P1 and P2:
(Rows and columns are expressed in percentage format.)

         ROWS: PER_1     COLUMNS: PER_2

                 0        10        20        30        40        50        60        70
    0  0.020000  0.001000  0.001000  0.001000  0.001000  0.001000  0.001000  0.001000
   10  0.001000  0.020000  0.001000  0.001000  0.001000  0.001000  0.001000  0.001000
   20  0.001000  0.001000  0.020000  0.001000  0.001000  0.001000  0.001000  0.001000
   30  0.001000  0.001000  0.001000  0.020000  0.001000  0.001000  0.001000  0.001000
   40  0.001000  0.001000  0.001000  0.001000  0.020000  0.001000  0.001000  0.001000
   50  0.002000  0.002000  0.002000  0.002000  0.002000  0.020000  0.001000  0.001000
   60  0.004000  0.006000  0.006000  0.003000  0.002000  0.005000  0.020000  0.001000
   70  0.010000  0.020000  0.020000  0.012000  0.005000  0.005000  0.005000  0.020000
   80  0.040000  0.034000  0.034000  0.032000  0.020000  0.010000  0.008000  0.006000
   90  0.040000  0.034000  0.034000  0.032000  0.016000  0.010000  0.008000  0.006000
 1000.040000  0.040000  0.040000  0.030000  0.021000  0.010000  0.008000  0.006000
```

```
          80        90       100
  0  0.001000  0.001000  0.001000
 10  0.001000  0.001000  0.001000
 20  0.001000  0.001000  0.001000
 30  0.001000  0.001000  0.001000
 40  0.001000  0.001000  0.001000
 50  0.001000  0.001000  0.001000
 60  0.001000  0.001000  0.001000
 70  0.001000  0.001000  0.001000
 80  0.024000  0.001000  0.001000
 90  0.004000  0.030000  0.001000
100  0.004000  0.001000  0.040000

       CELL CONTENTS --
                POST:DATA
```

TYPE 'Y' AND RETURN TO SEE A GRAPH OF THE POSTERIOR DISTRIBUTION:

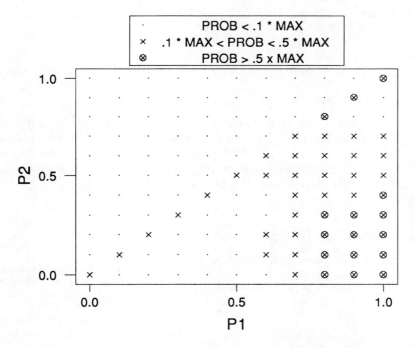

TYPE 'Y' AND RETURN TO SEE A TABLE OF THE POSTERIOR DISTRIBUTION OF THE DIFFERENCE IN PROBABILITIES P2-P1:
y
(The values of the difference are expressed in percentage form.)

(*continued*)

```
ROWS: PCT_DIFF

         POST
          SUM

-100  0.040000
 -90  0.080000
 -80  0.114000
 -70  0.108000
 -60  0.111000
 -50  0.086000
 -40  0.059000
 -30  0.036000
 -20  0.030000
 -10  0.027000
   0  0.254000
  10  0.010000
  20  0.009000
  30  0.008000
  40  0.007000
  50  0.006000
  60  0.005000
  70  0.004000
  80  0.003000
  90  0.002000
 100  0.001000
 ALL  1.000000
```

In some situations, one is interested in testing whether the proportions p_1 and p_2 are equal. In Section 8.2 a prior is constructed that places a fixed probability on the proportion pairs (p_1, p_2), where p_1 and p_2 are equal. In the text *Bayesian Computation Using Minitab*,[5] the program 'pp_disct' can be used to construct a flat-type prior and perform a Bayesian test that the proportions are equal.

Chapter Notes

1. Jared Diamond, *Natural History* (February 1989): 8–18.

2. Reported by R. H. Bartlett, et al., *Pediatrics* 76 (1985): 479–487.

3. J. H. Ware and M. F. Epstein, *Pediatrics* 76 (1985): 849–851.

4. J. H. Ware, *Statistical Science* 4 (1990): 298–305.

5. J. Albert, *Bayesian Computation Using Minitab* (Belmont, California: Duxbury Press, 1996).

6. A. Tversky and T. Gilovich, *Chance* 2 (1989): 16–21.

7. H. B. Muss, A. Thor, D. A. Berry, et al. "c-*erb*B-2 expression and S-phase activity predict response to adjuvant therapy in women with node positive early breast cancer," *New England Journal of Medicine* 330 (1994): 1260–1266.

8. L. Hardell, "Relation of soft-tissue sarcoma, malignant lymphoma and colon cancer to phenoxy acids, chlorophenols and other agents," *Scandinavian Journal of Work, Environment and Health* 7 (1981): 119–130.

9. L. Garrido of Massachusetts General Hospital, as reported by S. Blakeslee in *The New York Times*, August 10, 1994.

10. G. W. Snedecor and W. G. Cochran, *Statistical Methods*, 6th ed. (Ames, Iowa: Iowa State University Press, 1967), 218.

11. D. K. Snyder of Texas A & M University and colleagues, as reported by B. Bower in *Science News* 139 (February 23, 1991): 118.

9 Densities for Two Proportions

MANY of the calculations in Chapter 8 are cumbersome. Just as Chapter 7 made the calculations of Chapter 6 easier, so this chapter makes those of the previous chapter easier. It does so in the same way—using beta densities. The setting is more complicated than that of Chapter 7 because now there are two samples, and therefore two population models to worry about. It would be most convenient if we could simply use two beta densities, doubling the work of problems in Chapter 7—indeed that is the tack of this chapter.

The separation of the data and the prior information continues to be fundamental. The evidence in the data is carried by the likelihood. As in Chapter 8, when the samples are independent, the joint likelihood is the product of the individual likelihoods. As usual, posterior probabilities equal the product of prior and likelihood.

An assumption made throughout this chapter is that the two proportions are independent on the basis of prior information. This assumption is not always appropriate. One may suspect that if one proportion is big, the other is likely to be small. Or perhaps if one proportion is big, the other is likely to be big as well. A special case of the latter is when there is a positive probability on the null hypothesis of equality (such as in Section 8.2). Calculations for prior densities in which a positive prior probability is assigned to the null hypothesis require a computer and are carried out in the Bayesian statistics software package, *Bayesian Computation Using Minitab,* by James Albert (Duxbury Press, 1996). Section 3 of this chapter considers an alternative way to test null hypotheses.

9.1 Multiplying Likelihoods for Two Proportions

We know how to incorporate data into our probability judgments: We use likelihoods. In this chapter, a population proportion can take any value between 0 and 1. We know how to multiply two likelihoods when the samples are independent. But we have not addressed the issue of multiplying two likelihoods when the proportions take any values between 0 and 1. That is the purpose of this section.

EXAMPLE 9.1
 Malaria and sickle cells—open-minded assessor. Recall from the examples of Section 8.1 that 2 of 15 carriers and 14 of 15 noncarriers contracted malaria. Just as in those examples, letting $s = 2$ and $f = 13$ gives this likelihood for the carrier model:

$$P(D_C \mid p_C) = p_C^2(1 - p_C)^{13}$$

Letting $s = 14$ and $f = 1$ gives the likelihood for the noncarrier model:

$$P(D_N \mid p_N) = p_N^{14}(1 - p_N)$$

The bar charts in Example 8.1 showing these likelihoods are restricted to the values of p_C and p_N equal to 0, .1, .2, .3, ... , 1 because these 11 models were the only ones considered there. But the likelihoods are defined for all values of p between 0 and 1. Figure 9.1 shows the two likelihoods over the entire range. You will get a perfect fit if you place the bar charts in Figure 8.1 onto these figures.

Figure 9.1 Likelihoods for models in Example 9.1: (a) Carrier (b) Noncarrier

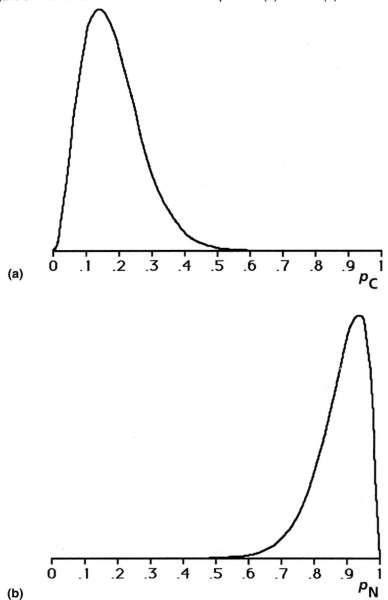

Sec. 9.1 / Multiplying Likelihoods for Two Proportions

For our purposes, we have to combine the two likelihoods. Models are now pairs of models—that is, one value for p_C and another for p_N. Figure 9.2 shows the two likelihoods on the same plot.

Figure 9.2 Likelihoods for both population proportions

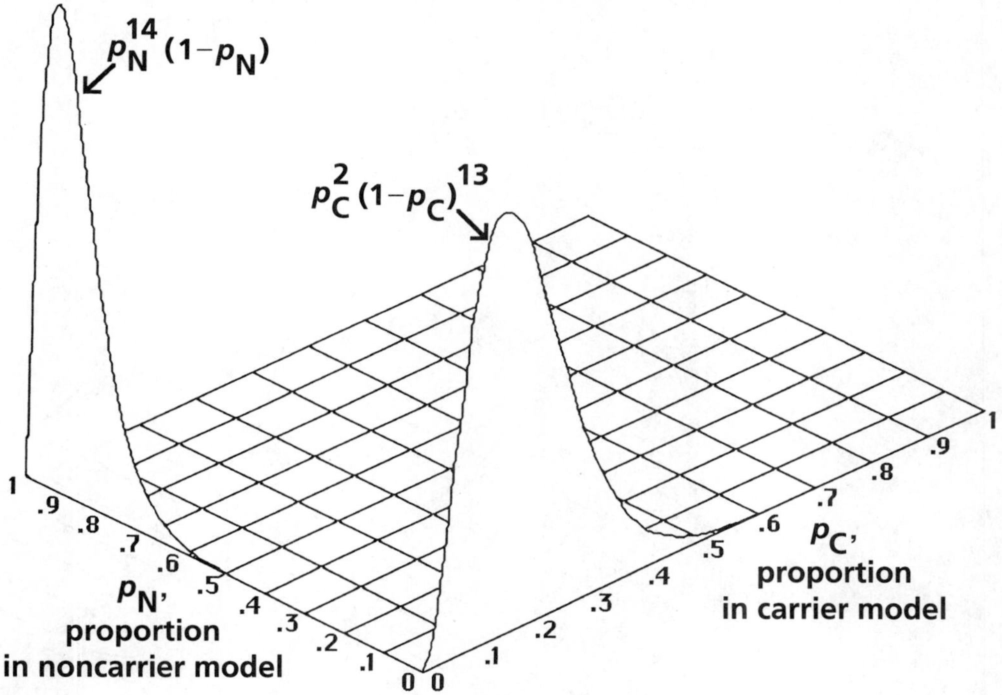

$p_N^{14}(1-p_N)$

$p_C^2(1-p_C)^{13}$

p_N, proportion in noncarrier model

p_C, proportion in carrier model

Recall from the previous chapter that likelihoods multiply when samples are independent. This still holds:

> **Multiply likelihoods for independent samples.**

Figure 9.3 (page 280) shows schematically the result of multiplying the two likelihoods.

To calculate posterior probabilities, multiply the likelihood by the prior. The bar chart in Figure 8.5 (page 246) shows Opey's probabilities—they are spread equally on the various pairs of models. As discussed in that example, this joint distribution is the product of Opey's prior probabilities for p_C and p_N separately. That is, the two models are independent a priori for Opey. The corresponding density for p_C and p_N separately (see Chapter 7) is the beta(1, 1).

Figure 9.3 To show the product of two likelihoods requires three dimensions

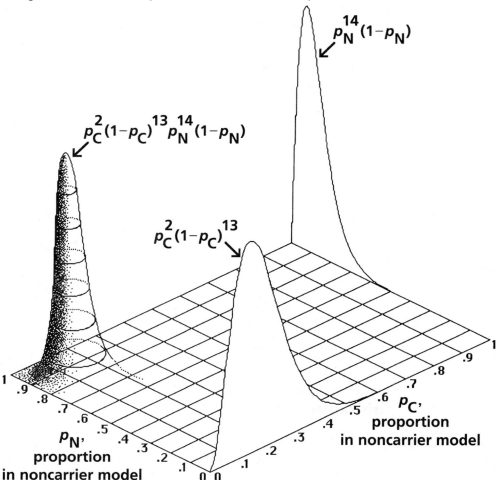

Just as when dealing with bar charts, the product of two beta(1, 1) densities gives the same weight to each pair of proportions—one from the carrier model and the other from the noncarrier model. This is represented by the flat surface in Figure 9.4. Using Bayes' rule to find the posterior density, the likelihood is multiplied by the prior density. Since the height of the prior density is the same at each point, the product is the likelihood itself (just as it was for bar charts). So the product of likelihoods pictured in Figure 9.3 is also the posterior density.

How to calculate probabilities of interest? First, Chapter 7 applies to probabilities of separate models and so these are easy. The data for the carrier sample were $s = 2$ and $f = 13$. So Opey's new density for the carrier proportion is beta($a + s, b + f$) = beta(1 + 2, 1 + 13) = beta(3, 14). For the noncarrier sample, $s = 14$ and $f = 1$, and so $(a + s, b + f) = (1 + 14, 1 + 1) = (15, 2)$. The means of these two densities

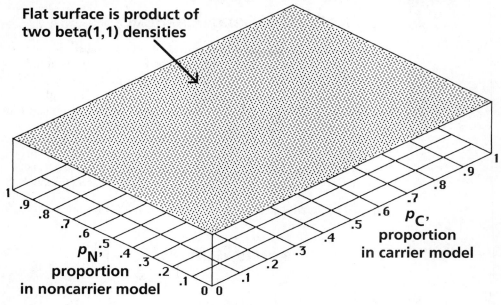

Figure 9.4 Opey's prior density in Example 9.1: Product of two flat densities is flat

(which are also the predictive probabilities of a case of malaria in the population in question) follow from the formula for the mean of a beta(a, b) density. For carriers:

$$\frac{a + s}{a + b + s + f} = \frac{1 + 2}{1 + 1 + 2 + 13} = \frac{3}{17} = .176$$

For noncarriers:

$$\frac{a + s}{a + b + s + f} = \frac{1 + 14}{1 + 1 + 14 + 1} = \frac{15}{17} = .882$$

These compare with .180 and .859 calculated in Example 8.2.

Other probabilities are more involved. We will return in subsequent examples to discuss them. ◁

To summarize, we have:

1. Observed a sample from each of two populations.
2. Assumed exchangeability within each population (but not across populations).
3. Calculated individual likelihoods.
4. Multiplied to find the joint likelihood.
5. Multiplied the joint likelihood by the joint prior to find the posterior.

Let's also be clear where we are not. We do not yet know how to apply our knowledge to calculate interesting posterior probabilities, such as the probability that carriers are protected. We will learn how to do this in the next section.

9.2 Applications of Densities: PdAL

There are several types of probabilities of interest when dealing with two unknown proportions. One is the probability that the $d = p_N - p_C$ is at least a particular value, say .6 (PdAL.6). To calculate PdAL.6 we add probabilities in the region of the bar chart (or table) in which the difference is at least .6. Example 9.4 will illustrate this calculation. Another interesting event is that p_C is in some interval *and* p_N is in some other interval. This is the subject of the next example.

EXAMPLE 9.2
▷ **Probabilities of rectangles.** In the setup of Example 9.1, let's calculate Opey's posterior probability that p_C is less than .3 *and* p_N is greater than .8. Since the two sets

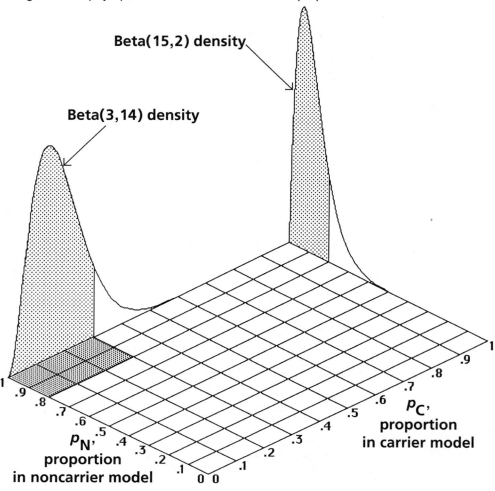

Figure 9.5 Opey's posterior densities for individual proportions

of models are independent (for Opey), the answer is a simple application of the methods in Section 7.5. Namely, we can calculate areas under the two densities in turn, then multiply. Figure 9.5 shows this schematically. We want to find Opey's probability in the shaded region of the (p_C, p_N) space—this shaded region is the one that looks like a doormat for the beta(3, 14) density, a doormat made by sewing together six rectangular pieces.

We will use the methods of Section 7.5 to find the shaded area in each of the two densities in the figure. For the p_C density, $a = 3$ and $b = 14$. The probability to the left of .3 for the beta(3, 14) density is .901. For the p_N density, $a = 15$ and $b = 2$. The probability to the right of .8 for the beta(15, 2) density is .859. Opey's probability that p_C is less than .3 *and* p_N is greater than .8 is the product of these two probabilities:

$$.901 \times .859 = .774 \text{ or } 77\%. \triangleleft$$

This procedure works for any **rectangle** provided the two population models are independent. A rectangle is formed by an interval of values on one proportion *and* an interval of values of the other proportion.

The most important region for us is a **triangle**—and not just any triangle. We want to find PdAL's, where $d = p_N - p_C$. PdALx is the probability that p_C is bigger than p_N by at least x—that is, the probability that $p_N - p_C \geq x$.

EXAMPLE 9.3
▷ **Probabilities of triangles.** Consider Opey's probability of difference due to the sickle-cell gene of at least .6. This is the probability of the shaded region in Figure 9.6 (page 284). It is not as easy to calculate the probability of a triangle as it is to calculate the probability of a rectangle. There is no way to make individual calculations—one for the carrier density and one for the noncarrier density—and multiply them.

One method of calculation is an approximation using the standard normal table. Just as in Section 7.5, we assume that the a's and b's of the individual beta densities are large. We need to find a z-score. First, calculate the values of r, r^+, and t for both densities:

$$r = \frac{a}{a+b} \qquad r^+ = \frac{a+1}{a+b+1} \qquad t = \sqrt{r(r^+ - r)}$$

For the noncarrier density:

$$r = \frac{15}{15+2} = \frac{15}{17} \qquad r^+ = \frac{15+1}{15+2+1} = \frac{16}{18}$$

$$t = \sqrt{\frac{15}{17}\left(\frac{16}{18} - \frac{15}{17}\right)} = \sqrt{\frac{5}{867}} = .0759$$

For the carrier density:

$$r = \frac{3}{3+14} = \frac{3}{17} \qquad r^+ = \frac{3+1}{3+14+1} = \frac{4}{18}$$

$$t = \sqrt{\frac{3}{17}\left(\frac{4}{18} - \frac{3}{17}\right)} = \sqrt{\frac{7}{867}} = .0899$$

Figure 9.6 Opey's posterior densities—shaded triangle has difference $d \geq 6$

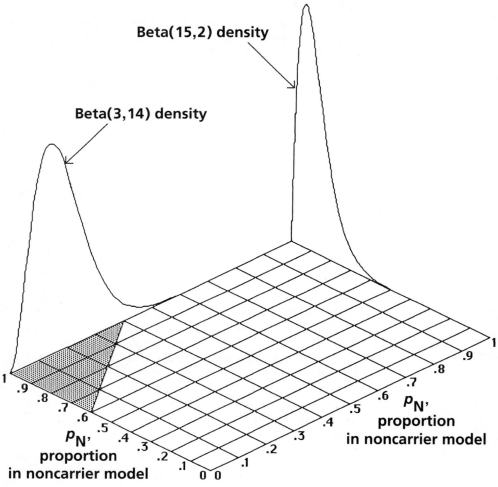

There are two proportions; we will append subscripts N and C to the r's and t's so that we can tell which is **Noncarrier** and which is **Carrier**:

$$r_N = \frac{15}{17} \quad \text{and} \quad t_N = \sqrt{\frac{5}{867}}$$

$$r_C = \frac{3}{17} \quad \text{and} \quad t_C = \sqrt{\frac{7}{867}}$$

The z-score for a 60% difference between noncarrier and carrier is

$$z = \frac{.6 - (r_N - r_C)}{\sqrt{t_N^2 + t_C^2}}$$

Sec. 9.2 / Applications of Densities: PdAL

In using this expression, be sure you have the labels in the right order. Since $d = p_N - p_C$, this z-score is for addressing the increase in proportion of malaria of N over C. (If you want the increase of C over N, then replace $r_N - r_C$ in this expression with $r_C - r_N$.) Substituting the numbers we calculated previously into this z-score gives

$$z = \frac{.6 - \left(\frac{15}{17} - \frac{3}{17}\right)}{\sqrt{\frac{5}{867} + \frac{7}{867}}} = \frac{-9/85}{2/17} = -\frac{9}{10}$$

The standard normal table gives the probability that z is less than $-.90$ as .1841. The PdAL.6 is the probability *greater* than $z = -.90$, namely, $1 - .1841 = .8159$. This compares with .8791 that we calculated in Example 8.4. The discrepancy is substantial, but easy to explain. Consider Figure 8.7 (page 253) from Example 8.4. There are three bars in the figure that sit on top of the line with difference in proportions equal to .6. In Example 8.4, their total weight counts fully in the calculation that the difference is at least .6. But when they are smoothed out, as in the current example, part of their weight (a little less than half of it) spills over outside the triangle. So the smoothing process of this section serves to decrease the probability of the triangle.

I used the preceding method to calculate Opey's PdAL for all differences from 0 to 1. These PdAL's are shown as a solid curve in the graph in Figure 9.7. The dashed line shown is taken from Figure 8.8 on page 255. The dashed curve is usually higher than the solid curve. It is higher at differences of .6, .7, and .8 because of the smoothing

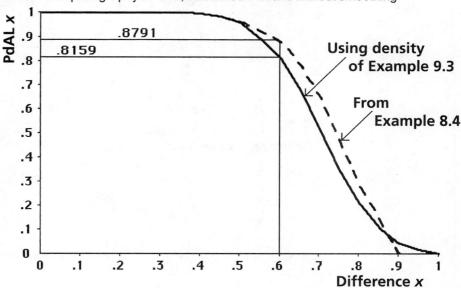

Figure 9.7 Comparing Opey's PdAL, calculated with and without smoothing

process, just as indicated for PdAL.6. The two PdAL's at .6 are highlighted using horizontal lines on this figure. ◁

EXAMPLE 9.4
▷ **Effectiveness of semipersonal letters.** A researcher[1] was interested in whether and how the type of transmittal letter affected response in a mail survey. He sent 2,040 letters to people, randomly selecting between a semipersonal letter and a form letter. Of the 1,018 people who received a semipersonal letter, 325 responded, while of the 1,022 people who received a form letter, 225 responded. How much of an improvement in response rate is due to using semipersonal letters?

These sample sizes are rather large. This has two effects. One relates to calculations—I will get to that shortly. The other concerns the role of the prior density. I have prior notions about whether someone is less likely to respond to a form letter, but the sample sizes are so large that the data will swamp my opinions. So what I choose as a prior density matters little. And my conclusion will be about the same as those of any reasonably open-minded person.

Consider two population models, one for semipersonal letters and the other for form letters. Suppose they are independent a priori and that the proportion of responses in each has a beta(1, 1) density. So the posterior density for the semipersonal-letter proportion is a beta(326, 694) and for the form-letter proportion is beta(226, 798).

For the semipersonal-letter proportion:

$$r = \frac{326}{326 + 694} = \frac{326}{1,020} = .3196 \qquad r^+ = \frac{326 + 1}{326 + 694 + 1} = \frac{327}{1,021} = .3203$$

$$t = \sqrt{\frac{326}{1,020}\left(\frac{327}{1,021} - \frac{326}{1,020}\right)} = .01459$$

For the form-letter proportion:

$$r = \frac{226}{226 + 798} = \frac{226}{1,024} = .2207 \qquad r^+ = \frac{226 + 1}{226 + 798 + 1} = \frac{227}{1,025} = .2215$$

$$t = \sqrt{\frac{226}{1,024}\left(\frac{227}{1,025} - \frac{226}{1,024}\right)} = .01295$$

Since S and F are too easily confused with success and failure, I will use subscripts 1 and 2 for semipersonal and form, respectively:

$$r_1 = \frac{326}{1,020} \qquad t_1 = .01459$$

$$r_2 = \frac{226}{1,024} \qquad t_2 = .01295$$

The z-score for calculating the probability that the difference $d = p_1 - p_2$ is less than 0 is

Sec. 9.2 / Applications of Densities: PdAL

$$z = \frac{0 - (r_1 - r_2)}{\sqrt{t_1^2 + t_2^2}} = \frac{0 - \left(\frac{326}{1,020} - \frac{226}{1,024}\right)}{\sqrt{2.130\text{E}-4 + 1.678\text{E}-4}} = -5.07$$

Such a small value of z is not in the standard normal table; the probability to the left of $z = -5.07$ is .0000, to four-decimal accuracy. So PdAL0 (the probability of a difference of at least 0) is 1.0000.

To calculate PdAL.05 and PdAL.10, we need the z-scores

$$z = \frac{.05 - (r_1 - r_2)}{\sqrt{t_1^2 + t_2^2}} = \frac{.05 - \left(\frac{326}{1,020} - \frac{226}{1,024}\right)}{\sqrt{2.130\text{E}-4 + 1.678\text{E}-4}} = -2.506$$

$$z = \frac{.10 - (r_1 - r_2)}{\sqrt{t_1^2 + t_2^2}} = \frac{.10 - \left(\frac{326}{1,020} - \frac{226}{1,024}\right)}{\sqrt{2.130\text{E}-4 + 1.678\text{E}-4}} = .056$$

corresponding to PdAL.05 of .994 and PdAL.10 of .522. So the probability of an additional response rate of 5% using semipersonal letters as opposed to form letters is 99%, and the probability of a response rate of 10% better than form letters is 52%.

The graph in Figure 9.8 shows the PdAL curve for differences from 0–20%.

Figure 9.8 PdAL for semipersonal letters over form letters in Example 9.4

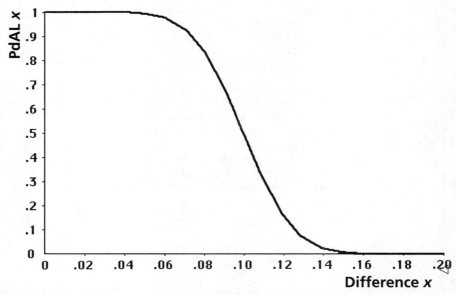

Conclusions such as those in this section require care in interpreting and applying. For example, suppose you decide to use semipersonal letters because an additional 5%

response is important to you, and the previous example indicates a very high probability of a response at least this large. If your addressees are not exchangeable with those in the study, then such an extrapolation may be wrong. If your semipersonal letters and your form letters are not the same as those in the study, again it may be wrong. If the letters and the people are "reasonably similar," you may get a similar increase in response, but if they are not, you could get an increase greater than 20% or you could even get a lower response rate than had you used a form letter—you may never know which. Judging whether they are "reasonably similar" is obviously subjective.

EXERCISES

9.1 During the 1940s and 1950s, Ted Williams played baseball for the Boston Red Sox and Joe DiMaggio played for the New York Yankees. They were regarded as among the best players of their day, and the question of who was better was the subject of many heated arguments among baseball aficionados. Williams had the better batting record. Was he in fact a better hitter? Their major league career statistics are given in the table that follows. Assume that at-bats are exchangeable for each player. (This assumption is only approximately correct for me. Players learn, change their batting style from time to time, and get older. Moreover, they do not always play in the same ball park, against the same pitcher, in the same weather, the style of ball may change, and so on. But the assumption may not be far off. And if we do not assume something, we will not be able to draw any conclusions!)

Data for Exercise 9.1

Player	At-bats	Hits	Batting average	Home runs	Home run average
Ted Williams	7,706	2,654	.3444	521	.0676
Joe DiMaggio	6,821	2,214	.3246	361	.0529

(a) First consider hits. My prior probabilities concerning these two players' "true proportions" of hits are given by independent beta(a, b) densities, where $a = 105$ and $b = 245$.

[The values of a and b we assume do not matter much. Both Williams and DiMaggio had so many at-bats that we would get about the same answer even if we were to assume $a = b = 1$. I decided on the given values by asking what I would have thought about a highly-touted rookie who came into the major leagues in the late 1930s. I assessed my probability of a hit on his first at-bat (or on any other given at-bat) to be $r = .3$, and assuming he got a hit on his first at-bat I would increase this a little to about $r^+ = .302$. Solving for a and b, as in Section 7.4, gives

$$a = \frac{r(1 - r^+)}{r^+ - r} = \frac{.3(.698)}{.302 - .3} = 105$$

$$b = \frac{(1 - r)(1 - r^+)}{r^+ - r} = \frac{.7(.698)}{.302 - .3} = 245$$

Find my posterior probability (given the preceding career statistics) that Ted Williams was a better hitter than Joe DiMaggio. (Interpret *better* to mean that there was a higher proportion of hits in the Williams model.)

(b) Now consider home runs. Find my probability that Williams was a better home-run hitter than DiMaggio, assuming that my prior probabilities are given by independent beta(19, 460) densities. (The corresponding probabilities of a home run on first at-bat and on second at-bat, given a home run on the first, are $r = .04$ and $r^+ = .042$.)

9.2 Exercises 7.18 and 7.19 (page 226) give the following data concerning high school graduation rates for children who had low and high exposure rates to lead:

	Exposure to lead	
	Low	High
Graduated	25	22
Not graduated	2	7

Assume the prior densities given in those exercises—beta(3, 1) for the low exposure group (L) and beta(1, 1) for the high exposure group (H). Calculate PdAL0 and PdAL.1 for the population proportions of graduates, where $d = p_L - p_H$.

9.3 Refer to the ECMO study considered in Exercises 8.2 and 8.3 (pages 256–257). Suppose your prior densities for the ECMO and CVT proportions are both beta(1, 1) and are independent. Redo parts (b) and (c) of Exercise 8.2 [repeated in parts (a) and (b) here] under this assumption. (Your answers should be close to those of Exercise 8.3.)

(a) Calculate your posterior densities of p_E and p_C. Use these data: 28 ECMO infants survived and 1 died; 6 CVT infants survived and 4 died.

(b) Using your posterior density from part (a), calculate your PdALx where $d = p_E - p_C$ and $x = 0, .2, .4,$ and $.6$.

9.4 A major nationwide study called CAST (Cardiac Arrhythmia Suppression Trial)[2] was started in 1987 and was to run until 1992. The purpose of the study was to see whether three drugs that had been shown to prevent cardiac arrhythmias actually improved one's chances of survival. The study involved patients who had experienced at least one heart attack. Patients were randomized to receive drug or placebo. The study was stopped in 1989 when interim results for two of the drugs (two that are very similar to each other) pointed in the opposite direction from the one expected: Of 730 patients given the drugs, 56 had died, while there were only 22 deaths of 730 patients in the placebo group. (Patients in a third group were assigned a drug with properties different from these two and continued in the study.) Assume independent population models and beta(1, 9) prior densities for both drug and placebo models. Calculate the posterior probability (given these data) that these drugs increase the death rate among the type of patients in the study. That is, find PdAL0, where $d = p_T - p_C$ and proportions refer to deaths.

9.5 A 1986 study[3] compared the drug AZT with placebo in the treatment of AIDS. Here is some background. AZT was synthesized in 1964 as a possible treatment for cancer, but it turned out to be ineffective at reducing tumors. Scientists discovered in 1984 that AIDS was caused by a retrovirus; therefore, it depended on an enzyme known as reverse transcriptase to reproduce itself. They realized the characteristics of AZT might enable it to block this enzyme. In 1985, scientists at the National Cancer Institute used cultures of AIDS-infected human cells to show that while AZT did not kill the AIDS virus, it did stop the virus from multiplying.

Later in 1985, doctors treated 11 patients with AIDS and eight with AIDS-related complex for a period of 6 weeks. The results were very promising. Lab tests showed that the immune systems of 15 of the 19 patients started to function better on the drug. Two patients had chronic fungus infections that cleared up. Six patients stopped running fevers at night. Cell cultures from some of the patients showed no trace of the AIDS virus. During the 6-week period, the patients gained an average of 5 pounds each; untreated AIDS patients tend to waste away. They continued giving the drug to these 19 patients beyond the 6-week period; after about one year, 16 of them were still alive.

Scientists regarded these results as less than conclusive. The U.S. Food and Drug Administration helped the drug's manufacturer, Burroughs Wellcome Co., to set up a double-blind clinical study to compare AZT and placebo in treating AIDS patients. The study was stopped in October 1986, which was earlier than originally planned, with 16 deaths among 137 patients on placebo and only 1 death among 145 patients given AZT. (While it is only approximately true, assume that all patients were admitted simultaneously and the study was stopped 6 months later.)

This study was controversial. Many AIDS patients claimed the drug should have been made available without a placebo-controlled study or, at the very least, it should have been stopped sooner. At the other end of the spectrum, some scientists claimed the evidence was still too scant to claim that AZT was effective in prolonging life. What do you think?

Back up your conclusion with an analysis. Use the background information in any way you like to assess prior beta(a, b) densities for the two population proportions. Be sure not to use the study information in your prior densities. Also, do not use information you have obtained about AZT since the trial. (If you do not trust your ability to set these latter two types of information aside when assessing your prior densities, then use open-minded densities.) Then calculate your PdALx for various x.

9.6 Eyewitnesses are not very reliable. Example 3.20 (page 89) deals with the accuracy of eyewitnesses and, in particular, with the susceptibility of eyewitnesses to false suggestions. The example deals with the average number of demonstrators reported. In this example, we will address the same question using a different experiment, one where the response is yes/no. Psychologist Loftus reported[4] an experiment in which she showed 150 students a short videotape of an automobile accident. She immediately administered a questionnaire of 10 questions designed to find out whether the students could be affected by an inaccurate suggestion. Nine of the questions were fillers and one was critical. She asked 75 of the students this question: "How fast was the white sports car going when it passed the barn while traveling along the country road?" The other 75 were asked a similar but distinctly different question: "How fast was the white sports car going while traveling along the country road?" In fact, there was no barn in the videotape.

One week later, all 150 students were asked 10 new questions, ostensibly to test their recall after the delay. But again nine of the questions were irrelevant to Loftus's purpose. The last one was, "Did you see a barn?" The students answered yes or no. I want to assess beta densities for the two proportions of yes responses. I know that eyewitnesses are unreliable, and I have a strong opinion about the power of suggestion. But I would like a conclusion that is not much affected by my opinion and so I will take the two densities to be the same—namely, both a's equal to 1 and both b's equal to 3.

These were the results: In the group given the suggestion that there was a barn (group T), 13 of the 75 answered yes and in the other group (C), 2 of the 75 answered yes. Assume independent population proportions and use my prior densities to calculate my posterior probability (given these data) that the question containing the false presupposition of a barn increases the propensity of students to answer yes. That is, find PdAL0, where $d = p_T - p_C$ and the proportions are of yes responses. (Example 12.2 gives the results of a similar experiment.)

9.3 Probability Intervals for Differences

In Section 7.6, we calculated probability intervals for population proportions. The same procedures apply to the individual proportions of this chapter. But they are easy to extend to differences in proportions. That is the purpose of this section. Again we will use subscripts to identify the two populations: p_T is beta(a_T, b_T) and p_C is beta(a_C, b_C). The necessary calculations are as usual:

$$r_T = \frac{a_T}{a_T + b_T} \qquad r_T^+ = \frac{a_T + 1}{a_T + b_T + 1} \qquad t_T = \sqrt{r_T(r_T^+ - r_T)}$$

$$r_C = \frac{a_C}{a_C + b_C} \qquad r_C^+ = \frac{a_C + 1}{a_C + b_C + 1} \qquad t_C = \sqrt{r_C(r_C^+ - r_C)}$$

As in the previous section, we will use z-scores for differences in proportions, that is, for $d = p_T - p_C$:

$$z = \frac{d - (r_T - r_C)}{\sqrt{t_T^2 + t_C^2}}$$

The following table gives the z-score, z_{perc}, used for constructing probability intervals containing perc% probability and is repeated from Section 7.6:

68% probability interval: $z_{68} = 1.00$
80% probability interval: $z_{80} = 1.28$
90% probability interval: $z_{90} = 1.65$
95% probability interval: $z_{95} = 1.96$
98% probability interval: $z_{98} = 2.33$
99% probability interval: $z_{99} = 2.58$

> **Constructing a Probability Interval for a Difference $d = p_T - p_C$ Using Normal Curves**
>
> For large a_T, b_T, a_C, b_C, a perc% probability interval for d is
>
> $$r_T - r_C \pm z_{\text{perc}} \sqrt{t_T^2 + t_C^2}$$
>
> where z_{perc}, the r's, and t's are as given above.

The question of how large is large arises once more. The rule of thumb we used in Chapter 7 was that the a's and b's should be at least 5, and preferably 10 or more. Again, the approximation will be better when the r's are near $\frac{1}{2}$, that is, when a_T and b_T are nearly equal and when a_C and b_C are nearly equal. The approximation can still be applied if this criterion is not met, but the approximation is not as good. In the next two examples, the a's and b's easily meet the criterion.

EXAMPLE 9.5

▷ **Effectiveness of semipersonal letters (revisited).** In Example 9.4, we considered data from a researcher who sent 2,040 letters to people, randomly selecting between a form letter and a semipersonal letter. Of the 1,022 people who received a form letter, 225 responded, while of the 1,018 people who received a semipersonal letter, 325 responded. What is a 99% posterior probability interval for $d = p_1 - p_2$ (using labels from Example 9.4)?

Suppose the two proportions are independent a priori and have beta(1, 1) densities. Again, the posterior density for p_1 is beta(326, 694) and for p_2 is beta(226, 798), and

$$r_1 = \frac{326}{1,020} \qquad t_1 = .01459$$

$$r_2 = \frac{226}{1,024} \qquad t_2 = .01295$$

$$r_1 - r_2 = .0989 \quad \text{and} \quad \sqrt{t_1^2 + t_2^2} = .01951$$

For a 99% probability interval, $z_{\text{perc}} = 2.58$, so

$$r_1 - r_2 \pm z_{\text{perc}} \sqrt{t_1^2 + t_2^2} = .0989 \pm .0503$$

or about 5% to 15%. This means that there is a 99% probability of improving the response rate by 5 to 15 percentage points when using semipersonal letters. ◁

In Chapters 6 and 8, we tested null hypotheses by finding their probabilities. Having a positive probability for $d = 0$ when using beta densities requires a computer (see Albert[5]). An alternative way to test null hypotheses using posterior probability intervals was suggested in Section 7.6 for proportions. It is developed below for differences between proportions.

> **Testing a null hypothesis means finding its posterior probability.**

> **Alternative Test of Null Hypothesis $d = 0$**
>
> Decide whether the 95% posterior probability interval for d contains 0. If yes, then the null hypothesis is supported; otherwise, it is not.
>
> Equivalently, calculate
>
> $$z = \frac{0 - (r_T - r_C)}{\sqrt{t_T^2 + t_C^2}}$$
>
> from the posterior beta densities. If $-1.96 < z < 1.96$, then the null hypothesis is supported; otherwise, it is not.

Again, I suggest using the following language. When 0 is not in the interval: "The null hypothesis is not supported since $d = 0$ is not in my 95% posterior probability interval." Or, "A difference between the two groups is supported." When 0 is in the interval: "The null hypothesis is supported since $d = 0$ is in my 95% posterior probability interval." To repeat from Chapter 7, the phrase "is supported" refers to posterior probabilities, which makes hypothesis testing subjective—a null hypothesis may be supported for one person but not for another. ["Testing a hypothesis" is a standard procedure in a frequentist approach to statistics; an approximate connection is the following: A null hypothesis is "rejected" if it is not supported by someone for whom both prior densities are beta(1, 1).]

EXAMPLE 9.6

▷ **Effectiveness of condoms in preventing AIDS transmission.** Researchers[6] tested female prostitutes in Zaire for the AIDS virus (HIV) and recorded their recollections of how frequently they had used condoms in the previous year. The numbers of prostitutes who converted from seronegative to seropositive are shown in Table 9.1, as adapted by a Mariposa Foundation paper.[7]

Table 9.1
Condom use and HIV-seroconversion among 373 Kinshasa, Zaire prostitutes

	None	1–25%	26–49%	50–74%	≥ 75%
Sero+	74 (26%)	19 (35%)	7 (32%)	0 (0%)	0 (0%)
Sero−	214	36	15	2	6
Total	288	55	22	2	6

It is difficult to interpret the results of this study. First, this is an observational study and several types of biases are possible. Perhaps prostitutes who use condoms regularly have a different type of clientele from those who do not. Similarly, a prostitute's activity may be related to frequency of condom use. Even if condoms are perfectly protective, a prostitute who uses them 25% of the time has the same exposure as one who never uses them but is only 75% as active.

Suppose there are no such biases. If condom use prevents transmission, then there should be a greater proportion of seropositivity with lower frequencies of condom use. Using curious logic, the original researchers dropped the "None" category and compared prostitutes who said they used condoms less than 50% of the time with those who said they used them at least 50% of the time:

Comparing moderate with high use

	1–50%	≥ 50%
Sero+	26 (34%)	0 (0%)
Sero−	51	8
Total	77	8

The Mariposa authors criticized this comparison ("Only by ignoring three-quarters of their data, as Mann et al. did, can a benefit be claimed for condom use.") and regrouped as follows:

Comparing no use with some use
(Mariposa analysis)

	None	Some
Sero+	74 (26%)	26 (31%)
Sero−	214	59
Total	288	85

This regrouping suggests that condoms make HIV transmission easier! But combining prostitutes who use condoms 1% of the time with those who use them all the time hardly seems appropriate. A more appropriate comparison would explicitly consider the level of use and would not combine categories. Such an analysis is beyond the level of this text.

The most appropriate comparison in which categories are combined into two groups may be < 50% vs ≥ 50%. Another may be **Low** vs **High**, as follows:

Comparing no or low condom use
with high use

	< 75%	≥ 75%
Sero+	100 (27%)	0 (0%)
Sero−	267	6
Total	367	6

Consider this last breakdown for the purposes of this example. The question of interest is whether high condom use (the condoms used by prostitutes in Zaire) is effective in preventing transmission of HIV. Assume independent beta(1, 1) priors for both proportions of seropositivity and calculate a 95% posterior probability interval for $d = p_L - p_H$.

The posterior density for p_L is beta(101, 268) and for p_H is beta(1, 7). Then,

$$r_L = \frac{101}{369} \qquad t_L = .0231793$$

$$r_H = \frac{1}{8} \qquad t_H = .1102396$$

$$r_L - r_H = .149 \quad \text{and} \quad \sqrt{t_L^2 + t_H^2} = .113$$

For a 95% probability interval, $z_{\text{perc}} = 1.96$, so

$$r_L - r_H \pm z_{\text{perc}} \sqrt{t_L^2 + t_H^2} = .149 \pm .221$$

or about -7% to 37%. Since this interval contains 0, the null hypothesis of no benefit of condom use (those used by Zairean prostitutes) is supported.

Equivalently,

$$z = \frac{0 - (r_L - r_H)}{\sqrt{t_L^2 + t_H^2}} = \frac{-.149}{.113} = -1.32$$

which is between -1.96 and $+1.96$.

A 68% probability interval ($z_{\text{perc}} = 1$) is $15\% \pm 11\%$ or from 4% to 26%, which does not contain 0. So while the null hypothesis has some support, it does not have very much. ◁

EXERCISES

In all these exercises, assume independent population proportions and independent samples.

9.7 In Exercise 9.1(a), you found my posterior probability that Ted Williams was a better hitter than Joe DiMaggio. These are the data:

Player	At-bats	Hits	Average
Ted Williams	7,706	2,654	.3444
Joe DiMaggio	6,821	2,214	.3246

Again assume beta(105, 245) densities.

(a) Find my 95% posterior probability interval for the increase in the hitting success proportion for Williams over DiMaggio.

(b) Test the null hypothesis of no difference between population proportions of hits.

9.8 Exercise 9.2 considered these data:

	Exposure to lead	
	Low	High
Graduated	25	22
Not graduated	2	7

As in that exercise, assume the prior density is beta(3, 1) for the population graduation rate for **L**ow and beta(1, 1) for **H**igh.

(a) Find the 95% probability interval for $d = p_L - p_H$.

(b) Test the null hypothesis of no effect of lead.

9.9 A randomized study in 59 centers in the United States addressed the effectiveness of the drug AZT in preventing transmission of the AIDS-associated virus, HIV, to newborns from infected mothers. The study was stopped on February 18, 1994. The interim results were deemed dramatic enough to warrant a *clinic alert*. Of newborns in the AZT group, 13 of 157 tested positive for HIV. Of newborns in the placebo group, 40 of 157 tested positive. [AZT was eventually approved by the U.S. Food and Drug Administration (FDA) for this indication.] Consider the proportions of those who test positive in the corresponding populations: AZT (T) and placebo (C), and the difference $d = p_T - p_C$. Now, $d < 0$ indicates a beneficial treatment effect. Find 95% posterior probability intervals using the three different sets of prior densities that follow.

(a) Beta(1, 1) for both p_T and p_C (These open-minded densities ignore prior evidence that led to organizing this large trial.)

(b) Beta(1, 1) for p_T and beta(25, 75) for p_C (The latter is consistent with available evidence concerning the rate of transmission with no treatment.)

(c) Beta(2, 18) for p_T and beta(25, 75) for p_C (The former is based on evidence from pre-clinical studies that AZT had a beneficial effect.)

9.10 The U.S. FDA has been criticized for requiring too much evidence before approving products. Repeat Exercise 9.9, but now address the possibility that the study might have stopped sooner by supposing that with about half the numbers of newborns, the rates were the same as when the study was actually stopped. (This is hypothetical—I do not know that the rates were the same at the halfway point.) In particular, suppose that 7 of 84 tested positive in the AZT group and 21 of 82 tested positive on placebo. Find 95% posterior probability intervals for d using the same three sets of prior densities in Exercise 9.9.

(a) Beta(1, 1) for both p_T and p_C

[*Check:* The interval in part (a) should be about 40% wider than the one in part (a) of Exercise 9.9 because n is halved and $\sqrt{2}$ is about 1.4.]

(b) Beta(1, 1) for p_T and beta(25, 75) for p_C

(c) Beta(2, 18) for p_T and beta(25, 75) for p_C

9.11 Between 1980 and 1989, 58 abstracts dealing with maternal cocaine use and fetal health problems were submitted for presentation at annual meetings of the Society for Pediatric Research.[8] Nine of the 58 reported no adverse effects on babies of cocaine users and the other 49 described reproductive risks linked to cocaine use. Peer reviewers accepted only one of the first nine for presentation, while they accepted 28 of the latter 49. Is this sufficient information to infer that an abstract's conclusion affects its acceptability to peers in this profession? For the purpose of this exercise, suppress your prior opinion and assume beta(1, 1) densities for acceptance proportions p_N (**Negative**) and p_A (**Affirmative**).

(a) Find the 90% posterior probability interval for $d = p_A - p_N$.

(b) Test the null hypothesis that in this field an abstract's conclusion has no impact on its acceptability.

(One should investigate the affirmative and negative studies to see whether the latter tended to be poorly designed or carried out in a shoddy fashion. In fact, they tended to have larger control groups and the researchers in these nine studies were more likely to verify cocaine use than were those in the 49 affirmative studies.)

9.12 In an article[9] in *Natural History,* Jared Diamond addressed the possibility of a relationship between smallpox (a disease that has been eradicated) and ABO blood type. He reports on an observational study completed in India in 1966. Two scientists (F. Vogel and M. R. Chakravartti) identified 415 unvaccinated smallpox cases. Of these, 261 carried type A blood (that is, their blood types were either A or AB) and 154 were noncarriers (that is, they were blood type O or B). Of the carriers, 201 had severe cases of the disease and the other 60 had mild cases. Of the noncarriers, 82 had severe cases of the disease and the other 72 had mild cases. (So now, as opposed to the malaria/sickle-cell example, *noncarriers* seem to be partially protected.) Define p_A to be the proportion of severe cases among all smallpox cases for type **A** carriers and p_N to be the proportion of severe cases for **Noncarriers**. Take beta(1, 1) for both p_A and p_N. Let $d = p_A - p_N$.

(a) Find the 90% posterior probability interval for d.

(b) Calculate PdAL0%, PdAL25%, and PdAL50%.

9.13 The following table is repeated from Exercise 7.4 (page 210). It gives differences between miles walked while on an experimental drug and miles walked while off the drug for 32 congestive heart-failure patients:

.00	+.56	+3.27	−2.55	+8.42	+1.07	−1.31	+3.19
−.59	+10.75	+11.73	−.05	+1.65	−3.42	+1.73	−1.44
+6.04	+12.21	+4.97	+1.68	+2.28	−6.57	−2.11	+.75
−.96	+1.68	+8.85	+7.45	−.59	+2.91	.00	+5.40

In Exercise 7.4, you used the fact that there are 20 increases (+'s) and 10 decreases (−'s). But congestive heart-failure patients may walk farther in a second week even without drugs. For comparison purposes, a second group of 31 patients (assumed to be exchangeable with those in the first group, except possibly for treatment) were administered placebo instead of drug in their second week. Their differences were as follows:

+.69	+12.76	−.20	−1.09	−1.07	−4.51	−1.32	−1.78
+.08	+1.91	−1.81	−2.20	−1.41	−5.33	−2.44	−2.48
−.71	+1.44	−.84	−3.16	−.90	+.37	−2.86	+1.55
+.30	−1.23	−3.83	−6.31	−3.97	+.10	−.27	

So there were 9 +'s and 22 −'s in the placebo group. Assume independent beta(5, 5) prior densities for p_T and p_C for the population proportions of improvement (+) in the **T**reatment and **C**ontrol groups. For differences of 0%, 10%, and so on up to 100%, where positive difference means distance improves more on drug, calculate the posterior PdAL.

9.14 Exercise 6.11 (page 191) indicates that a biologist identified 25 burrowing owl nests that had been lined with dung and observed two that were raided. He also identified 24 nests that had not been lined with dung and found that 13 had been raided. Take the prior densities of the proportions of raids in the **L**ined (p_L) and **U**nlined (p_U) populations to be beta(1, 2). Define $d = p_U − p_L$.
(a) For differences of 0%, 10%, and up to 70%, calculate the posterior PdAL.
(b) Find the 95% posterior probability interval for d.
(c) Test the null hypothesis of no effect of lining with dung.

9.15 Exercise 7.20 (page 226) considered what proportion of freshmen college women are too thin. As indicated in that exercise, the main purpose of the survey was to assess whether underweight women were more likely to subscribe to fashion or beauty magazines. Of the 36 underweight women, 16 or 44% did subscribe to one or more of *Mademoiselle, Young Miss, Vogue, Cosmopolitan, Seventeen, Glamour,* and *Sassy.* Of the 39 women who were not underweight, 14 or 36% subscribed to one or more of these magazines. Regard these to be random samples from the populations of **U**nderweight and **N**ot underweight freshmen women. Take the prior densities of the proportions of subscribers to be independent beta(1, 2) for both p_U and p_N. Define $d = p_U − p_N$.
(a) For differences of 0%, 10%, and 20%, calculate the posterior PdAL.
(b) Find the 90% posterior probability interval for d.
(c) Test the null hypothesis that $d = 0$.
[If there is a relationship here, it is not clear which is the cause and which is the effect (see also Section 3.2). Perhaps thin people are more likely to read fashion magazines or perhaps reading these magazines persuades people to become thin—or perhaps some of both.]

9.16 Tardive dyskinesia is a serious movement disorder that is apparently caused by antipsychotic drugs. Vitamin E had been used, with mixed results, to treat this condition. A study[10] compared

vitamin E with placebo. Symptoms decreased in 9 of 16 psychiatric patients given vitamin E and only in 1 of 12 patients given placebo. Assume independent beta(3, 3) prior densities for the population proportions of improvement (decrease in symptoms) in the **T**reatment and **C**ontrol groups, p_T and p_C. Define $d = p_T - p_C$.
   (a) Calculate the posterior PdALx for x = 0%, 10%, 20%, and 30%.
   (b) Find the 68% posterior probability interval for d.

9.17 Exercise 3.5 (page 72) describes a study that investigated the relationship between having experienced physical or sexual abuse as a child and developing post-traumatic stress disorder (PTSD) as a soldier. Of 38 Vietnam combat veterans seeking help for PTSD, 11 reported experiencing childhood physical or sexual abuse. Of 28 Vietnam combat veterans seeking medical care unrelated to PTSD, only 2 reported incidences of abuse. The rate of child abuse among U.S. combat veterans of Vietnam has for me a beta(1, 3) prior density—this is so both for those seeking help for PTSD and for those not.
   (a) Find an 80% posterior probability interval for the difference in rates of child abuse in the two groups (difference is PTSD group minus other group).
   (b) Test the null hypothesis of no difference between the two population proportions.

9.18 In a study[11] different from the one in the previous exercise, researchers assessed the mental statuses of 22 combat veterans and 22 POWs of the Korean War. Two of the first group and 19 of the second group suffered from PTSD. Assume independent beta(1, 1) prior densities for the population proportions of PTSD in the **C**ombat and **POW** groups, p_C and p_W, respectively. Define $d = p_W - p_C$.
   (a) Calculate the posterior PdALx for x = 40%, 60%, and 80%.
   (b) Find the 80% posterior probability interval for d.

9.19 Reconsider the violence-on-television data from Exercises 8.1 and 8.7 that follow. Assume beta(1, 1) prior densities for the population proportion of **M**en (p_M) and also of **W**omen (p_W) who would answer "yes." (See Example 3.12 for a discussion of the appropriate population.) Define $d = p_W - p_M$.

	Too much TV violence?	
	Yes	No
Men	8	12
Women	17	5

   (a) Find the posterior probability that in this population more women than men would answer "yes."
   (b) Calculate PdALx for x = 0, .3, and .6.
   (c) Find the 99% posterior probability interval for d.
   (d) Test the null hypothesis of no difference between men and women on this question.

9.20 Do people with a family history of alcoholism avoid alcohol? Two Duke students[12] thought so. In April 1993, they asked 68 other Duke students whether they had a family history (including parents, siblings, grandparents, aunts, uncles, and cousins) of alcoholism and how many alcoholic drinks they had per week. Of the 29 who said they had a family history, 15 abstained from alcohol ($\frac{15}{29}$ = 51.7%). Of the 39 who did not have a family history, 17 abstained ($\frac{17}{39}$ = 43.6%). Assume these are random samples from larger populations. Assume beta(3, 3) prior distributions for both p_H and p_N, the proportions of abstainers among those with family **H**istories and with **N**o family histories, and let $d = p_H - p_N$.
   (a) Find the posterior probability of $d > 0$.
   (b) Find the 95% posterior probability interval for d.

9.21 Example 3.7 (page 68) describes a procedure called gastric freezing for treating ulcers and the results of a randomized trial (see table). Take the perspective of someone who is persuaded by prior anecdotal evidence and has a beta(9, 1) density for the population proportion of successes for **G**astric freezing (p_G)—that equates to a 90% probability of success—and a beta(5, 5) density for the corresponding proportion on **C**ontrol (p_C). Define $d = p_G - p_C$. Test the null hypothesis $d = 0$ at:
(a) 6 weeks.
(b) 24 weeks.

Randomized trial of Example 3.7 for Exercise 9.21

After	Gastric Freezing		Control	
	Total	Success	Total	Success
6 weeks	82	64 (78%)	78	53 (68%)
24 weeks	71	39 (55%)	70	43 (61%)

9.22 Three Duke students[13] were interested in whether basketball players are more effective when under less pressure. They considered the three-point shots attempted by Duke's 1992–1993 basketball team in the **F**irst half vs **S**econd half of games, thinking there would be more pressure in the second half. Regard the following as random samples from larger populations: Of the 211 three-point shots attempted in the first half, 71 or 33.6%, were successful; of the 255 attempted in the second half, 90 or 35.3% were successful. Assuming beta(1, 1) prior densities for both success proportions, find the 99% posterior probability interval for $d = p_F - p_S$.

9.23 The three students cited in Exercise 9.22 also collected data on free throws, wondering whether first or second attempts were more successful—and by how much. Duke's 1992–1993 basketball team had 345 opportunities to shoot two free throws. They made 267 (77%) of the **F**irst and 250 (72%) of the **S**econd. Regard these as independent random samples from larger populations. [They are paired (success then success, success then failure, and so on), but you are to ignore the pairings—I have not told you this information in any case.] Assume beta(1, 1) prior densities for both success proportions and let $d = p_F - p_S$.
(a) Find the 90% posterior probability interval for d.
(b) Find PdAL0, PdAL.05, and PdAL.1.

9.24 Patients with advanced breast cancer were evaluated for their response to **H**ormonal therapy and to **C**hemotherapy in two separate studies. (Patients were not randomized to therapy, but assume that they are exchangeable except for therapy.) Consider the proportions of favorable responses (improvement in disease symptoms) in the two populations (the populations of patients receiving hormonal therapy and those receiving chemotherapy). Assume beta(1, 1) prior densities for both p_H and p_C.

There were 99 patients in each study, with 30 favorable responses to hormonal therapy and 32 favorable responses to chemotherapy. Find your posterior probability that chemotherapy is better than hormonal therapy for advanced breast cancer.

9.25 Exercise 7.14 (page 225) addresses the question of whether opposites attract. The same Duke students included males as well as females in their survey, with the results given in the table at the top of page 300. There seems to be little difference between **F**emales and **M**ales, but the sample sizes are small. Their sizes result in a moderately large probability interval for $d = p_F - p_M$, where these are proportions of yes responses. Find a 95% probability interval for d, assuming beta(1, 1) prior densities for both proportions.

	Attracted to Opposites?	
	Yes	No
Females	8	20
Males	9	19

9.26 Exercise 8.10 (page 266) considered whether agricultural workers who experience short-term exposure to phenoxy acid or chlorophenols have an increased risk of cancer. These were the study results:

	Exposed	Unexposed	Total
Cancer	11	16	27
Not cancer	9	44	53
Total	20	60	80

In Exercise 8.10, I assessed my prior probabilities for p_U (population proportion of cancer among those **Unexposed**) alone, as follows:

p_U	.1	.2	.3	.4	.5	.6	.7	.8	.9
Probability	.20	.25	.25	.15	.07	.04	.02	.01	.01

The mean and standard deviation of this distribution are $r = .295$ (add up the products of the two rows in the table) and $t = .166$. Assuming a beta(a, b) density, to find a and b, use the following formulas from Chapter 7:

$$a = \frac{r(1 - r^+)}{r^+ - r} \qquad b = \frac{(1 - r)(1 - r^+)}{r^+ - r} \qquad t = \sqrt{r(r^+ - r)}$$

The formula for t gives $r^+ = t^2/r + r = .389$. Then $a = 1.92$ and $b = 4.59$. So use beta(1.92, 4.59) prior densities for both p_U and p_E (for **Exposed**). (Having independent prior densities does not capture the greater probability of the null hypothesis $p_U = p_E$ that is present in the prior assignment for Exercise 8.10.) Find the 80% posterior probability interval for $p_E - p_U$.

9.27 Exercise 3.25 (page 103) describes a randomized 2 × 2 factorial study of the drugs deprenyl and vitamin E for the treatment of patients with Parkinson's disease. The results given here[14] are based on an average of 14 months of follow-up. Assume a beta(1, 1) prior distribution for each of the population proportions.

(a) Consider the "okay" rates of **Deprenyl** vs Control ($d = p_D - p_C$). Find PdAL0, PdAL.1, and PdAL.2.

(b) Consider the "okay" rates of vitamin **E** vs Control ($d = p_E - p_C$). Find PdAL0, PdAL.1, and PdAL.2.

(c) Consider the "okay" rates of **Both** vitamin E and deprenyl vs vitamin **E** alone ($d = p_B - p_E$). Find PdAL0, PdAL.1, and PdAL.2.

	Deprenyl	Vitamin E & Deprenyl	Vitamin E	Control (Placebo)
Okay	122 (60%)	123 (62%)	93 (46%)	86 (43%)
Disabled	80	74	109	113
Total	202	197	202	199

9.28 Fatigue, heightened irritability, and a sense of demoralization make up something psychologists call "vital exhaustion."[15] Apparently, it is associated with poor health. Researchers interviewed 127 adults who had undergone angioplasty to widen at least one severely restricted coronary artery; 43 reported experiencing vital exhaustion and 84 did not. In the vitally **E**xhausted group, 14 (33%) suffered one of the following cardiac reactions: chest pains, heart attack, or sudden death; only 14 (17%) in the **O**ther group experienced one of these reactions. Assume beta(1, 3) distributions for both proportions of cardiac reactions.

(a) Find a 99% posterior probability interval for $d = p_E - p_O$.
(b) Test the hypothesis that $d = 0$.

9.29 Example 9.6 addresses the prevention of man-to-woman sexual transmission of HIV by the use of condoms. Data from a subsequent study[16] are summarized in the following table. The participants were couples in which the man had tested positive for HIV and the woman had tested negative; 71 of the couples never used condoms, 55 sometimes did, and 171 always did. A total of 19 women converted to seropositive, as shown in the table. Combine the categories "Never" and "Sometimes" into "Not always" as in the second table, and test the null hypothesis that condom use does not matter. Take the two proportions to have independent beta(1, 1) prior densities.

Condom use and HIV-seroconversion among 305 serologically discordant couples for Exercise 9.29

	Never	Sometimes	Always
Conversions	8 (10%)	8 (15%)	3 (2%)
Not converted	71	47	168
Total	79	55	171

	Not always	Always
Conversions	16 (12%)	3 (2%)
Not converted	118	168
Total	134	171

9.30 Following up on Exercise 9.29, different brands of condoms are not equally effective. A study[17] shows that some brands leak viruses during simulated coitus. There may even be variability for different styles within brands, as the following table suggests. Assume independent beta(1, 1) densities for both population proportions.

(a) Find the 95% posterior probability interval for the increase in leakage rate of TNR over TR.
(b) Test the null hypothesis of no difference between leakage rates of TNR and TR.

Viral leakages for Trojan Naturalube Ribbed (TNR) vs Trojan Ribbed (TR) condoms for Exercise 9.30

	TNR	TR
Leaked	21 (23%)	8 (9%)
Not leaked	71	81
Total	92	89

9.31 Some women whose sex partners will not use condoms resort to spermicide to avoid transmission of HIV. A study[18] addressed the effectiveness of a vaginal sponge containing a spermicide called nonoxynol 9. Seronegative prostitutes in Nairobi, Kenya, were instructed in the use of the sponge and were randomized to a sponge plus spermicide and to a placebo cream or suppository. All were instructed on the importance of using condoms as well. The data are shown in the table. Assume independent beta(1, 1) densities for both population proportions and test the null hypothesis of no difference between seroconversion rates of sponge and placebo.

Sponge with spermicide vs placebo and HIV-seroconversion among 116 serologically negative prostitutes in Kenya for Exercise 9.31

	Spermicide	Placebo
Sero+	27 (45%)	20 (36%)
Sero−	33	36
Total	60	56

(The seroconversion rate was actually worse in the spermicide group. Apparently, the participants were not blinded as to treatment assignment. Blinding is important for separating out treatment effect. It is possible that prostitutes assigned to the sponge were less likely to use condoms, feeling some confidence in the spermicide. Such an effect can mask any benefit of the spermicide.)

9.32 Does retaining students (holding them back to repeat a grade) affect whether they eventually drop out of school? A researcher[19] tracked 100 students in three California counties. All 100 had failing grades in the seventh grade—47 were retained in that grade and 53 were promoted to eighth grade. After 4 years, 32 (68%) of the 47 and 21 (39%) of the 53 had dropped out of school.

(a) Regard these as random samples from the two populations and assume independent beta(1, 1) prior densities for the two population proportions of dropping out within 4 years. Find the posterior probability that the retained population has a higher dropout rate than does the promoted population.

(b) Part (a) does not address an interesting comparison because it does not mean that promoting failing students will decrease their rate of dropping out. Why not? How would you design a study that does address the possibility that retention affects dropout rate? (Refer to Chapter 3, especially Section 3.2.)

9.33 A company is concerned about an annoying, but not limiting, defect in one of its products. The defect cannot be found on the assembly line, but must be tested in real or simulated conditions. In attempting to isolate the cause, a total of 168 items from five different manufacturing lots were tested with the following results:

	Lot of manufacture				
	1	2	3	4	5
Defective	11 (28%)	11 (55%)	23 (82%)	14 (22%)	1 (6%)
Good	29	9	5	50	15
Total	40	20	28	64	16

(The data are real, but I am not identifying the product or the defect.) The differences across lots suggest a lot-specific cause of the defect. Assuming independent beta(1, 1) prior densities for all proportions, test the hypothesis that $d = 0$ in comparing the following lots.

(a) Lots 1 and 2
(b) Lots 1 and 3
(c) Lots 1 and 4
(d) Lots 1 and 5

9.34 A report[20] indicated that U.S. military enlisted men were treated more harshly than officers for bringing home illegal war trophies after the Gulf War—most of the trophies were AK-47 rifles. The report indicated that court-martial was "a punishment the lower ranks could expect to face far more often than officers. Twenty-three percent of the enlisted Army, Air Force and Marine members were court-martialed. Only 14 percent of the officers were." To be specific, of 159 enlisted men charged, 37 were court-martialed and of 43 officers charged, six were court-martialed. (Especially in view of the small numbers of cases, "far more often" seems an exaggeration.) Consider these to be random samples from larger populations and assume independent beta(1, 1) densities for both proportions of court-martial.

(a) Find the posterior probability that the court-martial proportion is greater for enlisted personnel than for officers.

(b) Calculate the 99% posterior probability interval for the difference between the two population proportions.

(c) Test the hypothesis that the population proportions of court-martial are the same for both officers and enlisted personnel.

9.35 Hyperactive boys seem to graduate from high school at a lower rate than do normally active boys. But more may grow up to own their own businesses. A study[21] involved 91 men (average age: 26) who were hyperactive as boys. Sixteen of them owned their own small business; this compares with only five small-business owners of 95 normally active men in a control group. (Examples of small businesses include bicycle store, automobile repair shop, burglar alarm business, cookie franchise, limousine service, dog-training school.) Assume these are random samples from the corresponding populations. (A reservation: Effecting this is difficult. Most selection procedures either underrepresent or overrepresent small-business owners. The researchers reviewed charts of white males of the same age and socioeconomic status at the Long Island Jewish Medical Center, selecting those who had no behavioral problems at school, as reported by parents. I suspect that this underrepresents small-business owners who are likely neither to have medical insurance nor to have charts at LIJMC.) Test the hypothesis of no difference between population proportions assuming independent beta(1, 5) prior densities.

9.36 A study[22] carried out on macaques (a kind of Asian monkey) concluded that immune function is boosted by long-term social ties; but the focus of this exercise is on another aspect of the study. Forty-three male macaques were housed for 14 months in unchanging groups of four or five members. Twenty-two continued in their groups for another 26 months. (These were selected randomly; presumably, five groups were selected randomly from 10 groups in all.) The other 21 individuals were kept in "unstable" groups—with three or four members leaving each month to live with another group. An observation from the study was that 14 of the 21 in unstable groups displayed "affiliative" behavior (friendly touching or grooming another animal) while only 8 of the 22 in stable groups displayed such behavior. Test the hypothesis of no difference in population proportions assuming independent beta(1, 1) prior densities.

[As usual, you assumed independent observations within the samples. This may not be correct. Affiliative behavior may be contagious: "You scratch my back and I'll scratch yours." Depending on the way in which affiliative behavior is displayed and affected by the behavior of

9.37 Is alcoholism hard-wired? Does it come with our genes or is it conditioned by environment? Substantial evidence exists that there is a genetic component of alcoholism for men. A study[23] of twins addressed the issue for women. Of 1,180 women in 590 identical twinships, 61 were alcoholics (using the most stringent of definitions). These represented a total of 53 twinships; in 8 of them, both twins were alcoholics. (For the 61 alcoholics, 16 of their cotwins were also alcoholics for a concordance rate of $\frac{16}{61} = 26.2\%$.) Of 880 women in 440 fraternal twinships, 67 were alcoholics. They represented a total of 63 twinships; in 4 of them, both twins were alcoholics. (For the 67 alcoholics, 8 of their cotwins were also alcoholics for a concordance rate of $\frac{8}{67} = 11.9\%$.) Researchers assessed the level of alcoholism and whether twins were identical while blinded as to the other factor. Summarizing:

Sample proportions of twinships in which both are alcoholics: Data for Exercise 9.37

	Identical	Fraternal
Both	8 (15%)	4 (6.3%)
One	45	59
Total	53	63

Assume beta(1, 1) prior densities for the population proportions of "both." Identical twins are genetically identical, while fraternal twins share about half of each other's genes. If the population proportions of "both" are equal, then alcoholism does not have a genetic explanation. Test the hypothesis that the two proportions are equal.

(The researchers used a different analysis than the preceding one, and indeed they do not provide the information in the form of the table given here. They say their analysis is best and conclude that "genetic factors play a major etiologic role in alcoholism in women." While there is some evidence in their study that alcoholism in women is hard-wired, most women with the same genes as their alcoholic cotwins are not themselves alcoholic. "Major etiologic role" is a major exaggeration.)

9.38 Is sexual orientation genetic? A study[24] addressed this possibility for men. (Exercise 9.39 deals with a similar study for women and Exercise 12.17 deals with quite a different study of the same question for men.) By advertising in gay magazines, the researchers identified 167 gay or bisexual men who were identical twins or who had adoptive brothers. In each case, they ascertained the sexual orientation of the man's twin or adoptive brother and found the proportions of gay relatives given in the following table.

	Identical	Fraternal	Adoptive	Nontwin Brother
Gay[†]	29 (52%)	12 (22%)	6 (11%)	13 (9.2%)
Straight	27	42	51	129
Total	56	54	57	142

[†]Homosexual or bisexual.

The 110 twin probands (as psychologists call the individuals in the original sample) reported 142 nontwin biological brothers about whose sexual orientation their brothers were virtually certain. The last column gives the results for nontwin brothers.

(a) In trying to estimate the relative contributions of genetics and environment on the basis of these results, there are enormous difficulties in adjusting for biases due to the sampling process. (The researchers recognize this.) But if the population proportion of identical twins who are gay is different from that of fraternal twins, then there *is* a genetic component. In the sampling process, any biases (such as willingness to volunteer for the study related to the sexual orientation of one's cotwin) are likely to be present in both the identical and fraternal group. So assuming random sampling is probably not unreasonable when comparing these two groups. Also, assume independent beta(1, 1) prior densities and test the hypothesis that the proportions of gays for identical and fraternal twins are equal.

(b) There may be a differential bias in comparing fraternal twins with nontwin brothers. The genetic relationship is the same and so the population proportions of gays should be the same. Assuming random samples and beta(1, 1) prior densities, test the hypothesis that they are the same. (An explanation for any difference is that the nontwin brothers are different in age and so may not as readily confide in their brothers. Another explanation related to age is that there is an environmental difference even within families. The former seems more likely—the authors cite a separate study in which 22% of the nontwin brothers of gays were gay.)

9.39 The researchers cited in Exercise 9.38 conducted a similar study[25] among women. The information provided in the original article is not specific: Participants included "115 twin pairs, about equally divided between identical and fraternal twins." In addition, "Homosexuality and bisexuality occurred among both sisters in nearly half of the identical twin pairs" and "drops to one-quarter of the fraternal twin pairs and one in six of adoptive-sister pairs." To be specific, we estimate the data to be as indicated in the following table:

	Identical	Fraternal	Adoptive
Lesbian[†]	27 (47%)	14 (25%)	5 (16%)
Straight	31	43	27
Total	58	57	32

[†]Homosexual or bisexual.

If there is no genetic component in sexual orientation of women (and barring biases in the sampling, some of which are mentioned in the previous exercise), then the corresponding population proportions are equal.

(a) Repeat your analysis of part (a) in Exercise 9.38. That is, assume independent beta(1, 1) prior densities and test the hypothesis that the proportions of lesbians for identical and fraternal twins are equal.

(b) Assume independent beta(1, 1) prior densities and test the hypothesis that the proportions of lesbians for fraternal twins and adoptive sisters are equal.

Appendix: Using Minitab with Densities for Two Proportions

The Minitab program **'pp_beta'** will summarize the posterior distribution for the difference between two proportions when independent beta prior distributions are used.

Consider Example 9.1, where one wishes to learn about the difference $p_N - p_C$, where p_N and p_C are the probabilities that noncarriers and carriers, respectively, contract malaria. The open-minded assessor (Opey) uses independent beta(1, 1) distributions to reflect his prior beliefs about the two proportions. With the data $(s_1, f_1) = (2, 13)$, $(s_2, f_2) = (14, 1)$, the prior distributions for p_N and p_C are independent with p_N being beta(3, 14) and p_C being beta(15, 2). Suppose that one is interested in graphing the posterior density for the difference in probabilities $p_N - p_C$. Also, one wishes to compute the probability PdAL that $p_N - p_C$ exceeds each of the values 0, .5, .6, .7, .8, .9. Run this program by typing

exec 'pp_beta'

This program uses a slightly different method of computing the posterior distribution of $p_N - p_C$ than the one described in this chapter. This program simulates 1,000 values of the probability p_C from a beta(3, 14) distribution and 1,000 values p_N from a beta(15, 2) distribution. The 1,000 pairs of simulated (p_C, p_N) values are displayed on a scatterplot. For each simulated (p_C, p_N) value, the program computes the difference $p_N - p_C$. The 1,000 simulated values of this difference in proportions is displayed using a dot plot. Most of the dots are located in the range .45 to .90—this means that Opey is pretty certain that the difference $p_N - p_C$ falls in the interval from .45 to .90. This program also lets one compute the probability PdAL that $p_N - p_C$ exceeds particular values of interest. On the Minitab 'DATA' line, input the values of possible improvement. The output of this program consists of the columns 'PdALx' and 'sim_se'. For a given value x, the value of 'PdALx' gives the probability that $p_N - p_C$ exceeds a value x and the value in 'sim_se' is an estimate of the error due to simulation in the computation of this probability. The probability that $p_N - p_C$ exceeds .6 is .814 with a corresponding simulation error of .012. So the exact value of this probability is .814 give or take the error of .012. (Example 9.3 finds this probability to be .816.)

```
MTB > exec 'pp_beta'

FOR PROPORTION P1,
ENTER VALUES OF BETA PARAMETERS A1 AND B1:
DATA> 3 14

FOR PROPORTION P2,
ENTER VALUES OF BETA PARAMETERS A2 AND B2:
DATA> 15 2

HOW MANY VALUES OF (P1, P2) DO YOU WISH TO SIMULATE?
DATA> 1000

TYPE 'y' TO SEE A PLOT OF THE JOINT DISTRIBUTION
OF P1 AND P2:
y
```

Ch. 9 / Appendix: Using Minitab with Densities for Two Proportions   307

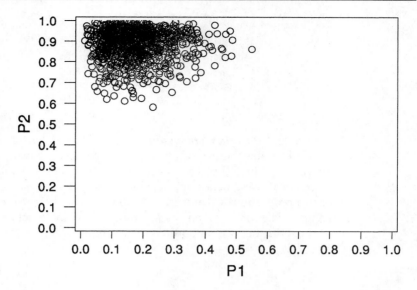

```
TYPE 'y' TO SEE A PLOT OF THE DISTRIBUTION OF
THE DIFFERENCE IN PROPORTIONS P2-P1:
y

Each dot represents 4 points


    Each dot represents 4 points


                                    ::
                                 .:. :: :
                              .  :::.::.::.
                             . :: :::::::::: .
                            : :.:::::::::::::.:
                          :..: ::::::::::::::::::
                        ..:::::::::::::::::::::::. .
           .      ...  :::::::::::::::::::::::::::::.
          -------+---------+---------+---------+---------+---------+---------P2-P1
                0.30      0.45      0.60      0.75      0.90      1.05

TYPE 'y' to COMPUTE PROBABILITIES OF IMPROVEMENT
FOR P2-P1:
    ----------------------------------------------------------------
Input values of possible improvement.
The output is the probability PdALx that P2-P1 exceeds
each improvement value x. The column sim_se gives
simulation standard deviations for the estimated probabilities.
    ----------------------------------------------------------------
y
DATA> 0 .5 .6 .7 .8 .9
DATA> end
```

ROW	x	PdALx	sim_se
1	0.0	1.000	0.000
2	0.5	0.934	0.008
3	0.6	0.814	0.012
4	0.7	0.570	0.016
5	0.8	0.249	0.014
6	0.9	0.024	0.005

The text *Bayesian Computation Using Minitab* by James Albert (Duxbury Press, 1996) illustrates the use of other density models for two binomial proportions. The program 'pp_bet_t' performs a test of the hypothesis that the two proportions p_1 and p_2 are equal using beta prior densities. The program 'pp_exch' illustrates inference for two proportions when the two proportions are believed exchangeable. This model is appropriate for use when you are unsure about the exact locations of the proportions, but think that the two proportions are of similar size.

Chapter Notes

1. M. J. Matteson, *Journal of Applied Psychology* 59 (1974): 535–536.
2. Preliminary report by the Cardiac Arrhythmia Suppression Trial (CAST) Investigators, *New England Journal of Medicine* 321 (1989): 406–412.
3. D. M. Barnes, "Promising results halt trial of anti-AIDS drug (AZT)," *Science* 234 (October 3, 1986): 15–16.
4. E. F. Loftus, "Leading questions and the eye-witness report," *Cognitive Psychology* 7 (1975): 560–572.
5. J. Albert, *Bayesian Computation Using Minitab* (Belmont, Calif.: Duxbury Press, 1996).
6. J. Mann, T. C. Quinn, P. Piot, et al., "Condom use and HIV infection among prostitutes in Zaire," *New England Journal of Medicine* 316 (1987): 345.
7. B. Voeller, J. Nelson, and C. Day, "Viral leakage through selected brands of condoms," Mariposa Occasional Paper #19 (Topanga, Calif.: The Mariposa Foundation, 1993).
8. G. Koren, *Lancet* (December 16, 1989). Excerpted in *Science News* 137 (January 6, 1990): 13.
9. Jared Diamond, *Natural History* (February 1990): 26–30.
10. Conducted by Lenard A. Adler and reported in *Science News* 141 (May 23, 1992): 351.
11. P. B. Sutker and colleagues at New Orleans Veterans Administration Medical Center, reported in *Science News* 139 (February 2, 1991): 68.
12. I am indebted to Tamara Gehris and Megan Wilson for these data.
13. I am indebted to Darriel Hoy, William Horwath, and Demian Gutierrez for these data.
14. The Parkinson Study Group, "Effects of tocopherol and deprenyl on the progression of disability in early Parkinson's disease," *New England Journal of Medicine* 324 (January 21, 1993): 176–183.
15. Reported by B. Bower, *Science News* 146 (August 6, 1994): 87. Based on an article by W. J. Kop et al. in *Psychosomatic Medicine* (1994).
16. A. Sarracco, M. Musicco, A. Nicolosi, et al., "Man-to-woman sexual transmission of HIV: Longitudinal study of 343 steady partners of infected men," *Journal of AIDS* 6 (1993): 497–502.
17. B. Voeller, J. Nelson, and C. Day, "Viral leakage through selected brands of condoms," Mariposa Occasional Paper #19 (Topanga, Calif.: The Mariposa Foundation, 1993).
18. "Spermicides may not offer HIV protection," *Science News* 142 (July 25, 1992): 54.
19. Reported by S. Finucane, *Santa Barbara News-Press* (August 14, 1994). Based on doctoral dissertation research of C. Mackey (1991).
20. R. R. Smith, "Lower ranks met harsher fates for Gulf War trophy infractions," *The News & Observer* (Raleigh, N.C., November 28, 1992): 1A.
21. S. Mannuzza, R. G. Klein, A. Bessler, et al., "Adult outcome of hyperactive boys," *Archives of General Psychiatry* 50 (1993): 565–576.
22. "Social ties boost immune function," *Science News* 142 (October 10, 1992): 237.

23. K. S. Kendler, A. C. Heath, M. C. Neale, et al. "A population-based twin study of alcoholism in women" *Journal of the American Medical Association* 268 (1992): 1877–1882.

24. J. M. Bailey and R. C. Pillard, "A genetic study of male sexual orientation," *Archives of General Psychiatry* 48 (1991): 1089–1096.

25. Conference presentation of R. C. Pillard and J. M. Bailey as reported by B. Bower, "Genetic clues to female homosexuality," *Science News* 142 (August 22, 1992): 117.

10 General Samples and Population Means

CHAPTERS 6–9 dealt with making inferences about proportions. Observations were of two types, sometimes called success and failure. When the observations are exchangeable, the data can be summarized using just the number s of successes and the number f of failures. We used probabilities to represent available knowledge of the process producing the data, and we used Bayes' rule to update these probabilities as information accumulated. An important derivation was the (predictive) probability of a success on the next observation.

This chapter and the next two chapters treat general types of observations. A general setting is a little more complicated than one with only two types of observations. But the statistical questions we consider here are not very different from those addressed in earlier chapters. In particular, we still want to give intervals to represent our uncertainty and we want to predict the next observation or set of observations.

Section 10.1 generalizes Chapter 6, which dealt with proportions only. Section 10.2 deals with normal models and with means and standard deviations. It addresses inferences when the population is assumed to be a normal model—a subject that will be addressed more extensively in Chapter 11. Chapter 12 considers two samples and two populations. Chapter 13 describes an alternative approach to inferences that does not rely on assuming that the population is a normal model.

General Population Models

Just as we did for proportions, we will deal with observations that we regard to be exchangeable. Such observations comprise a **sample,** as defined in Section 1.3. In the previous four chapters, samples contained only two types of observations: S and F. For example, in Section 6.1, we considered this sample of 25 observations:

SFFSS FFFSF SFFFS SSSFS SFSFS

The next example gives data that are both more complicated and more interesting. Its context is somewhat frivolous, but it serves to convey the ideas.

EXAMPLE 10.1
▷ **Darts.** I once watched a man play a carnival game in which he tossed darts at a board. The board was divided into a grid of 396 (= 18 × 22) very small squares. Each

square was labeled with one of the numbers 1, 2, 3, 4, 5, and 6. A dart that landed in one of these squares scored the number labeled on that square. (When the dart missed the board altogether or fell out upon landing, the player threw another until scoring one of these six numbers.) I watched him throw 30 darts, getting these scores:

$$65314 \quad 44531 \quad 44535 \quad 34343 \quad 41333 \quad 14355$$

I will call this sample D, for data. (The game's operator kept trying—unsuccessfully—to get me to play, and was visibly annoyed when I started to write things down.)

Since there are more than two types of observations, we cannot use the methods we developed for proportions. But just as we did for proportions, we would like to use the data and any other available information to predict the next observation—the player's score on his next toss. We do not have to predict a single number. Rather, we seek predictive probabilities for each of the numbers. For example, if we decide that the six numbers are equally likely, then it will be sufficient to say that. We can use these predictive probabilities to answer such questions as whether playing the game is silly. We can also use them to decide whether we would like more information—paying for it, perhaps—before concluding that the game is silly. (To come to such decisions, you would have to know how much a player must pay to play the game, what the player has to score in order to win, and what the prizes are. I will describe the game further in Exercises 10.1 and 11.5. However, being the worldly person that you are, as soon as I said "carnival game," you had an inkling as to the game's reasonableness!)

To predict the next observation, we need a way to incorporate information from the sample into what we know about the game. We proceed as usual to assess our prior information in terms of population models. Now there are six types of chips: 1, 2, 3, 4, 5, 6. The models specified in an application should reflect one's opinions concerning the process in question. To keep the calculations manageable, I will consider only the following three models—labeled M1, M2, and M3—and I will associate probability $\frac{1}{3}$ with each:

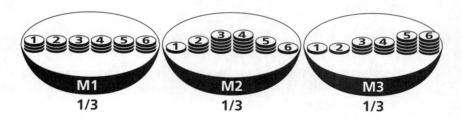

There are many alternatives to these models, and it is hardly reasonable to suppose that anyone's knowledge of a process can be adequately represented with such a small number of possibilities. But as I said, I want to keep things manageable. (As you will see in Exercise 10.1, these models and probabilities do not correspond with my actual prior knowledge in this dart example.)

Choosing to have 18 chips in each model was arbitrary. Since we replace chips between selections in a population model, any two models with the same proportions of the various numbered chips are equivalent. For example, both of these models are equivalent to model M1 (see figure at the top of page 312):

In table form, the proportions of chips in the three models are as follows:

| | \multicolumn{6}{c}{Result} |
Model	1	2	3	4	5	6
M1	3/18	3/18	3/18	3/18	3/18	3/18
M2	1/18	3/18	5/18	5/18	3/18	1/18
M3	1/18	1/18	3/18	3/18	5/18	5/18

If we knew that model M1, say, corresponded with the dart throwing of this player, then the entries in the table would be the predictive distribution for the next observation. In this example, we will assume that the correct model is one of these three, but we do not know which it is. To find the predictive distribution, we will average over the three as usual, using the law of total probability from Section 5.3.

We will write the event that the next observation is 1, for example, simply as 1. Also, we will let M1, M2, and M3 be the events that the observations are being selected from the model with that number. The (predictive) probability of 1 is

$$P(1) = P(1 \mid M1)P(M1) + P(1 \mid M2)P(M2) + P(1 \mid M3)P(M3)$$
$$= (\tfrac{3}{18})(\tfrac{1}{3}) + (\tfrac{1}{18})(\tfrac{1}{3}) + (\tfrac{1}{18})(\tfrac{1}{3}) = \tfrac{5}{54}$$

The rest of the probabilities are found similarly:

$$P(2) = (\tfrac{3}{18})(\tfrac{1}{3}) + (\tfrac{3}{18})(\tfrac{1}{3}) + (\tfrac{1}{18})(\tfrac{1}{3}) = \tfrac{7}{54}$$
$$P(3) = (\tfrac{3}{18})(\tfrac{1}{3}) + (\tfrac{5}{18})(\tfrac{1}{3}) + (\tfrac{3}{18})(\tfrac{1}{3}) = \tfrac{11}{54}$$
$$P(4) = (\tfrac{3}{18})(\tfrac{1}{3}) + (\tfrac{5}{18})(\tfrac{1}{3}) + (\tfrac{3}{18})(\tfrac{1}{3}) = \tfrac{11}{54}$$
$$P(5) = (\tfrac{3}{18})(\tfrac{1}{3}) + (\tfrac{3}{18})(\tfrac{1}{3}) + (\tfrac{5}{18})(\tfrac{1}{3}) = \tfrac{11}{54}$$
$$P(6) = (\tfrac{3}{18})(\tfrac{1}{3}) + (\tfrac{1}{18})(\tfrac{1}{3}) + (\tfrac{5}{18})(\tfrac{1}{3}) = \tfrac{9}{54}$$

How do we learn about the correct model as we make observations? Take one step at a time. Consider the first observation, a 6, and call this data D1. Bayes' rule in Section 5.4 says

$$P(M1 \mid D1) = \frac{P(D1 \mid M1)P(M1)}{P(D1 \mid M1)P(M1) + P(D1 \mid M2)P(M2) + P(D1 \mid M3)P(M3)}$$
$$= \frac{(3/18)(1/3)}{(3/18)(1/3) + (1/18)(1/3) + (5/18)(1/3)} = \frac{3}{9}$$

The denominator is simply the probability of D1 (a 6), as already calculated:

$$P(D1) = P(6) = \tfrac{9}{54}$$

The numerator is the first of the three terms in the calculation of $P(6)$: $(\tfrac{3}{18})(\tfrac{1}{3}) = \tfrac{3}{54}$. Similarly,

$$P(M2 \mid D1) = \frac{(1/18)(1/3)}{(3/18)(1/3) + (1/18)(1/3) + (5/18)(1/3)} = \frac{1}{9}$$

$$P(M3 \mid D1) = \frac{(5/18)(1/3)}{(3/18)(1/3) + (1/18)(1/3) + (5/18)(1/3)} = \frac{5}{9}$$

As they must, these three posterior probabilities sum to 1.

The probabilities of D1, given the various models, are likelihoods. From Chapter 6,

> **The likelihood of a model is the probability of the data calculated assuming that model.**

Since there is no constraint on the likelihood that its total must be 1, likelihoods are easy to handle. We need worry about making the total equal to 1 only after multiplying likelihoods by prior probabilities:

> **Bayes' Rule**
>
> **Calculate the posterior probability of a model by multiplying its prior probability and its likelihood. To make the total probability 1, divide by the sum of the products over all models.**

The likelihoods of the models are determined by the proportions of chips similar to the one observed. In this case we observed a 6-chip:

So the likelihoods are $\frac{3}{18}$, $\frac{1}{18}$, and $\frac{5}{18}$ for models M1, M2, and M3. Since the prior probabilities are equal, the posterior probabilities are proportional to 3, 1, and 5 (the common factor $\frac{1}{18}$ is irrelevant). Divide each by their sum $(3 + 1 + 5 = 9)$ to obtain the posterior probabilities. These are shown in Figure 10.1 with the 6-chips darkened. The probabilities of the chips change in proportion to the number of 6-chips (that is, the observation) in the bowl.

Figure 10.1 Updated probabilities of models are proportional to numbers of 6-chips

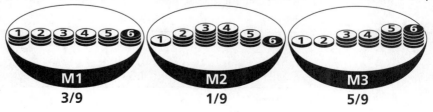

Another way of viewing this calculation (if you aren't weary of it yet!) is presented in the tree diagram in Figure 10.2. The posterior probabilities (given that the first result is 6) of M1, M2, and M3 are proportional to the probabilities along the branches of this tree that correspond to chips labeled 6; these are shown in the boxes.

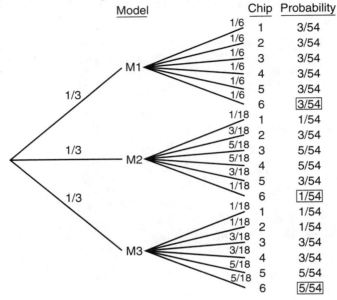

Figure 10.2 Tree diagram showing Bayes' rule calculation when a 6-chip is observed

Now consider the first *two* observations, 6 and 5, and call these data D2. Just as when we considered proportions, we will assume that the observations are independent when sampling from a particular model. So again likelihoods multiply. The likelihoods for the three models are therefore as follows:

Model	Likelihood of 6	Likelihood of 5	Likelihood for D2
M1	3	3	9
M2	1	3	3
M3	5	5	25

Then we apply Bayes' rule for data D2:

Model	Likelihood for D2	P(Model)	Product	P(Model | D2)
M1	9	1/3	9/3	9/37 = .24
M2	3	1/3	3/3	3/37 = .08
M3	25	1/3	25/3	25/37 = .68
Sum			37/3	

These posterior probabilities are also pictured in Figure 10.3.

Figure 10.3 Probabilities change in proportion to the product for two observations, 5 and 6

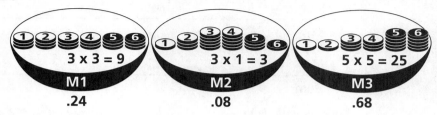

Now we look at the full data, D, repeated and reordered here to make counting easier:

$$11113 \quad 33333 \quad 33334 \quad 44444 \quad 44455 \quad 55556$$

From the law of total probability,

$$P(D) = P(D \mid M1)P(M1) + P(D \mid M2)P(M2) + P(D \mid M3)P(M3)$$

Consider $P(D \mid M2)$. This is the probability of D given model M2. But it is also the likelihood of M2 for the data D. Because there is only one 1-chip and only one 6-chip in M2, each has probability $\frac{1}{18}$ when conditioning on M2. So $P(D \mid M2)$ has a factor of $\frac{1}{18}$ each time there is a 1 or a 6 in D. Similarly, each occurrence of 2 or 5 brings in a factor of $\frac{3}{18}$ and each 3 or 4 brings in a factor of $\frac{5}{18}$. Since there are four 1's and one 6 (total of 5), no 2's and six 5's (total of 6), and ten 3's and nine 4's (total of 19), for data D the likelihood of model M2 is

$$P(D \mid M2) = \left(\frac{1}{18}\right)^5 \left(\frac{3}{18}\right)^6 \left(\frac{5}{18}\right)^{19} = 1^5 \, 3^6 \, 5^{19} \left(\frac{1}{18}\right)^{30}$$

Similarly,

$$P(D \mid M1) = \left(\frac{3}{18}\right)^{30} = 3^{30} \left(\frac{1}{18}\right)^{30}$$

and

$$P(D \mid M3) = \left(\frac{1}{18}\right)^4 \left(\frac{3}{18}\right)^{19} \left(\frac{5}{18}\right)^7 = 1^4 \, 3^{19} \, 5^7 \left(\frac{1}{18}\right)^{30}$$

Since the factor $(\frac{1}{18})^{30}$ appears in all three, we can drop it throughout:

$$\begin{aligned}
\text{Likelihood of M2:} &\quad 1^5 \, 3^6 \, 5^{19} = 1{,}390\text{E}+13 \\
\text{Likelihood of M1:} &\quad 3^{30} \phantom{\, 3^{19} \, 5^7} = 21\text{E}+13 \\
\text{Likelihood of M3:} &\quad 1^4 \, 3^{19} \, 5^7 = 9\text{E}+13
\end{aligned}$$

All that matters are likelihoods in relation to the other likelihoods, so we can also drop the E+13 from all three. (As before, anything canceled has to be the same in every likelihood—it would not be correct to write the likelihood of M2 as 1.39E+16 and drop E+16 from it and E+13 from the others.) The posterior probabilities are found as follows (considering 1 first):

Model	Likelihood for D2	P(Model)	Product	P(Model ∣ D2)
M1	21	1/3	21/3	21/1,420 = .015
M2	1,390	1/3	1,390/3	1,390/1,420 = .979
M3	9	1/3	9/3	9/1,420 = .006
Sum			1,420/3	

So, *assuming that the correct model is one of these three,* M2 is very likely to be it.

We can find the predictive distribution of the 31st observation given the first 30 as follows (considering 1 first):

$$\begin{aligned}P(1 \mid D) &= P(1 \mid M1, D)P(M1 \mid D) + P(1 \mid M2, D)P(M2 \mid D) \\ &\quad + P(1 \mid M3, D)P(M3 \mid D) \\ &= P(1 \mid M1)P(M1 \mid D) + P(1 \mid M2)P(M2 \mid D) \\ &\quad + P(1 \mid M3)P(M3 \mid D) \\ &= (\tfrac{3}{18})(.015) + (\tfrac{1}{18})(.979) + (\tfrac{1}{18})(.006) = .058\end{aligned}$$

Here are all the predictive probabilities in table form:

Observation	1	2	3	4	5	6
P(Observation ∣ D)	.058	.169	.281	.281	.171	.060

Because model M2 has almost all the posterior probability, these are very similar to the proportions of chips in M2. ◁

EXERCISES

10.1 Consider the dart game of Example 10.1. Assuming the player is sufficiently far from the board (or that he is a sufficiently bad shot!), I would put some credence in the model—call it M4—in which each of the 396 squares on the dart board has the same probability. With this model in mind, I counted the squares on the dart board corresponding to each of the six numbers. The proportions for M4 are shown in the following table, along with those of the three models considered in Example 10.1 (repeated here for convenience):

Model/Result	1	2	3	4	5	6
M1	3/18	3/18	3/18	3/18	3/18	3/18
M2	1/18	3/18	5/18	5/18	3/18	1/18
M3	1/18	1/18	3/18	3/18	5/18	5/18
M4	57/396	29/396	142/396	114/396	27/396	27/396

Consider models M1, M2, M3, and M4 in the same way I considered M1, M2, and M3 in Example 10.1. Namely, give each of the four the same prior probability ($\tfrac{1}{4}$) and calculate their posterior probabilities based on data D.

10.2 Reconsider the dart game of Example 10.1 and Exercise 10.1. Again consider models M1, M2, M3, and M4. But now you have more data. In addition to the 30 observations given in the

Sec. 10.1 / General Population Models

example, consider a second set of 30 dart throws observed from the same player. All 60 results are listed here, with the first row repeated from the example:

```
6 5 3 1 4    4 4 5 3 1    4 4 5 3 5    3 4 3 4 3    4 1 3 3 3    1 4 3 5 5
3 1 4 3 4    3 3 3 4 2    3 3 4 6 5    3 4 3 1 2    4 6 3 3 3    6 3 4 6 3
```

(In the 60 throws, there are six 1's, two 2's, twenty-four 3's, sixteen 4's, seven 5's and five 6's.) Again, give each of the four models the same prior probability ($\frac{1}{4}$) and calculate their posterior probabilities.

10.3 Tossing two coins will result in either 0, 1 or 2 heads. Two of your friends, Jon and Sue, disagree about the corresponding probabilities. Jon is convinced that they are equal, each being $\frac{1}{3}$. Sue is also convinced, but she thinks that 0 heads will result one-quarter of the time, 1 head half the time, and 2 heads one-quarter of the time. You are wisely open-minded about this issue, and for now give equal probability to each of your friends being right. The two models are shown with your probabilities indicated. Having a scientific bent, you try to settle the issue by carrying out an experiment. Select two coins and toss them a total of 50 times. (The total number of coin tosses is 100, but you must observe them in pairs.) Keep track of the number of times in the 50 tosses that you get 0 heads, 1 head, and 2 heads.

Models for Exercise 10.3

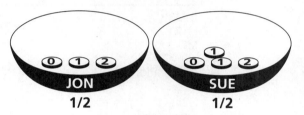

(a) Of the 50, how many times did you get 0 heads? 1 head? 2 heads?
(b) Calculate your posterior probabilities of the two population models. That is, find $P(\text{Jon} \mid \text{your data})$ and $P(\text{Sue} \mid \text{your data})$.
(c) Given your data, what is your predictive probability of observing 1 head (and 1 tail) in the next toss of the two coins?

10.4 Consider families with four children. Not many have all girls or all boys. If the sexes of different children are independent and if each is a boy with probability $\frac{1}{2}$, then the proportion of families having all girls is $(\frac{1}{2})^4 = \frac{1}{16}$. The same holds for all boys. For the probability of 1, 2, or 3 girls, one can proceed by counting the number of possibilities as in Section 7.7. These probabilities are shown in the model labeled INDEP. But maybe the sexes are not independent. Maybe some couples are good at having girls and others are good at having boys. If all couples were perfect in this regard, then there would be only single-sex families, and this would be the model:

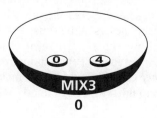

The label MIX3 indicates that the corresponding population is an equal mixture of subpopulations of couples; one subpopulation can have only girls and the other can have only boys. The 0 below the model indicates my prior probability (I know there are families with 1, 2, and 3 girls). You entertain two other models, called MIX1 and MIX2, as shown in the picture and table. These are not as extreme as MIX3, but they are consistent with some couples having a probability of a girl greater than $\frac{1}{2}$ and others having a probability less than $\frac{1}{2}$.

Model proportions for Exercise 10.4

Number of girls	0	1	2	3	4
INDEP	1/16	4/16	6/16	4/16	1/16
MIX1	2/20	5/20	6/20	5/20	2/20
MIX2	1/8	2/8	2/8	2/8	1/8

You do some research and find 100 families of four children, with the following distribution of number of girls:

Number of girls	0	1	2	3	4
Frequency	9	24	30	26	11

[There are 206 girls (51.5%) of the 400 children, but this fact is irrelevant here.]

(a) Calculate your posterior probabilities of the three population models. That is, find $P(\text{INDEP} \mid \text{above data})$, $P(\text{MIX1} \mid \text{above data})$, and $P(\text{MIX2} \mid \text{above data})$.

(b) Given the data, what is your predictive probability of observing 2 girls (and 2 boys) in the next family with four children selected from this same population?

Normal Models

The methods of Example 10.1 were more complicated than those used for proportions in earlier chapters. On the other hand, Example 10.1 was easier than many examples in earlier chapters in the sense that it involved only three models. Yet this characteristic makes it unrealistic. There may be settings in which there are just a few competing theories; but in most applications of statistics (and in dart throwing, in particular), there are many possible models. To make the example more realistic, we should consider a large enough number of models that we can anticipate at least one being close to correct.

Specifying a large number of models is not easy because there are so many possibilities to consider. In Chapters 6–9, we had only to worry about the probability of success (the probability of failure being 1 minus it). But in the dart game example, we have to consider the probabilities of 1, 2, 3, ..., and 6. Specifying models gets even harder when there are more than six possible observations. Before we can handle problems with many possibilities, we need to greatly improve the efficiency of our calculations. This section takes a step in that direction.

When we introduced densities in Chapter 7, we increased the number of models by making a finer division in the proportion of S-chips. But the models themselves continued to have only two types of chips: S and F. In this section, we again consider densities of population models. But now the form of the data is more complicated than data that consist of two types. It is also more complicated than the six possible outcomes of Example 10.1. Now we are also going to smooth out the composition of the models themselves. That is, we are going to smooth *within* models, as well as *across* models.

The method we develop will be quite easy to use, but developing it is not easy. You may want to skim through the development first so you can appreciate how easy it is once you reach the end. Then come back and read more carefully when you see that traveling the road is indeed worthwhile.

In Section 2.5, we introduced data histograms as a pictorial method of showing a sample distribution. If we had the entire population we could make a **population histogram** in the same way. If we know the shape of the population histogram, sampling from it is not very interesting because we learn nothing from the sample—just as tossing a coin that we know is fair is not very interesting. But imagine taking a sample from a known population distribution. When the sample is very large, the data histogram will look like the population histogram. This fact is really the law of large numbers of Section 6.2, where it was stated for proportions. We are not dealing with proportions in this chapter, but we can apply the results to other types of observations, as in the next example.

EXAMPLE 10.2

▷ **Piglet weight gains.** Example 2.17 considers the histogram of weight gains for $n = 100$ young pigs who had been fed a certain diet. Suppose a "success" is a weight gain of 20–24 lb. When $n = 10$, the proportion of weight gains that fall in this interval is $\frac{2}{10} = 20\%$; after $n = 20$, it is $\frac{6}{20} = 30\%$; after $n = 50$, it is $\frac{8}{50} = 16\%$; and after $n = 100$, it is $\frac{12}{100} = 12\%$. The law of large numbers (Section 6.2) means that these sample proportions will stabilize, and that they tend to the population proportion. The same is true for the proportions in every other interval. The histograms in Figure 10.4 (page 320) show the first 10, 20, 50, and 100 weight gains. (To show proportions accurately, the vertical heights have been adjusted to make the total shaded area equal to 1.) The individual proportions fluctuate initially but eventually stabilize, just as for the coin tosses in Section 6.2. Moreover, the histograms themselves become smoother as the sample distribution tends to look more and more like the population distribution. No one knows what the population histogram looks like.

Figure 10.5 shows a normal curve (or density) superimposed on the $n = 100$ histogram, suggesting that this type of curve may be the limit.

Figure 10.4 Data histograms for piglet weight gains: Different sample sizes for Example 10.2

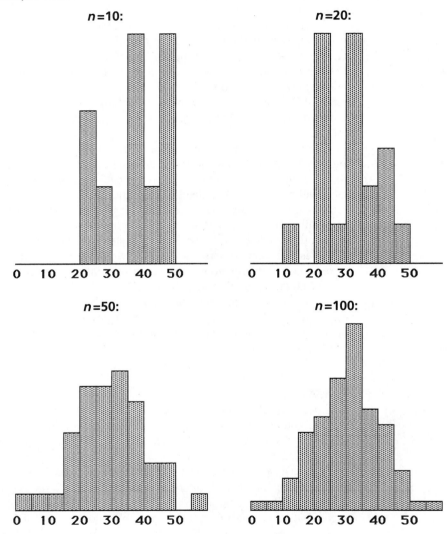

Figure 10.5 Normal density superimposed on $n = 100$ histogram from Example 2.17

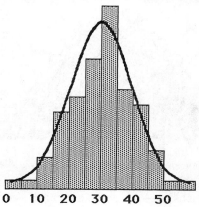

◁

> **Law of Large Numbers**
> **For large samples, a data histogram is similar to the population density.**

Data histograms sometimes have the shape of a normal density, as in the previous example. The reason for this shape may be that the population is itself a bell or normal density. Assuming that the population is normal leads to a rich theory in which calculations are not very difficult.

> **Population distributions sometimes have the shape of a normal density.**

In calculating likelihoods, think of the density in the same way you are used to thinking about chips in bowl models, where the height of a stack of chips labeled with the observation is the height of the density at that observation. Figure 10.6 (page 322) is an example, where the observation in question is a weight gain. The chips are not labeled in the figure, but think of each chip as being labeled with the weight gain it is sitting on.

To consider models for populations that are normal densities requires learning something about the mathematical description of these densities. We encountered normal curves in Section 7.5 when calculating areas under beta(a, b) densities. We used z-scores. For example, the z-score of proportion .5 is

$$z = \frac{.5 - r}{t}$$

where $r = a/(a + b)$ is the center or mean of the beta(a, b) density and t is its standard deviation. We are not discussing proportions or beta densities in this chapter, but we can still talk about the center or mean of a density, as well as about the standard deviation of a density.

Figure 10.6 Normal density with chips indicating corresponding weight gain

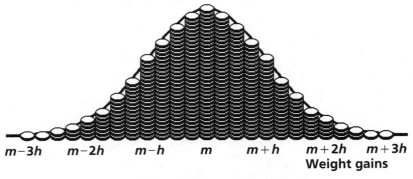

$$m-3h \quad m-2h \quad m-h \quad m \quad m+h \quad m+2h \quad m+3h$$
Weight gains

If m is the mean of a density—say, of weight gains—and h is the standard deviation, then the **z-score** of any particular weight gain is

$$z = \frac{\text{Weight gain} - \text{Mean weight gain}}{\text{Standard deviation of weight gain}}$$

$$= \frac{\text{Weight gain} - m}{h}$$

Figure 10.7 shows how a normal density depends on the z-score and how z-scores depend on the original scale of the data. A normal density has the following mathematical form, given in terms of the z-score:*

$$\text{Height of normal density} = e^{-z^2/2}$$

where $e = 2.718$.†

Figure 10.7 Normal density showing weight gains and z-scores

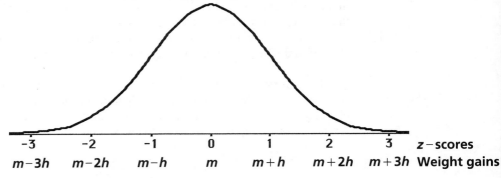

---

*In other books, you may see this formula with a multiplicative constant, $1/\sqrt{2\pi}$; this constant serves only to make the total area under the curve equal to 1. I have not included it because we are interested in likelihoods—heights of the bell curve—and so constant multiples are irrelevant.

†The number e is the base of the natural logarithm. More accurately, $e = 2.718281828459\ldots$.

The normal density is symmetric about $z = 0$, since the value of z^2 is the same for $z < 0$ as it is for $z > 0$. If your calculator has an e^x key, you can use it to calculate the height of the normal. To calculate $-z^2/2$ use this sequence: $x^2, +/-, \div, 2, =$. Then, press e^x to calculate $e^{-z^2/2}$. Some calculators use **INV** and then **ln** x in place of e^x. If your calculator does not have an e^x key, but does have a y^x key, then input **2.718**, followed in turn by y^x and $-z^2/2$ and, finally, $=$. If your calculator has none of these niceties, you can interpolate from Table 10.1, a short table of *heights* of the normal curve—*not areas* under the curve (the latter are given in the Standard Normal Table inside the front cover of this text).

Table 10.1
Heights of normal density for z-scores

z	Normal	z	Normal	z	Normal	z	Normal
0	1	1.0	.6054	2.0	.1353	3.0	.0111
.1	.9950	1.1	.5461	2.1	.1103	3.1	.0082
.2	.9802	1.2	.4868	2.2	.0889	3.2	.0060
.3	.9560	1.3	.4296	2.3	.0710	3.3	.0043
.4	.9231	1.4	.3753	2.4	.0561	3.4	.0031
.5	.8825	1.5	.3247	2.5	.0439	3.5	.0022
.6	.8353	1.6	.2780	2.6	.0340	3.6	.0015
.7	.7827	1.7	.2357	2.7	.0261	3.7	.0011
.8	.7261	1.8	.1979	2.8	.0198	3.8	.0007
.9	.6670	1.9	.1645	2.9	.0149	3.9	.0005

The mean of z-scores is 0 and the standard deviation is 1. Consider the scale of original measurements rather than the z-scale. A normal density centered at m and having standard deviation h is a **normal(m, h) density**.

In the next example we use the sample data to update information about weight gains of pigs.

EXAMPLE 10.3
▷ **Piglet weight gains (revisited).** What population model is appropriate for the data in Example 10.2? Let's start slowly by considering just three models, as shown in Figure 10.8 (page 324). All three are normal densities with the same shape ($h = 10$ in all three cases), but with different means: $m = 20$, $m = 30$, and $m = 40$. Each model has its mean as its name. The figure shows my prior probability of $\frac{1}{3}$ for each model.

A model in Chapters 6–9 is characterized by its proportion of successes. Now it is handy to think of models as being characterized by m-values. Figure 10.9 puts the densities on the same scale, one that is defined by m-values.

As usual, our goal is to update probabilities of models using sample data. We proceed as in Example 10.1, where the likelihood of a model equals the height of the stack of chips of the type observed in that model. Now the likelihood is the height of normal density at the observed data point.

Figure 10.8 Three normal models considered in Example 10.3

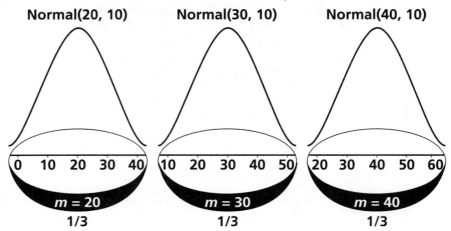

Figure 10.9 Normal models of Example 10.3: Centered at *m*-values 20, 30, and 40

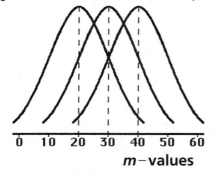

> **A likelihood is the height of the model density at the value observed in the experiment.**

The model that best explains a small observation (25 or less) is the one with the highest likelihood, $m = 20$. The model with smallest likelihood, $m = 40$, is the least well supported by small observations. On the other hand, large observations (35 or more) lend more credence to $m = 40$ and middle-sized observations (between 25 and 35) support $m = 30$. Also, the combined effect of one small observation (say, 10) and one large observation (say, 50) lends more support to $m = 30$ than it does to the other two models. We will quantify the phrase "lends more support" by updating probabilities of the models using Bayes' rule.

As in Example 10.1, first consider a single observation. The first weight gain was 33. The heights of the three densities at weight gain = 33 are shown in Figure 10.10.

Sec. 10.2 / Normal Models

The normal-density heights were calculated from Table 10.1 using the three z-scores:

$$z = \frac{33 - \mathbf{20}}{10} = 1.3 \qquad z = \frac{33 - \mathbf{30}}{10} = .3 \qquad z = \frac{33 - \mathbf{40}}{10} = -.7$$

The means in boldface type (**20, 30, 40**) are the only differences in these three expressions. To find the height of the normal density, calculate z^2. For the first ($m = \mathbf{20}$), $z^2 = 1.3^2 = 1.69$. The exponent is $-1.69/2 = -.845$ and $e^{-.845} = .4296$. The other two heights are $e^{-.09/2} = e^{-.045} = .9560$ and $e^{-.49/2} = e^{-.245} = .7827$.

Figure 10.10 Likelihoods for observing 33 for models considered in Example 10.3

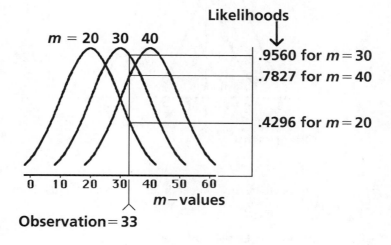

Calculating Likelihood of m

Find $z = \dfrac{\text{Observation} - m}{h}$ and then $e^{-z^2/2}$.

Proceed as in the previous section:

Model	Likelihood of 33	P(Model)	Product	P(Model | 33)
$m = 20$	.4296	1/3	.4296/3	4,296/21,683 = .198
$m = 30$	.9560	1/3	.9560/3	9,560/21,683 = .441
$m = 40$	.7827	1/3	.7827/3	7,827/21,683 = .361
Sum			2.1683/3	

For the observed value 33, the $m = 30$ density is the highest of the three, and so its updated probability is the largest of the three.

The second observation was 25. As usual, we will assume that the observations are independent, given any particular model (but they are not independent without conditioning on a model). The likelihoods are indicated in Figure 10.11 and correspond to z-scores of .5, $-.5$, and -1.5 (.8825 for the first two and .3247 for the third).

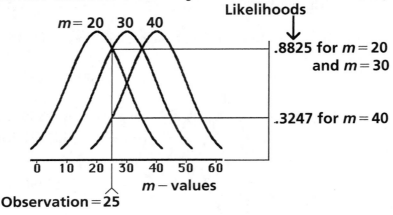

Figure 10.11 Likelihoods for observing 25 for models considered in Example 10.3

To calculate my probabilities as updated after both observations, proceed as follows:

Model (mean)	Likelihood of 25	P(Model | 33)	Product	P(Model | 33, 25)
20	.8825	.198	.175	175/681 = .257
30	.8825	.441	.389	389/681 = .571
40	.3247	.361	.117	117/681 = .172
Sum			.681	

At this rate, it will take forever to get through 100 observations. Such deliberation is hardly worthwhile because this example is not very interesting—its only purpose is to show you how to carry out calculations. [Even if you are into weight gains of piglets, assuming that the population mean is one of three values (20, 30, and 40) is quite artificial.] Luckily, there are ways to avoid applying Bayes' rule separately for each data point. I will give one here, and we will consider an even better method in the next example.

Consider the previous pair of calculations. Likelihoods of individual data points multiply (assuming independence within models). So we can multiply the likelihoods for the two observations and then apply Bayes' rule just once, as follows:

Sec. 10.2 / Normal Models

Model	Likelihood Of 33	Of 25	Product	P(Model)	Product	P(Model | 33, 25)
$m = 20$	.4296	.8825	.3791	1/3	.3791/3	3,791/14,769 = .257
$m = 30$	.9560	.8825	.8437	1/3	.8437/3	8,437/14,769 = .571
$m = 40$	.7827	.3247	.2541	1/3	.2541/3	2,541/14,769 = .172
Sum					1.4769/3	

The last column is the same as in the previous table—as it must be, since the only difference is that we multiplied in a different order.

To calculate posterior probabilities after seeing all 100 weight gains, we can multiply all 100 likelihoods for each model. That may be easier than using Bayes' rule 100 times, but it is still an unwieldy task. It would be unwieldy even if we could look up 100 numbers in the table of normal heights and multiply them. Luckily, there is an easier way.

Multiplying 100 likelihoods turns out to be the same as the likelihood for a *single* normal density, but one different from that of the population model. Showing this is a matter of algebra, and it is not very difficult. But it is too much math for the level of this course and so I will omit it. The pertinent result is that the appropriate z-score is

$$z = \sqrt{100}\,\frac{\bar{x} - m}{h}$$

The general expression is

$$z = \sqrt{n}\,\frac{\bar{x} - m}{h}$$

where n is the sample size. In this example, $h = 10$, which cancels with $\sqrt{100} = 10$, giving

$$z = \bar{x} - m$$

In Example 2.20, we calculated the sample mean to be $\bar{x} = 30$. The z-scores for $m = 20, 30,$ and 40 are, therefore,

$$z = 30 - 20 = 10 \quad z = 30 - 30 = 0 \quad z = 30 - 40 = -10$$

The corresponding likelihoods are

$$e^{-10^2/2} = e^{-50} = 2\text{E}-22 \quad e^{-0^2/2} = e^0 = 1 \quad e^{-(-10)^2/2} = e^{-50} = 2\text{E}-22$$

In view of the relative sizes of these z-scores, applying Bayes' rule is easy:

Model	Likelihood	P(Model)	Product	P(Model | D)
$m = 20$	2E−22	1/3	7E−23	0
$m = 30$	1	1/3	1/3	1
$m = 40$	2E−22	1/3	7E−23	0
Sum			1/3	

So, I am now overwhelmingly convinced that the mean of pigs' weight gains in this population is $m = 30$. (The probability of model $m = 30$ is not exactly 1. A more accurate calculation is $1 - 4E-22$, which is .9999999999999999999996.) How can I be so sure that $m = 30$? I can be sure only in comparison with $m = 20$ and $m = 40$. These two values are so far from $\bar{x} = 30$ that their likelihoods are essentially 0. If I were to consider alternative models closer to $m = 30$, then I would be much less convinced that $m = 30$.

Assuming that m is either 20, 30, or 40, the prediction problem is easy: Base all predictions on the normal(30, 10) density. ◁

Restricting the mean to $m = 20$, 30, and 40 in the previous example was unrealistic. The next example lifts this restriction. It involves nothing conceptually different from the previous example, but its formulation and conclusion point the way to a conceptual improvement: smoothing *across* models. This improvement will be carried out in Chapter 11 and will lead to a satisfying solution of the main problem treated in this section.

EXAMPLE 10.4
▷ **Piglet weight gains (yet again).** Reconsider the setting of Example 10.3, but now enlarge the set of models. Suppose I am willing to assert that the population mean, m, is between 0 and 100, and all intervals of m-values between 0 and 100 have the same probability as every other interval of the same width. (My probabilities are not really as specified, but you will see that the conclusions are quite insensitive to this assumption.) We will consider 1,000 models, one for each of these values of m (in lb): 0, .1, .2, .3, . . . , 99.9, with each having probability 1/1,000. Three of these 1,000 were the models considered in Example 10.3.

I cannot show you a picture of 1,000 models. But since each of these 1,000 values of m corresponds to a point on the interval from 0 to 100, I could show you a bar chart with 1,000 bars of height 1/1,000. Even this is too difficult, but I can show you a small slice of this bar chart. The bar chart in Figure 10.12 is restricted to the range of m-values from 26.6 to 33.4. The reason I selected this particular slice will be clear shortly.

Figure 10.12 Prior probabilities of selected models in Example 10.4

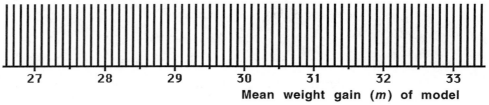

Recall that

likelihood of m: $e^{-z^2/2}$

Sec. 10.2 / Normal Models

where the z-score is

$$z = \sqrt{100}\,\frac{\bar{x} - m}{h} = \bar{x} - m = 30 - m$$

The likelihoods are given in Table 10.2 in which the range is restricted to the right half of the bar chart in Figure 10.12. The likelihoods are symmetric because m enters through z^2. Since, for example, $(+2)^2 = (-2)^2$, z^2 and therefore the likelihoods are the same at 28 as they are at 32. (Because there are many more m-values than considered in earlier examples, I have strung them vertically instead of horizontally in this table, using three columns.) These likelihoods—for m both larger and smaller than 30—are shown also as bars in Figure 10.13.

**Table 10.2
Likelihoods for Example 10.4**

m	Likelihood	m	Likelihood	m	Likelihood
30.0	1.0000	31.2	.4868	32.4	.0561
30.1	.9950	31.3	.4296	32.5	.0439
30.2	.9802	31.4	.3753	32.6	.0340
30.3	.9560	31.5	.3247	32.7	.0261
30.4	.9231	31.6	.2780	32.8	.0198
30.5	.8825	31.7	.2357	32.9	.0149
30.6	.8353	31.8	.1979	33.0	.0111
30.7	.7827	31.9	.1645	33.1	.0082
30.8	.7261	32.0	.1353	33.2	.0060
30.9	.6670	32.1	.1103	33.3	.0043
31.0	.6065	32.2	.0889	33.4	.0031
31.1	.5461	32.3	.0710		

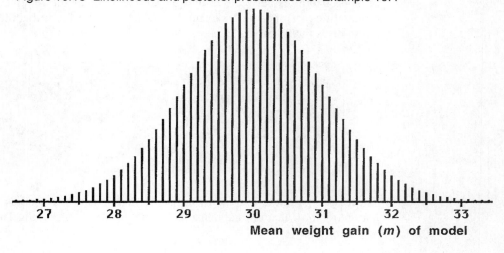

Figure 10.13 Likelihoods and posterior probabilities for Example 10.4

Since prior probabilities are equal on the various m-values, it is simple to find the posterior probabilities—just add up the likelihoods and divide each individual likelihood by the sum. I have not done this because I am interested only in showing the bar chart of posterior probabilities, which is the same as for the bar chart of likelihoods. However, I will tell you the posterior probability of $m = 30$:

$$P(30 \mid \text{Data}) = .040$$

This is a far cry from the answer in Example 10.3, where my probability of $m = 30$ was essentially 1. As you can see from Figure 10.13, the probability of $m = 30$ is larger than that of any other value of m. The reason it is so small is that there are many values of m near 30 that easily could have accounted for the 100 weight gains observed.

The striking thing about this bar chart of posterior probabilities is that it is essentially itself a *normal curve*! Indeed, if you measure the heights of the bars, you will find that it is a perfect normal shape (where $z = m - 30$). In Sections 11.1 and 11.3, we will exploit this characteristic when making inferences about the population mean, m, and when predicting future observations.

I want to address one last aspect of this example. We have assumed that my probabilities were equal for all 1,000 values of m. I suggested that they really were not equal. I was pretty sure that young pigs that eat a reasonable amount (of almost anything!) over a 20-day period will gain at least some weight. Also, while I am far from an expert, I would have thought it most unlikely that any pig could gain as much as 90 pounds in 20 days. Moreover, because I am not an expert, I would have assigned about the same probability to all values of m between 26 and 34. Since this is the range with greatest likelihoods, these are the only values of m whose prior probabilities affect my posterior probabilities. The only characteristics of these prior probabilities that matter are their relative sizes; in particular, my total prior probability of m's between 26 and 34 is nearly irrelevant—provided it is not 0. Since the relative sizes of my actual prior probabilities between 25 and 35 are the same as the relative sizes of the prior probabilities assumed in this example, my actual posterior probabilities are essentially the same as those calculated here. ◁

> **For a normal population with standard deviation h and a sample of size n with mean $\bar{x}$, calculate the likelihood of the population mean, m, as $e^{-z^2/2}$, where z is the z-score**
>
> $$z = \sqrt{n}\,\frac{\bar{x} - m}{h}$$

It is important to know how to use normal densities as models. But it is also important to know that normal densities are not always appropriate. The next example is a case in point.

EXAMPLE 10.5

▷ **Treating hearts that beat out of sync.** Some people with heart disease, as well as some with hearts that are normal otherwise, experience frequent premature heart beats. A premature beat is not efficient because it occurs before the heart is filled with blood and so the heart pumps less blood. The problem may be mechanical or electrical.

An experimental drug for treating this condition was evaluated in 12 patients in a clinic. (It seems unlikely that premature beats do the individual any good, but whether they should be treated is open to question.) Investigators counted these patients' premature beats during a 1-minute period, administered the drug, waited a short time, and then counted premature beats during another 1-minute period. [A better design would have been a crossover design (Section 3.6), except that patients receiving first drug and then placebo may have continued to have a drug effect at the time of placebo administration. Another possibility is a parallel design (Section 3.5), but that would have required more patients. This was a preliminary study that would eventually be confirmed by parallel studies.] The results are shown in Table 10.3.[1] The sample mean and standard deviation are $\bar{x} = 11.50$ and $s = 13.68$.

Table 10.3
Premature beats per minute before and after drug for Example 10.5

Patient Number	1	2	3	4	5	6	7	8	9	10	11	12
Before	6	9	17	22	7	5	5	14	9	7	9	51
After	5	2	0	0	2	1	0	0	0	0	13	0
Difference	1	7	17	22	5	4	5	14	9	7	−4	51

The dot plot in Figure 10.14 shows the last row of this table. It indicates that patient 12 is very far from the rest of the patients. Such an observation is called an **outlier**—see Chapter 13. Without patient 12, the sample mean and standard deviation are $\bar{x} = 7.91$ and $s = 7.03$. This 30% decrease in the mean is large, but the nearly 50% decrease in standard deviation is even larger. Consider the population that produced this sample. Among normal densities, the one that fits the sample data best is the normal($\bar{x}$, s). The best-fitting normals are normal(11.50, 13.68) and normal(7.91, 7.03). These are shown in Figure 10.15 (page 332) for the two cases: with and without the outlying patient 12. The fact that one observation affects the best-fitting normal curve so dramatically makes the normal shape questionable as a model for these data.

Figure 10.14 Dot plot for decrease in premature beats/minute for Example 10.5

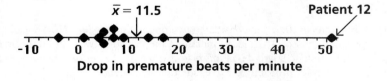

Drop in premature beats per minute

◁

Example 10.5 shows that it is handy to be able to assume that the true model is a normal density. We will frequently make this assumption for that reason. Predictions

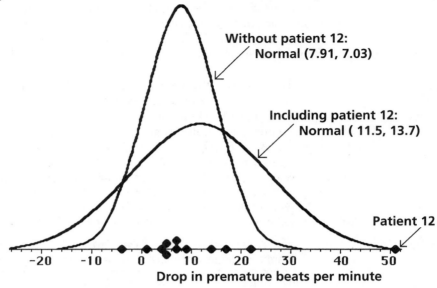

Figure 10.15 Normal densities drawn to fit the data of Example 10.5

and other conclusions depend heavily on this assumption. (Statisticians say that a method that gives one answer under one assumption and a very different answer under others is not very **robust**.) If the population density is not far from being a normal one, then conclusions may not be bad. But when there are outlying observations as in the sample of Example 10.5, conclusions may be quite wrong. There is no foolproof way to check the assumption of a normal population. Whenever you assume a model has a normal shape, worry that you might be wrong and consider alternatives, such as those given in Chapter 13.

EXERCISES

10.5 In Exercise 2.24 (page 52), you calculated $\bar{x} = 2.406$ for the mean increase in miles walked by 32 congestive heart-failure patients while using an experimental drug. Consider only two possible models—both normal densities. In one, the mean is $m = 0$, corresponding to no difference between the 2 weeks. In the other, the mean is $m = 3$, corresponding to a mean improvement of 3 miles walked in the second week. In both, the population standard deviation is $h = 5$ (miles). Assign prior probability $\frac{1}{2}$ each to $m = 0$ and $m = 3$. Find the posterior probabilities of these two models.

10.6 In Exercise 10.5, you considered two population models. Now expand that exercise by considering a total of 16 models, including the previous two. All are normal densities and all have population standard deviation $h = 5$ (miles). The means are $m = -5, -4, -3, \ldots, +10$. Assign prior probability $\frac{1}{16}$ to each of these models.
    (a) Calculate your posterior probabilities of these 16 models.
    (b) Draw a bar chart to show your posterior probabilities.
    (c) What is your posterior probability that m is greater than 0?

10.7 In Exercise 2.26 (page 52), you found the mean of $n = 50$ spot-weld strengths to be $\bar{x} = 404.2$. Consider five models with normal densities and population means of 395, 400, 405, 410, 415, and standard deviation $h = 20$ in each case. Assume equal prior probabilities and calculate your posterior probabilities of these five models.

10.8 A method for weighing extremely light objects gives the following results (in micrograms, μg) for nine weighings of a particular specimen:[2]

$$114 \quad 129 \quad 121 \quad 98 \quad 140 \quad 134 \quad 122 \quad 133 \quad 125$$

Suppose the specimen actually weighs m μg and that the population from which this sample is drawn is normal(m, 10). Assume prior probabilities of $\frac{1}{3}$ for each of the three m-values: 110, 120, and 130. Calculate the posterior probabilities for these three m-values.

10.9 Exercise 2.1 (page 13), described an experiment that Darwin carried out to learn whether cross- or self-pollination would produce more vigorous seeds. In Exercise 2.22 (page 52), you calculated the sample mean of the differences in the 15 pots, cross- vs self-pollinated plants.

(a) Suppose the model that produced these differences is a member of a set of 40 normal densities, all having standard deviation $h = 30$ and with one normal density for each mean, $m = -100, -95, -90, \ldots, 95$. Assign each of these models prior probability $\frac{1}{40}$. Calculate your posterior probabilities for these 40 models. [To ease your calculating burden, recognize that only a small number of these models need be considered—you can ignore those with small likelihoods. The likelihood goes down as m gets farther from the sample mean of 20.93; once the likelihood reaches less than $\frac{1}{1,000}$ of the greatest likelihood, stop calculating.]

(b) Draw a bar chart showing your posterior probabilities, but restrict the picture to the interval of m-values having probabilities of at least .001.

(Your dot plot in Exercise 2.1 draws into question the normality assumption for the population model. Exercises 13.10 and 13.11 will use different assumptions.)

10.10 A criminal spilled his blood at a crime scene and was observed going into a house. Police found two men, Smith and Jones, in the house; regard them as equally likely to be the criminal. A laboratory analysis of the blood found that a particular fragment of the criminal's DNA has a molecular weight of 944. But there is error in this test and the error has a normal density. [This assumption has strong support in laboratory procedures that use restriction fragment length polymorphism (RFLP) analysis.] The actual weight has mean 944 and standard deviation $h = 8$. Both are tested; Smith's DNA fragment has a molecular weight of 940 and Jones's weighs 964. (These, too, are subject to error, but for simplicity assume they are known exactly. Recognizing errors in both suspect and crime sample assessments leads to an analysis very similar to the one you will use in this exercise.) Who is the criminal? Answer by calculating the posterior probabilities of Smith and Jones. [*Hint:* For Smith, use z-score = $(940 - 944)/8$ and for Jones, use z-score = $(964 - 944)/8$.]

10.11 This is a continuation of Exercise 10.10. Now suppose the laboratory analyzes a different fragment of DNA from each of the three samples. The crime sample's fragment weighs 1,120, Smith's weighs 1,128, and Jones's weighs 1,136. Again assume normal error with $h = 8$ for the crime sample and regard this new evidence as independent of the first. Update the posterior probabilities you found in the previous exercise based on this new evidence.

Appendix: Using Minitab with General Samples and Population Means

The Minitab program **'m_disc'** will compute posterior probabilities for a normal mean, m, when the models of interest can be listed. In Example 10.3, one is interested in learning about the average weight gain, m, on the basis of 100 observations. Suppose that the seven values 27, 28, 29, 30, 31, 32, 33 are plausible values for m and each is assumed to have the same (prior) probability. You then observe a sample mean of 30. To compute posterior probabilities in this case, you need to know the standard deviation of the population of weight gains. We will assume that this standard deviation is equal to 10. Before this program is run we place the values of the mean in the column 'm' and the prior probabilities in the column 'prior.' To compute the posterior probabilities, type

```
exec 'm_disc'
```

This program asks for two inputs: the value of the known population standard deviation and the observed data. If the data are in the Minitab worksheet, then one inputs the number of the column that contains the data. In this particular case, the data are available in summary form and we input on the 'DATA' line the sample mean and sample size. The output of the program is a table of the posterior probabilities, the values of the prior and posterior means of m, and a spike chart plot of the values of the posterior probabilities. On the basis of this sample, you are fairly confident (with approximate probability .88) that the value of the mean, m, is either 29, 30, or 31.

```
MTB > name c1 'm' c2 'prior'
MTB > set 'm'
DATA> 27 28 29 30 31 32 33
DATA> end
MTB > set 'prior'
DATA> .14 .14 .14 .14 .14 .14 .14
DATA> end
MTB > exec 'm_disc'
INPUT POPULATION STANDARD DEVIATION:
DATA> 10

OBSERVED DATA IN WORKSHEET? (TYPE 'y' OR 'n'.)
 IF YES, INPUT NUMBER OF COLUMN.
 IF NO, INPUT OBSERVED SAMPLE MEAN AND SAMPLE SIZE:
n

DATA> 30 100

MEAN
 30

COUNT
 100
```

```
ROW   m     prior    M_X_PRIO    LIKE   PRODUCT      POST   M_X_POST
  1  27  0.142857      3.8571   11109      1587  0.004433     0.1197
  2  28  0.142857      4.0000  135335     19334  0.054006     1.5122
  3  29  0.142857      4.1429  606531     86647  0.242036     7.0191
  4  30  0.142857      4.2857 1000000    142857  0.399050    11.9715
  5  31  0.142857      4.4286  606531     86647  0.242036     7.5031
  6  32  0.142857      4.5714  135335     19334  0.054006     1.7282
  7  33  0.142857      4.7143   11109      1587  0.004433     0.1463
  8                   30.0000                                30.0000
```

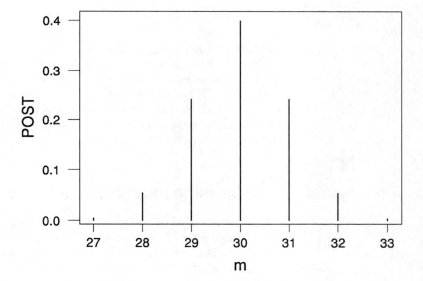

Chapter Notes

1. D. A. Berry, *Biometrics* 43 (1987): 439–456.
2. G. E. P. Box and G. C. Tiao, *Bayesian Inference in Statistical Analysis* (New York: Wiley, 1973), 118.

11
Densities for Means

SECTION 3.5 compared parallel and paired designs. A parallel design results in two samples and a paired design gives a single sample—perhaps of differences between twins or of treatment measurement divided by control. The analysis of two samples is the subject of Chapter 12 and this chapter deals with a single sample. The single sample may be the result of a paired experiment or it may be the result of making any type of measurement on each experimental unit, such as those considered in the previous chapter. So this chapter is a continuation of Chapter 10.

11.1 Prior Densities and Normal Models

Chapter 7 introduced the notion of smoothed histograms or densities. In that chapter, and again in Chapter 8, we smoothed *over population models*. So we were using densities to reflect prior information. Section 10.2 introduced smoothing *over observations*. In this section and in the next, we will combine these two types of smoothings. We will assume the population has a normal density (with unknown mean, m, and standard deviation, h) and we will also specify prior probabilities for the population mean using a normal density.

EXAMPLE 11.1
▷ **Piglet weight gains.** Consider the mean of the population from which the 100 piglet weight gains of Example 2.15 represent a sample. What is my posterior probability that the population mean is between 28 and 32 lb? The most detailed prior distribution we considered in Chapter 10 was in Example 10.4. There were 1,000 models, differing from each other only in the mean observation, m. It is difficult to show all 1,000 models in a bar chart so I settled for the small slice in Figure 11.1.

Figure 11.1 Prior probabilities of selected models in Example 10.4

Showing all 1,000 bars from 0 to 100 in that example gives Figure 11.2.

Figure 11.2 Prior probabilities of all models in Example 10.4

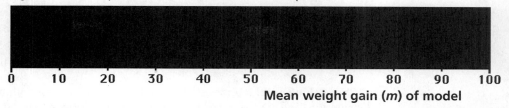

This suggests a density—for obvious reasons I will call it *flat*. Our style has been to save ink and leave the density open, as in Figure 11.3.

Figure 11.3 Flat prior density of *m* for Example 11.1

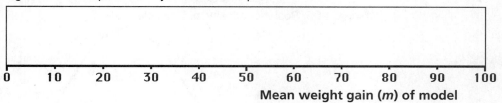

After rather extensive calculations, my posterior distribution of *m*-values in Example 10.4 was the bar chart shown in Figure 11.4. The heights of these bars are the same as the likelihoods because the prior probabilities of the various models were all the same. I restricted the range to a narrower interval (27 to 33 or so) because this is where the posterior probabilities are greatest. Restricted to the same interval, the prior density is shown in Figure 11.5.

Figure 11.4 Posterior probabilities from Example 10.4

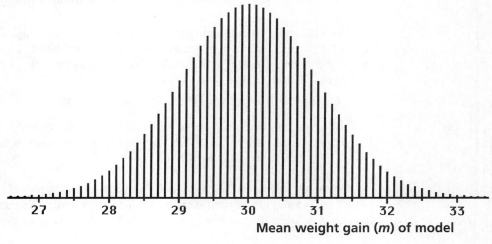

Figure 11.5 Flat prior density of *m* for Example 11.1, restricted to *m* from 27 to 33

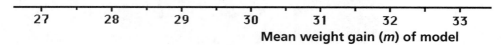

Mean weight gain (*m*) of model

Smoothing out the prior distribution has the effect of smoothing out the posterior distribution. The posterior density that goes with the prior density shown in Figure 11.5 is obtained by connecting the tops of the bars in Figure 11.4 with a curve and dropping the bars, as shown in Figure 11.6. This is a picture of a normal(30, 1) density. As I indicated in Example 10.4, the *z*-score is $z = m - 30$. So the mean weight gain has a normal density centered at 30 and with standard deviation equal to 1.

Figure 11.6 Posterior density for Example 11.1

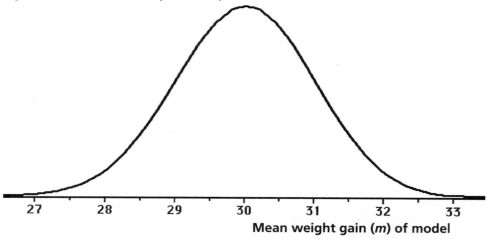

Mean weight gain (*m*) of model

We can easily calculate posterior probabilities of intervals of *m*-values as areas under the normal curve using *z*-scores and the Standard Normal Table inside the front cover of this text. I started this example asking about my posterior probability that *m* is between 28 and 32. The two *z*-scores are

$$z = 28 - 30 = -2 \quad \text{and} \quad z = 32 - 30 = 2$$

According to the Standard Normal Table, the area to the left of $z = -2$ is .0228 and the area to the left of $z = 2$ is .9772. So the area between them is the difference: .9772 − .0228 = .9544, about 95%. ◁

Suppose *n* observations are made from a population model, and suppose every model considered is a normal density. Assume the standard deviation of each normal density is *h*. We will continue to use *m* to indicate the mean of the true model, which is then a normal(m, h) density. We are assuming a normal(m, h) density—so what is the purpose of sampling? The answer is that we do not know m (and sometimes we do not know h), and we are sampling to get more information about it.

We have to formulate our prior information into probabilities. Suppose the prior density is flat over a substantial region of m-values. The posterior density of m-values turns out to be a normal one with mean $\bar{x}$ and standard deviation $h/\sqrt{n}$.

> **Updating Rule for Normal Models**
>
> **For n observations from a normal(m, h) density, if the prior density of m-values is flat, then the posterior is a normal($\bar{x}$, $h/\sqrt{n}$) density.**

To calculate the z-score for any particular value of m, proceed as usual. First, subtract the mean of m—in this case, it is $\bar{x}$—and then divide by the standard deviation of m, which is $h/\sqrt{n}$:

$$z = \frac{m - \bar{x}}{h/\sqrt{n}}$$

Consider predicting the next observation. If we knew m, the next observation would have a normal(m, h) density, just like any other observation. But we do not know m. Instead, m itself has a normal density, as already indicated. The center of the normal density is $\bar{x}$. The two sources of variability (h and $h/\sqrt{n}$) add, but standard deviations do not add; *squares* of standard deviations add:

$$h^2 + \left(\frac{h}{\sqrt{n}}\right)^2 = h^2\left(\frac{n+1}{n}\right)$$

So the standard deviation of the next observation is $h\sqrt{(n+1)/n}$. When the sample size is large, the posterior density of m has a small standard deviation and, in fact, $h/\sqrt{n}$ tends to 0 as n gets large. But the standard deviation of the predictive density tends to h as n gets large. To calculate probabilities for the next observation, use the Standard Normal Table and the z-score:

$$z = \frac{\text{Obs.} - \bar{x}}{h\sqrt{\frac{n+1}{n}}}$$

> **Prediction for Normal Models**
>
> **For n observations from a normal(m, h) density, if the prior density of m-values is flat, then for prediction, the next observation has a normal($\bar{x}$, $h\sqrt{(n+1)/n}$) density.**

Because of the extra variability in the predictive density of the next observation, it is much more spread out than the posterior density of the population mean—unless n

is very small, such as 1. Figure 11.7 shows this for the piglet weight gains. So you can compare them better, the area under the density is the same for both.

Figure 11.7 Predictive vs posterior density in piglet weight gain example

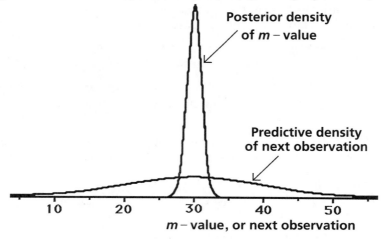

Thus far, we have considered population models that have the same standard deviation, h, with h known. In practice, it is unusual to know the standard deviation of a population but not know its mean. Allowing for different standard deviations as well as different means makes for calculational issues that are too advanced for this course.[1]

The law of large numbers (Section 10.2, page 321) comes to the rescue. This law means that the shape of the sample histogram approximates the shape of the population density when the sample is sufficiently large. In particular, the sample standard deviation, s, will be close to the population standard deviation, h. So for large samples, we can use s in place of h and proceed as though we knew h to be s. (I will discuss how large is "sufficiently large" shortly.) The standard deviation of the posterior density of m-values is $h/\sqrt{n}$, which is then approximately $s/\sqrt{n}$. The standard deviation of the predictive density of the next observation is approximately

$$s\sqrt{\frac{n+1}{n}}$$

which is $\sqrt{n+1}$ times the standard deviation of the posterior density of m. This is itself close to s, especially for large n.

EXAMPLE 11.2
▷ **Power generated by manufactured devices.** A new type of device operates on electricity. A company experimenting with manufacturing this type of device made 40 of them for testing. They measured steady-state power (in watts). These measurements are given here, in order of time of manufacture:

18 18 19 17 25   18 18 17 18 18   22 20 20 17 20   18 18 20 15 17
18 17 19 17 17   17 19 18 18 18   16 17 18 18 18   17 17 22 16 20

Consider these 40 to be a sample from the larger population of such devices yet to be manufactured and consider the next such device to be manufactured. What is the probability that its steady-state power is greater than 20 watts? Assume a normal population density with mean m and a flat prior density for m. To find the posterior density, proceed as in Example 10.4. First find

$$\bar{x} = 18.25 \quad \text{and} \quad s = 1.799$$

The posterior density of m is therefore normal(18.25, 1.799/$\sqrt{40}$). The standard deviation of m is $s/\sqrt{n} = 1.799/\sqrt{40} = .284$. The z-score for any particular m is

$$z = \frac{m - \bar{x}}{.284}$$

The posterior density is shown in Figure 11.8, with the z-scale added. If we wanted to find the probability of $m > 20$, say, we would calculate the z-score of $m = 20$:

$$z = \frac{m - \bar{x}}{.284} = \frac{20 - 18.25}{.284} = 6.16$$

This value of z is out of the range of the Standard Normal Table. So the probability to the right of $m = 20$ is essentially 0. But the question deals with the next observation and not with m.

Figure 11.8 Posterior density for m-values in Example 11.2

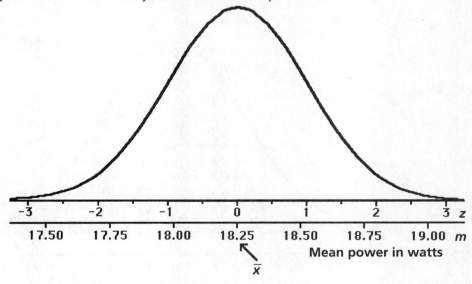

The predictive density of the steady-state power of the next device tested (that is, the 41st) is also normal with the same mean, $\bar{x} = 18.25$, and with standard deviation $s\sqrt{(n + 1)/n} = 1.799\sqrt{41/40} = 1.821$. (Since $\sqrt{41/40}$ is close to 1, this is not very different from the sample standard deviation, s.) This is much larger than the standard deviation of the posterior density of m. The reason is that most of the variability in predicting is in the process of observation and not because of our uncertainty in m. In fact, we know m pretty well, but that has a limited benefit in prediction.

The predictive standard deviation of 1.821 indicates that our best guess of 18.25 watts could easily be off by a few watts. The predictive density is shown in Figure 11.9, where the z-score is now

$$z = \frac{\text{Obs.} - \bar{x}}{1.821}$$

The dots on this figure correspond with the currently available observations. Obviously, the predictive density is similar to—and in fact, is a smoothed version of—the sample histogram. These dots should make it clear why there is much more variability in prediction than in statements about m: If the next observation were in fact one of the previous observations, then its standard deviation is about 1.8.

The z-score corresponding to 20 watts is

$$z = \frac{20 - 18.25}{1.821} = .96$$

which is shown in the figure. According to the Standard Normal Table, the probability to the right of $z = .96$ is about 17%, which is the area shown shaded in the figure. The probability that the next observation is greater than 20 is 17%.

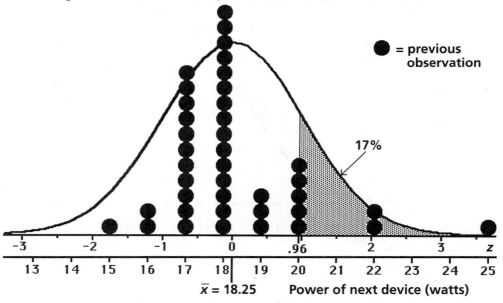

Figure 11.9 Predictive density for the next observation in Example 11.2

When is n large enough that h can be replaced by s? Bigger is better. Any choice of boundary is arbitrary, but I will adopt the convention that "large" means $n > 30$. When $n \leq 30$, the histogram of m-values is not well approximated by a normal curve. Moreover, it cannot be made into a normal curve by a simple change of scale (as we have changed to a z-scale, say). Most statistics books develop and use the exact mathematical curve, called a Student t distribution. This exact development is complicated.

Moreover, making this one aspect of an answer exact seems not to be worth the effort in the face of other uncertainties in empirical research and given the questionable nature of assuming a normal population model in the first place. We will use a simple adjustment to the z-score. This gives posterior probabilities of intervals that are only approximately correct, but the discrepancy is usually within .02 of the exact answer given by the Student t distribution.*

The reason the normal curve does not work when n is small is that the denominator $s/\sqrt{n}$ in the z-score,

$$z = \frac{m - \bar{x}}{s/\sqrt{n}}$$

may not be close to $h/\sqrt{n}$. The problem is most severe when s happens to underestimate h. So we will approximate the actual curve with a normal curve by introducing a **small-n correction factor,** k. Namely, we will inflate the sample standard deviation, multiplying it by the factor

$$k = 1 + \frac{20}{n^2}$$

and use sk in place of h in the z-score, estimating h to be

$$sk = s\left(1 + \frac{20}{n^2}\right)$$

> **When the sample size, n, is 30 or less and the population standard deviation, h, is unknown, estimate h to be**
>
> $$h = sk = s\left(1 + \frac{20}{n^2}\right)$$

> **When the sample size, n, is 30 or less, use this as the z-score for the posterior density of m-values:**
>
> $$z = \frac{m - \bar{x}}{sk/\sqrt{n}}$$

> **When the sample size, n, is 30 or less, use this as the z-score for the predictive density of the next $[(n + 1)\text{st}]$ observation:**
>
> $$z = \frac{\text{Obs.} - \bar{x}}{sk\sqrt{(n + 1)/n}}$$

* Some instructors may prefer to hand out and explain Student t tables in lieu of the modification used in this text.

The small-n correction factor applies for any value of n, large or small. If $n = 40$, as in the previous example, $k = 1.0125$, and so the change in s is only about 1%. In any application, a 1% change in standard deviation is negligible. On the other hand, using the small-n correction factor for moderate n is not very important. For example, if $n = 20$, then $k = 1.05$ and a 5% change in s usually does not matter much.

Having to correct the standard deviation when n is small is an additional penalty. A greater penalty comes from the lack of information in a small sample. The posterior standard deviation is $sk/\sqrt{n}$. The factor $1/\sqrt{n}$ is halved when n is quadrupled. While k is decreased when n is increased, this effect is tiny compared with the decrease due to $1/\sqrt{n}$.

The form of the denominator of the z-score varies. The denominator is the standard deviation of the numerator, making the z-score a standardized score. There are four reasons for different denominators: (1) The available information may be prior or posterior, and the prior and posterior standard deviations are different. (2) If the numerator is the mean of the population, then the standard deviation has a factor $1/\sqrt{n}$; if the numerator is the next observation, then there is no such factor in the standard deviation. (3) If the standard deviation of the population is known, then it appears in the denominator, but if it is not known then it is estimated from the sample. (4) In the case that the population standard deviation is not known, and when the sample size is small, the estimate is inflated to account for the additional uncertainty.

EXERCISES

11.1 Example 11.2 gives the steady-state power of the first 40 devices manufactured by a company. The steady-state power measurements for the next 28 devices were as follows:

$$25 \ 20 \ 25 \ 18 \ 24 \quad 26 \ 24 \ 24 \ 24 \ 24 \quad 18 \ 18 \ 17 \ 17$$
$$21 \ 15 \ 13 \ 12 \ 25 \quad 19 \ 18 \ 24 \ 25 \ 18 \quad 25 \ 24 \ 24 \ 25$$

Ignore the 40 measurements given in Example 11.2 and treat these 28 as a sample from a second population. Assume a normal density for the population and a flat prior density for m, the mean of the population.

(a) Find the posterior probability that m is greater than 20 watts.

(b) Calculate the probability that the next (i.e., the 29th) observation will be greater than 20 watts.

[These answers are different from those calculated in Example 11.2 (.00 and .17). The first 40 observations seem not to be exchangeable with the second 28. In Chapter 12, we will consider the differences between the two populations.]

11.2 Example 2.7 described a study that evaluated a drug called amiloride in patients who were suffering from cystic fibrosis. The example gives the patients' forced vital capacities after 6 months of using an aerosol spray containing amiloride and after 6 months of using a spray without amiloride (vehicle only). The following table repeats those data and also includes a column of differences (positive differences indicate a better response with the drug than without). Consider only the column of differences. Assume normal models and a flat prior density for the population mean.

(a) Calculate the posterior probability that the population mean, m, of the differences is greater than 0.

(b) Predict the next patient's response to the drug. In particular, calculate the probability that the next patient responds more favorably on amiloride than on vehicle alone (that is, has a positive difference).

Patient Number	Forced vital capacity (in ml)		Difference (Amiloride − Vehicle)
	Vehicle	Amiloride	
1	2,925	2,760	−165
2	4,190	4,490	+300
3	5,067	5,617	+550
4	2,588	2,543	−45
5	3,934	3,810	−124
6	3,952	3,985	+33
7	2,547	2,392	−155
8	4,108	3,880	−228
9	2,646	2,732	+86
10	3,635	3,758	+123
11	2,890	2,960	+70
12	3,125	3,387	+262
13	3,805	4,048	+243
14	1,741	1,787	+46

(The sample mean and standard deviation of differences are $\bar{x} = 71.1$ and $s = 207.1$.)

11.3 Geologists and climatologists are interested in the age of glacial moraines. One of the methods they use to date such moraines involves measuring the accumulation of chlorine. In one study,[2] researchers reported on the amounts of chlorine in several moraines in Bloody Canyon, California, which is in the eastern Sierra Nevada range. These were the amounts of chlorine (in parts per million) in five samples taken from a moraine called Older Tahoe:

$$73 \quad 75 \quad 49 \quad 76 \quad 115$$

Find the predictive density of the amount of chlorine in the next sample; give the predictive probability that the next sample's chlorine will be greater than 150 ppm. Assume a normal density for the population and a flat prior density for the population mean.

11.4 The following data[3] are the percentages of ammonia lost during 21 days of operation of a plant that oxidized ammonia to nitric acid, given in order from smallest to largest:

$$7\ 8\ 8\ 8\ 9 \quad 11\ 12\ 13\ 14\ 14 \quad 15\ 15\ 15\ 18\ 18 \quad 19\ 20\ 28\ 37\ 37 \quad 42$$

Regard this as a random sample from some population. Find the predictive density of a 22nd day's percentage of ammonia lost. Assume a normal density for the population and a flat prior density for the population mean.

11.5 Exercise 10.2 gives the results of 60 dart tosses. These actually represented 10 games played by the tosser, who paid $2 to toss six darts in each game ($20 total). He would win a large stuffed animal if his total score for the six tosses was 29 or more, and a small stuffed animal if his total score was 14 or less. If you add up the tosses in groups of six, you will see that his 10 scores were these:

$$23\ 21\ 23\ 18\ 21 \quad 18\ 18\ 25\ 19\ 25$$

He has not yet won a stuffed animal. Should he spend another $2 to play the game an 11th time? Address this question by calculating the probability that his total score on the next (11th) game is the following:

(a) 29 or more
(b) 14 or less

11.6 The study described in Example 3.24 compared the drugs formoterol (F) and salbutamol (S) in aerosol solutions with a placebo (P) solution in 30 patients suffering from exercise-induced asthma. The following table repeats the original data, as well as three columns of differences: $F - P$, $S - P$, and $F - S$. (The column of $F - P$ is the sum of $F - S$ and $S - P$.) Consider only formoterol vs salbutamol—that is, the last column, $F - S$. Ignore any possible effect of the sequence of administration. Assume a flat prior for the mean of the population from which this column of differences represents a sample.

Forced expiratory volume (in ml) after F, S, or P in aerosol and differences in volume between drugs

Patient Number	Sequence	Period 1	Period 2	Period 3	F − P	S − P	F − S
1	FSP	35	32	29	6	3	3
10	FSP	34	28	22	12	6	6
17	FSP	23	22	17	6	5	1
21	FSP	23	13	14	9	−1	10
23	FSP	30	24	18	12	6	6
2	FPS	31	18	24	13	6	7
11	FPS	28	16	22	12	6	6
14	FPS	31	16	14	15	−2	17
19	FPS	23	15	22	8	7	1
25	FPS	30	17	26	13	9	4
28	FPS	31	21	28	10	7	3
3	SFP	21	32	10	22	11	11
12	SFP	16	23	16	7	0	7
18	SFP	16	14	8	6	8	−2
24	SFP	31	32	10	22	21	1
27	SFP	28	31	20	11	8	3
4	SPF	22	11	26	15	11	4
8	SPF	28	20	28	8	8	0
16	SPF	24	17	34	17	7	10
6	PFS	22	25	24	3	2	1
9	PFS	22	32	33	10	11	−1
13	PFS	8	14	10	6	2	4
20	PFS	9	13	15	4	6	−2
26	PFS	17	26	24	9	7	2
31	PFS	14	25	22	11	8	3
5	PSF	9	19	29	20	10	10
7	PSF	15	26	20	5	11	−6
15	PSF	12	22	27	15	10	5
22	PSF	24	26	38	14	2	12
30	PSF	19	27	28	9	8	1

(Assume a normal density for the population and a flat prior density for the population mean. Then use these 10 observations to find the predictive density of the 11th. You do not have to refer back to the earlier example and you should not consider the 60 scores of the individual dart tosses.)

(a) Calculate the posterior probability that the mean difference is positive—that is, that F is more effective than S.

(b) Calculate the posterior probability that the mean difference (F − S) in forced expiratory volume is greater than 2, 4, 6, and 8 ml—that is, that F is more effective than S by 2, 4, 6, and 8 ml.

11.7 Repeat parts (a) and (b) of Exercise 11.6 for the column F − P of the table.

11.8 Repeat parts (a) and (b) of Exercise 11.6 for the column S − P of the table.

11.2 Choosing a Normal Density As a Prior

The results of the previous section were limited to the case in which there is very little prior information concerning m. The prior probabilities are spread out in such a way that every interval of m-values gets the same prior probability as every other interval of the same width. A person with such an opinion would be indifferent between wagering that m is in the interval from −10 to 0 and that it is from 1,000 to 1,010. One usually has more information than this concerning the value of m. Even if one's prior probability density is flat, the posterior is no longer flat. This posterior is the prior for the next experiment (if the observations in the two are deemed to be exchangeable), and so it will not be flat. So the results of Section 11.1 will not carry us very far. We need to handle other types of prior information. The following example shows how to assess one's prior density if it is bell shaped.

EXAMPLE 11.3
▷ **How much do I weigh?** To find out how much I weigh, I use a scale that is not perfectly accurate (see Example 11.4 in the next section). Think of the population of measurements as based on a normal model with chips labeled with measurements of my weight. Stepping on the scale and observing the weight shown is analogous to selecting a chip. Assume that the mean of the true model is m, my actual body weight, so the scale is correct *on average*.

What is my prior density for m? It might be reasonable to assume a flat prior as in the previous section if you are the assessor and because you cannot see me, but it is not reasonable if I am the assessor. I think m is about 174 lb. But if it is as small as 170 or as large as 180, I would not be very surprised. I am willing to assume that my prior density is normal shaped—say, centered at m_0 and having standard deviation h_0. The value of m_0 is my best guess; that is $m_0 = 174$. Finding h_0 is more involved.

To find my prior standard deviation h_0 of m, I can use the methods of Section 4.4 to assess my probability that m is greater than 180 lb, say. I have done this and conclude that this prior probability is 10%. According to the Standard Normal Table, the value of z corresponding to 10% probability to the right (or 90% of the probability to the left) is 1.28. So

$$z = \frac{180 - 174}{h_0} = 1.28$$

Solving this for my prior standard deviation gives

$$h_0 = \frac{6}{1.28} = 4.69$$

I should check that this value reflects my opinions by examining some of its implications. One is that my probability to the left of 170 lb is the proportion of the normal curve less than the z-score of

$$\frac{170 - 174}{4.69} = -.85$$

According to the Standard Normal Table, this is about .20, which is consistent with my opinions. If it were not the case that 20% of my probability is to the left of 170 lb, then I would have to revise h_0. I have examined other implications of $m_0 = 174$ and $h_0 = 4.69$ and find all to be consistent with my opinions about m. ◁

> Assess a prior density for population mean m by focusing first on the prior mean, m_0, of m and then on the prior standard deviation, h_0, of m.

Suppose an implication of the assessment process in the previous example had been inconsistent with my opinions and suppose it could not be corrected without creating other inconsistencies. Then my prior density is not bell shaped. In that case, I cannot use the methods developed in this chapter. Chapter 13 presents methods that apply somewhat more generally.

EXERCISES

11.9 I have assessed my information about the drop in blood pressure that will result from using a particular drug that is now in development. For people whose diastolic blood pressure is 95 mmHg, my estimate for the mean drop, m, is 6 mmHg and my probability that m is greater than 12 mmHg is 15%. Assuming a normal(m_0, h_0) prior density for m, find m_0 and h_0.

11.10 I have assessed my information about the increase in IQ score that will result for fourth graders given an intensive course in spatial thinking. My prior probability that the mean increase, m, is greater than 20 points is 20% and my prior probability that m is greater than 10 points is 50%. Assuming a normal(m_0, h_0) prior density for m, find m_0 and h_0.

11.11 Consider the time, m, from the present (in years) that the next major earthquake hits Southern California. I would like to assume a normal(m_0, h_0) prior density for m. My assessment gives $m_0 = 9$ and my probability that m is bigger than 20 is 25%. Obviously, m cannot be negative and so my probability of $m < 0$ should be 0. Why does a normal density not apply, at least not exactly?

11.12 How old are children when they begin to speak? Consider a child's mean age, m (in months), at which they speak their first word. Assume a normal(m_0, h_0) prior density and assess your values of m_0 and h_0. Describe each step of your process. Consider at least one check of h_0 to see that it is consistent with your opinions.

Rule of Means

Suppose your prior density for a mean m of normal(m, h) densities is normal(m_0, h_0). Then your posterior density is also normal—let's call it normal(m_1, h_1). We want to find m_1 and h_1. These values will depend on the observed data and also on m_0 and h_0. It seems reasonable to expect that m_1 should be between $\bar{x}$ and m_0, and it is. Moreover, if n is larger, then m_1 should be closer to $\bar{x}$ because the data then have more credibility. I will not derive the following formulas, except to say that m_1 has the characteristics just mentioned. In particular, it is a weighted average of the prior mean, m_0, and the sample mean, $\bar{x}$.

The weights in this weighted average are proportional to their **precisions,** which are defined as follows:

$$\text{Prior precision:} \quad c_0 = 1/h_0^2$$

$$\text{Sample precision:} \quad c = n/h^2$$

Think of precision as measuring *how well* one knows m, based on the corresponding information. Precisions add. So the posterior precision is easy to calculate (and is the reason for considering the inverse of the square of the standard deviation in the first place):

$$\text{Posterior precision:} \quad c_1 = c_0 + c$$

Prior precision c_0 indicates how accurately one knows m before obtaining the sample. If the prior standard deviation, h_0, is small (high prior precision), then the prior density is tightly concentrated about its mean, m_0. But if h_0 is large (low prior precision), then little information is available about m.

Similarly, the information about m present in the sample is n/h^2. If the standard deviation, h, is small or the sample size, n, is large, then the sample precision is high and the data are very informative about m. The formula for posterior precision reflects the intuition that how well m is known after the sample is the sum of how well it was known before the sample and what was learned from the sample.

I have indicated that, in calculating the posterior mean, the weights of the prior and sample means are proportional to precisions. This is the rule of means.

Rule of Means

$$\text{Posterior mean:} \quad m_1 = \frac{c_0}{c_1} m_0 + \frac{c}{c_1} \bar{x}$$

In view of the formula $c_1 = c_0 + c$, the weights of m_0 and of $\bar{x}$ (c_0/c_1 and c/c_1, respectively) do indeed sum to 1.

The posterior standard deviation of m is trivial once the posterior precision, c_1, is known:

> **Posterior standard deviation:** $h_1 = 1/\sqrt{c_1}$

In the extreme case in which there are no data at all, and so $n = 0$, these formulas give obvious answers: $c = 0$, $c_1 = c_0$, $h_1 = h_0$, and $m_1 = m_0$. In the other extreme in which the sample size is very large, the data completely determine the posterior mean: $c_1 \approx c$ (which is very large), $h_1 \approx 0$, and $m_1 \approx \bar{x}$. When a standard deviation is close to 0, there is little variability. So this result is consistent with the law of large numbers (see box in Section 10.2): A large sample implies that the sample average is close to the population mean.

We have generalized the result given in Section 11.2 (which applies when the prior density of m-values is flat):

> **Updating Rule for Normal Models**
>
> For n observations from a normal(m, h) density, if the prior density of m is normal(m_0, h_0), then the posterior density of m is normal(m_1, h_1), where m_1 and h_1 are as given in the preceding boxes.

The following generalizes the corresponding flat-prior result for making predictions. The standard deviation actually looks a bit simpler here than in the previous section, but it is more complicated because c_1 incorporates n (as well as both c and c_0).

> **Prediction for Normal Models**
>
> For n observations from a normal(m, h) density, if the prior density of m is normal(m_0, h_0), then the predictive density of the next observation is a normal$(m_1, \sqrt{h^2 + 1/c_1})$, where m_1 and c_1 are as given in the preceding boxes.

The preceding formulas will be used repeatedly in this chapter and in the next two chapters. The result of the previous section (in which the prior density is flat) is the special case in which h_0 is infinite (and so c_0 is 0, which means that m_0 does not enter into the posterior density). In that case, the formulas become

$$\text{Prior precision:} \quad c_0 = 1/h_0^2 = 0$$

$$\text{Sample precision:} \quad c = n/h^2$$

$$\text{Posterior precision:} \quad c_1 = c_0 + c = c$$

$$\text{Posterior mean:} \quad m_1 = \frac{c_0}{c_1} m_0 + \frac{c}{c_1} \bar{x} = \frac{0}{c} m_0 + \frac{c}{c} \bar{x} = \bar{x}$$

$$\text{Posterior standard deviation:} \quad h_1 = 1/\sqrt{c_1} = 1/\sqrt{c} = h/\sqrt{n}$$

When the population standard deviation, h, is unknown, we proceed as in the previous section, using an inflated version of the sample standard deviation, s, in place of h:

$$h = s\left(1 + \frac{20}{n^2}\right)$$

Again, conclusions for small n will not generally be very satisfactory. When n is small, it is especially important to carefully address the assessment of the prior information about m. Larger sample sizes allow one the luxury of being somewhat cavalier in the assessment process.

EXAMPLE 11.4

▷ **How much do I weigh? (revisited).** My bathroom scale gives what seem to be quite variable readings. I have stepped on it several times before getting two like readings. Maybe it is defective, but its readings do have a consistent pattern. I weighed myself 10 times in succession, with these results:

$$182 \quad 172 \quad 173 \quad 176 \quad 176 \quad 180 \quad 173 \quad 174 \quad 179 \quad 175$$

I calculate $\bar{x} = 176$ and $s = \sqrt{10} = 3.16$.

Think of this sample as selections from a population model, one with mean m. If the scale is not biased in the sense that its readings are neither consistently too big nor too small—a big assumption—and the mean measurement error is 0, then the mean m is my actual weight. What information does the sample provide about the population? And what does the sample suggest that the next (and 11th) reading will be?

In Example 11.3, I assessed my prior density to be normal(m_0, h_0), where

$$m_0 = 174 \quad \text{and} \quad h_0 = 4.69$$

Correct the sample standard deviation for the small sample size:

$$h = s\left(1 + \frac{20}{n^2}\right) = \sqrt{10}\left(1 + \frac{20}{100}\right) = \sqrt{10}\,(1.2) = 3.79$$

The formulas give

$$\text{Prior precision:} \quad c_0 = \frac{1}{h_0^2} = .0455$$

$$\text{Sample precision:} \quad c = \frac{n}{h^2} = \frac{10}{14.4} = .694$$

$$\text{Posterior precision:} \quad c_1 = c_0 + c = .740$$

$$\text{Posterior mean:} \quad m_1 = \frac{c_0}{c_1}m_0 + \frac{c}{c_1}\bar{x} = \frac{.0455}{.740}(174) + \frac{.694}{.740}(176)$$

$$= .061(174) + .939(176) = 175.9$$

$$\text{Posterior standard deviation:} \quad h_1 = 1/\sqrt{c_1} = 1/\sqrt{.740} = 1.163$$

The posterior mean is much closer to the sample mean of 176 than it is to the prior mean of 174. This is because the sample precision is much larger than the prior precision—about 15 times as large. Even though my scale is not very accurate, using it several times increases its accuracy.

Recall from Example 11.3 that my prior probability of my weight being less than 170 lb was about 20%. To find the corresponding posterior probability, calculate the z-score:

$$z = \frac{170 - m_1}{h_1} = \frac{170 - 175.9}{1.163} = -5.07$$

This is way off the scale in the Standard Normal Table, so the posterior probability is less than .0001.

The density of the next reading on the scale is a normal$(m_1, \sqrt{h^2 + 1/c_1})$, that is, normal(175.9, 3.97). The z-score corresponding to 170 lb is

$$z = \frac{170 - 175.9}{3.97} = -1.49$$

According to the Standard Normal Table, my probability that the next reading is less than 170 lb is .0681, about 7%. The predictive normal density and z-score for 170 lb are shown in Figure 11.10. My probability that the next reading is less than 170 lb is shown as a shaded area on the figure. The figure also shows the 10 previous observations as filled circles. The predictive density is based mostly on these previous observations and only somewhat on prior information (since the prior precision is much less than the sample precision). So it is not surprising that these 10 observations look like they could be a sample from the predictive density.

Figure 11.10 Predictive density of next observation in Example 11.4

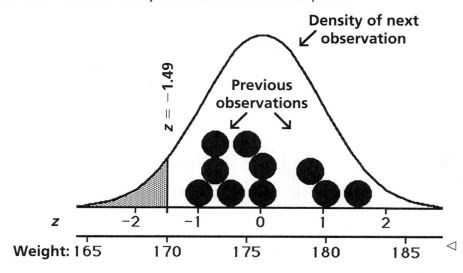

EXAMPLE 11.5

▷ **Drop in blood pressure.** A small pilot study[4] evaluated the effectiveness of an experimental drug. The (sitting diastolic) blood pressures of 16 patients who had mild to moderate hypertension were measured before and after receiving the drug. (There was a placebo group in the study as well; see Exercise 12.2.) The changes in blood pressure (in mmHg), where negative numbers are decreases while on the drug and positive numbers are increases, are as follows:

+3.7	−6.7	−10.5	−6.1	−17.6	+2.3	−7.9	−8.9
−4.5	−7.7	−9.4	−10.4	−10.9	−9.3	−16.7	−7.2

I find $\bar{x} = -7.99$ and $s = 5.33$.

My prior density for the mean change in blood pressure is normal(m_0, h_0). My prior opinion perhaps should reflect the fact that the drug was designed to lower blood pressure. But to minimize the effect of my opinions on the posterior probabilities, I will take my estimate of the mean effect of the drug to be $m_0 = 0$. I associate some probability that the drug causes a dramatic average decrease, such as 10 mmHg, and also a dramatic increase, and so my prior standard deviation is $h_0 = 10$.

These are the calculations needed to find my posterior density:

Population standard deviation: $\quad h = s\left(1 + \dfrac{20}{n^2}\right) = 5.33\left(1 + \dfrac{20}{256}\right)$

$$= 5.33(1.078) = 5.74$$

Prior precision: $\quad c_0 = \dfrac{1}{h_0^2} = .010$

Sample precision: $\quad c = \dfrac{n}{h^2} = \dfrac{16}{33.0} = .485$

Posterior precision: $\quad c_1 = c_0 + c = .495$

Posterior mean: $\quad m_1 = \dfrac{c_0}{c_1}m_0 + \dfrac{c}{c_1}\bar{x} = \dfrac{.010}{.495}(0) + \dfrac{.485}{.495}(-7.99) = -7.83$

Posterior standard deviation: $\quad h_1 = 1/\sqrt{c_1} = 1/\sqrt{.495} = 1.42$

My posterior density is normal(−7.83, 1.42). As in the previous example, the posterior mean is much closer to the sample mean (−7.99) than it is to the prior mean (0).

Of interest is the posterior probability that the drug causes a mean drop in blood pressure of at least 5 mmHg. Calculate the z-score at −5:

$$z = \dfrac{-5 - m_1}{h_1} = \dfrac{-5 - (-7.83)}{1.42} = 1.99$$

According to the Standard Normal Table, the probability of a z-score less than 1.99 is .9767, or about 98%.

If we are interested in a particular patient's response and not the mean response of the population of patients, then there is more variability. The standard deviation of the predictive density is $\sqrt{h^2 + (1/c_1)} = \sqrt{5.74^2 + 2.02} = \sqrt{35.0} = 5.92$. The predictive z-score for a change of -5 is

$$z = \frac{-5 - m_1}{\sqrt{h^2 + 1/c_1}} = \frac{-5 - (-7.83)}{5.92} = .48$$

So the probability that a *particular patient's* blood pressure drops on drug by at least 5 mmHg is only about 68%. ◁

The examples in this section assume normal prior densities while those in Section 11.1 assume a flat prior density—that is, zero prior precision. For flat priors the posterior density is proportional to the likelihood. If the prior precision is not 0, but is small relative to the sample precision, then the posterior density is not much changed by setting it to 0. In other words, when there is not very much prior information and the sample size n is moderately large, then the posterior density is approximately the same as when using a flat prior.

> **Principle of Flat Priors**
>
> **Use flat priors as an approximation when the prior precision is small in comparison with the sample precision.**

In which direction does the approximation go—does it make the conclusions more or less convincing? The answer depends on whether the prior evidence is consistent with the data. If it is, then the conclusions will be slightly weakened by using flat priors (to an extent that depends on the relative sizes of the precisions) because flat priors make the posterior standard deviation as large as possible; that is, flat priors are then more conservative. If the prior evidence is inconsistent with the data, then the conclusions will tend to be more convincing than is appropriate.

> **Using flat priors is conservative when the actual prior evidence is consistent with the sample data.**

EXAMPLE 11.6

▷ **Drop in blood pressure (revisited).** In Example 11.5, we considered a sample with $\bar{x} = -7.99$ and $s = 5.33$. My prior density for the population mean change in blood pressure was normal(0, 10). My posterior density turned out to be normal(-7.83, 1.42) and my probability of a mean drug effect greater than 5 mm is about 97.7%.

Suppose I had used a flat prior instead. Then my posterior density would have been normal(-7.99, 1.44). The z-score for a 5-mm drop is

$$z = \frac{-5 - (-7.99)}{1.44} = 2.08$$

and the probability to the left is 98.1%.

The difference is minimal because the prior precision contributes only $.01/.495 = 2\%$ to the posterior precision. In such a case, a flat prior seems quite acceptable. But the conclusion is a little stronger when using the flat prior. That is because the prior was centered at 0 and the likelihood is centered near -8. ◁

EXERCISES

11.13 One of the measurements made in a study[5] of 21 children was their age in months at the time they spoke their first word. These ages are as follows:

15 26 10 9 15   20 18 11 8 20   7 9 10 11 11   10 12 42 17 11   10

In Exercise 11.12, you assessed your normal(m_0, h_0) prior density for the mean age, m, in months at which children speak their first word. For the purposes of this exercise, assume $m_0 = 12$ and $h_0 = 6$.
(a) Find the posterior density of m based on the given data.
(b) Find the probability that a child from this population (but not one of the 21 in the study) has still not uttered a word at age 24 months.

11.14 The observed breaking strengths (in g) of 20 pieces of yarn taken from spinning machines in a certain production area[6] are as follows:

46 58 40 47 47   53 43 48 50 55   49 50 52 56 49   54 51 50 52 50

Assume that this is a sample from a normal density with unknown mean m. Also assume a normal(60, 15) prior density for m. Find the following:
(a) Posterior density of m.
(b) Predictive probability that the next (that is, the 21st) piece of yarn will be stronger than any of the 20 pieces in the sample (that is, stronger than 58 g).

11.15 In Exercise 10.5, you made some calculations for data involving 32 congestive heart failure patients. The data are increases in miles walked while on an experimental drug as compared with miles walked in the previous week while off drug:

.00	+.56	+3.27	−2.55	+8.42	+1.07	−1.31	+3.19
−.59	+10.75	+11.73	−.05	+1.65	−3.42	+1.73	−1.44
+6.04	+12.21	+4.97	+1.68	+2.28	−6.57	−2.11	+.75
−.96	+1.68	+8.85	+7.45	−.59	+2.91	.00	+5.40

Assume that this is a sample from a bell-shaped population density with unknown mean m. Also assume a normal(0, 10) prior density for m. Find the predictive probability that the next (that is, the 33rd) patient from this population will have a beneficial response (that is, the increase will be positive). [You will first have to calculate the posterior density of m.]

Normal Densities for Means of Large Samples

In the previous sections, I claimed that the normal density is the posterior density of the population mean, m, when the population is a normal model. Assuming normal models for populations is not necessary when the sample size, n, is large. This follows from a result in probability theory called the **central limit theorem.** In our setting, we can simply change some of the words from the result given in the previous section.

The most important change is in the phrase "from a normal model" to "from *any* population":

> **Central Limit Theorem**
>
> **For a large sample of n observations from *any* population, if the prior density of m is normal(m_0, h_0), then the posterior density of m is normal(m_1, h_1).**

The values of m_1 and h_1 are as before:

$$m_1 = \frac{c_0}{c_1} m_0 + \frac{c}{c_1} \bar{x} \quad \text{and} \quad h_1 = 1/\sqrt{c_1}$$

where the prior precision is $c_0 = 1/h_0^2$, the sample precision is $c = n/s^2$, and the posterior precision is $c_1 = c_0 + c$.

The **posterior standard deviation, h_1,** is the same as in the previous section. Because the sample size is large, we can use the sample standard deviation s in place of the population standard deviation h without the small-n correction factor. With this change, all formulas are the same as in the previous section.

The other main result in the previous section deals with predictive densities. The central limit theorem does not help in this regard: For predictive densities to be bell shaped requires that the population density itself be bell shaped—an assumption not made in this section. When the sample is large, however, the sample distribution may be taken as the population distribution (or model) and also as the predictive distribution of the next observation. The important question is, how large is "large"? Sometimes a sample large enough for one use is not large enough for another. The next example is a case in point.

EXAMPLE 11.7
▷ **Molecular weights of fragments of DNA.** Example 2.24 gives a histogram of 3,092 molecular weights of DNA fragments at locus D2S44 (reproduced in Figure 11.11). There are four blips in the right-hand tail that are barely visible and that represent outlying observations at 1,652, 1,762, 1,874, and 2,094.

The sample size is large enough to give some information about the shape of the population. It seems to be a mixture of perhaps four populations centered near 800, 1,100, 1,220, and 1,520. (Actually, biological considerations suggest that it is a mixture of a larger number of populations.) Each component of the mixture may be a normal, but the population itself is obviously not a normal.

Since the sample size is large, we can appeal to the law of large numbers (Section 6.2) and consider using the sample histogram as the predictive density. It serves well for some purposes. For example, the interval from 1,200 to 1,240 contains 462 observations. So the probability that the next observation is between 1,200 and 1,240 is about $462/3{,}092 = 14.94\%$. This is a problem of estimating proportions as in Chapter 7. Using a beta$(1, 1)$ prior density, this probability is $463/3{,}094 = 14.96\%$, which is essentially the same as using the observed proportion ($462/3{,}092 = 14.94\%$).

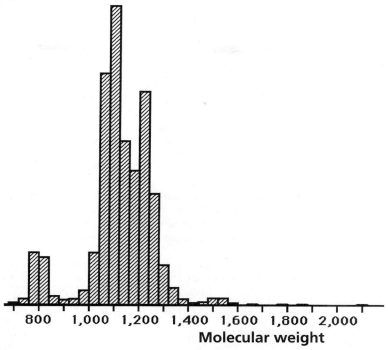

Figure 11.11 Histogram of molecular weights of DNA, locus D2S44 (Caucasians), for Example 11.7

However, the sample histogram does not serve well as the predictive density for the interval from 2,000 to 2,040, say. The frequency in this interval is 0; that is, none of the 3,092 allele weights were in this interval. The predictive probability assuming a beta(1, 1) prior density gives a different answer: $1/3{,}094 = .03\%$. In a sense, this is close to 0. But it is far from 0 in legal applications: If a blood sample left at a crime scene measures 2,020 and a suspect's blood is found to also measure 2,020, using a probability of 0 for the interval from 2,000 to 2,040 would be taken (incorrectly) to mean that no one else in the world matches the crime sample. Using a probability of 1/3,094 would be taken to mean that this allele weight is rare. In the law, there is a big difference between being rare and being unique. ◁

When the sample size is large, as in the previous example, the sample histogram may serve as the predictive density. It serves well in some circumstances. But it serves poorly when the sample size is small—unless the population is assumed to be a normal. It also may serve poorly in regions where the observations are sparse.

> **The predictive distribution can be approximated by the sample histogram when the sample size is large.**

EXERCISES

11.16 In Exercise 2.26, you calculated the sample mean, $\bar{x}$, and sample standard deviation, s, for the following strengths of 50 spot welds (in psi):

$$\begin{array}{ccccc ccccc}
400 & 395 & 398 & 421 & 445 & 389 & 372 & 408 & 398 & 401 \\
399 & 386 & 423 & 364 & 394 & 414 & 390 & 412 & 398 & 363 \\
388 & 431 & 392 & 438 & 411 & 399 & 399 & 408 & 390 & 420 \\
400 & 389 & 430 & 426 & 388 & 406 & 431 & 411 & 404 & 424 \\
450 & 416 & 397 & 404 & 388 & 405 & 392 & 405 & 379 & 419
\end{array}$$

(a) Assume a flat prior density for the population mean, m, and find its posterior density.

(b) Assume a normal(m_0, h_0) prior density with $m_0 = 1{,}000$ and $h_0 = 100$. Find the posterior density of the mean.

[Your answers to parts (a) and (b) should be the same: Taking h_0 to be very large gives the same answers as assuming a flat prior density.]

(c) Now assume that the population density is itself bell shaped. Using your result in either part (a) or (b), find the predictive probability that the next spot weld has strength greater than 425 psi. [Seven of the 50 strengths were greater than 425 psi; if your answer is very different from $7/50 = .14$, either it is wrong or assuming a bell shape for the population is inappropriate.]

11.17 Example 2.25 considers test scores of 223 prospective Chicago city engineers competing for 15 jobs. Ignoring the 15 scores of questionable authenticity leaves 208 test scores. These are given in the following table and histogram. (I inserted an X in the table where a questionable score occurs in the table of Example 2.25.) Consider one of these 15 questionable cases. Suppose there was no fraud and that such a person's score was exchangeable with the 208 legitimate scores. Using the approximation suggested in this section and a normal(60, 20) prior density, find the predictive probability that this person's score will be as follows:

(a) Greater than 70.

(b) In the top 14 scores. (The top 15 would involve ties.)

(c) In the top 14 scores, given that it is greater than 70.

Test scores of 208 (legitimate) applicants for Exercise 11.17

61	48	58	27	57	39	37	57	47	60	55	67	51	42	50
84	61	53	X	X	33	49	80	74	52	71	74	55	45	83
68	35	43	30	45	83	60	36	63	61	37	82	58	51	58
62	64	26	37	84	41	40	54	56	46	43	X	45	39	69
60	37	53	84	29	54	80	34	47	45	31	44	56	66	X
59	49	59	33	31	44	46	76	43	69	47	61	X	60	32
X	33	42	37	69	39	56	58	48	27	49	81	84	46	47
50	48	60	57	27	27	32	42	X	54	65	58	82	36	61
57	62	69	34	51	55	44	62	61	43	51	42	X	80	56
37	36	39	58	47	30	30	67	59	54	34	63	51	73	52
66	52	42	69	84	43	67	83	44	83	80	44	56	78	33
47	X	39	48	58	58	30	75	48	76	53	48	54	44	46
46	66	45	37	69	72	46	33	X	84	X	X	84	53	81
53	X	52	45	31	43	43	45	75	68	39	43	71	81	48
59	X	35	60	74	67	48	52	39	69	69	X	49		

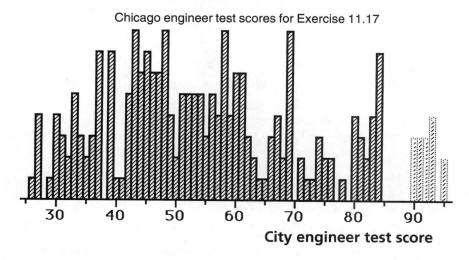
Chicago engineer test scores for Exercise 11.17

11.5 Probability Intervals for Population Means

Section 7.6 described percentiles and their relationship to probability intervals calculated for normal densities. The same relationship and calculations apply in the general setting of this chapter. Now, m_1 and h_1 play the roles of r and t, respectively. Although this section starts from scratch, you should refresh your memory about Section 7.6 before reading on.

EXAMPLE 11.8
▷ **Drop in blood pressure (revisited).** In Example 11.5, we considered the mean change in blood pressure (measured in mmHg) among patients who took an experimental drug. We found that the posterior density of m-values was normal(-7.83, 1.42). We calculated the posterior probability that the population mean was less than -5 by calculating the z-score for -5,

$$z = \frac{-5 - (-7.83)}{1.42} = 1.99$$

and then used the Standard Normal Table to find the probability of a z-score less than 1.99 to be .9767. Thus, -5 is about the 98th percentile of m, or, more accurately, the 97.67th percentile.

Suppose that we wanted to find the 97.67th percentile of the density of m-values. Then we would reverse the order of the calculation. We would go to the Standard Normal Table first and find the 97.67th percentile of the standard normal curve: $z = 1.99$. So we know that the value of m we are looking for satisfies

$$z = \frac{m - m_1}{h_1} = 1.99$$

Solving for m gives
$$m = zh_1 + m_1$$
Substituting for z, m_1 and h_1 gives
$$m = (1.99)(1.42) - 7.83 = -5.00$$
which is, therefore, the 97.67th percentile of the density of m-values.

To give an example where we do not know the answer in advance, consider the 90th percentile of m-values. According to the Standard Normal Table, the 90th percentile of z-scores is about 1.28. Proceeding as before, we set
$$m = (1.28)(1.42) - 7.83 = -6.01$$
Similarly, the 10th percentile of the standard normal density is -1.28. So the 10th percentile of m-values is
$$m = (-1.28)(1.42) - 7.83 = -9.65$$
Therefore, the probability between $m = -9.65$ and $m = -6.01$ is 90% − 10% = 80%. This is shown in Figure 11.12.

Figure 11.12 80% probability interval in Example 11.8

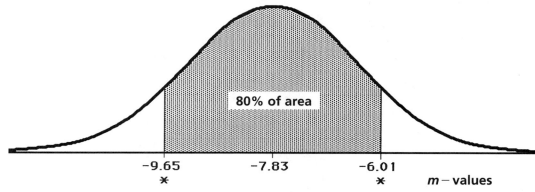

The interval of m-values from -9.65 to -6.01 (indicated by asterisks) is an 80% probability interval. This interval can be expressed as being from -9.65 to -6.01 or simply as -7.83 ± 1.82, where $1.82 = zh_1 = (1.28)(1.42)$. ◁

In the previous example, we found an 80% probability interval going from $z = -1.28$ to $z = 1.28$. According to the Standard Normal Table, the 5th percentile of the standard normal density is -1.65 and the 95th percentile is 1.65. Since 95% − 5% = 90%, going from $z = -1.65$ to $z = 1.65$ would give a 90% probability interval of m-values. The following short table was constructed in this way from the Standard Normal Table. It gives the z-score, z_{perc}, used for constructing probability intervals containing perc% probability.

68% probability interval: $z_{68} = 1.00$
80% probability interval: $z_{80} = 1.28$
90% probability interval: $z_{90} = 1.65$

95% probability interval: $z_{95} = 1.96$
98% probability interval: $z_{98} = 2.33$
99% probability interval: $z_{99} = 2.58$

> **Constructing a Probability Interval for m**
>
> Find the posterior mean, m_1, and standard deviation, h_1, as on pages 349–350. A perc% probability interval for m is
>
> $$m_1 \pm z_{\text{perc}} h_1$$
>
> where z_{perc} is given in the preceding table.

A perc% probability interval assuming a flat prior density is the same as a perc% *confidence interval* in a frequentist statistical approach.

EXAMPLE 11.9
▷ **Water pollution.** A report on water pollution[7] gave the amounts of chemical oxygen demands (mg/l) for 18 water samples:

$$\begin{array}{cccccccccc} 580 & 674 & 512 & 540 & 616 & 298 & 960 & 570 & 640 \\ 588 & 556 & 588 & 582 & 844 & 574 & 420 & 696 & 620 \end{array}$$

Assume that this is a random sample from a population with mean m. The value of m is a measure of the degree of pollution in the water being sampled. An important problem is to understand how well m is known from the data; we seek a 95% posterior probability interval for m.

I find that $\bar{x} = 603.2$ and $s = 138.5$. Assume that the population is normal. As a check to see whether this is reasonable, I sketched a normal(603.2, 138.5) curve on a dot plot of the data, as shown in Figure 11.13. The middle 14 observations would be fit better by a density with smaller standard deviation than the curve in the figure. The other four observations—especially 298 and 960—seem rather far from the middle (relative to the standard deviation, 138.5) and so are more consistent with a larger standard deviation. However, the fit is not horrible and we will proceed to find probability intervals for m assuming a normal population. (See Chapter 13 for cases in which you do not want to make this assumption.)

Figure 11.13 Dot plot with possible normal population density for Example 11.9

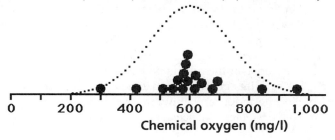

Take the prior density of m to be flat (as in Section 11.1). So

$$m_1 = \bar{x} = 603.2$$

$$h = s\left(1 + \frac{20}{n^2}\right) = 138.5\left(1 + \frac{20}{324}\right) = 138.5(1.062) = 147.1$$

$$h_1 = h/\sqrt{n} = 147.1/\sqrt{18} = 34.66$$

Since $z_{95} = 1.96$ and $(1.96)(34.66) = 67.9$, the 95% probability interval is 603.2 ± 67.9, or from 535.3 to 671.1. To find a 99% probability interval, use $z_{99} = 2.58$. Since $(2.58)(34.66) = 89.4$, this interval is 603.2 ± 89.4, or from about 514 to 693. (You may wonder how it is possible that five of the 18 data points can be outside this interval; the interval refers to the population mean and not to the observations.) ◁

As in Chapters 7 and 9, we cannot test null hypotheses by finding probabilities of particular values of d without a computer. As before, use the following alternative:

Alternative Test of Null Hypothesis $m = 0$

Decide whether the 95% posterior probability interval contains 0. If yes then the null hypothesis is supported; otherwise, it is not.

Equivalently, calculate $z = \dfrac{0 - m_1}{h_1}$ using the posterior values m_1 and h_1. If $-1.96 < z < 1.96$, then the null hypothesis is supported; otherwise, it is not.

As I've pointed out before, a null hypothesis may be "supported" for one person but not for another. An approximate connection with the frequentist approach to statistics is that the null hypothesis that $m = 0$ is "rejected" if it is not supported by someone whose prior density of m is flat.

EXAMPLE 11.10
▷ **Drop in blood pressure (revisited).** In earlier examples, we considered a sample with $\bar{x} = -7.99$ and $s = 5.33$. My posterior density turned out to be normal(-7.83, 1.42). My 80% probability interval (from Example 11.8) is -9.65 to -6.01. My 95% probability interval is $-7.83 \pm (1.96)(1.42)$, or from about -10.61 to -5.05. Since this does not contain the null hypothesis mean of 0, the null hypothesis is not supported. So the evidence suggests that the drug is effective in lowering blood pressure.

My 99% posterior probability interval is from about -11.49 to -4.17, which is still far from 0, so the evidence against the null hypothesis is quite strong. ◁

In this chapter, we considered models for populations and, using a sample of observations, we addressed finding the posterior density of a population mean and predicting future observations. But we considered one sample from one population—including observations from a paired experiment in which we considered differences between two measurements. Parallel designs (Section 3.5) are more common in research because researchers usually obtain a second sample to serve as a control. In Chapters 8 and 9, we considered samples from two populations that contain only two types of observations, such as success and failure. In the next chapter, we consider a sample from each of two populations, where the populations are of the general type that we have considered in the present chapter.

EXERCISES

11.18 Exercise 10.8 gives the following nine measured weights (in μg) for a particular specimen:

$$114 \quad 129 \quad 121 \quad 98 \quad 140 \quad 134 \quad 122 \quad 133 \quad 125$$

Again, suppose the specimen actually weighs m μg and that the population is normal with mean m. Now the population standard deviation is unknown. Assume a prior density for m that is normal(100, 20).
(a) Find the posterior density for m.
(b) Find a 90% posterior interval for m.
(c) Find a 95% posterior interval for m.
(d) Find a 99% posterior interval for m.
[Hint: The intervals get wider from parts (b) through (d).]

11.19 Example 11.2 and Exercise 11.1 give the steady-state power of 40 "earlier" and 28 "later" devices, respectively. Using the posterior densities of the two m-values, find the 99% posterior probability intervals for the following:
(a) The m of the earlier devices
(b) The m of the later devices
[These intervals do not intersect. But it is not clear how to interpret this. Chapter 12 gives better ways of addressing the difference between two means.]

11.20 In Exercise 11.2, you found the posterior density of m, the population mean difference of forced vital capacity in ml. Use this density to do the following.
(a) Find the 95% probability interval for this mean difference.
(b) Test the hypothesis that $m = 0$.

11.21 Exercise 2.23 gave the following 13 measurements of zinc concentration (mg/kg):

$$13.5 \quad 23.8 \quad 23.3 \quad 20.9 \quad 23.8 \quad 29.0 \quad 20.9 \quad 24.4 \quad 16.4 \quad 18.3 \quad 17.6 \quad 25.4 \quad 23.3$$

For this sample, assume a normal population and a flat prior for its mean, m. Find a 99% posterior probability interval for m.

11.22 In an experiment to test observational learning,[8] 16 octopuses watched other octopuses that had been trained to select red balls when offered a choice between red and white. Each octopus was then given five opportunities in the same setting. The mean number of reds selected was 4.31 with standard deviation $s = .98$. Convert these to proportions: $\bar{x} = .862$ and $s = .196$. Assume these 16 are a random sample from the population of octopuses allowed to make similar observations and selections, and that the mean proportion of correct selections in this population is m. Assume a flat prior and find the 95% posterior probability interval for m.

11.23 The experiment described in Exercise 11.22 employed a control group of 18 octopuses that had not observed other octopuses selecting a ball. Again, each was given five opportunities to choose between red and white. The mean proportion of reds was .422 with $s = .24$. Again, assume a flat prior for the m of this population (which now includes octopuses that were not allowed to watch others selecting red). Find the 95% posterior probability interval for m.

11.24 Is the brain like a muscle—use it and it grows? Researchers[9] selected pairs of rats from the same litters and placed them in different environments. One rat in each pair was selected at random for communal living with 11 other rats and given new toys to play with each day. The other rat received the same food and drink, but lived in isolation without toys. The animals were sacrificed at the same age and their cortexes were weighed. (There is experimental variability and possibility for bias in the weighing process; the investigators coded the animals, arranged littermates in random order, and the technicians, who dissected the brains, were blinded as to treatment group.)

The experiment was repeated many times; the results of the first five experiments are given in the table below. The treatment animals (T) within each experiment lived together. (Apparently, one of the 12 pairs in experiment 1 was invalid, perhaps one member of the pair died early.) The treatment animal's cortex weight is on the left within each pair and its littermate (C) is on the right.

Assess the effect of treatment by considering the increase in treatment over control. This increase is measured in milligrams, which is difficult to interpret. So consider percentage increase. These are shown in the table at the top of the next page, given in the same order as the results in the first table. The sample mean is 5.72% and the sample standard deviation is 5.09%. So rats' cortexes in the treatment group were about 6% heavier on average, which in turn suggests that communal living and playing with toys enhances the ability to think. (A factorial design as described in Section 3.7 would allow for deciding the relative contribution of these two factors.)

Cortex weights (mg) of treatment and control animal pairs for Exercise 11.24

Exp. 1		Exp. 2		Exp. 3		Exp. 4		Exp. 5	
T	C	T	C	T	C	T	C	T	C
689	657	707	669	690	668	700	662	640	641
656	623	740	650	701	667	718	705	655	589
668	652	745	651	685	647	679	656	624	603
660	654	652	627	751	693	742	652	682	642
679	658	649	656	647	635	728	578	687	612
663	646	676	642	647	644	677	678	653	603
664	600	699	698	720	665	696	670	653	593
647	640	696	648	718	689	711	647	660	672
694	605	712	676	718	642	670	632	668	612
633	635	708	657	696	673	651	661	679	678
653	642	749	692	658	675	711	670	638	593
		690	621	680	641	710	694	649	602

Sec. 11.5 / Probability Intervals for Population Means

Percentage increase in cortex weights of treatment over control animals for Exercise 11.24

		Experiment		
1	2	3	4	5
4.9	5.7	3.3	5.7	−.2
5.3	13.8	5.1	1.8	11.2
2.5	14.4	5.9	3.5	3.5
.9	4.0	8.4	13.8	6.2
3.2	−1.1	1.9	26.0	12.3
2.6	5.3	.5	−.1	8.3
10.7	.1	8.3	3.9	10.1
1.1	7.4	4.2	9.9	−1.8
14.7	5.3	11.8	6.0	9.2
−.3	7.8	3.4	−1.5	.1
1.7	8.2	−2.5	6.1	7.6
	11.1	6.1	2.3	7.8

Dot plot (a) suggests that control weights vary with the experiment (and the treatment weights do as well)—see Exercise 12.29. The other dot plot suggests that the percentage increases are quite comparable across experiments—see Exercise 12.30. So combining these percentages seems reasonable. Assume a flat prior for the population mean, m, of percentage increases. The histogram shown at the top of page 366 draws into question the assumption of a normal population, but the central limit theorem of Section 11.4 justifies assuming that the posterior density is approximately normal.

(a) Find the posterior density of m.
(b) Use the posterior density to find the 95% probability interval for m.
(c) Test the hypothesis that $m = 0$.

Dot plots of control and percentage change of treatment for Exercise 11.24

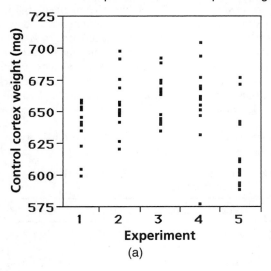
(a)

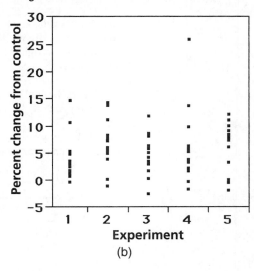
(b)

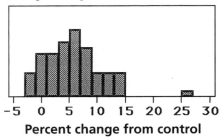

Histogram of percentage change of treatment to control for Exercise 11.24

Percent change from control

11.25 Example 2.18 gives birth weights (in lb) of babies depending on the smoking habits of their mothers. Consider only those birth weights of babies whose mothers have never smoked (repeated here). Regard this as a sample from a bell-shaped population with mean m.

Birth weights of children of nonsmoking mothers for Exercise 11.25

3.3	6.1	6.6	6.9	7.3	7.8	7.9	8.5	9.2
5.3	6.1	6.6	7.1	7.4	7.8	8.3	8.6	9.2
5.6	6.1	6.6	7.1	7.4	7.8	8.3	8.6	10.9
5.6	6.5	6.7	7.1	7.6	7.8	8.3	8.6	
5.6	6.6	6.7	7.1	7.8	7.9	8.4	8.8	

(a) Assume a normal(7.5, 2.0) prior density for m. Find the 90% posterior probability interval for m.

(b) Assume a flat prior density for m. Find the 90% posterior probability interval for m.

[The interval in part (b) should be wider than that in part (a). Since the prior precision in part (b) is 0, if the two answers turn out to be close, then the prior precision in part (a) is small in comparison with the sample precision.]

11.26 Can very elderly people increase muscle strength by exercising? Apparently so. A randomized study[10] with a factorial design (Section 3.7) evaluated the effects of exercise and a nutrition supplement in people whose average age was 87. The exercise involved high-intensity resistance training of the knee and hip extensors 3 days a week, 45 minutes per day, for 10 weeks. Various measures of muscle performance were used, including left- and right-knee strength and press and left- and right-hip strength. All show marked improvements for exercise. I used stair-climbing power (as a percentage increase from prestudy values) in the table shown because it is an important measure of a person's well-being. [I estimated the sample sizes and standard deviations using the limited information given in the article—including the total sample size (83) for this measurement and the sample size allocation of the 100 subjects in the entire study (25:24:25:26).]

Consider only the exercise group and give a 90% posterior probability interval for the population mean percentage change; use a flat prior.

Percentage increase (from prestudy) in stair-climbing power for Exercise 11.26

	Exercise	Supplement	Both	Control
Mean	33.8	12.5	23.0	−5.2
Std. dev.	42	54	45	44
n	21	20	21	21

11.27 Before they were found to cause cancer in animals, DDT and PCBs were used extensively in many countries as insecticides and as fluid insulators of electrical components. Humans ingested them and breathed them. But humans do not metabolize them well and so they remain in human tissue essentially indefinitely. Are they harmful to humans? A study[11] compared blood levels of DDT (actually, its main metabolite DDE) and PCBs in 58 breast cancer patients. For each patient, researchers selected controls matching the patient's age, menopausal status, number and dates of blood donations, and, for premenopausal patients, the day of the menstrual cycle on which the blood was drawn. They selected two controls for each of the 34 postmenopausal patients and four controls for each of the 24 premenopausal patients. (I do not know the reason for the difference.) Also, seven patients had one extra control each. The controls provide a pairing; they subtracted the average of the controls from the patient's blood level. The appropriate analysis involves these differences. The article gives the following table of sample means and standard deviations:

DDT and PCBs in blood (ng/ml) of breast cancer patients and controls for Exercise 11.27

	DDT		PCBs	
	BC	Controls	BC	Controls
Mean	11.0	7.7	8.0	6.7
Std. dev.	9.1	6.8	4.1	6.7
n	58	171	58	171

This information would be relevant if this were a two-sample problem, as in Chapter 12. But it involves a single sample. The article does not give the standard deviations of differences, but I inferred them from other calculations provided:

Difference (ng/ml) in DDT and PCBs in blood of breast cancer patients vs controls for Exercise 11.27

	DDT	PCBs
Mean	2.7	1.0
Std. dev.	9.2	3.9
n	58	58

[The reason $11.0 - 7.7 \neq 2.7$ is that the patients had different numbers of controls.] For DDT and also for PCBs, consider the differences to be a random sample from some population. (They are not really. The researchers would not make the data available to me, but the larger number of controls for premenopausal patients should make their standard deviation smaller. This kind of difference in standard deviation should not have much effect on the conclusions.) Assume flat prior densities for the population means, m_{DDT} and m_{PCB}.

(a) Test the hypothesis that $m_{DDT} = 0$.
(b) Test the hypothesis that $m_{PCB} = 0$.

Appendix: Using Minitab with Densities for Means

The Minitab program **'m_cont'** will compute the posterior distribution for a normal mean when a normal prior density is assumed. In Example 11.4, we wish to learn about my true weight, m, on the basis of 10 measurements on my bathroom scale. My opinions about m are modeled using a normal distribution with mean 174 and standard deviation 4.69. First, we place the observed measurements (182, 172, 173, 176, 176, 180, 173, 174, 179, 175) in the column 'data'. Then type

```
exec 'm_cont'
```

The program asks if we wish to use an informative prior for the mean. We answer yes by typing 'y' and input the mean and standard deviation of the normal prior on the 'DATA' line. We indicate next that the data are contained in the worksheet and input the number of the column. The output of the program is the mean and standard deviation of the normal posterior density for the true weight, m. In addition, the program gives the mean and standard deviation of the normal distribution which can be used to predict the value of the next measurement on the scale. To summarize the results, we are fairly confident (with probability .95) that my true weight is 175.86 plus or minus 1.96×1.22. Also, we believe (with probability .95) that the next measurement will be 175.86 plus or minus 1.96×4.18.

```
MTB > name c1 'data'
MTB > set c1
DATA> 182 172 173 176 176 180 173 174 179 175
DATA> end
MTB > exec 'm_cont'
DO YOU WISH TO USE AN INFORMATIVE NORMAL PRIOR FOR M? (TYPE 'y' OR
   'n'.) IF YES, INPUT PRIOR MEAN AND PRIOR STANDARD DEVIATION.
y
DATA> 174 4.69

PR_MEAN
   174

PR_STD
   4.69

OBSERVED DATA IN WORKSHEET? (TYPE 'y' OR 'n'.)
   IF YES, INPUT NUMBER OF COLUMN.
   IF NO, INPUT OBSERVED SAMPLE MEAN, STANDARD DEVIATION, AND
   SAMPLE SIZE.
y
DATA> 1

data
      182     172     173     176     176     180     173     174     179     175
```

```
MEAN         =     176.00
COUNT        =     10
ST. DEV.     =     3.3333
```

THE POSTERIOR DENSITY FOR M IS NORMAL
WITH MEAN AND STANDARD DEVIATION:

```
MEAN     STD
175.864  1.221
```

THE PREDICTIVE DENSITY OF THE NEXT OBSERVATION
IS NORMAL WITH MEAN AND STANDARD DEVIATION:

```
MEAN     STD
175.864  4.182
```

This program will give slightly different posterior standard deviations from the ones computed in Chapter 11. This program uses the common "divide by sample size minus 1" definition of the sample standard deviation to correspond to the definition used in Minitab. For most examples, the program's formula will give a posterior standard deviation very close to the standard deviation computed in the text.

The book *Bayesian Computation Using Minitab* by James Albert (Duxbury Press, 1996) gives other programs useful in making inferences about a normal mean. The program '**normal_s**' finds the particular normal curve that matches two probability statements about the prior density. For an arbitrary normal curve, the program '**normal**' plots the density and will compute cumulative probabilities and percentiles. The program '**m_nchi**' produces the "exact" normal–chi-square inference about the mean and standard deviation of a normal distribution using a noninformative prior distribution. The program '**m_norm_t**' performs a test that the mean, *m*, is equal to a particular value using a normal prior density.

Chapter Notes

1. Interested students should refer to the book *Probability and Statistics,* 2nd ed., by M. H. DeGroot (Reading, Mass.: Addison-Wesley, 1986).
2. F. M. Phillips et al., *Science* 248 (1990): 1529–1532.
3. K. A. Brownlee, *Statistical Theory and Methodology in Science and Engineering* (New York: Wiley, 1965).
4. J. D. Goldberg and K. J. Koury, in *Statistical Methodology in the Pharmaceutical Sciences,* ed. D. A. Berry (New York: Marcel Dekker, 1989), 220.
5. M. R. Mickey, O. J. Dunn, and V. Clark, *Computers and Biomedical Research* 1 (1967): 105–111.
6. G. E. P. Box and G. C. Tiao, *Bayesian Inference in Statistical Analysis* (New York: Wiley, 1973), 83.
7. K. J. Shapland, "Industrial effluent treatability—A case study," *Water Pollution Control* 85 (1986): 75–80.
8. G. Fiorito and P. Scotto, "Observational learning in *octopus vulgaris,*" *Science* 256 (April 24, 1992): 545–547. This article reports other types of training, for example, octopuses that observe other octopuses selecting white balls.
9. M. Rosenzweig, E. L. Bennett, and M. C. Diamond, "Brain changes in response to experience," *Scientific American* (February 1964): 22–29. Reported in D. Freedman, R. Pisani, and R. Purves, *Statistics* (New York: W. W. Norton, 1978), 451–453.
10. M. A. Fiatarone, E. F. O'Neill, N. D. Ryan et al., "Exercise training and nutritional supplementation for physical frailty in very elderly people," *The New England Journal of Medicine* 330 (1994): 1769–1775.
11. M. S. Wolff, P. G. Toniolo, E. W. Lee, M. Rivera, and N. Dubin, "Blood levels of organochlorine residues and risk of breast cancer, *Journal of the National Cancer Institute* 85 (1983): 648–652.

12 Comparing Two or More Means

CHAPTER 11 addressed problems in which there is a single sample of observations. Problems with one sample arise from paired designs and from other, simpler designs as well. This chapter addresses problems with two or more samples—one each from two or more distinct populations.

EXAMPLE 12.1
▷ **Mothers' smoking habits and birth weights (revisited).** Example 2.18 gives birth weights of 72 children categorized by whether their mothers smoked, used to smoke but quit, and never smoked. These data are repeated in Table 12.1. These conditions define three populations. The motivation for the investigators collecting such data is presumably to address whether a mother's smoking affects a child's birth weight. They are interested in comparing the sample distributions of birth weights in the three groups. In this chapter, we will focus on comparing population means using sample means. Moreover, we will consider problems with two samples. This example qualifies if, as in Example 2.18, we combine the samples of mothers who smoke and mothers who quit into one and compare it with that of mothers who never smoked.

Table 12.1
Birth weights of children based on mothers' smoking habits
for Example 12.1

Smokes			Quit	Never smoked			
4.5	6.9	9.9	5.4	3.3	6.6	7.3	7.9	9.2
5.4	6.9		6.6	5.3	6.6	7.4	8.3	9.2
5.6	7.1		6.8	5.6	6.6	7.4	8.3	10.9
5.9	7.1		6.8	5.6	6.7	7.6	8.3	
6.0	7.2		6.9	5.6	6.7	7.8	8.4	
6.1	7.5		7.2	6.1	6.9	7.8	8.5	
6.4	7.6		7.3	6.1	7.1	7.8	8.6	
6.6	7.6		7.4	6.1	7.1	7.8	8.6	
6.6	7.8			6.5	7.1	7.8	8.6	
6.6	8.0			6.6	7.1	7.9	8.8	

◁

The next example concerns two samples with no collapsing.

EXAMPLE 12.2

▷ **Eyewitness testimony.** Exercise 9.6 addresses the accuracy of eyewitnesses and, in particular, the susceptibility of eyewitnesses to false suggestions. In that exercise, the response type is yes/no. Example 3.20 deals with the same question for a different experiment and different response type. Namely, 40 students were shown a 3-minute videotape of a classroom lecture being disrupted by 8 demonstrators and immediately filled out a questionnaire. One of the questions asked of half the students was whether the leader of the "4 demonstrators" was a male, and the other 20 students were asked whether the leader of the "12 demonstrators" was a male. A week later the instructor asked all 40 students how many demonstrators they had seen. Those who had been fed the "4 demonstrators" line reported seeing an average of 6.40 demonstrators, whereas those who were asked about the "12 demonstrators" reported seeing an average of 8.85. (Deleting the two students who said exactly "4" from the first group and the two who said exactly "12" from the second still gave a substantial difference between the two groups.)

You will analyze these data in Exercise 12.1. ◁

Chapters 10 and 11 deal with single samples from general types of populations; they address models for making predictions of future observations and for making inferences about population means. This chapter deals with two (or more) populations with one sample available from each. We will assume the samples are independent (see Chapters 5, 8, and 9). This assumption may not be perfectly apt in every problem we consider, but it holds at least approximately in all of them. Moreover, as I have indicated before, if you do not make assumptions, you will get nowhere. Make assumptions, and worry about whether they are appropriate.

Problems with two samples are common in statistics. They arise in many settings. These include the analysis of studies having parallel designs. Our goal in this chapter is to compare the means of the populations. For convenience, we will call one population "treatment" and the other "control." However, many problems do not involve actual "treatments."

Chapters 8 and 9 convey some of the importance of two-sample problems in scientific research. These earlier chapters dealt with proportions—the populations have only two types of members—but the arguments extend to the general case considered in this chapter. In many ways, two-sample problems are just two one-sample problems. So the ideas developed in Chapters 10 and 11 are important in this chapter.

Normal Densities for Differences

By now you know of many roles played by normal densities. We used them to approximate areas under beta densities (Section 7.5), as models of population distributions, and as prior densities (and also posterior densities) for population means. In this chapter, they will serve as models for the difference between two population means. The next example begins the development.

EXAMPLE 12.3

▷ **Comparing two sets of devices.** Example 11.2 gave the steady-state powers (in watts) of 40 experimental devices, as follows:

18 18 19 17 25   18 18 17 18 18   22 20 20 17 20   18 18 20 15 17
18 17 19 17 17   17 19 18 18 18   16 17 18 18 18   17 17 22 16 20

In that example, we found

$$\bar{x} = 18.25 \quad \text{and} \quad s = 1.799$$

Exercise 11.1 gave the steady-state powers for the next 28 devices manufactured, as follows:

25 20 25 18 24   26 24 24 24 24   18 18 17 17
21 15 13 12 25   19 18 24 25 18   25 24 24 25

As indicated in Exercise 11.1, the first 40 observations do not seem to be exchangeable with the remaining 28. The two samples are different in several ways. One is that the numbers in the second set tend to be bigger and another is that they are more dispersed:

$$\bar{x} = 21.14 \quad \text{and} \quad s = 4.033$$

Are the respective populations really different and, if so, how different are they? The two probability intervals you calculated in Exercise 11.19 do not satisfactorily address this question.

Discussion is difficult when there are two different $\bar{x}$'s and two different s's. If I say $\bar{x}$, you will not know which one I am talking about. To distinguish between these two sets of sample means and standard deviations, I will use the subscript E for the Earlier set and L for the Later set. So

$$\bar{x}_E = 18.25 \quad \text{and} \quad s_E = 1.799$$

$$\bar{x}_L = 21.14 \quad \text{and} \quad s_L = 4.033$$

Similarly, there are two population means, call them m_E and m_L, and two population standard deviations, h_E and h_L. Also, there are two sample sizes: $n_E = 40$ and $n_L = 28$. In other examples and in the exercises, I will use similar labeling schemes. For example, I use the subscripts T and C for treatment and control. But I have run into something of a notational bind. In Chapter 11, I used subscripts 0 and 1 in the prior and posterior densities of the population mean: normal(m_0, h_0) and normal(m_1, h_1). I will continue to use 0 and 1 for this purpose, but now I need two sets of subscripts. So m_T has prior normal(m_{T0}, h_{T0}) and posterior normal(m_{T1}, h_{T1}) densities, and m_C has prior normal(m_{C0}, h_{C0}) and posterior normal(m_{C1}, h_{C1}) densities.

In Example 11.2, we assumed a flat prior density and found the posterior density for m_E to be a normal(18.25, .284). In Exercise 11.1, you assumed a flat prior density and found the posterior density for m_L to be normal(21.14, .762). These two densities are shown in Figure 12.1. (The one for m_E may look different from the density shown in Example 11.2, but it is not. The height of the m_E curve above each value of m is identical with the height in the previous picture. It looks different only because I changed the horizontal scale so that the density for m_L would fit on the same figure. The areas under the two densities are equal.)

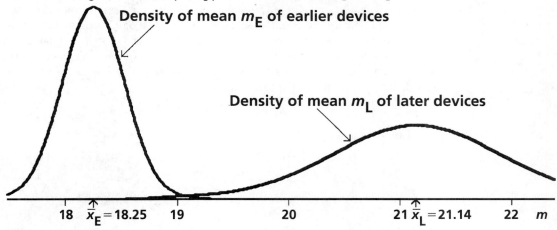

Figure 12.1 Comparing posterior densities of m_E and m_L

Two aspects of this figure are important. First, the peakedness of the density on the left suggests that we have more information about m_E than about m_L. Indeed, the posterior precision of m_E is $1/h_{E1}^2 = 1/.284^2 = 12.4$, which is more than 7 times as big as the posterior precision of m_L: $1/h_{L1}^2 = 1/.762^2 = 1.7$. The additional information about m_E comes from two sources: (1) the greater size of the earlier sample (40 vs 28) and (2) the smaller sample standard deviation (1.799 vs 4.033). Source (1) accounts for about 20% of the seven-fold increase: $40/28 = 1.4$, and source (2) accounts for the rest: $4.033^2/1.799^2 = 5.0$.

The second important aspect in this figure is that the two densities are separated. Clearly, some of the m_E-values are farther from the m_L-values than are others. The distance between the two densities itself has a distribution. Finding this distribution is one way to address the earlier question in this example: Are the respective populations really different and, if so, how different are they? Example 12.4 will address this question, but first we will consider densities of differences. ◁

Densities of Differences

The previous example dealt with one sample from each of the treatment and control populations. Under our assumptions, the posterior densities of the treatment and control population means, m_T and m_C, also have normal densities. Think about the distance between the two densities as follows. A density is a pictorial way of showing a distribution assuming a particular model. Select a number from the first density and another number from the second density. Do this in such a way that the two selections are independent. Subtract the second number from the first and write down the answer.

Repeat this until you have many differences written down. Make a stem-and-leaf diagram of the differences. This stem-and-leaf diagram approximates the density of the distance between the first and second densities.

Call a difference between the two means d, so $d = m_T - m_C$. It seems reasonable that the mean of the density of d is the difference of the posterior means—in this case, $\bar{x}_T - \bar{x}_C$. Since both m_L and m_E have normal densities, it seems reasonable that d also has a normal density. Both statements are correct. The only thing that remains to find is the posterior standard deviation of the normal density of d. Advanced statistics books[1] show that the square of the standard deviation of the difference is the sum of the squares of the individual standard deviations: in general, $h_{T1}^2 + h_{C1}^2$. Because the prior densities were flat in the example, we have

$$h_{T1}^2 + h_{C1}^2 = s_L^2/n_L + s_E^2/n_E$$

where $n_L = 28$ and $n_E = 40$. Summarizing:

Rule of Differences

If the posterior densities of m_T and m_C are normal(m_{T1}, h_{T1}) and normal(m_{C1}, h_{C1}), respectively, then the difference $d = m_T - m_C$ has a normal$(m_{T1} - m_{C1}, \sqrt{h_{T1}^2 + h_{C1}^2})$ density.

The next example continues from the previous example and applies these formulas.

EXAMPLE 12.4

▷ **Comparing two sets of devices (continued).** In Example 12.3, we were interested in finding the density of the difference between the mean steady-state powers of devices manufactured early and those manufactured late. The posterior density of m_E is a normal(18.25, .284) and the posterior density of m_L is a normal(21.14, .762).

I carried out the experiment mentioned in that example based on the chip-in-bowl analogy. Namely, I selected a chip from the model whose density is that of m_L and a chip from the model whose density is that of m_E, and subtracted the second from the first. I repeated this experiment 50 times. This is called **simulation**.[2] The results are shown in Table 12.2 (I rounded off the differences *after* subtraction). Lest there be confusion, I want to stress that these are *not* selections from the populations of steady-state powers of individual Later and Earlier devices, but rather selections from the densities of the corresponding population *means*. Had we sampled from the predictive distribution of devices, the variability in the observations would be much greater in all the columns.

Figure 12.2(a) is a stem-and-leaf diagram of the differences in means—the numbers in the Difference column in Table 12.2. To make this stem-and-leaf diagram, I used three stems for each units digit. The topmost stem within each triple consists of numbers ending in .00 to .33, the middle stem has numbers ending in .34 to .66, and the bottom stem consists of numbers ending in .67 to .99. Had I used one stem per digit, the numbers would have bunched up in the 2- and 3-stems and the picture would

Table 12.2
50 simulations from m_L and m_E models for Example 12.4

m_L	m_E	Difference	m_L	m_E	Difference
21.24	17.61	3.62	20.34	19.56	.78
21.25	18.73	2.52	21.13	17.39	3.74
20.82	17.76	2.52	20.81	18.55	2.26
21.22	19.67	1.56	20.81	17.66	3.15
21.83	19.26	2.58	21.51	18.47	3.04
21.49	19.03	2.45	21.20	19.64	1.56
21.38	19.17	2.21	21.19	18.74	2.44
20.57	17.83	2.75	21.63	18.58	3.05
21.28	17.71	3.56	20.88	18.14	2.74
20.49	19.26	1.22	21.42	17.89	3.53
21.93	18.70	3.23	21.11	18.90	2.21
21.37	19.18	2.19	21.28	17.89	3.39
21.08	17.07	4.01	21.46	18.43	3.03
21.13	17.42	3.71	21.66	19.24	2.42
21.10	17.86	3.24	21.37	18.59	2.78
21.19	18.27	2.92	20.76	17.81	2.95
21.41	17.27	4.14	21.09	17.82	3.27
21.09	18.70	2.39	21.32	17.02	4.30
21.19	18.41	2.77	21.17	16.92	4.25
21.10	18.53	2.57	21.08	19.00	2.07
20.69	16.85	3.84	21.17	17.52	3.66
20.89	18.80	2.10	21.05	16.57	4.48
21.34	17.68	3.66	21.29	18.40	2.90
21.47	18.99	2.47	20.79	17.82	2.98
21.11	18.39	2.72	21.40	18.14	3.26

Figure 12.2 (a) Stem-and-leaf diagram for Example 12.4; (b) Stem-and-leaf diagram rotated and normal curve superimposed

```
0 8
1 2
1 66
1
2 112223
2 444555566
2 778889900
3 00022233
3 456677
3 778
4 0133
4 5
   (a)
```

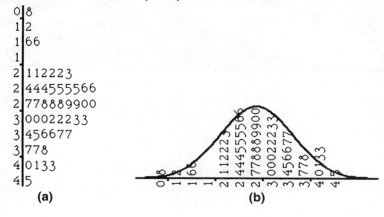

(b)

not have conveyed the shape very well. When listing the leaves, I rounded off to the nearest tenth of a watt. That means that while the number 2.98 gets listed in the bottom-most 2-stem, it rounds off to 3.0 and so its leaf is a 0.

As I have suggested, the stem-and-leaf diagram shows a distribution that has a bell shape. [Also shown in Figure 12.2(b) is the diagram turned on its side with a normal curve superimposed.] Do not read too much into the hole on the left side or the small bump on the right side—these are quite consistent with sampling and simulation variability. The rule of differences indicates that the density of $d = m_L - m_E$ is a normal with mean $21.14 - 18.25 = 2.89$ and standard deviation $\sqrt{.762^2 + .284^2} = .813$. So d has a normal(2.89, .813) density.

Treating the 50 differences as a sample, their sample mean is 2.90 and the sample standard deviation is .781. The law of large numbers of Section 10.2 says that these sample values will tend to their theoretical values as the sample size gets larger. While 50 is only moderately large, both values are close to their respective theoretical values of 2.89 and .813.

The relationship given for the standard deviations of m_L, m_E, and d is the Pythagorean relationship of the sides of right triangles: $.762^2 + .284^2 = .813^2$. Figure 12.3 shows the three densities, all in the same scale and laid out on a right triangle. The early and late densities are replicas from Figure 12.1 in Example 12.3, except that they are shaded. The sides of the triangle are in the same proportion as the standard deviations: $762:284:813$. The center of the m_L density is 21.14, that of m_E is 18.25, and that of d is $2.89 = 21.14 - 18.25$. This figure shows that there is more variability in the difference than in either of the components.

Figure 12.3 Standard deviation of d follows the Pythagorean theorem

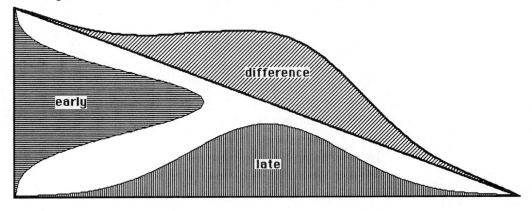

Knowing that d has a normal density, we can calculate the probability of any interval of values of d using z-scores. For example, the z-score of $d = 0$ is

$$z = \frac{0 - 2.89}{.813} = -3.55$$

So, the probability of $d > 0$ is the area under the standard normal curve to the right of -3.55—about $1 - .0002 = .9998$, according to the Standard Normal Table inside the front cover of this book. Such a large probability is consistent with the fact that all 50 differences calculated in Table 12.2 are positive, the smallest being .78. It is *possible* to get a number from the "earlier mean" model that is bigger than the number selected from the "later mean" model, but it should happen roughly once in every $1/.0002 = 5,000$ times in my simulations. (I set my computer the task of selecting from the two models repeatedly until getting a negative difference. It took 5,866 tries. On the 5,866th, the observations were $m_L = 21.01$ and $m_E = 21.37$, for a difference of $d = -.36$.)

Consider another calculation. To find the probability of $d > 5$ watts, we need the z-score corresponding to $d = 5$:

$$z = \frac{5 - 2.89}{.813} = 2.60$$

The Standard Normal Table indicates that the area to the right of $z = 2.60$ is $1 - .9953 = .0047$. This again seems reasonable in light of the 50 differences calculated: None of the 50 was greater than 5, and the biggest difference was 4.48. (It should happen once per $1/.0047 = 210$ times. I programmed my computer to select from the two models repeatedly until getting a difference in means greater than 5. On the 104th try, it got $m_L = 21.19$, $m_E = 16.10$, $d = 5.09$.) ◁

EXERCISES

In these exercises, assume independent population means and independent samples.

12.1 Consider the results of the experiment discussed in Example 12.2: $n_4 = n_{12} = 20$, $\bar{x}_4 = 6.40$, $\bar{x}_{12} = 8.85$, where the subscripts stand for the students who were fed the "4 demonstrators" and "12 demonstrators" lines. In the paper,[3] the author did not give the sample standard deviations separately. Rather, she indicated the standard deviation of the difference (saving you this calculation!): .98.

Assuming flat prior densities for the two means, d ("12" group minus "4" group) has a normal(d_1, .98) posterior density, where $d_1 = \bar{x}_{12} - \bar{x}_4 = 8.85 - 6.40 = 2.45$. What is the posterior probability that $d > 0$?

12.2 Example 11.5 cites a small pilot study[4] and gives changes in blood pressure for 16 patients who were given an experimental drug. The changes, given as after-drug minus before-drug—so that a negative number means a drop in blood pressure—were as follows:

+3.7	−6.7	−10.5	−6.1	−17.6	+2.3	−7.9	−8.9
−4.5	−7.7	−9.4	−10.4	−10.9	−9.3	−16.7	−7.2

A second set of 10 patients in the study served as a control group and were given a placebo instead of the drug. Their changes (after- minus before-placebo) were as follows:

−.7	−2.2	+.2	−7.4	−1.0	−3.1	−2.3	+1.8	−11.7	+2.3

In Example 11.5, I assumed a normal(0, 10) density for the mean drug effect in the population of patients and found that the corresponding posterior density was a normal(−7.83, 1.42). Placebo patients may experience a drop because, for example, the patients may have been more rested for the second measurement. But I think a large mean change is less likely among placebo

patients and so my prior density for mean change among placebo patients is a normal(0, 5). Consider the difference d = mean change on drug − mean change on placebo.
  (a) Find the posterior density of d given the data.
  (b) Calculate the posterior probability of $d > 0$.

PdAL and Probability Intervals for Differences

In Chapters 8 and 9 we calculated the **P**robability that difference **d** is **At L**east a particular value x, calling this PdAL, sometimes adding the value x to the end: PdALx. In the earlier chapters, d was the difference between two population proportions, for example, $d = p_T - p_C$. In this chapter, d stands for the difference between two population means: $d = m_T - m_C$. This coincidence in notation is intentional. In fact, labeling success as 1 and failure as 0, the population mean *is* the proportion of successes. In this section, we again calculate PdAL. We will calculate probability intervals for d-values, a notion developed in Section 11.5 for m-values.

EXAMPLE 12.5
▷ **Comparing two sets of devices (continued).** In Example 12.4, we found that the probability of $d > 0$ is close to 1. In other words, the probability of d at least 0 is PdAL0 ≈ 1. This means that we are nearly convinced that the mean steady-state power of the later devices is greater than that of the earlier devices. The probability of $d > 5$ is close to 0 and it means that PdAL5 ≈ 0. This indicates that we are nearly convinced that the mean steady-state power of the later devices is not more than 5 watts greater than that of the earlier devices.

Proceeding as in Section 11.5, we can calculate a 99% probability interval for the mean difference between these two populations of devices. To find the lower limit, use $z = -2.58$:

$$z = \frac{d_{\text{lower}} - 2.89}{.813} = -2.58$$

which gives

$$d_{\text{lower}} = 2.89 - (2.58)(.813) = 2.89 - 2.10 = .79$$

To find the upper limit, use $z = +2.58$:

$$z = \frac{d_{\text{upper}} - 2.89}{.813} = +2.58$$

which gives

$$d_{\text{upper}} = 2.89 + 2.10 = 4.99$$

So a 99% probability interval for d is 2.89 ± 2.10, or from about .8 to 5.0.

This interval does not contain $d = 0$, the null hypothesis value corresponding to no difference between the two population means. The same is therefore true for a 95%

probability interval. Our convention from previous chapters is to regard this to be lack of support for the null hypothesis of no difference between the two population means. ◁

To find probability intervals, use this table of z-scores repeated from Chapter 11:

68% probability interval: $z_{68} = 1.00$
80% probability interval: $z_{80} = 1.28$
90% probability interval: $z_{90} = 1.65$
95% probability interval: $z_{95} = 1.96$
98% probability interval: $z_{98} = 2.33$
99% probability interval: $z_{99} = 2.58$

Constructing a Probability Interval for $d = m_T - m_C$

Find the posterior means, m_{T1} and m_{C1}, and standard deviations, h_{T1} and h_{C1}.

A perc% probability interval for d is

$$m_{T1} - m_{C1} \pm z_{\text{perc}} \sqrt{h_{T1}^2 + h_{C1}^2}$$

where z_{perc} is given in the preceding table.

As in Chapters 7, 9, and 11, we cannot test null hypotheses by finding probabilities of particular values of d without a computer. As before, use the following alternative for testing $d = 0$:

Alternative Test of Null Hypothesis $d = 0$:

Decide whether the 95% posterior probability interval for d contains 0. If yes, then the null hypothesis is supported; otherwise, it is not.

Equivalently, calculate

$$z = \frac{0 - (m_{T1} - m_{C1})}{\sqrt{h_{T1}^2 + h_{C1}^2}}$$

If $-1.96 < z < 1.96$, then the null hypothesis is supported; otherwise it is not.

As I have pointed out before, a null hypothesis may be "supported" for one person but not for another. An approximate connection with the frequentist approach to statistics is that the null hypothesis that $d = 0$ is "rejected" if it is not supported by someone for whom the prior densities of both m_T and m_C are flat.

Example 12.5 used the results of preceding examples. The next example starts from the fundamentals and employs the most complicated set of conditions we have considered: unknown population standard deviations, small sample sizes, and normal prior densities. Various steps can be eliminated when the corresponding conditions do not apply. For example, if one of the sample sizes were greater than 30, then you could drop the corresponding small-n correction factor. The example also shows that probability intervals for d are closely related to PdAL's.

EXAMPLE 12.6
▷ **Does birth weight increase when a mother quits smoking?** Examples 2.18 and 12.1 give birth weights (in lb) of 72 children according to their mothers' smoking habits. The data for mothers who smoke and for mothers who once smoked but quit are repeated in Table 12.3. I will use Q for those who Quit and, because the letter S means standard deviation, I will use C for those who still smoke Cigarettes. What is the benefit of quitting? Consider $d = m_Q - m_C$. If this is positive, then there is a benefit. The probability that d is positive is PdAL0. The probability that the benefit is at least .5 lb is PdAL.5.

Table 12.3
Birth weights of children based on mothers' smoking habits for Example 12.6

Smokes					Quit	
4.5	6.1	6.9	7.5	9.9	5.4	7.2
5.4	6.4	6.9	7.6		6.6	7.3
5.6	6.6	7.1	7.6		6.8	7.4
5.9	6.6	7.1	7.8		6.8	
6.0	6.6	7.2	8.0		6.9	

I used the data from a study[5] to be considered in Exercise 12.23 to assess my prior densities for both m_Q and m_C. Both are normal densities and have the same mean and standard deviation—I do not think quitting helps for this purpose. I took the mean to be on the low side of the data in the study considered in Exercise 12.23 and I chose a standard deviation that makes it reasonably open-minded: normal(7.7, .7). Since one standard deviation to the left of the mean is $7.7 - .7 = 7.0$, my prior probability of a mean birth weight that is less than 7 lb is the value in the Standard Normal Table for $z = -1$: about 16%. This is also my probability of a birth weight more than one standard deviation to the right of the mean, which is 8.4 lb.

The sample sizes are $n_Q = 8$ and $n_C = 21$. I calculate these sample means and standard deviations:

$$\bar{x}_Q = 6.800 \qquad s_Q = .589$$
$$\bar{x}_C = 6.824 \qquad s_C = 1.093$$

To find the posterior means and standard deviations of m_Q and m_C, we proceed as in Section 11.3.

First, for m_Q:

Population standard deviation of m_Q:

$$h_Q = s_Q\left(1 + \frac{20}{n_Q^2}\right) = .589\left(1 + \frac{20}{64}\right) = .589(1.31) = .774$$

Prior precision: $\quad c_{Q0} = \dfrac{1}{h_{Q0}^2} = \dfrac{1}{.7^2} = 2.04$

Sample precision: $\quad c_Q = \dfrac{n_Q}{h_Q^2} = \dfrac{8}{.599} = 13.35$

Posterior precision: $\quad c_{Q1} = c_{Q0} + c_Q = 15.39$

Posterior mean: $\quad m_{Q1} = \dfrac{c_{Q0}}{c_{Q1}} m_{Q0} + \dfrac{c_Q}{c_{Q1}} \bar{x}_Q$

$$= \frac{2.04}{15.39}(7.7) + \frac{13.35}{15.39}(6.8) = 6.92$$

Posterior standard deviation: $\quad h_{Q1} = 1/\sqrt{c_{Q1}} = 1/\sqrt{15.39} = .255$

So the posterior density of m_Q is normal(6.92, .255).

Now for m_C:

Population standard deviation of m_C:

$$h_C = s_C\left(1 + \frac{20}{n_C^2}\right) = 1.093\left(1 + \frac{20}{441}\right) = 1.093(1.05) = 1.142$$

Prior precision: $\quad c_{C0} = \dfrac{1}{h_{C0}^2} = \dfrac{1}{.7^2} = 2.04$

Sample precision: $\quad c_C = \dfrac{n_C}{h_C^2} = \dfrac{21}{1.304} = 16.10$

Posterior precision: $\quad c_{C1} = c_{C0} + c_C = 18.14$

Posterior mean: $\quad m_{C1} = \dfrac{c_{C0}}{c_{C1}} m_{C0} + \dfrac{c_C}{c_{C1}} \bar{x}_C$

$$= \frac{2.04}{18.14}(7.7) + \frac{16.10}{18.14}(6.824) = 6.92$$

Posterior standard deviation: $\quad h_{C1} = 1/\sqrt{c_{C1}} = 1/\sqrt{18.14} = .235$

So the posterior density of m_C is normal(6.92, .235). The only difference between this density and that of m_Q is the slightly smaller standard deviation.

According to the rule of differences in the previous section, the posterior density of $d = m_Q - m_C$ is a normal(6.92 − 6.92, $\sqrt{.255^2 + .235^2}$) = normal(0, .347). According to the abbreviated table of z-scores (page 379), $z_{95} = 1.96$. Since

$(1.96)(.347) = .680$, the 95% probability interval for d is $0 \pm .68$, or from $-.68$ to $.68$. So, I am pretty sure—95% sure—that the mean difference in children's birth weight between quitting and not quitting is less than .68 lb (about 11 oz).

Since the individual densities have the same mean (6.92), the probability of $d > 0$ is $\frac{1}{2}$. The following table gives various other values of PdAL, including the value of z corresponding to a birth weight difference of x (using the formula $z = (x - 0)/.347$). Figure 12.4 shows the PdAL curve from -1 lb to 1 lb.

PdAL for various differences in birth weights for Example 12.6

x	-1	$-.75$	$-.5$	$-.25$	0	.25	.5	.75	1
z	-2.88	-2.16	-1.44	$-.72$	0	.72	1.44	2.16	2.88
PdALx	.9980	.9846	.9251	.7642	.5000	.2358	.0749	.0154	.0020

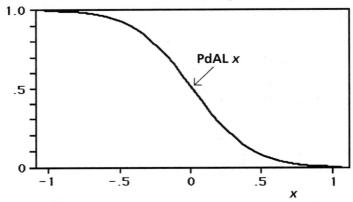

Figure 12.4 PdAL for differences in birth weights in Example 12.6

Consider PdAL.68. The corresponding z-score is $.68/.347 = 1.96$. So PdAL.68 is .025 and PdAL$(-.68)$ is .975. Subtracting the first from the second gives .95; this is the probability between $-.68$ and $.68$. So using PdAL is a way of finding a probability interval—in this case, a 95% probability interval.

This interval contains $d = 0$ and so we take this as support for the null hypothesis of no difference between population means. Equivalently, the observed value $z = 0$ is between -1.96 and 1.96. It comes as no surprise that these data support the null hypothesis $d = 0$ because the observed mean difference is 0. ◁

EXERCISES

In these exercises, assume independent population means and independent samples.

12.3 Example 3.25 and Exercise 8.8 consider a study[6] designed to see if children with cancer who blow on a party blower are distracted from the pain and distress of needle pricks. The exercise asked you to compare the proportions of children in the "intervention" and control groups who required restraining. The investigators also rated the "observed distress" of each of the 13 patients in the **I**ntervention group and each of the 10 patients in the **C**ontrol group. Larger numbers mean greater distress. The results were as follows: $\bar{x}_\text{I} = 6.07$, $s_\text{I} = 4.11$, $\bar{x}_\text{C} = 8.60$, $s_\text{C} = 2.41$. Assume flat prior densities for both population means. Define $d = m_\text{I} - m_\text{C}$.

(a) Find the posterior probability that intervention is effective (this is the same as $1 - \text{PdAL0}$).

(b) Find the 90% posterior probability interval for d.

(c) Test the hypothesis that $d = 0$.

12.4 The children in Exercise 12.3 were asked to assess their own pain after being pricked by the needles. (Larger numbers mean greater pain.) Repeat parts (a), (b), and (c) of Exercise 12.3 for these sample means and standard deviations of "self-reported pain:" $\bar{x}_\text{I} = 2.27$, $s_\text{I} = 1.19$, $\bar{x}_\text{C} = 4.22$, $s_\text{C} = 1.39$. Assume flat prior densities for the two population means.

12.5 In Exercise 11.6, you found the probabilities of differences of at least 0, 2, 4, 6, and 8 ml in forced expiratory volume for formoterol (F) over salbutamol (S) in aerosol solutions using 30 patients suffering from exercise-induced asthma. In this exercise, you will repeat this analysis, but assume that the study had a parallel design—see the accompanying data on F and S, given in the same order as in Exercise 11.6. This assumption of a parallel design is wrong because the patients were paired in a crossover design. The analysis in Exercise 11.6 is correct (although it ignored any effect of sequence of administration) and the analysis in this exercise is wrong. I do not usually ask you to do an analysis I know to be wrong. However, its purpose is to demonstrate that it is not only wrong, but also that it is weaker in the sense suggested in Section 3.5: Not considering interpatient variability decreases the precision of the posterior density. Assume a flat prior for each population mean difference and calculate the posterior PdALx for $x = 0, 2, 4, 6$, and 8. Compare these with your answers for Exercise 11.6.

F: 35 34 23 23 30   31 28 31 23 30   31 32 23 14 32
    31 26 28 34 25   32 14 13 26 25   29 20 27 38 28

S: 32 28 22 13 24   24 22 14 22 26   28 21 16 16 31
    28 22 28 24 24   33 10 15 24 22   19 26 22 26 27

12.6 Exercise 11.3 contained the following data concerning the amounts of chlorine (in parts per million) in five samples taken from a moraine in Bloody Canyon, California called Older Tahoe (population T):

$$73 \quad 75 \quad 49 \quad 76 \quad 115$$

Seven samples taken from a second moraine in Bloody Canyon called Younger Tahoe (population C) contained the following amounts of chlorine:[7]

$$74 \quad 31 \quad 64 \quad 74 \quad 100 \quad 38 \quad 90$$

As in Exercise 11.3, assume that the mean, m_T, of the Older Tahoe population has a flat prior density, and now assume that the mean, m_C, of the Younger Tahoe population also has a flat prior density. Define $d = m_\text{T} - m_\text{C}$.

(a) Find the posterior density of d.

(b) Find the posterior probability of $m_\text{T} > m_\text{C}$.

(c) Find PdAL0, PdAL5, PdAL10.

(d) Find a 68% posterior probability interval for d.

(e) Test the null hypothesis that $d = 0$.

12.7 In Exercise 11.18, you found the posterior distribution for the actual weight of a specimen with the following nine measured weights (in μg):

$$114 \quad 129 \quad 121 \quad 98 \quad 140 \quad 134 \quad 122 \quad 133 \quad 125$$

A second specimen was measured in the same experiment, with the following two measured weights (in μg):

$$109 \quad 85$$

Suppose the first specimen actually weighs m_1 μg and the second weighs m_2 μg. Also assume that the populations are normal(m_1, h) and normal(m_2, h). The prior density for m_1 is normal(100, 20) and that for m_2 is also normal(100, 20). Define $d = m_1 - m_2$.
(a) Find the posterior density of d.
(b) Find PdAL0 and PdAL10.
(c) Find a 95% posterior probability interval for d.
(d) Test the null hypothesis that $d = 0$.

12.8 Exercise 2.11 considered pressures (in mmHg) generated by two heart pumps. The sample sizes are $n_A = 40$ and $n_B = 26$. Consider the diastolic pressures of the two pumps, repeated here:

Pump A	6	7	9	9	10	11	11	12	12	13	15	16	16	16	17
	17	18	19	19	21	23	23	24	30	30	31	31	32	33	39
	44	44	45	53	55	59	62	65	66	69					
Pump B	1	2	3	4	8	11	12	14	15	16	17	18	20	26	29
	30	31	31	32	35	44	45	54	60	63	64				

Assume flat prior densities for population means m_A and m_B. Define $d = m_A - m_B$.
(a) Find PdAL0, PdAL(-5), PdAL5.
(b) Find a 90% posterior probability interval for d.
(c) Test the null hypothesis that $d = 0$.

12.9 Exercise 2.3 gives the following pressures required to remove cannulae from animal hearts for two cannula types:

Cannula A	.30	.30	.32	.43	.44	.47	.52	.59	.70	.77
	.79	.81	.95	1.33	1.43	1.54				
Cannula B	.04	.14	.19	.21	.25	.26	.32	.35	.40	.40
	.42	.53	.53	.56	.57	.68				

Are these two types of cannula equally effective? A way of answering is to see if $d = 0$ is in a posterior probability interval for $d (= m_A - m_B)$. So find the 95% posterior probability interval for d and indicate whether it contains 0. Assume m_A and m_B have flat prior densities.

12.10 Example 3.23 described a study among people with gingivitis concerning the improvement of oral hygiene that resulted from using an oxygen gel (T) as opposed to a placebo gel (C). The sample sizes are $n_T = 30$ and $n_C = 34$. The two sets of improvements in oral hygiene index are repeated here:

Oxygen gel	10	15	6	10	11	3	8	8	3	13	10	9	8	9	8
	4	10	15	11	5	14	7	8	8	2	13	6	2	7	3
Placebo gel	5	6	4	3	3	5	6	4	4	2	0	7	0	3	2
	2	3	6	0	3	-3	1	6	6	8	2	12	24	5	3
	3	3	13	4											

Assume normal(4, 6) prior densities for both m_T and m_C. Let d stand for the mean improvement from oxygen gel as compared with placebo gel: $d = m_T - m_C$.
(a) Find the posterior PdAL0 and PdAL5.
(b) Find the 95% posterior probability interval for d.
(c) Find the 99% posterior probability interval for d.
(d) Test the null hypothesis that $d = 0$.

12.11 To compare two machines that fill 5.65-liter bottles, bottles are sampled from their production—20 from machine A and 25 from machine B.[8] The data are as follows:

Machine A	5.63	5.65	5.62	5.63	5.62	5.68	5.66	5.68	5.62	5.65
	5.61	5.65	5.61	5.68	5.60	5.64	5.61	5.61	5.61	5.63

Machine B	5.68	5.69	5.66	5.62	5.63	5.64	5.60	5.63	5.64	5.60
	5.60	5.60	5.60	5.65	5.64	5.61	5.64	5.61	5.66	5.65
	5.67	5.60	5.65	5.65	5.63					

Assume normal(5.65, .1) prior models for both m_A and m_B. Find the 95% posterior probability interval for $d = m_A - m_B$. [*Option:* Because m_{A0} and m_{B0} are equal (their common value being 5.65), the number 5.65 will disappear from the problem. You can handle the data as given if you like, but to ease your calculational burden, you may want to subtract 5.65 (or any other number) from each observation and proceed. For example, machine A: $-.02, 0, -.03, -.02$, and so on.]

12.12 A study[9] addressed whether a diet supplement of calcium would lower blood pressure. The blood pressures of 75 healthy men were measured once a week for 4 weeks. Then, for weeks 5 to 16 of the study, 37 of the men (chosen randomly) received calcium and the other 38 received placebo; their blood pressures were measured every other week during this period. The table on page 386 shows the diastolic blood pressures (taken while lying down) before treatment and after treatment. The Pre column is the average of the first 4 weeks of the study and the Post column is the average of study weeks 10, 12, 14, and 16. The table also gives the changes from Pre to Post, as well as the means and standard deviations of the changes for the two groups. The changes are also shown in the histograms in the figure below.

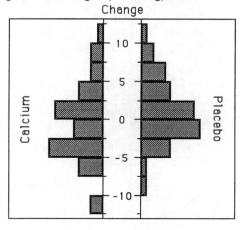

Histograms of changes (in mmHg) for Exercise 12.12

Exercise 12.12: Diastolic blood pressure
(in mmHg) before and after treatment

Calcium			Placebo		
Pre	Post	Change	Pre	Post	Change
71.50	67.75	−3.75	74.75	81.75	7.00
82.50	80.00	−2.50	67.50	65.00	−2.50
80.50	77.25	−3.25	68.75	79.25	10.50
81.25	81.75	.50	63.00	65.50	2.50
67.00	69.25	2.25	67.50	66.25	−1.25
65.50	73.00	7.50	76.25	75.75	−.50
63.25	62.75	−.50	69.75	71.25	1.50
73.75	69.25	−4.50	78.50	81.25	2.75
58.25	69.00	10.75	76.50	67.75	−8.75
59.00	64.75	5.75	75.75	77.75	2.00
68.50	69.75	1.25	62.50	71.75	9.25
71.25	75.00	3.75	70.50	71.25	.75
86.00	81.25	−4.75	70.00	68.00	−2.00
72.50	67.50	−5.00	82.50	83.25	.75
85.50	75.25	−10.25	72.50	74.75	2.25
65.50	67.75	2.25	76.75	75.50	−1.25
79.50	69.00	−10.5	71.75	76.00	4.25
77.50	74.75	−2.75	67.75	73.00	5.25
82.75	81.00	−1.75	72.25	65.50	−6.75
66.00	68.75	2.75	68.50	64.25	−4.25
76.25	71.00	−5.25	68.00	69.50	1.50
72.75	72.25	−.50	75.75	72.25	−3.50
76.00	71.25	−4.75	69.00	69.00	.00
85.75	83.25	−2.50	81.75	77.00	−4.75
71.25	65.25	−6.00	78.75	77.00	−1.75
84.25	77.00	−7.25	67.25	71.25	4.00
72.50	77.75	5.25	85.50	88.25	2.75
76.00	77.50	1.50	92.75	88.00	−4.75
69.50	77.00	7.50	89.00	84.75	−4.25
75.25	71.50	−3.75	73.75	72.00	−1.75
73.50	74.25	.75	73.50	71.00	−2.50
64.25	68.25	4.00	70.75	69.50	−1.25
87.00	82.00	−5.00	84.00	84.75	.75
77.00	71.00	−6.00	75.00	74.50	−.50
68.75	71.25	2.50	62.75	68.50	5.75
59.50	60.50	1.00	70.50	80.00	9.50
60.75	60.75	.00	77.00	78.50	1.50
			70.25	77.50	7.25
Sample mean:		−.845	Sample mean:		.776
Standard deviation:		4.84	Standard deviation:		4.42
Sample size:		37	Sample size:		38

These histograms suggest that the populations are reasonably bell shaped.

There are two controls in this study. One is the subject's own pretreatment blood pressure and the other is the placebo group. Part (a) exploits the first and part (b) considers both.

(a) Consider the calcium group only and use the methods of Chapter 11 to find the posterior probability that the mean change is less than 0 (calcium lowers blood pressure). Assume a flat prior for the population mean m_C.

(b) Consider both samples of changes and find the posterior probability that the mean change on calcium is less than the mean change on placebo (calcium lowers blood pressure). That is, where $d = m_C - m_P$, find the probability that $d < 0$. Assume flat priors for population means m_C and m_P.

[Which analysis is more appropriate, that of part (a) or (b)? The latter accounts for the possibility that blood pressure decreases over the study period even without calcium. But the blood pressure of the placebo recipients actually increased, and subtracting the positive sample mean for the placebo group increases the apparent effectiveness of calcium. I do not know which is the better comparison, but since I do not think that increases over time are normal—even for subjects in a study—I lend more credence to part (a).]

12.13 Exercise 11.22 deals with octopuses trained to select red balls and Exercise 11.23 deals with a similar experiment using untrained octopuses. The following table gives proportions correct for the two groups. Perhaps the discrepancy between the two groups is merely a chance observation. Assume flat priors for both means.

Proportion of reds selected by trained and control octopuses for Exercise 12.13

	Trained	Controls
Mean	.862	.422
Std. dev.	.196	.240
n	16	18

(a) Find the probability that the mean of the trained population is greater than that of the controls.

(b) Find the 68% posterior probability interval for the improvement in proportion effected by the training.

(c) Test the null hypothesis of no difference between trained octopuses and controls.

12.14 An experiment[10] was conducted to learn to what extent batch-to-batch variation in a certain raw material was responsible for variation in the final product yield. Five samples were taken from each of six batches of raw material. (There was no pairing across batches and so these were independent samples.) The product yields of dyestuff in grams of standard color for the last two batches are as follows:

| Batch X | 1,595 | 1,630 | 1,515 | 1,635 | 1,625 |
| Batch Y | 1,520 | 1,455 | 1,450 | 1,480 | 1,445 |

Consider the difference $d = m_X - m_Y$, where m_X and m_Y are the means of the populations of yields. Assume flat prior densities for both means.

(a) Find the posterior probability that $d > 0$, that is, that the mean yield for batch X is bigger than that for batch Y.

(b) Test the null hypothesis $d = 0$.

12.15 How does background music affect worker productivity? The Middlesex (New Jersey) County Board of Social Services wanted to remove a music system from a branch office because they believed that it distracted workers. The union contended that the music had no effect and agreed to a productivity test by an outside firm.[11] Nine workers in each of two units were timed while completing 20 food stamp applications. One unit worked with music and the other without. They did not know they were being observed and timed. The results (in minutes) are as follows:

Music	35.0	36.8	40.2	46.6	50.4	64.2	83.0	87.6	89.2
No music	28.2	28.6	33.0	34.8	45.4	50.8	52.6	66.4	67.8

Using flat priors for both means, test the null hypothesis that the difference in population means is 0.

[I worry that the design of this study does not produce random samples from the appropriate populations and so the calculations you made are therefore themselves inappropriate. The workers in the study were not randomized to the two units. Work units are cultures in and of themselves, with work speed and attitude "rubbing off" from one worker to another. There is an aspect of this type of design for which the experimental unit is the work unit and not the worker. If that were true in the extreme, this study provides a sample of size 1 from each population of work units, where the total time to complete 180 food stamp applications is 533.0 minutes for the music unit and 407.6 minutes for the music-free unit. Conclusions are weak when the sample sizes equal 1, unless the prior information is convincing.]

12.16 An experiment[12] was designed to test pattern recognition abilities of kittens. The experiment involved nine males and six females. The numbers of trials required for the kittens to give at least 27 correct responses out of 30 are as follows:

Males (δ)	40	76	89	106	120	130	150	155	382
Females ($\female$)	66	69	94	103	117	391			

Use flat prior densities for both mean numbers of trials and find the 95% posterior probability interval for the difference in means, $d = m_\delta - m_\female$.

[You may question whether the assumption of a normal model is appropriate. This assumption will be dropped in Exercises 13.8 and 13.20.]

12.17 An important question in the study of human behavior is the relative contribution of genetic and environmental factors in determining sexual orientation. (Exercise 9.38 deals with quite a different study of the same question.) Various physical differences between heterosexuals and homosexuals have been identified. Some are related to reproductive function. One not directly related to reproductive function is the midsagittal area of the anterior commissure in the brain. This is a nerve bundle connecting the left and right sides of the brain and may be related to lateralization, the degree to which the two sides of the brain perform different functions. This nerve bundle tends to be larger in women than in (heterosexual) men, which may explain a tendency for women to have better language skills. Researchers[13] measured the bundle's midsagittal area for 30 gay men and 30 age-matched straight men. These were men who died in Southern California between 1983 and 1991 and had autopsies. (Some died of AIDS, but AIDS has little or no effect on the nerve bundle's area.) Appropriately, measurers were blinded as to the subjects' sexual orientation. The results are shown in the following table. (I approximated the table entries from dot plots given in the original article; as a check, the mean and standard deviation of my estimates agree with the ones the researchers calculated.) Use flat prior densities for the means m_G and m_S of the gay and straight populations and define $d = m_G - m_S$.

(a) Find the 95% posterior probability interval for d.
(b) Test the hypothesis that $d = 0$.

Midsagittal area of the anterior commissure (mm²)
of 30 gay and 30 straight men for Exercise 12.17

Gay			Straight		
10.1	13.1	15.1	5.0	9.9	11.9
10.4	13.2	15.1	6.7	10.2	12.4
10.5	13.4	15.3	6.7	10.2	12.5
10.9	13.6	15.4	6.9	10.2	12.6
11.1	13.6	15.5	7.0	10.4	12.9
11.2	13.6	15.7	8.6	10.5	13.4
11.4	14.2	16.4	9.0	10.6	14.3
11.4	14.3	18.3	9.3	11.0	14.6
12.3	14.4	24.0	9.4	11.0	14.7
12.6	14.4	25.6	9.5	11.0	16.0

12.18 Edward Morley determined the density of oxygen using potassium chlorate ($n=7$) and also electrolysis ($n=10$).[14] If there were no impurities arising in either method, then the two populations being sampled would have the same mean—the true density of oxygen. The values he obtained are shown in the following table and dot diagram. Assume flat priors for both population means and test the hypothesis that the means of the two populations are the same (which would happen if there were no impurities arising in either method).

Oxygen from Potassium Chlorate	Electrolytic Oxygen
1.42920	1.42932
1.42860	1.42908
1.42906	1.42910
1.42957	1.42951
1.42910	1.42933
1.42930	1.42905
1.42945	1.42914
	1.42849
	1.42894
	1.42886

Dot diagram for Exercise 12.18

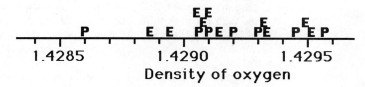

12.19 Lord Rayleigh determined the density of nitrogen by obtaining it from nitric oxide, nitrous oxide, and ammonium nitrite ($n=8$) and also by obtaining it from air whereby the oxygen was withdrawn by either hot copper, hot iron, or ferrous hydrate ($n=8$).[15] If there were no impurities arising in either method, then the two populations being sampled would have the same mean: the true density of nitrogen. The values he got are shown in the following table and dot diagram.

Nitrogen from chemical compounds	Nitrogen from air
2.30143	2.31026
2.29890	2.31017
2.29816	2.30986
2.30182	2.31003
2.29869	2.31007
2.29940	2.31024
2.29849	2.31010
2.29889	2.31028

Dot diagram for Exercise 12.19

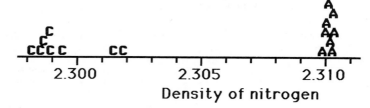

In view of the dot diagram, perhaps no further analysis is necessary. But assume flat priors for both population means and carry out the following analysis.

(a) Find the 95% posterior probability interval for the difference between the population mean of nitrogen from air and that of nitrogen from chemical compounds; $d = m_A - m_C$.

(b) Test the hypothesis that the means of the two populations are the same.

[You should find a difference. Rayleigh did. He may not have made the calculation you made, but may simply have used a dot diagram. His observation led him to discover the element argon, which was present in his extractions from air but not from the chemical compounds. This discovery won him a Nobel prize in 1904.]

12.20 Exercise 12.19 makes it clear that some errors are within experiments and others can apply to the whole experiment. An across-experiment error suggests that the quantities being measured in the two experiments are different. At a conference in July 1993,[16] two groups presented their measurements of the density of Pluto's moon, Charon. A group from NASA's Jet Propulsion Laboratory reported 1.30 g/cm³ with "an uncertainty" of .23 g/cm³. A group from MIT reported 2.35 g/cm³ with an uncertainty of .02. I do not know what "an uncertainty" means, but I believe it is the sample standard deviation divided by the number of measurements. Taking flat priors makes the posterior standard deviation of the difference d equal to $\sqrt{.23^2 + .02^2} = .23$.

Test the hypothesis that the two population means are equal—that is, that the two groups are measuring the same quantity (including the possibility that both are measuring the wrong quantity).

[Updating as usual, assuming the population means are equal, would lead us to conclude that the density of Charon is between 1.30 and 2.35 g/cm³. But I suspect one group got it right and the other has a consistent bias in measurement (although perhaps both groups got it wrong). It is likely that Charon is made of ice, since it is probably a (rather large) chip knocked off Pluto by a comet or some other large space traveler at some time during Pluto's lifetime. Pluto's outer layers are ice and that probably has been the case for much of its life. My posterior density for Charon's density is shown in the following figure. I arrived at this using a more complicated application of Bayes' rule than we have used in this book.]

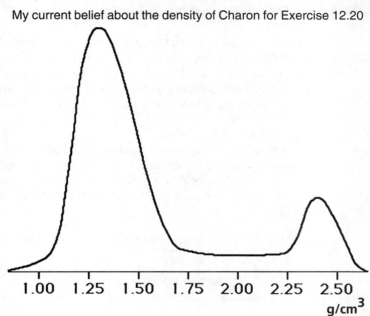

My current belief about the density of Charon for Exercise 12.20

12.21 A women's health trial showed that women could change their dietary habits and lower their fat intake—specifically, from 40% of their total energy intake to 26% of their energy intake, which compares to a control group change from 39% to 38%. Moreover, the 26% level increased by only about 1% one year after the study's completion. Researchers[17] also wanted to know how the women's dietary changes affected their husbands' fat intake. The corresponding percentages for their husbands at 1 year after the study are shown in the following table.

Percentages of fat intake and of polyunsaturates in total fat for husbands of women in health trial for Exercise 12.21

Group	Percentage fat intake		Percentage polyunsaturates	
	Intervention	Control	Intervention	Control
Mean	32.9	36.9	54	58
Std. dev.	8.2	7.9	20	22
n	156	148	156	148

(a) Test the hypothesis that the two population means are the same. Assume independent normal(40, 10) prior densities for both population means.

(b) The table also shows the percentage of polyunsaturated fat in the husband's total fat intake. Test the hypothesis that the two population means are the same. Assume independent normal(60, 10) prior densities for both population means.

12.3 Comparing Several Means

The previous two sections dealt with two populations. This section treats settings in which there are more than two. There are two basic types of settings: those in which the populations are ordered (so it is clear how to arrange them in a list, including which one of any three populations lies between the other two) and those which are not ordered. The next example is one in which the populations are ordered.

EXAMPLE 12.7
▷ **Alcohol vs caffeine consumption (revisited).** Example 2.2 considered a possible relationship between daily caffeine consumption and per sitting alcohol consumption (see Figure 12.5). The four populations are ordered by number of drinks per sitting: 0, 1–2, 3–5, and 6 or more. There are several possible analyses, as will be described in this section.

Figure 12.5 Dot plots from Example 2.2 for Example 12.7

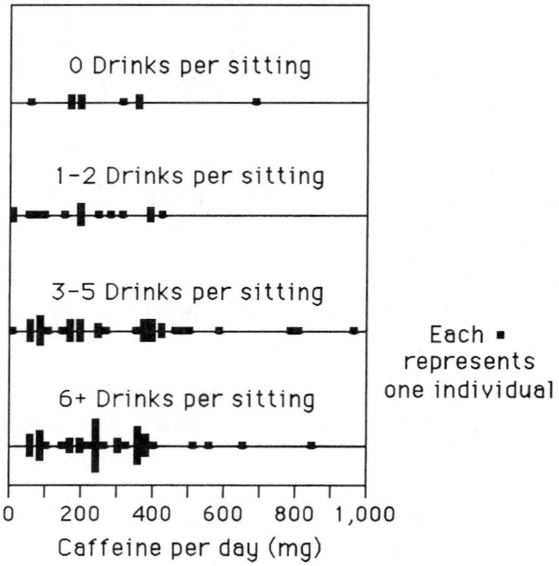

◁

The next example represents one in which the populations are not ordered.

EXAMPLE 12.8
▷ **Can fish oil prolong pregnancy?** Chapter 1 refers to an article[18] in *Science News* with the same heading as this example. In the study, 533 healthy Danish women in

their last trimester of pregnancy were randomly assigned to a treatment or a control group. The treatment group received daily capsules of fish oil. There were two control groups: One received capsules of olive oil and the other took no capsules and were simply observed. The motivation for the study was stated as follows: "The high birth weights and long duration of pregnancy in the Faroe Islands led us to suggest that a high intake of marine-fat-derived n-3 fatty acids might prolong pregnancy by shifting the balance of production of prostaglandins involved in parturition." The original journal article[19] gives this information on the gestation period (in days):

	Fish oil	Olive oil	No oil
Number	266	136	131
Mean	283.3	279.4	281.7
Std. dev.	11.1	13.1	11.6

There is no obvious best order in listing the three groups. Perhaps the fish oil group should appear on the left or right since it is supposed to contrast with the other two, but the order of olive oil and no oil is not clear.

This study will be analyzed in Examples 12.10 and 12.11 and Exercise 12.22. ◁

The next example deals with a different study of pregnant women. Now the populations are partially ordered in that the appropriate arrangement is clear for some populations but not for others.

EXAMPLE 12.9
▷ **Mother's exercise and child's birth weight.** Table 12.4 gives the summary data from Example 2.22 of children's birth weights based on mothers' exercise levels while pregnant:

Table 12.4

Exercise level	n	Birth weight (g) $\bar{x}$	s
None	185	3,389.3	487.9
Changing pattern	213	3,451.4	502.4
Low/Moderate	49	3,554.4	382.1
Heavy	15	3,713.8	298.4

The level "Changing pattern" properly falls between the levels "None" and "Heavy." Also, "Low/Moderate" falls between "None" and "Heavy." But, as discussed in Example 2.22, the order of the two middle categories is not clear: Which should be closer to None?

This study will be analyzed further in Example 12.12. ◁

The following list gives five approaches for analyzing data from more than two populations. Each way can be criticized, but each may be appropriate in certain circumstances.

1. Compare particular pairs of populations as in Sections 12.1 and 12.2. For example, compare caffeine consumption for people having no drinks per sitting with those having six or more. Another is Example 12.6 in which we restricted the analysis to mothers who smoke and mothers who quit; we did not consider mothers who never smoked.

   Advantages: You know how to do it; it may be the appropriate comparison in a particular problem.
   Disadvantages: Does not consider the data in the other samples; multiplicities abound—can choose from among many pairs of populations for comparisons (seemingly harmless options create multiplicity problems as discussed in Section 3.4).

2. Combine samples into just two and then compare the combined populations as in Sections 12.1 and 12.2. For example, compare caffeine consumption for people having 0–2 drinks per sitting with those having 3 or more. (Actually, the populations in this example are already combinations, with, for example, people having 1 and 2 drinks per sitting combined into the 1–2 category.)

   Advantages: You know how to do it; increases sample sizes giving greater precision; uses all the data; lessens the effect of extreme observations (so it is *robust* in the sense of Chapter 13).
   Disadvantages: Combines distinct (although perhaps closely related) populations, which may muddy the comparison; the person choosing the combination can select the one that makes a particular point as strongly as possible ("Smoking's bad and let's make sure that comes across").

3. This is a combination of the first two. Compare pairs of neighboring populations and if they are sufficiently similar, combine them for an eventual comparison with some other combination. For example, compare 0 drinks with 1–2 drinks and, if they are similar, combine them. Do the same for 3–5 drinks and 6+ drinks. Then compare 0–2 drinks with 3+ drinks, or with 6+ drinks, if the second combination did not seem reasonable.

   Advantage: Partially overcomes the disadvantages of approaches 1 and 2.
   Disadvantage: Difficult to decide whether any particular combination is appropriate.

4. Model the relationship between the type of population (alcohol use, say) and the measurement of interest (caffeine consumption). Such a model incorporates the possibility that increasing alcohol use a little increases caffeine consumption a little, and increasing it more increases caffeine consumption still more. Regression analysis, the subject of Chapter 14, is an example of such a model.

   Advantages: Explicitly considers the order and possible relationship among the populations; uses all the data; allows for interpolation (in number of drinks, say) and extrapolation.
   Disadvantages: You do not know how to do it—not yet anyway; conclusions depend on the model assumed.

Sec. 12.3 / Comparing Several Means 395

5. Ask whether the mean of any particular population or subset of populations is different from the mean of the other populations. This leads to a method considered in many texts (but not in this one) and called analysis of variance (ANOVA).

   Advantages: Considers the four samples separately; addresses problems of multiplicities explicitly (see Section 3.4).
   Disadvantages: Does not consider the order of the populations (which is obviously not a disadvantage when there is no order); asking whether any mean is different (without asking which, and by how much) may not be realistic; provides a single way to handle multiplicities, and no one way is appropriate for every setting.

We will use approaches 1 and 2 in this section and approach 4 in Chapter 14. We will not consider the other two approaches further. Approach 3 has promise, and many practitioners use it, as I have myself. But it is unwieldy and not very precise. More importantly, the only good advice I have concerning its use is to be sure the combination you use was not dictated by a desire to get a particular answer. If you are worried that it was, compare your conclusions with those of other combinations. Approach 5 (ANOVA) may be the most commonly used statistical procedure. It is assuredly the most commonly misused statistical procedure! The reasons for its use and for its misuse are related to the disadvantages mentioned above. Many texts[20] address ANOVA.

The next example uses approach 1.

EXAMPLE 12.10
▷ **Can fish oil prolong pregnancy? (revisited).** Example 12.8 gives the following data on gestation periods (in days):

	Fish oil	Olive oil	No oil
Number	266	136	131
Mean	283.3	279.4	281.7
Std. dev.	11.1	13.1	11.6

Before the experiment, I would be a little concerned that any effect of fish oil is shared by olive oil. So no oil seems to be an appropriate control. I know that the mean human gestation period is about 280 days, and I would be surprised if any of these groups has a mean more than 5 days in either direction. I will take normal(280, 5) as the prior density for both the fish- and no-oil population means, m_F and m_N.

So $n_F = 266$ and $n_N = 131$. The sample means and standard deviations are, from the table,

$$\bar{x}_F = 283.3 \qquad s_F = 11.1$$
$$\bar{x}_N = 281.7 \qquad s_N = 11.6$$

To find the posterior means and standard deviations of m_F and m_N, we proceed as in Section 11.3. First, for m_F:

Population standard deviation of m_F (no small-n correction factor): $h_F = s_F = 11.1$

$$\text{Prior precision:} \quad c_{F0} = \frac{1}{h_{F0}^2} = \frac{1}{5^2} = .04$$

Sample precision: $c_F = \dfrac{n_F}{h_F^2} = \dfrac{266}{123.2} = 2.159$

Posterior precision: $c_{F1} = c_{F0} + c_F = 2.199$

Posterior mean: $m_{F1} = \dfrac{c_{F0}}{c_{F1}} m_{F0} + \dfrac{c_F}{c_{F1}} \bar{x}_F$

$= \dfrac{.04}{2.199}(280) + \dfrac{2.159}{2.199}(283.3) = 283.2$

Posterior standard deviation: $h_{F1} = 1/\sqrt{c_{F1}} = 1/\sqrt{2.199} = .67$

So, the posterior density of m_F is normal(283.2, .67).

Now for m_N:

Population standard deviation of m_N (no small-n correction factor): $h_N = s_N = 11.6$

Prior precision: $c_{N0} = \dfrac{1}{h_{N0}^2} = \dfrac{1}{5^2} = .04$

Sample precision: $c_N = \dfrac{n_N}{h_N^2} = \dfrac{131}{134.6} = .973$

Posterior precision: $c_{N1} = c_{N0} + c_N = 1.013$

Posterior mean: $m_{N1} = \dfrac{c_{N0}}{c_{N1}} m_{N0} + \dfrac{c_N}{c_{N1}} \bar{x}_N$

$= \dfrac{.04}{1.013}(280) + \dfrac{.973}{1.013}(281.7) = 281.6$

Posterior standard deviation: $h_{N1} = 1/\sqrt{c_{N1}} = 1/\sqrt{1.013} = .99$

So, the posterior density of m_N is normal(281.6, .99).

According to the rule of differences, the posterior density of $d = m_F - m_N$ is normal(283.2 − 281.6, $\sqrt{.67^2 + .99^2}$) = normal(1.61, 1.2). The z-score for finding PdAL0 is $z = (0 - 1.61)/1.20 = -1.34$. The PdAL0 is the probability to the right of this, which according to the Standard Normal Table is about 91%.

A 95% probability interval for d captures most of the uncertainty in d. According to the table in Section 12.2, $z_{95} = 1.96$. Since $(1.96)(1.20) = 2.35$, the 95% probability interval for d is 1.61 ± 2.35, or from about −.7 to 4.0. While there is some evidence that fish oil extends gestation (over no oil), $d = 0$ is in this interval and so the null hypothesis is supported. I am not persuaded that there is much difference between the means in these two populations. ◁

The next example—or, more precisely, its comparison with the previous example—shows how difficult inferential issues can be when there are several populations. It employs approach 2, the device of combining similar populations.

EXAMPLE 12.11
▷ **Can fish oil prolong pregnancy? (revisited).** The purpose of the study described in Examples 12.8 and 12.10 was to address the effect of fish oil. The other two groups were controls. Those assigned to olive oil were similar to the fish-oil group in that both took capsules, while the no-oil group did not. So in this sense, of the two controls, the olive-oil group provides the more appropriate comparison. You will compare fish oil and olive oil in Exercise 12.22. (Choosing a subscript for olive oil in that exercise and in this example is a bit of a problem. The letter O is too similiar to the number 0, and so I will use V, for ol**v**e oil.) Another possibility is to combine the two groups. The investigators may have had this in mind because the total number of subjects in the two control groups is about the same as the number in the fish-oil group. I will take this tack in this example.

There is a bit of a problem in combining two control groups because we do not have the full data. We have to combine means and standard deviations. The mean is easy; it is the following weighted average, where the weights are $n_V/(n_V + n_N) = 136/(136 + 131) = .51$ and $131/(136 + 131) = .49$:

$$\bar{x}_C = .51\bar{x}_V + .49\bar{x}_N = .51(279.4) + .49(281.7) = 280.5$$

Because the sample sizes are nearly equal, the weights are close to $\frac{1}{2}$; so, the combined mean is approximately the simple average of the two sample means. The standard deviation of the combined group is more complicated and I will simply tell you that the answer is 12.6. Summarizing:

	Fish oil	Controls
Number	266	267
Mean	283.3	280.52
Std. dev.	11.1	12.6

The calculations for the posterior density of m_F are the same as in the previous example: It is normal(283.2, .67).

These are the calculations for m_C:

Population standard deviation of m_C (no small-n correction factor): $h_C = s_C = 12.6$

$$\text{Prior precision:} \quad c_{C0} = \frac{1}{h_{C0}^2} = \frac{1}{5^2} = .04$$

$$\text{Sample precision:} \quad c_C = \frac{n_C}{h_C^2} = \frac{267}{158.8} = 1.682$$

$$\text{Posterior precision:} \quad c_{C1} = c_{C0} + c_C = 1.722$$

$$\text{Posterior mean:} \quad m_{C1} = \frac{c_{C0}}{c_{C1}} m_{C0} + \frac{c_C}{c_{C1}} \bar{x}_C$$

$$= \frac{.04}{1.722}(280) + \frac{1.682}{1.722}(280.52) = 280.5$$

$$\text{Posterior standard deviation:} \quad h_{C1} = 1/\sqrt{c_{C1}} = 1/\sqrt{1.722} = .76$$

So the posterior density of m_C is normal(280.5, .76).

According to the rule of differences, the posterior density of $d = m_F - m_C$ is normal$(283.2 - 280.5, \sqrt{.67^2 + .76^2}) = $ normal$(2.73, 1.02)$. The z-score for finding PdAL0 is $z = (0 - 2.73)/1.02 = -2.68$. The PdAL0 is the probability to the right of this. According to the Standard Normal Table, PdAL0 = 99.6%, which is much closer to 100% than the 91% of the previous example. A 95% posterior probability interval for d is 2.73 ± 2.00, or from about .7 to 4.7. Since this interval does not contain 0, the null hypothesis of no difference between fish oil and controls is not supported.

In contrast with the previous example, the evidence now seems reasonably compelling that fish oil extends gestation. ◁

Why do these two examples come to different conclusions? First, combining the olive-oil group decreased the average gestation time in the controls from 281.7 in Example 12.10 to 280.5 in Example 12.11. Second, the sample size of the control group doubled, thereby decreasing the posterior standard deviation of m_C. The first factor had more of an effect on the PdAL0, but both served to increase it.

Does fish oil prolong gestation or not? The answer depends on how credible you consider the various comparisons. My probability that it prolongs gestation as compared with no oil is not quite as high as the 99.6% of the second example, but it is closer to 99.6% than it is to the 91% of the first example. (You will get still another answer in Exercise 12.22.)

The next example considers the data from Example 12.9 in which the populations are only partially ordered.

EXAMPLE 12.12

▷ **Mother's exercise and child's birth weight.** Examples 2.22 and 12.9 describe a study relating children's birth weights to the mothers' exercise level while pregnant, with the results given in Table 12.4 (page 393). Does exercise matter? The issue is complicated by not knowing how to relate the levels "Changing pattern" and "Low/Moderate." Three possibilities seem reasonable: (1) Compare "None" with "Heavy" and ignore the other two; (2) compare "Heavy" with the combination of the other three; and (3) compare "None" with the combination of the other three. The first two possibilities have the important drawback that the "Heavy" group stands alone in both and its sample size is quite small. So I will adopt option (3), which I believe to be reasonable and to have trustworthy conclusions.

These are the combined data:

Exercise?	n	Birth weight (g)	
		$\bar{x}$	s
No	185	3,389.3	487.9
Yes	277	3,483.8	478.8

I will assume flat priors for both m_Y and m_N. The calculations follow the usual pattern. This time, I will combine the two sets of calculations—a flat prior makes them easier:

Population standard deviations of m_Y and m_N (no small-n correction factors):
$$h_Y = s_Y = 478.8 \quad \text{and} \quad h_N = s_N = 487.9$$

Prior precisions: $c_{Y0} = c_{N0} = 0$ (due to flat priors)

Sample precisions: $c_Y = \dfrac{n_Y}{h_Y^2} = \dfrac{277}{229{,}300} = 1.208\text{E}-3$

$$c_N = \dfrac{n_N}{h_N^2} = \dfrac{185}{238{,}000} = 7.772\text{E}-4$$

Posterior precisions: $c_{Y1} = c_{Y0} + c_Y = 1.208\text{E}-3$
$c_{N1} = c_{N0} + c_N = 7.772\text{E}-4$

Posterior means: $m_{Y1} = \bar{x}_Y = 3{,}484$
$m_{N1} = \bar{x}_N = 3{,}389$

Posterior standard deviations: $h_{Y1} = 1/\sqrt{c_{Y1}} = 1/\sqrt{1.208\text{E}-3} = 28.8$
$h_{N1} = 1/\sqrt{c_{N1}} = 1/\sqrt{7.772\text{E}-4} = 35.9$

So, the posterior densities of m_Y and m_N are normal(3,484, 28.8) and normal(3,389, 35.9).

According to the rule of differences, the posterior density of $d = m_Y - m_N$ is normal(3,484 − 3,389, $\sqrt{28.8^2 + 35.9^2}$) = normal(94.5, 46.0). The z-score for finding PdAL0 is $z = (0 - 94.5)/46.0 = -2.05$. The PdAL0 is the probability to the right of this; according to the Standard Normal Table, PdAL0 ≈ 98%. A 95% probability interval for d is 94.5 ± (1.96)(46.0) = 94.5 ± 90.2, or from 4.3 g to 184.7 g, which is about 0 oz to 7 oz. The null hypothesis is on the boundary of being supported and not supported.

So, I am pretty convinced that women who exercise have bigger babies, but I do not think that they are a lot bigger. Also, there is at least a suggestion that more intensive exercise makes for even bigger babies. (However, I have the usual reservations that this is not a randomized study. Even though I am willing to believe that women who *do* exercise have larger babies, it takes a leap of faith to also conclude that a sedentary woman can increase the weight of her baby if she starts to exercise. Someone willing to leap should know that they did.) ◁

> **When there are more than two samples, either compare them in pairs or combine and then compare in pairs.**

Comparing Means May Not Be Appropriate

You know how to compare population means, but just because you know how does not mean you should! Comparing population means is not always appropriate. The next example is a case in point.

EXAMPLE 12.13

▷ **Forecasting tornadoes (revisited).** Example 2.9 addressed the role of two variables, umax and mda, for predicting which thunderstorms will produce tornadoes—see the dot plots from that example on page 22. The following table gives sample means and standard deviations of the three populations: those storms producing no, weak, and strong tornadoes.

Tornado?	n	umax + mda	
		$\bar{x}$	s
No	123	17.1	11.3
Weak	28	35.1	7.6
Strong	24	49.9	15.7

These three populations are ordered, with "Weak" falling between "No" and "Strong." It might seem reasonable to combine the "Weak" and "Strong" populations and compare the mean of the combination with that of "No." But it is not. The scientific issue is not one of distinguishing between the means of the groups. Just as in Exercise 3.14 (which dealt with screening for prostate cancer), means are nearly irrelevant. Just as in that exercise, the question is one of prediction: Given a value of umax + mda, will the storm spawn a tornado? So, it is not appropriate to compare means as is typical of other examples in this chapter. ◁

EXERCISES

In these exercises, assume independent population means and independent samples.

12.22 Reconsider the study of Examples 12.8, 12.10, and 12.11. Now compare fish oil with olive oil—use V for olive oil. These are the data:

	Fish oil	Olive oil
Number	266	136
Mean	283.3	279.4
Std. dev.	11.1	13.1

Let $d = m_F - m_V$ and use a normal(280, 5) prior density for both population means.
(a) Calculate PdAL0.
(b) Find a 95% probability interval for d.
(c) Test the null hypothesis $d = 0$.

12.23 Example 12.1 gives the birth weights of 72 children by whether their mothers smoked. These data are repeated here, but with the mothers who smoke combined with those who quit, as in the histograms of Example 2.18. Call the "smokes or quit" group treatment T and the "never smoked" group control C. The sample sizes are $n_T = 29$ and $n_C = 43$. (Compare to Exercise 11.25.) Another study[21] reports that babies of smokers weigh about .4 lb less. So the prior density of m_T is normal(7.1, 2.0) and that of m_C is normal(7.5, 2.0). Let $d = m_T - m_C$.

Birth weights of children based on mothers' smoking habits for Exercise 12.23

Smokes or quit				Never smoked			
4.5	6.9	9.9	5.4	3.3	6.6	7.3	7.9	9.2
5.4	6.9		6.6	5.3	6.6	7.4	8.3	9.2
5.6	7.1		6.8	5.6	6.6	7.4	8.3	10.9
5.9	7.1		6.8	5.6	6.7	7.6	8.3	
6.0	7.2		6.9	5.6	6.7	7.8	8.4	
6.1	7.5		7.2	6.1	6.9	7.8	8.5	
6.4	7.6		7.3	6.1	7.1	7.8	8.6	
6.6	7.6		7.4	6.1	7.1	7.8	8.6	
6.6	7.8			6.5	7.1	7.8	8.6	
6.6	8.0			6.6	7.1	7.9	8.8	

(a) Find PdAL0.
(b) Find a 90% probability interval for d.
(c) Test the null hypothesis $d = 0$.

12.24 Researchers[22] addressed the impact of the acorn crop in the Southern Appalachians on the local deer population. They measured various characteristics of harvested deer and related them to the acorn crop of the current and previous years. The categorization of acorn crop into poor, fair, and good was somewhat arbitrary and was based on an oak mast index. One of the measures was the length of antler beam (which is the antler's main stem) of bucks. Means, standard deviations, and sample sizes are given in the table at the top of page 402 for three categories of the previous year's acorn crop. To show the consistent effect of both age of deer and size of crop, I have plotted the mean lengths—using symbols having areas proportional to sample sizes.

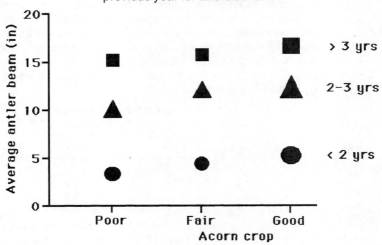

Mean antler beams for three ages of male deer depending on size of acorn crop in previous year for Exercise 12.24

Antler beam length (in.) for Exercise 12.24

Age/Acorn crop	Poor			Fair			Good		
	$\bar{x}$	s	n	$\bar{x}$	s	n	$\bar{x}$	s	n
< 2 yr	3.45	.95	18	4.55	.93	16	5.38	.79	29
2–3 yr	10.14	1.36	17	11.35	1.61	15	11.66	1.15	26
> 3 yr	15.13	1.40	17	15.78	1.82	16	16.65	1.48	27

The appropriate analysis is not clear. The original paper analyzes data within each age group and compares pairs of populations defined by acorn crop. More appropriate would be an overall conclusion in which the effects of acorn crop and deer age are modeled. Such an approach is beyond the scope of this book. Instead, we will analyze within age category and combine the fair and good acorn crops into a "fair-to-good" category for comparison with the poor crop.

The next table shows the fair and good crops combined (again, I have not told you the algebra necessary for calculating the standard deviation of the combined data). Use a flat prior for each population mean antler beam length. Test the null hypothesis that there is no difference in the population means after fair-to-good than after poor crops for the following age groups.

(a) Male deer less than 2 years old.
(b) Male deer between 2 and 3 years old.
(c) Male deer more than 3 years old.

Antler beam length (in.) for Exercise 12.24

Age/Acorn crop	Poor			Fair-to-Good		
	$\bar{x}$	s	n	$\bar{x}$	s	n
< 2 yr	3.45	.95	18	5.08	.93	45
2–3 yr	10.14	1.36	17	11.55	1.34	41
> 3 yr	15.13	1.40	17	16.33	1.67	43

12.25 Exercise 11.26 describes a factorial study evaluating the effects of exercise and a nutrition supplement in very elderly people. The study considered four populations. Subjects taking the supplement gained weight (averaging 1 kg in the 10-week study period), but it had little, if any, effect on muscle. So it seems reasonable to combine the supplement and nonsupplement groups. The following table ignores the supplement and combines the two exercise groups (Exercise plus Both from Exercise 11.26) into one and the two no-exercise groups (Supplement plus Control from Exercise 11.26) into one. Assume flat priors for the population means, m_Y and m_N, and let $d = m_Y - m_N$.

Percentage in stair-climbing power for Exercise 12.25

Exercise	Yes	No
Mean	28.4	3.6
Std. dev.	44	50
n	42	41

(a) Find the posterior probability that the mean percentage increase for exercise is greater than that for no exercise. That is, find the probability that $d > 0$.
(b) Find the 90% posterior probability interval for d.
(c) Test the null hypothesis that $d = 0$.

[Just as in Exercise 12.12, it is not clear how best to use the randomized control group in the presence of the prestudy comparison. There is no question of its value as a check that nothing unusual is happening in the population or in the experiment. In this exercise, the average 3.6% increase is not very different from 0%, which serves to validate the prestudy comparison as a legitimate control. Whether it should be subtracted from the treatment average increase—as I asked you to do—is open to question.]

12.26 Aging is as mysterious as life itself. One theory of aging is that it is due to oxidative-stress: Reactive oxygen causes molecular damage. This and the following exercise deal with separate experiments designed to test this theory. Researchers[23] assigned 200 houseflies to a large cage where they would live out their lives. These "high-activity" flies were able to fly and did so. The researchers also assigned 47 flies to live in small individual jars. These "low-activity" flies could walk but did not have enough room to fly. All flies were fed sucrose and water. The low-activity flies lived longer on average—see the table. Assume normal(30, 20) densities for both population means and find a 90% posterior probability interval for the (mean) advantage of low exercise.

Lifetimes (in days) of houseflies
for Exercise 12.26

Activity	High	Low
Mean	20.6	48.1
Std. dev.	6*	14.2
n	200	47

*The article gives only one-digit accuracy.

(You assumed random samples from the two populations in doing this problem. But the lifetimes of 200 flies in the cage may not be independent of each other. For example, harmful bacteria may have been present in the cage, the flies may have passed around a communicable disease, or a low level of toxin may have leaked into the cage. Perhaps continuously exposing flies to other flies causes stress and shortens life. The flies are experimental units, but so is the cage—and there was only one cage. My reservations would be eased by a more extensive experiment with more large cages and with some of them housing single flies.)

12.27 The researchers in Exercise 12.26 also evaluated the effect on housefly lifetime of exposure to 100% oxygen. The table shows the results. The "Perm. O_2" group were continually exposed to oxygen. They started dying on day 4 and were all dead by day 11. The "Temp. O_2" group received 100% oxygen for 3.5 days—"just before heavy mortality would set in"—and then were placed in normal air. The control group was in normal air throughout. Let m_T be the population mean lifetime of the Temp. O_2 group and m_C that of the control group. Define $d = m_T - m_C$. If d is less than 0, then the oxygen has caused damage and this would support the oxidative-stress theory of aging. Assume independent normal(30, 20) prior densities for m_T and m_C.

Lifetimes (in days) of houseflies
for Exercise 12.27

Group	Perm. O_2	Temp. O_2	Control
Mean	5.1	25.9	17.6
Std. dev.*	2	8	6
n	200	200	200

*The article gives only one-digit accuracy.

(a) Find PdAL0.
(b) Test the hypothesis that $d = 0$.

[The researchers report that the metabolic rate of the Temp. O_2 group was less than that of the controls. Perhaps being nearly dead before the oxygen was removed lowered their ability to exercise and their lives were extended by the resulting low activity just as for the flies in Exercise 12.26.

My reservations about the design of this experiment are similar to those in the previous exercise. I do not know how many cages were used for each group, but I think there was only one. Cage variability should be assessed. Doing so requires multiple cages. Also, the effect of oxygen may depend on the number of flies in the cage, which therefore should itself be varied. Moreover, the results are sufficiently nonintuitive that there may have been a handling difference of the Temp. O_2 and control cages that would cause a reversal of the expected results. That the researchers can explain their paradoxical results may be reassuring, but it is far from convincing; the human mind can explain any observation—even those that are wrong!]

12.28 The definitions of the **C**ontrol group from Exercise 12.27 and the **H**igh-activity group from Exercise 12.26 seem identical. (The data are repeated here.) If they were treated differently, the article does not say so. But they were different flies. The differences may be due to cage variability, or variability in sampling from the population of flies. To see whether the differences are consistent with sampling variability, test the hypothesis that $m_H = m_C$. Assume independent normal(30, 20) prior densities for m_H and m_C.

Lifetimes (in days) of houseflies for Exercise 12.28

Group	High Activity (Ex. 12.26)	Control (Ex. 12.27)
Mean	20.6	17.6
Std. dev.	6	6
n	200	200

12.29 Exercise 11.24 compared the cortex weights of rats that lived in large, communal cages and had daily toy replacements with littermates that were isolated and had no toys. I indicated that the cortex weights differed in the different experiments. Consider the controls of experiments 1, 3, and 5 repeated in the table here. Assuming flat priors for the corresponding population means, find the posterior probability that the control mean of experiment 3 is greater than that of:
(a) Experiment 1
(b) Experiment 5

Experiment	Control cortex weights									
1	657	623	652	654	658	646	600	640	605	635	642
3	668	667	647	693	635	644	665	689	642	673	675	641
5	641	589	603	642	612	603	593	672	612	678	593	602

12.30 (This exercise continues from the previous exercise.) In Exercise 11.24, I indicated that the percentage increase in cortex weight for treatment rat over control rat did not seem to differ by experiment. Consider the percentage increases for experiments 1, 3, and 5 repeated in the table here. Assuming flat priors for the corresponding population means, find the posterior probability that the percentage increase mean of experiment 3 is greater than that of:
(a) Experiment 1
(b) Experiment 5

Experiment	Percentage increases from control cortex weights
1	4.9 5.3 2.5 .9 3.2 2.6 10.7 1.1 14.7 −.3 1.7
3	3.3 5.1 5.9 8.4 1.9 .5 8.3 4.2 11.8 3.4 −2.5 6.1
5	−.2 11.2 3.5 6.2 12.3 8.3 10.1 −1.8 9.2 .1 7.6 7.8

12.31 Is psychotherapy for depression beneficial? Is computer-assisted therapy as good as therapist-assisted? A study[24] compared three treatment groups with the results indicated in the following table. Participants were recruited by a newspaper ad and had to meet criteria for major, minor, or intermittent depressive disorder. The first two groups were given identical cognitive-behavioral treatment in six weekly sessions, one administered by a clinical psychologist and the other by a computer. Therapy for the control group was delayed for the 2 months required for completing the study. The three groups had comparable Beck Depression Inventory scores before treatment. (The control group had a pretreatment mean of 22.92; the drop to 20.67 is an example of the regression effect described in Section 14.3.) Assume flat priors for the means of each of the three groups.

2-month follow-up Beck Depression Inventory for Exercise 12.31

Therapy	Therapist	Computer	Control
Mean	8.27	6.17	20.67
Std. dev.	8.84	5.37	9.89
n	11	12	12

(a) Test the hypothesis that the population means of the therapist- and computer-assisted groups are equal.
(b) Test the hypothesis that the population means of the therapist-assisted and control groups are equal.
(c) Test the hypothesis that the population means of the computer-assisted and control groups are equal.

Appendix: Using Minitab for Comparing Two or More Means

The program '**mm_cont**' will compute the mean and standard deviation for the difference in two means $m_1 - m_2$ when m_1 and m_2 have independent normal posterior distributions. In Example 12.6, one is interested in comparing the average birth weight, m_1, of mothers who smoke with the average birth weight, m_2, of mothers who have quit smoking. Suppose that before observing data, you believe that there is no effect of smoking on birth weights and you assign to each of m_1 and m_2 a normal prior distribution with mean 7.7 and standard deviation .7. The birth weights of the smoking mothers are placed in the column 'smoke' and the birth weights for the mothers who have quit smoking are placed in the column 'quit.'

First, we summarize the posterior density for the mean m_1 by executing the program '**mm_cont**'. We input the prior mean and standard deviation and the number of the column corresponding to 'smoke.' The program gives the mean and standard deviation of the normal posterior density for m_1. The program 'm_cont' is run again for the 'quit' data. The output is the mean and standard deviation for the mean m_2. To summarize the difference in means, $m_1 - m_2$, we run the program 'mm_cont.' We input the mean and standard deviation for m_1 and m_2 and the program gives the mean and standard deviation for the difference in means. In this example, the increase in birth weight due to quitting is approximately $-.007$, plus or minus a standard deviation of .361. Since this posterior distribution has a large probability on each side of 0, there does not appear to be any evidence in these data that smoking has an effect on birth weights.

```
MTB > name c1 'smoke' c2 'quit'
MTB > set 'smoke'
DATA> 4.5 5.4 5.6 5.9 6 6.1 6.4 6.6 6.6 6.6 6.9 6.9
DATA> 7.1 7.1 7.2 7.5 7.6 7.6 7.8 8 9.9
DATA> end
MTB > set 'quit'
DATA> 5.4 6.6 6.8 6.8 6.9 7.2 7.3 7.4
DATA> end
MTB > exec 'm_cont'

DO YOU WISH TO USE AN INFORMATIVE NORMAL PRIOR FOR M? (TYPE 'y' OR
'n'.)
   IF YES, INPUT PRIOR MEAN AND PRIOR STANDARD DEVIATION.
y

DATA> 7.7 .7

OBSERVED DATA IN WORKSHEET? (TYPE 'y' OR 'n'.)
   IF YES, INPUT NUMBER OF COLUMN.
   IF NO, INPUT OBSERVED SAMPLE MEAN, STANDARD DEVIATION, AND
SAMPLE SIZE.
y
DATA> 1

THE POSTERIOR DENSITY FOR M IS NORMAL
WITH MEAN AND STANDARD DEVIATION:

MEAN STD
   6.92677    0.23996

MTB > exec 'm_cont'

DO YOU WISH TO USE AN INFORMATIVE NORMAL PRIOR FOR M? (TYPE 'y' OR
'n'.)
   IF YES, INPUT PRIOR MEAN AND PRIOR STANDARD DEVIATION.
y

DATA> 7.7 .7

OBSERVED DATA IN WORKSHEET? (TYPE 'y' OR 'n'.)
```

```
IF YES, INPUT NUMBER OF COLUMN.
IF NO, INPUT OBSERVED SAMPLE MEAN, STANDARD DEVIATION, AND
SAMPLE SIZE.
Y
DATA> 2

THE POSTERIOR DENSITY FOR M IS NORMAL
WITH MEAN AND STANDARD DEVIATION:

MEAN       STD
 6.93373    0.26983


MTB > exec 'mm_cont'

INPUT MEAN AND STANDARD DEVIATION FOR NORMAL DISTRIBUTION FOR MEAN
M1:
DATA> 6.92677    0.23996

INPUT MEAN AND STANDARD DEVIATION FOR NORMAL DISTRIBUTION FOR MEAN
M2:
DATA> 6.93373    0.26983

THE POSTERIOR DENSITY FOR M1-M2 IS NORMAL
WITH MEAN AND STANDARD DEVIATION:

Row           mn              st
 1        -0.0069599      0.361094
```

The book *Bayesian Computation Using Minitab* by James Albert (Duxbury Press, 1996) illustrates a different method of summarizing the difference between means of normal distributions. The program **'mm_tt'** uses simulation to approximate the difference between two independent *t* distributions. This method may be more accurate than the approach taken in this text, since it is based on the exact posterior densities of m_1 and m_2.

Chapter Notes

1. For example: D. A. Berry and B. W. Lindgren, *Statistics: Theory and Methods* (Pacific Grove, Calif.: Brooks/Cole, 1990), 527.

2. D. A. Berry and B. W. Lindgren, *Statistics: Theory and Methods* (Pacific Grove, Calif.: Brooks/Cole, 1990), 378–381.

3. E. F. Loftus, *Cognitive Psychology* 7 (1975): 560–572.

4. J. D. Goldberg and K. J. Koury, in *Statistical Methodology in the Pharmaceutical Sciences,* ed. D. A. Berry (New York: Marcel Dekker, 1989), 220.

5. M. C. Hatch, X-O. Shu, D. E. McLean, et al., *American Journal of Epidemiology* 137 (1993): 1105–1114.

6. S. L. Mann, W. H. Redd, P. B. Jacobsen, K. Gorfinkle, O. Schorr, and B. Rapkin, *Journal of Consulting and Clinical Psychology* 58 (1990): 565–572.

7. F. M. Phillips et al., *Science* 248 (1990): 1529–1532.

8. A. H. Bowker and G. J. Lieberman, *Engineering Statistics*, 2nd ed. (Englewood Cliffs, N.J.: Prentice Hall, 1972).

9. R. M. Lyle, C. L. Melby, G. C. Hyner, J. W. Edmonson, J. Z. Miller and M. H. Weinberger, "Blood pressure and metabolic effects of calcium supplementation in normotensive white and black men," *Journal of the American Medical Association* 257 (1987): 1772–1776.

10. O. L. Davies, *Statistical Methods in Research and Production,* 3rd ed., (London: Oliver and Boyd, 1967), 105. *See also* G. E. P. Box and G. C. Tiao. *Bayesian Inference in Statistical Analysis* (New York: Wiley, 1973), 246.

11. I am indebted to William Stellmach for this application and for providing me with these data.

12. P. C. Dodwell, F. E. Wilkinson, and M. W. von Gruman, "Pattern recognition in kittens," *Perception* 12 (1983): 393–410.

13. L. S. Allen and R. A. Gorski, "Sexual orientation and the size of the anterior commissure in the human brain," *Proceedings of the National Academy of Sciences* 89 (1992): 7199–7202.

14. I. Freund, *The Study of Chemical Composition: An Account of Its Method and Historical Development* (Cambridge: Cambridge University Press, 1904; reprinted, New York: Dover Publications, 1968), 88.

15. I. Freund, *The Study of Chemical Composition: An Account of Its Method and Historical Development* (Cambridge: Cambridge University Press, 1904; reprinted, New York: Dover Publications, 1968), 89–91.

16. *Science News* 144 (July 10, 1993): 22.

17. A. L. Shattuck, E. White, and A. R. Kristal, "How women's adopted low-fat diets affect their husbands," *American Journal of Public Health* 82 (1992): 1244–1250.

18. *Science News* 141 (May 16, 1992): 334.

19. S. F. Olsen, et al., *Lancet* 339 (1992): 1003–1007.

20. For example: D. A. Berry and B. W. Lindgren, *Statistics: Theory and Methods* (Pacific Grove, Calif.: Brooks/Cole, 1990), Chapter 14.

21. M. C. Hatch, X-O. Shu, D. E. McLean, et al., *American Journal of Epidemiology* 137 (1993): 1105–1114.

22. J. M. Wentworth, S. Johnson, P. E. Hale and K. E. Kammermeyer, "Relationship of acorn abundance and deer herd characteristics in the Southern Appalachians," *Journal of American Forestry* (1992): 5–8.

23. R. S. Sohal, S. Agarwal, A. Dubey, and W. C. Orr, "Protein oxidative damage is associated with life expectancy of houseflies," *Proceedings of the National Academy of Sciences, U.S.A.* 90 (1993): 7255–7259.

24. P. M. Selmi, M. H. Klein, J. H. Greist, et al., "Computer-administered cognitive-behavioral therapy for depression," *American Journal of Psychiatry* 147 (1990): 51–56.

13 Data Transformations and Nonparametric Methods

CHAPTERS 6–9 dealt with proportions. The populations had two types of members and we focused on the proportion of one of them. Chapters 10–12 dealt with general types of populations and we focused on population means. In these latter chapters, we allowed for any type of population when the sample size was large. When the samples were small, we assumed that the population had a bell shape. However, assuming something does not make it so—and you will never really know a population's true shape.

All conclusions for small samples in Chapters 10–12 are tied to the assumption of normal populations. When the population density is other than normal, a conclusion may still be valid, at least approximately. Problem populations are those with **outliers**—values that are small in probability and far from the bulk of the population. Section 13.1 gives some examples. Samples may also contain outliers, as in Example 10.5. Sample outliers come from the population, of course, but they may or may not be outliers in the population.

> **Outliers are values in the population that have a small probability and are far from regions with the greatest concentration of density.**

A purpose of this chapter is to extend the previous three chapters to allow for small samples and for populations that may not be normal. The resulting procedures are called *robust*. The basic approach of the first four sections of this chapter is to manipulate or **transform** the data to pull in outliers (and other values as well). After transforming, proceed as usual. Transforming can make the population more consistent with the normal assumption. How small the sample should be before you use the methods of this chapter depends on the proportion and type of outliers in the population. It also depends on the accuracy required in making inferences. Usually, $n = 100$ is large enough to assume a normal. But in rare circumstances, $n = 1,000$ is still not big enough. Fortunately, the methods in this chapter apply for any sample size, small or large. Use them when you are concerned about the possibility of outliers in the population, as well as when you are not. Robust procedures are forgiving: If the population density happens to be a normal and you use a robust procedure, guarding against outliers that do not exist, your conclusions will likely be nearly as good as if you had assumed a normal model.

A mathematical operation—such as taking logarithms, square roots, or exponentials—is a transformation. Simple multiplication or addition does not change the shape of the population. While multiplying by a small number shrinks outliers, it shrinks every other observation proportionally. We need transforms that shrink the population *differentially*—larger observations more than smaller ones. An example is the square root, which will not work if any observations are negative. The most commonly used transformation is the **logarithm** (see Section 13.2). It is routinely applied to counts and to other positive numbers. Another common and useful way to transform data is to assign each observation its **rank** (see Section 13.4). Still another is to **categorize,** perhaps into big, small, and middle-sized (see Section 13.3). This chapter demonstrates the use of each and addresses some of the advantages and disadvantages of each.

Using a robust procedure is straightforward. First, decide which analysis method you plan to use (Chapter 11 for paired differences, say, or Chapter 12 for parallel designs), then transform the data and carry out the analysis as planned. Such a procedure requires the original data—it is not enough to know summary values such as n, $\bar{x}$, and s.

> **Robust procedure:** Transform the data to lessen the impact of outliers; then proceed with the analysis method of choice (as in previous chapters).

13.1 Outliers

Defining an outlier as a value in a population or sample that is far from the bulk of the other values is not rigorous because what is "far" is not specific. But this is consistent with the standard interpretation of the term: You might call something an outlier that I would not. I have two purposes in trying to define it at all. First, I want to make clear that populations can have outliers even though the sample does not. Second, I want to *exclude* something from the definition: An outlier is neither bad nor good—it just is. I stress this because of the widespread confusion concerning this issue. For example, some statistics computer packages automatically delete outliers from the sample. An outlier may be the most important observation in the sample: Would it be reasonable to delete Babe Ruth when evaluating the 1927 New York Yankees? If you ever find yourself using such a package, throw it away!

There are two main types of outliers: (1) those that reflect an aspect of the population that is important for addressing the scientific question of interest and (2) those that do not. An example of type (1) is a patient who dies after taking a headache remedy. Examples of type (2) are values that arise from the measurement or data collection process—say, a data recording mistake such as a misplaced decimal point. We should seek to understand as much as possible about the type (1) outliers, but the type (2) are a nuisance that we would like to avoid.

The way to handle an outlier depends on its type. If we knew that an extreme sample observation is a data recording mistake, we could simply ignore it. If we knew that a particular extreme observation is indicative of an important population characteristic, we would prize the information it contains. Consider Darwin's cross- vs self-pollination experiment described in Exercise 2.1 and Example 3.18. We should not ignore the cross-pollinated plant in pair 2 because it may have been stunted "by some larva gnawing at [its] roots." This plant might be suffering from a defect that results *because* it was cross-pollinated. Even if the larva theory is correct, perhaps larvae are more attracted to plants that are cross-pollinated—possibly because they are heartier! Does this manifest weakness or heartiness? The answer depends on the scientific question and, in this case, on what one means by "superior." Unfortunately, we may never know whether an outlier is type (1) or type (2)—as we will see in Example 13.1.

Outliers in the population have an impact on the population mean (and also on the standard deviation) that is disproportionate to their frequencies. If no outliers show up in the sample, then we would have a false idea of the population mean. If some are in our sample, they may lead us to think the population mean is more extreme than it really is. Outliers do not exist in bell-shaped populations—or to be more accurate, they are so improbable that their existence can be ignored. Some observations from a normal density are relatively big and some are relatively small, but they are not *unusually* big or *unusually* small. The fact that there are no surprises in a sample from a normal population means that learning takes place in a rather smooth way. But the learning process is sensitive to the presence of outliers—even when they are not observed.

Population densities with **heavy tails** have outliers that may or may not end up in the sample—Figure 13.1 has heavy tails on both sides. Heavy tails are problematic when the sample is small. In the figure, $n = 6$, and there is an outlier in the right-hand tail, but not in the left. The sample mean, $\bar{x}$, overestimates the population mean, m, because the outlier makes $\bar{x}$ large. (In an application you would not know m and you would not know the population density that is sketched in this figure.) The presence of the large observation makes s large and gives a wide 95% probability interval for m (assuming a normal population and a flat prior for m); this interval is indicated by asterisks on the figure.

Figure 13.1 Heavy-tailed population densities can give extreme observations with undue influence on analysis. Asterisks indicate the endpoints of a 95% probability interval.

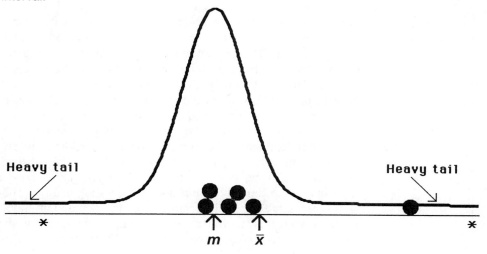

Figure 13.2, in contrast, shows a typical sample when the population has light tails. The asterisks delineating the 95% probability interval are now close together because there are no outliers to make s large.

Heavy tails house outliers.

Figure 13.2 Light-tailed population densities result in samples that do not cause problems in analysis, even for small samples. Asterisks indicate endpoints of a 95% probability interval.

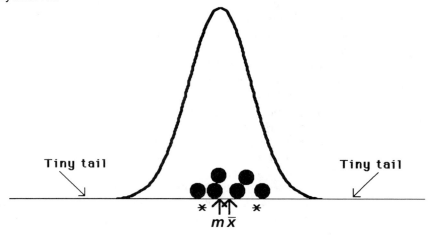

The population density shown in Figure 13.3 has a single, heavy tail and it too is problematic. The indicated sample shows another problem resulting from heavy tails. The tail makes the population mean large—I labeled the mean as m, but its exact value depends on what the density looks like to the right of where it has been cut in the figure. There are outliers in the heavy tail that did not show up in the sample. So the sample mean is too small. Look at the sample values only, which is all you could see in an application. They look quite consistent with a normal population. But m is outside the 95% posterior probability interval calculated assuming a normal population density and a flat prior for m. This interval is inappropriately narrow because the population is not normal. So you need to be concerned about outliers even if you do not see them! Like a species of mosquito in which the males buzz but do not bite and the females do not buzz but bite—you have to worry as much when you hear nothing!

The problem density in Figure 13.4 is a composite or **mixture,** in the sense that the population is made up of two subpopulations. In the figure, the smaller bump has only about 5% of the total density (i.e., area). The figure shows four observations as solid dots. The fifth observation (dot with question mark) is either an outlier (i.e., in the smaller bump) or in the main part of the population along with the other four observations. If it is an outlier, then the sample mean shifts to the left and the sample standard deviation makes a substantial jump. If it is among the other four observations, then both $\bar{x}$ and s stay about the same as they were. In the first case, $\bar{x}$ is too small and s is too large; in the second case $\bar{x}$ is too large and s is too small.

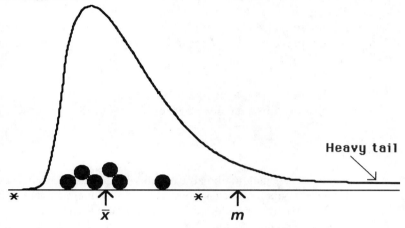

Figure 13.3 Population mean, m, is much larger than sample mean because of the heavy tail in the population with no sample values in the tail. Asterisks indicate a 95% probability interval.

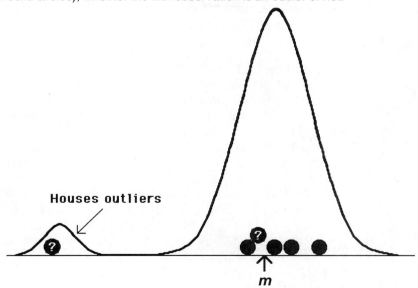

Figure 13.4 Populations that are mixtures are problematic for small n (here, $n = 4$, the four solid circles), whether the fifth observation is an outlier or not.

All such problems disappear for very large sample sizes. As is true generally, the law of large numbers (Section 6.2) means that, for very large n, the sample histogram will roughly approximate the density. In particular, about 5% of the observations will be in the leftmost subpopulation in the figure. So for large n, we will have a good idea of what the population looks like.

In addition, mixtures are not problematic if both subpopulations are at least moderately common, even for small n. Think of labeling observations in one subpopulation as a "failure" and in the other as a "success" and ignore the differences within subpopulations. (This is not unreasonable because the between-subpopulation variability dominates the within-subpopulation variability.) Then the inferential problem is comparable to that for proportions: Observe values in the subpopulations just as you observed successes and failures in Chapters 6–9, and draw conclusions about the relative sizes of the subpopulations.

> **Populations that are mixtures of very different subpopulations can give rise to outliers.**

Since we may not be able to tell which outliers are nuisances and which indicate important population characteristics, we need methods that minimize the damage from the first and yet provide some information in the second. Such methods are the focus of this chapter. As I have already indicated, they are especially important when the sample size is small.

Some natural approaches are robust. The next example shows that the impact of outliers is decreased by simply combining similar observations. Of course, the problem is to decide which observations are "similar."

EXAMPLE 13.1

▷ **Digoxin levels over time (revisited).** Example 2.12 describes a study in which 15 patients took digoxin (digitalis) throughout the study and an experimental drug for part of the time—the data are repeated in Table 13.1. The patients had been taking only digoxin prior to and including days A, B, E, and F. On days C and D (shown in boldface type), they had also been taking the experimental drug. The motivation for the study was to decide whether the digoxin levels were changed on days C and D, thus implicating the experimental drug. To see whether digoxin levels are increased while on the experimental drug, we might compare the mean of days C and D with the mean of days A, B, E, and F; these means are shown in Table 13.2.

Example 2.12 addressed two outlying digoxin levels: day A for patient 5 (143) and day E for patient 1 (121). When I first saw these data, I questioned the pharmacologists who made the measurements concerning these two observations, which they agreed were unusual. In particular, could the 1 at the hundreds digit have crept onto the sheet by accident? After checking and rechecking, they claimed that these were not recording errors, and they could not explain why they were so large.

These outliers make the means of days A, B, E, and F for patients 1 and 5 larger than those for the other patients, but averaging them with three other levels in each case gives results that are at least close to those of the other patients. This is an example

Sec. 13.1 / Outliers

of the advantage of approach 2 mentioned in Section 12.3, but in a paired setting: Averaging observations collected under similar conditions lessens the impact of outliers.

Table 13.1
Digoxin levels over time for Example 13.1

Patient Number	Plasma digoxin (ng/ml/100)					
	Day A	Day B	**Day C**	**Day D**	Day E	Day F
1	34	40	56	45	121	34
2	74	15	44	49	38	37
3	31	49	52	58	41	23
4	40	22	38	25	29	32
5	143	33	57	51	49	46
6	54	58	59	43	41	35
7	31	48	65	53	44	30
8	52	61	70	62	66	38
9	51	63	76	50	32	16
10	21	36	56	40	39	38
11	42	38	68	52	55	51
12	20	30	36	40	31	40
13	51	47	47	38	46	31
14	27	28	61	67	92	75
15	52	43	73	56	56	44

Table 13.2
Mean digoxin levels and differences for Example 13.1

Patient Number	Plasma digoxin (ng/ml/100)		
	Average **C, D**	Average A, B, E, F	Difference
1	50.5	57.25	−6.75
2	46.5	41	5.5
3	55	36	19
4	31.5	30.75	.75
5	54	67.75	−13.75
6	51	47	4
7	59	38.25	20.75
8	66	54.25	11.75
9	63	40.5	22.5
10	48	33.5	14.5
11	60	46.5	13.5
12	38	30.25	7.75
13	42.5	43.75	−1.25
14	64	55.5	8.5
15	64.5	48.75	15.75

Combining measurements uses all the data, and it decreases the impact of single measurements. But it may hide important aspects of an experiment. For example, understanding the variability of the measurement process may be important. Many reports and publications are guilty of combining data without informing readers that there was a combination. It is always safe to give the full data set, as collected in the study or experiment—as I have done in this example. An investigator who does not give the full data set must say so and must provide an assessment of the variability between any measurements that were combined.

> **Averaging replicated measurements can lessen the effect of outliers.**

We would like to find the posterior probability of $m > 0$, where m is the mean of the population of differences. We would also like to find a 95% probability interval for m. Assuming a flat prior density for m, and following Section 11.1, the posterior density is normal($\bar{x}$, $h/\sqrt{n}$), where $h = sk$, and the small-n correction factor, $k = (1 + 20/15^2) = 1.089$. The following table gives the required summary statistics:

Increases in digoxin level: On vs off experimental drug				
n	$\bar{x}$	s	sk	$sk/\sqrt{n}$
15	8.17	9.97	10.9	2.80

The z-score for $m = 0$ is

$$z = \frac{0 - \bar{x}}{sk/\sqrt{n}} = -2.91$$

and so the posterior probability of $m > 0$ is about .998. The 95% probability interval for m is $8.17 \pm (1.96)(2.80) = 8.17 \pm 5.49$, or from about 3 to 14.

This interval does not contain 0 and so the null hypothesis $m = 0$ is not supported. ◁

Transforming Data: Taking Logarithms

In this section, we will consider transformations and focus on logarithms. But first I want to address a disadvantage of transforming data: It is harder to think about the population and therefore harder to formulate prior opinions. For example, I may have a reasonable understanding of the distribution of heart rates (beats per minute) in a certain population. Suppose I think its mean is about 64. Considering the square roots of heart rates makes my understanding more remote. Someone with a heart rate of 64 becomes an 8 in the new scale. But 8 is no longer the mean in this new scale. For example, suppose a sample includes 45, 45, 45, 64, 121. The sample mean is 64. The square roots are 6.708, 6.708, 6.708, 8, 11; the mean of these five square roots is 7.825. Reverting to the original scale by squaring gives $7.825^2 = 61.23$, as opposed to 64.

To formulate prior opinions of transformed values, first assess them in the original scale. The changeover is easy if there are a finite number of population models. Figures 13.5 and 13.6 give an example. Each bar in the m-scale is moved to $\sqrt{m}$ when taking the square root in Figure 13.5. The same thing happens for densities, shown in Figure 13.6 (page 418). The square root squeezes the density (for m-values larger than 1), with the squeezing greater for larger m's. The effect when taking logarithms is similar.

Figure 13.5 Probabilities slide over when scale is transformed by square root.

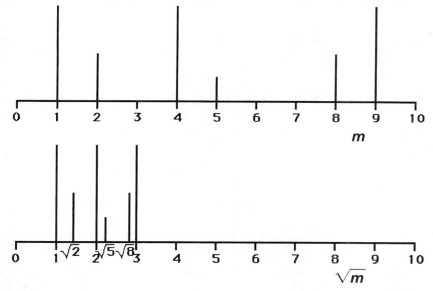

We will consider logarithms (logs, for short) in this section. Just as with square roots, logs can be applied only for positive numbers. They are frequently used for count data, such as number of heart beats, number of electrons, and so on. You will need a scientific calculator or computer to take logs. Logarithms change a multiplicative scale into an additive scale. For example, with a base of 10, the log of 10^x is just x itself, the number of 0's in the decimal form. So, for example, the log of 10,000 is 4 and the log of 1,000,000 is 6.

Since every logarithm is a constant multiple of every other, its base is irrelevant for our purposes. (However, do not use more than one base in the same problem.) The so-called natural logarithm has $e = 2.71828\ldots$ as its base. It is written "ln" on many calculators. To return to the original scale, use the antilog or inverse log or exponential (e^x) key. The logarithm with base 10 is written "log" on many calculators. To return to the original scale, use the antilog or 10^x key.

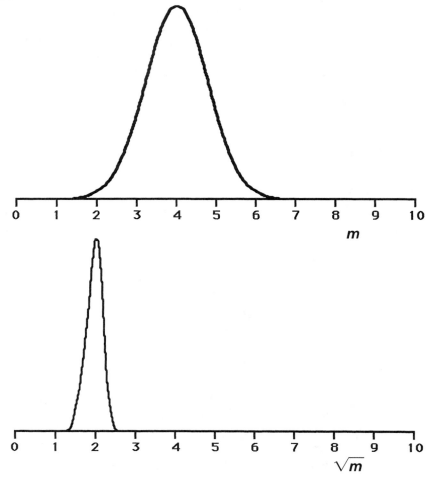

Figure 13.6 Density moves and becomes squeezed as scale is transformed by square root.

Examples of Log Scales

The Richter scale of earthquake intensity is a log scale having base 10. So an earthquake of magnitude 7 is 10 times as intense as one of magnitude 6, which is 10 times as intense as one of magnitude 5. An earthquake that is 6.5 on the Richter scale is 3.16 times as powerful as one that is a 6, since $10^{6.5-6} = 10^{.5} = \sqrt{10} = 3.16$.

The brightness of stars is also measured on a log scale—the Pogson scale. The base was chosen to fit roughly with the ancient classification of magnitude: First magnitude stars are about 100 times as bright as sixth magnitude stars and so the base is the fifth root of 100: $\sqrt[5]{100} = 2.51$. This means that a first magnitude star is 2.51 times as bright as a second magnitude star, $2.51^2 = 6.31$ times as bright as a third magnitude star, and $2.51^5 = 100$ times as bright as a sixth magnitude star. (The smaller the magnitude, the brighter the star.) The zero was also chosen to fit the ancient scheme; it is approximately that of Alpha Centauri, which has a magnitude of .06, and of Vega (Alpha Lyra) at a magnitude of .14. Our brightest star is Sirius (Alpha Canis Major) with a magnitude of 1.58, which means it is $2.51^{1.58+.14} = 2.51^{1.72} = 4.88$ times as bright as Vega. Planets have magnitudes calculated on the same scale: Venus's greatest magnitude is about -4.4, which means it is $2.51^{4.4-1.58} = 2.51^{2.82} \approx 13$ times as bright as Sirius.

Still another example for the use of the log scale is pH (*potential* of *H*ydrogen), which measures acidity or alkalinity. It is the logarithm of the reciprocal of the hydrogen ion concentration, expressed in gram atoms per liter of a solution. The base of the logarithm is 10.

I have to choose a base in order to make calculations. For no particular reason, I will use natural logarithms. Suppose the original number is 25. The natural log of 25 is 3.22. Taking the antilog of 3.22 gives $e^{3.22} = 25.0$. The natural log of .2 is -1.61; the antilog of -1.61 is $e^{-1.61} = .2$.

Interpreting the Richter and Pogson scales shows one advantage of using logs. Suppose a treatment increases the (natural) log of response from 2 to 2.3. The antilog of the difference is $e^{2.3-2} = e^{.3} = 1.35$. This indicates that the treatment increases the response by 35%. It is sometimes convenient and it may be natural to talk in terms of percentage change rather than difference in raw numbers.

Shrinking Outliers Using Logs

We seek to exploit the shrinkage resulting from taking logarithms. Logs shrink large values much more than small ones. So if a population has a heavy right-hand tail as shown in Figure 13.7, taking logs shrinks values in the tail and, we hope, converts the shape of the population distribution to something resembling a normal.

> **Taking logarithms can make populations more nearly normal.**

EXAMPLE 13.2
▷ **Cycles to failure of worsted yarn (revisited).** Example 3.27 described a factorial experiment relating the number of cycles of loading to failure of worsted yarn to the length of the test specimen, amplitude of the loading cycle, and load. All three of these factors seem to be important. Let's consider length, 250 mm vs 300 mm, and use the other two factors to pair the observations. So focus on the rows of the table in Example 3.27. For example, we find that for an amplitude of loading cycle equal to 8 mm and a load equal to 40 g, the numbers of cycles to failure are 674 and 1,414 for 250 mm vs 300 mm lengths, respectively. This pair is shown in the table on page 420, along with the other eight pairs and corresponding differences (1,414 − 674 = 740 cycles advantage when using 300 mm, in the first case).

Figure 13.7 Taking logarithms can convert a curve with a heavy tail (a) into a near-normal curve (b)

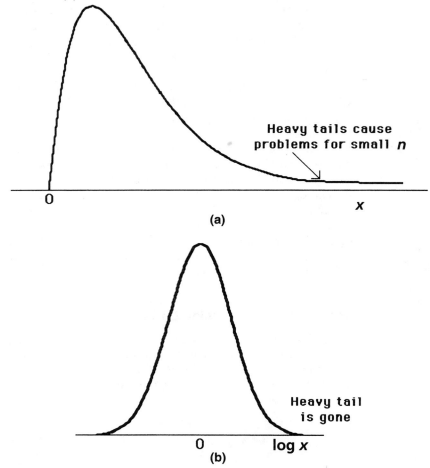

Length (mm)	Cycles								
300	1,414	1,198	634	1,022	620	438	442	332	220
250	674	370	292	338	266	210	170	118	90
Difference	740	828	342	684	354	228	272	214	130

The second table uses the same scheme as the first, except that it considers logs of number of cycles. In this instance, first I took logs and then I subtracted. This is not the same as taking the differences first and then the log; although that would also be better than analyzing the differences themselves, it does not make sense in general because the differences could be negative.

Length (mm)				Log cycles					
300	7.25	7.09	6.45	6.93	6.43	6.08	6.09	5.81	5.39
250	6.51	5.91	5.68	5.82	5.58	5.35	5.14	4.77	4.50
Difference	.74	1.18	.77	1.11	.85	.73	.95	1.04	.89

The dot plots in Figures 13.8 and 13.9 show the differences, in cycles and in log cycles. (The asterisks show the 95% posterior probability limits for the population means assuming flat priors.) There are several things to notice in comparing the two dot plots. First, the differences in log cycles are more tightly packed (in comparison with 0) and the sample seems somewhat more consistent with a normal population. More importantly, the second set of dots is much more closely packed relative to 0 (no difference between 250 mm and 300 mm). It seems much less likely that the second sample could have come from a population with 0 or negative mean than the first sample—see z-scores that follow.

Figure 13.8 Dot plot for Example 13.2: Cycles for lengths of 300 mm vs 250 mm

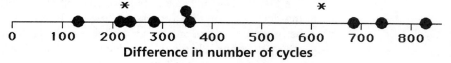

Difference in number of cycles

Figure 13.9 Dot plot for Example 13.2: Log cycles for lengths of 300 mm vs 250 mm

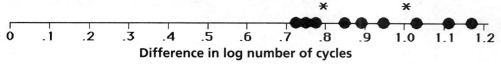

Difference in log number of cycles

The following table gives the summary statistics required for an analysis similar to those in Chapter 11 [where the small-n correction factor $k = (1+20/9^2) = 1.247$]:

Differences Between Lengths of 300- and 250-mm yarn

	n	$\bar{x}$	s	sk	$sk/\sqrt{n}$
Cycles	9	421.3	243.7	303.9	101.3
Log cycles	9	.9185	.1530	.1908	.0636

Assume a flat prior (as in Section 11.1) for the mean, m, of the corresponding population. To find the posterior probability that $m > 0$, we find the corresponding z-scores, first for cycles and then for log cycles:

$$z = \frac{0 - \bar{x}}{sk/\sqrt{n}} = -4.16$$

and

$$z = \frac{0 - \bar{x}}{sk/\sqrt{n}} = -14.44$$

So both probabilities that the population mean is positive are essentially 1. But the second z-score is much smaller than the first, indicating the power of taking logs.

The 95% probability intervals are as follows:

Cycles: $421.3 \pm (1.96)(101.3) = 421.3 \pm 198.5$, or from about 220 to 620

Log cycles: $.9185 \pm (1.96)(.0636) = .9185 \pm .1246$, or from about .79 to 1.04

Neither interval contains 0 and so the null hypothesis is not supported, whether one takes logs or not.

The lower and upper limits of these intervals are indicated in the corresponding dot plots using asterisks. (Remember that these intervals refer to population means and not to future observations.) Using antilogs, the second interval converts to a range of percentage increase for 300 over 250: namely, $e^{.79} = 2.2$, or a 120% increase, and $e^{1.04} = 2.8$, or a 180% increase.

(Fixing the levels of all factors but one to evaluate the effect of the one—in this case, fixing amplitude and load to assess length—is simple and reasonable. But it is neither the best nor the generally accepted form of analysis. As I indicated in Section 3.7, for generally acceptable methods and also for methods of assessing interactions of factors in a factorial design, see Chapter 12 of the text by Snedecor and Cochran.[1]) ◁

> **Taking logarithms gives a natural way to assess percentage change.**

EXERCISES

13.1 In Exercise 11.13, we considered the following sample of ages at which 21 children spoke their first word:

15 26 10 9 15   20 18 11 8 20   7 9 10 11 11   10 12 42 17 11   10

The histogram for these data shows at least one outlier. Now take logs and assume that the population (of logs) is normal with mean m. Assume that $m_0 = 2.5$ and $h_0 = 1.8$.

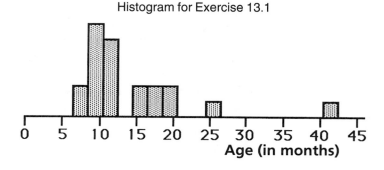

Histogram for Exercise 13.1

(a) Find the posterior density of m for the given data.

(b) In Exercise 11.13, you found the predictive probability that a child from this population (but not one of the 21 in the study) has still not uttered a word at age 24 months. On the log scale, 24 months is 3.18. Find the predictive probability that a selection from the log population is greater than 3.18. [This is analogous to your answer in Exercise 11.13(b).]

13.2 Using logarithms in Example 13.2, I calculated 95% probability intervals for the percentage increase in the number of cycles to failure in comparing lengths of 250 and 300 mm. Again take logs and find differences between logs. But now consider amplitudes of 8 and 9 and find the 95% probability interval for the percentage increase in the number of cycles to failure for Amp = 8 over Amp = 9. (To assist you, I have rearranged the table of data from Example 3.27 into pairs—see below.) Use a flat prior for the mean of the population of differences in logs. (It is okay to use a base of the logarithms different from the one I used.)

Length	Amp	Lead	Cycles	Length	Amp	Lead	Cycles	Length	Amp	Lead	Cycles
250	8	40	674	300	8	40	1,414	350	8	40	3,636
250	9	40	338	300	9	40	1,022	350	9	40	1,568
250	8	45	370	300	8	45	1,198	350	8	45	3,184
250	9	45	266	300	9	45	620	350	9	45	1,070
250	8	50	292	300	8	50	634	350	8	50	2,000
250	9	50	210	300	9	50	438	350	9	50	566

13.3 Example 13.1 dealt with digoxin levels over time and demonstrated that the effects of outliers could be reduced by averaging measurements taken under similar circumstances. Repeat the analysis of that example, but now take logs. The table of logs is given here. Proceed as in Example 13.1: For each patient, average days C and D and subtract the average of the other 4 days from it. Assume a flat prior density for the population mean, m, of the differences (of logarithms).

Logs of Plasma Digoxin for Exercise 13.3

Patient Number	Day A	Day B	Day C	Day D	Day E	Day F
1	3.526	3.689	4.025	3.807	4.796	3.526
2	4.304	2.708	3.784	3.892	3.638	3.611
3	3.434	3.892	3.951	4.060	3.714	3.135
4	3.689	3.091	3.638	3.219	3.367	3.466
5	4.963	3.497	4.043	3.932	3.892	3.829
6	3.989	4.060	4.078	3.761	3.714	3.555
7	3.434	3.871	4.174	3.970	3.784	3.401
8	3.951	4.111	4.248	4.127	4.190	3.638
9	3.932	4.143	4.331	3.912	3.466	2.773
10	3.045	3.584	4.025	3.689	3.664	3.638
11	3.738	3.638	4.220	3.951	4.007	3.932
12	2.996	3.401	3.584	3.689	3.434	3.689
13	3.932	3.850	3.850	3.638	3.829	3.434
14	3.296	3.332	4.111	4.205	4.522	4.317
15	3.951	3.761	4.290	4.025	4.025	3.784

(a) Calculate the posterior probability that digoxin levels are increased when on experimental drug.

(b) Find a 99% posterior probability interval for the percentage increase in digoxin level due to experimental drug.

(c) Test the null hypothesis that the percentage increase in digoxin level is 0.

13.4 Two electrodes were placed on subjects and their resistances measured.[2] The results are shown here. The measured resistance on electrode 2 for subject 15 seems large. The technician wrote the following note by this reading: "Hairy part of arm." But the technician did not indicate amounts of hair on the other 31 readings, and it is likely that had this reading been smaller, this note would never have appeared. So I am loath to do anything but consider it real. On the other hand, taking logs seems a reasonable way to decrease the effect of all the large readings. Take logs of these 32 resistances and do a pairwise analysis. Assume a normal(.3, 1.0) prior density for the population mean of difference in logs.

Resistance (kilo-ohms) for two electrodes, E1 and E2, on 16 subjects in Exercise 13.4

Subject Number	E1	E2	Subject Number	E1	E2	Subject Number	E1	E2
1	500	400	6	27	84	11	15	45
2	660	600	7	100	50	12	160	200
3	250	370	8	105	180	13	250	400
4	72	140	9	90	180	14	170	310
5	135	300	10	200	290	15	66	1,000
						16	107	48

(a) Find a 95% posterior probability interval for the percentage decrease in resistance from electrode 2 to electrode 1.

(b) Use the interval in part (a) to test the null hypothesis that the percentage decrease is 0.

13.5 Drugs taken orally are eliminated from the body gradually. A standard measure of elimination is plasma half-life: the time required to decrease the level of the drug in the plasma by 50%. Large half-lives mean that more of the drug accumulates and less frequent dosing is required. The drug verapamil, which is used to lower blood pressure, was administered to eight subjects and its plasma half-life measured. The subjects took another drug for a week and then again took verapamil and had the plasma half-life measured. The question was whether the half-life was sufficiently changed by the other drug as to require changing the verapamil dosing schedule—a change would be recommended if the increase was 30% or more. These were the half-lives of verapamil (V), in hours:

Subject	1	2	3	4	5	6	7	8
V alone	2.55	1.81	1.99	2.37	3.03	2.25	1.89	1.83
V plus other	3.15	2.07	3.22	2.67	2.90	2.47	1.31	2.68

Take logs of these 16 half-lives and do a pairwise analysis. Assume a flat prior density for the population mean difference in logs. Find the posterior probability that the half-life of verapamil is increased by more than 30% by the other drug. [This is the probability that the mean difference, (V plus other) − (V alone), in logs is greater than log(1.30).]

13.6 In an assay for a certain blood factor, two experiments (using different plates) were carried out to estimate the amount of factor present in each of 21 samples, and also to compare the reproducibility of the procedure. The following table gives the results and also shows the difference between the assays from experiments 1 and 2. The histogram of the differences is also shown and suggests that there is an outlier (sample 4). Take logs of both sets of results and calculate the differences in logs for the 21 samples. (The Difference column in the table is irrelevant for your calculations.) Assume a flat prior for the population mean.

Sample Number	Experiment 1	Experiment 2	Difference	Sample Number	Experiment 1	Experiment 2	Difference
1	12.77	13.16	−.39	11	2.24	2.79	−.55
2	6.54	7.73	−1.19	12	3.11	3.23	−.12
3	6.03	6.79	−.76	13	16.20	17.45	−1.25
4	12.60	16.85	−4.25	14	5.44	6.33	−.89
5	7.69	9.43	−1.74	15	3.90	4.15	−.25
6	5.69	5.18	.51	16	5.36	4.69	.67
7	1.23	1.74	−.51	17	2.27	2.49	−.22
8	3.98	4.12	−.14	18	1.57	2.25	−.68
9	2.56	2.92	−.36	19	.68	.95	−.27
10	3.19	3.83	−.64	20	2.18	1.90	.28
				21	8.85	8.42	.43

Histogram of differences in Exercise 13.6

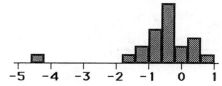

(a) Find the posterior density of the population mean.

(b) Find a 95% probability interval for the percentage increase for experiment 2 over experiment 1.

(c) Using the interval in part (b), test the hypothesis that the mean percentage increase is 0.

(d) Find the predictive density for the next (22nd) observation from the population of logs of differences. (Refer to Section 11.1 for predictive densities.) Calculate a 95% prediction interval. Transform this interval to percentage increases in experiment 2 over experiment 1.

13.7 Exercise 12.14 considers an experiment in which the product yields of dyestuff in grams of standard color for two batches are as follows:

| Batch X | 1,595 | 1,630 | 1,515 | 1,635 | 1,625 |
| Batch Y | 1,520 | 1,455 | 1,450 | 1,480 | 1,445 |

Reconsider the analysis, now taking logarithms. Where $d = m_X - m_Y$, take m_X and m_Y to be the means of the logs of the populations of yields. Assume flat prior densities for both means.

(a) Find the posterior probability that $d > 0$, that is, that the mean log of yield for batch X is bigger than that for batch Y.

(b) Test the null hypothesis that $d = 0$.

13.8 Reconsider the experiment described in Exercise 12.16. The numbers of trials required for kittens to give at least 27 correct responses out of 30 are given at the top of page 426.

Males (♂)	40	76	89	106	120	130	150	155	382
Females (♀)	66	69	94	103	117	391			

The largest observation in each group leads one to question the assumption that the populations are normal. Now take logs. Use flat prior densities for both means (of log number of trials). Assume these are independent random samples from the respective populations.

(a) Find the 95% posterior probability interval for the difference in means (of logs), $d = m_\delta - m_\varphi$.

(b) Convert this into an interval of percentage increase in mean number of trials required for males over females.

(c) Test the null hypothesis that $d = 0$.

13.9 Three drugs designed to reduce numbers of premature heart beats were considered in a three-period crossover design involving 12 patients.[2] The results (in premature beats per hour over a 2-day period) are shown in the table. (Drug C is the experimental drug considered in Examples 13.3 and 13.4.) We will ignore the crossover aspect of the study. A 0 in the table indicates that the patient had no premature beats on that drug. You cannot take the log of 0. So, add .1 to every one of the 36 observations and then take the log of each, as shown in the table. Carry out a pairwise analysis, assuming a flat prior density for the population mean of difference in logs for drug A vs drug C and drug B vs drug C. Find the posterior probability for the following:

(a) Drug A allows more premature beats than does drug C.
(b) Drug C cuts premature beats by 50% or more compared with drug A.
(c) Drug B allows more premature beats than does drug C.
(d) Drug C cuts premature beats by 50% or more compared with drug B.

Patient Number	Premature beats per hour			Logs(beats + .1)			Difference in Logs	
	Drug A	Drug B	Drug C	Log A	Log B	Log C	A vs C	B vs C
1	170	7	0	5.14	1.96	−2.30	7.44	4.26
2	19	1.4	6	2.95	.41	1.81	1.14	−1.40
3	187	205	18	5.23	5.32	2.90	2.34	2.43
4	10	.3	1	2.31	−.92	.10	2.22	−1.01
5	216	.2	22	5.38	−1.20	3.10	2.28	−4.30
6	49	33	30	3.89	3.50	3.40	.49	.10
7	7	37	3	1.96	3.61	1.13	.83	2.48
8	474	9	5	6.16	2.21	1.63	4.53	.58
9	.4	.6	0	−.69	−.36	−2.30	1.61	1.95
10	1.4	63	36	.41	4.15	3.59	−3.18	.56
11	27	145	26	3.30	4.98	3.26	.04	1.72
12	29	0	0	3.37	−2.30	−2.30	5.67	.00

13.3 Categorizing Data

Another type of transformation that lessens the effect of outliers is to specify a number of categories and assign the data to them. For example, we could use two categories and say the effect is big or small, positive or negative. Then we could analyze the observations as we did for proportions in Chapters 6–9. That is precisely what we did in Exercise 7.4. We decided that a success was a patient who walked farther while on the drug than off. We knew the actual differences in miles walked, but we reduced these into just two categories by considering only whether they were positive or negative. (In Exercise 10.5, we analyzed the actual miles walked.) Such a reduction may lose some important information, but it has the advantage that we do not have to assume a particular shape (such as the normal curve assumed in Exercise 10.5) for the population of differences.

The nonparametric methods of this section consider reductions of the data that are less dramatic than big/small. The next example reconsiders the experiment described in Example 10.5.

EXAMPLE 13.3
▷ **Percentage change in premature heart beats.** Example 10.5 considered reductions in the number of premature heart beats before and after taking an experimental drug. I want to make two modifications of that example. First, instead of considering the *number* of premature beats eliminated, consider the *percentage* of beats eliminated. This is

$$\text{Percentage decrease} = \frac{\text{Before} - \text{After}}{\text{Before}} \times 100$$

The percentage decrease in premature beats is considered by cardiologists to be more important than the actual number eliminated.

Appending percentage decreases to the table of premature beats of Example 10.5 gives Table 13.3. The dot plot in Figure 13.10 shows that this transformation makes the outlying value for patient 12 just another dot at 100% decrease. So simply consid-

Table 13.3
Premature beats per minute for Example 13.3

Patient Number	1	2	3	4	5	6	7	8	9	10	11	12
Before	6	9	17	22	7	5	5	14	9	7	9	51
After	5	2	0	0	2	1	0	0	0	0	13	0
Difference	1	7	17	22	5	4	5	14	9	7	−4	51
Decrease (%)	17	78	100	100	71	80	100	100	100	100	−44	100

Figure 13.10 Dot plot of % decreases in Example 13.3 with three categories of drug effect

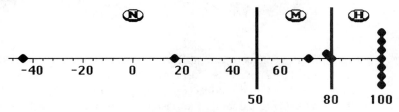

Percentage decrease in number of beats per minute

ering percentage change can be robust. However, now patient 11 seems to be an outlier. Handling this outlier is especially important because it is the only patient who experienced an increase. In any case, a normal density seems questionable as a model for the population.

In the dot plot, I have grouped percentage decrease into three categories. These categories agreed with the views of some cardiologists at the time of the experiment: A drug that caused 80% or greater decrease would be highly effective (H), from 50% to 80% decrease would be moderately effective (M), and less than 50% decrease would not be effective (N). So in the sample of 12 patients, there are eight patients in the H group and two in each of the other two groups.

We need models for the results. I will consider nine models and carry out an analysis along the lines of Chapter 10. The models considered ought to reflect a range of possibilities and ought to include models considered possible. As usual, our prior opinions depend on all the information at our disposal: animal studies, previous patients treated (if any), effectiveness of similar drugs (if any), molecular properties of the drug as compared with other drugs, and so on. The nine models shown in Figure 13.11 are reasonably comprehensive. At the time of the experiment, these nine models would have been approximately equally probable for me.

Figure 13.11 Nine possible models of drug effect for Example 13.3

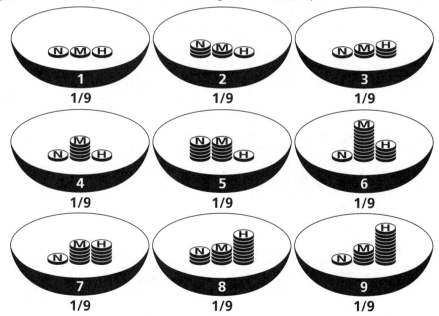

We proceed as in Example 10.1, but with nine models instead of three. The table in Figure 13.12 gives the likelihood (equal to the probability of the data for that bowl) for each of the nine models. I used Bayes' rule (in this case, dividing each likelihood by their sum) to calculate the posterior probabilities of the various models; these are given in the third column of the table. The last three columns give intermediate calcu-

Sec. 13.3 / Categorizing Data 429

lations for the predictive probabilities of the various categories. An entry is the product of the posterior probability (column 3) and the probability of the corresponding observation for that bowl. For example, the posterior probability of bowl 9 is .2565 and the probability of N in bowl 9 is $\frac{1}{15}$: $.2565(\frac{1}{15}) = .0171$.

Figure 13.12 Calculation of posterior and predictive probabilities for Example 13.3

Model	Likelihood		P(model | data)	$P(\text{N} \mid \text{model})P(\text{model} \mid \text{data})$	$P(\text{M} \mid \text{model})P(\text{model} \mid \text{data})$	$P(\text{H} \mid \text{model})P(\text{model} \mid \text{data})$
1:	$\left(\frac{1}{3}\right)^2\left(\frac{1}{3}\right)^2\left(\frac{1}{3}\right)^8$	= 1.88E-6	.0391	.0130	.0130	.0130
2:	$\left(\frac{3}{6}\right)^2\left(\frac{2}{6}\right)^2\left(\frac{1}{6}\right)^8$	= 1.65E-8	.0003	.0002	.0001	.0001
3:	$\left(\frac{1}{6}\right)^2\left(\frac{2}{6}\right)^2\left(\frac{3}{6}\right)^8$	= 1.21E-5	.2508	.0418	.0836	.1254
4:	$\left(\frac{1}{7}\right)^2\left(\frac{5}{7}\right)^2\left(\frac{1}{7}\right)^8$	= 1.81E-9	.0000	.0000	.0000	.0000
5:	$\left(\frac{5}{11}\right)^2\left(\frac{5}{11}\right)^2\left(\frac{1}{11}\right)^8$	= 1.99E-10	.0000	.0000	.0000	.0000
6:	$\left(\frac{1}{15}\right)^2\left(\frac{10}{15}\right)^2\left(\frac{4}{15}\right)^8$	= 5.05E-8	.0011	.0001	.0007	.0003
7:	$\left(\frac{1}{11}\right)^2\left(\frac{5}{11}\right)^2\left(\frac{5}{11}\right)^8$	= 3.11E-6	.0647	.0059	.0294	.0294
8:	$\left(\frac{3}{15}\right)^2\left(\frac{4}{15}\right)^2\left(\frac{8}{15}\right)^8$	= 1.86E-5	.3874	.0775	.1033	.2066
9:	$\left(\frac{1}{15}\right)^2\left(\frac{4}{15}\right)^2\left(\frac{10}{15}\right)^8$	= 1.23E-5	.2565	.0171	.0684	.1710
		4.81E-5	.9999	.1556	.2985	.5458

$P(\text{N} \mid \text{data})$
$P(\text{M} \mid \text{data})$
$P(\text{H} \mid \text{data})$

◁

Example 13.3 carries out detailed calculations as in Chapter 10. Let's now take advantage of the ease of calculation when dealing with normal models. The analysis is based on a very simple idea: Group the data into categories and assume that the observed categories come from a normal model. So we will calculate means and standard deviations of the category labels. First, we will adapt to the single-sample case of Chapter 11 and then consider two samples as in Chapter 12.

EXAMPLE 13.4

▷ **Categorizing premature heart beats.** Reconsider Example 10.5. Categorize (or transform) the differences as in Figure 13.13: 4 means the drug has a marked effect, decreasing the number of premature beats by 30 or more; 3 means it has a moderate effect; 2 means it has a mild effect; 1 means it has no effect or a negative effect. The categorized observations—there are $n = 12$ of them—are given here in order:

$$1 \quad 2 \quad 2 \quad 2 \quad 2 \quad 2 \quad 2 \quad 2 \quad 3 \quad 3 \quad 3 \quad 4$$

I calculate $\bar{x} = 2.33$ and $s = .745$. Adjusting for small sample size: $h = sk = s(1 + 20/n^2) = .745(1.139) = .848$. Assume a flat prior for the population mean, m. The posterior density of m is then normal$(2.33, .848/\sqrt{12})$ = normal$(2.33, .245)$.

Figure 13.13 Dot plot categorizing drug effect for Example 13.4

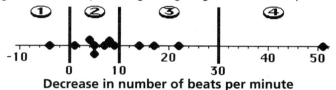

Decrease in number of beats per minute

To calculate the probability of a positive decrease means we have to undo the transformation. A positive decrease means $m = 2, 3$, or 4. To calculate the posterior probability of $m \geq 2$, find the z-score for $m = 2$:

$$z = \frac{2 - 2.33}{.245} = -1.35$$

According to the Standard Normal Table inside the front cover of this text, the probability to the right of $z = -1.35$ is about 91%. A positive decrease also means $m > 1$. The z-score for $m = 1$ is -5.43 and so the probability of $m > 1$ is 1.0000. Why the discrepancy? The mean, m, can be any number and is not restricted to integers 1, 2, and so on. So more appropriate than either $m > 1$ or $m \geq 2$ is to split the difference. Think about it this way: Category 2 is really the interval from 1.5 to 2.5 and so $m \geq 2$ includes all values down to 1.5. The z-score for $m = 1.5$ is -3.39, so the probability of a positive decrease is the probability to the right of this: .9997. ◁

Combining observations into categories lessens the effect of outliers.

There are several possible difficulties with the analysis in Example 13.4. First, a population that contains only 1's, 2's, 3's, and 4's cannot have a bell shape. However, a bell shape may not be a bad approximation, as the pictures in Figure 13.14 suggest. The pictures in Figure 13.15 show models in which the normal curve approximation will not be very good. However, when the sample size is large and we are interested in the population mean, the central limit theorem of Section 11.4 lets us conclude that the posterior distribution of m is close to bell shaped, even if the population is not. The

Sec. 13.3 / Categorizing Data 431

central limit theorem applies even for the model on the right in Figure 13.15, in which there are only two types of members of the population. Indeed, this is the setting in which we used normal curves to calculate probabilities when we dealt with proportions in Chapters 6–9.

Figure 13.14 Populations for which a normal density is a reasonable approximation

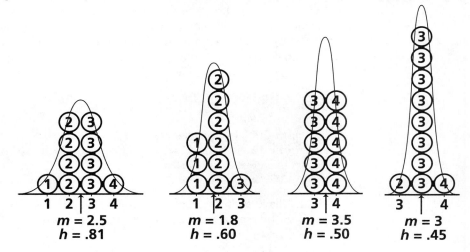

Figure 13.15 Populations for which a normal density may not be a good approximation, except as a posterior for population mean m and a large sample

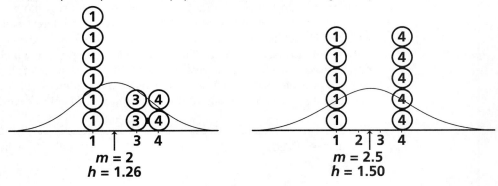

The central limit theorem does *not* mean we can use a normal model as a predictive density of the next observation. The predictive density depends very strongly on the population density. A larger sample size gives more information about the population model, but it does not make that model into a normal. So, before you use a normal density for prediction, worry about whether the shape of the population is consistent with it. If it is not, and if the sample size is large, use the approach of Example 11.7.

Another difficulty in the previous example is that the population mean cannot have a flat prior over all possible numbers. For example, restricting the observations to lie between 1 and 4 also restricts the population mean, m, to be between 1 and 4 (it restricts the population standard deviation h to lie between 0 and 1.5). This worry has little practical relevance. As I have indicated several times, the region in which the shape of the prior density matters is the vicinity of the data. So a flat prior density over all numbers is essentially the same as a flat (or open-minded) prior over numbers near $\bar{x}$.

Categorizing the observations is a very simple way to eliminate the undue influence of outliers. But you should be careful when assigning categories. For example, assigning 1, 2, 3, and 4 will give different results than will 1, 2, 3, and 5. Although it is not essential, try to have interpretations for the categories used—such as different levels of drug effect.

> **Robust Normal Model Approach**
>
> Group the data into categories having numerical labels. For several samples, use the same labels for all. Regard the labels as data and proceed as described in Chapter 11 or 12.

So far, I have addressed only the single-sample setting of Chapter 11. The next example is for two samples, as considered in Chapter 12.

EXAMPLE 13.5

▷ **Survival times.** To judge whether a particular "inhibitor" could slow the onset of cancer, 40 rats were injected with a known cancer-causing agent.[3] They were then randomized into two equal groups, with rats in the first group given the inhibitor (T) and those in the second serving as controls (C). Of interest was survival time and whether the rats had cancer. As rats died, postmortems were conducted to assess the presence or absence of cancer. The study lasted 192 days, at which time all surviving rats were sacrificed. The survival times are indicated in the following table. A plus sign, +, indicates a death unrelated to cancer and a minus sign, −, on day 192 means that the rat survived to the end of the study, but was found to have cancer.

Survival times for Example 13.5

Survival Times in Inhibitor Group				Survival Times in Control Group			
2+	55+	192+	192+	18+	73	106	171
2+	78	192+	192+	57	80	108	188
2+	78	192+	192−	63+	87	133	192−
2+	96	192+	192−	67+	87+	159	192−
2+	152	192+	192−	69	94	166	192−

In this example, the question is not so much the additional survival time afforded by the inhibitor, but whether it does indeed lengthen survival. The null hypothesis is that it does not. The + and − signs are problematic. A 192− does not mean that 192 is the day of death due to cancer, but only that cancer was present on day 192; however, it suggests that the rat would soon have died of cancer. A 192+ means that cancer was not present on day 192; when the rat would have died and whether death would have been due to cancer is not clear. Entries such as 67⁺ are particularly troublesome. The rat lived for 67 days and never contracted cancer; it might have eventually died of cancer, but it survived cancer free for about a third of the study.

One way to analyze these data is to consider the proportions of rats that died of cancer and then proceed as in Chapters 5 and 6. Assigning a "yes" or "no" is a categorization scheme—but how to categorize a 67⁺? A modified categorization is appropriate and appealing. (For an even more appealing but also more sophisticated analysis, see Snell[3] or Cox and Oakes.[4]) Various modifications are possible and reasonable, and there is nothing special about the one I will use.

I will simply delete rats that died of causes other than cancer before the first cancer death; these are the 2⁺, 18⁺, and 55⁺ rats. The 192⁺ rats had the best response and I will give them a 6 (an arbitrary assignment). Rats who died of other than cancer in the range of 50–100 days rate a middle-sized value—say, 4. Those that had cancer at the end of the study get a 3. (Perhaps this group should rate higher than the previous group. But rats normally live about 2 years, and while I do not know how old these rats were at the start of the study, having cancer after 6 months seems like a poor response.) Rats who died of cancer between 150 and 192 days rate a 2; 100 to 150 days, a 1; and before 100 days, a 0. These are shown in the following table, indicated in the same relative positions in the table as were the corresponding survival times in the previous table.

Modified survival times for Example 13.5

Categories for Inhibitor Group (T)					Categories for Control Group (C)			
—	—	6	6		—	0	1	2
—	0	6	6		0	0	1	2
—	0	6	3		4	0	1	3
—	0	6	3		4	4	2	3
—	2	6	3		0	0	2	3

I calculate

$$n_T = 14 \qquad \bar{x}_T = 3.79 \qquad s_T = 2.43$$

$$n_C = 19 \qquad \bar{x}_C = 1.68 \qquad s_C = 1.45$$

Assume a normal density for the inhibitor population and a flat prior for its mean, m_T. The posterior mean of m_T is the sample mean 3.79 and the posterior standard deviation is $h_1 = h/\sqrt{n} = [1 + 20/14^2](2.43/\sqrt{14}) = .71$. So the posterior distribution for m_T is a normal(3.79, .71) density. Also assume a normal model for the control population and a flat prior for the population mean m_C. The posterior mean of m_C is 1.68 and the posterior standard deviation is $[1 + 20/19^2](1.45/\sqrt{19}) = .35$; so the posterior distribution for m_C is a normal(1.68, .35) density. Since $3.79 - 1.68 = 2.10$ and $.71^2 + .35^2 = .80^2$, the difference in population means, $d = m_T - m_C$, has a normal(2.10, .80) density. The z-score for finding PdAL0 is

$$z = \frac{0 - 2.10}{.80} = -2.64$$

According to the Standard Normal Table, PdAL0 is about .996. So for someone who had flat prior densities for both m's, the inhibitor is very likely to be effective on average.

A 95% posterior probability interval for d is

$$2.10 \pm 1.96(.80) = 2.10 \pm 1.56, \text{ or from } .54 \text{ to } 3.66$$

This interval does not include 0 and so the null hypothesis is not supported.

These values refer to the scoring system described, and so are difficult to interpret. Over part of the range of the scoring system, a unit increase in score corresponds to about 50 days of remaining free of cancer. Since $.54 \times 50 = 27$ and $3.66 \times 50 = 183$, a possible interpretation of this interval is that the benefit of inhibitor is (very roughly) from 1 to 6 months. ◁

EXAMPLE 13.6
▷ **Experienced vs inexperienced analyst.** The accuracies of two analysts were compared[5] in their assay of carbon in a mixed powder. The first two columns in Table 13.4 give the differences in percentage carbon: assayed value minus correct value.

Table Table 13.4
Accuracies in assayed carbon (%)
for Example 13.6

Differences		Scoring system	
Exp.	Inexp.	Exp.	Inexp.
−8	−10	1	1
−3	16	0	2
20	−8	2	1
22	9	3	1
3	5	0	0
5	−5	0	0
10	5	1	0
14	−11	2	2
−21	25	3	3
2	22	0	3
7	16	1	2
8	3	1	0
16	40	2	3
	0		0
	−5		0
	16		2
	30		3
	−14		2
	25		3
	−28		3

The two right-hand columns of the table are transformations of the actual differences into scored deviations. If the assayed percentage minus actual percentage is between -5 and 5, then the analyst was off by 5 percentage points or less and I rate that a 0. The complete scoring system that I used follows:

$$\begin{array}{ll}\text{Off by 5\% or less:} & 0 \\ \text{Off by 6–10\%} & 1 \\ \text{Off by 11–20\%:} & 2 \\ \text{Off by more than 20\%:} & 3\end{array}$$

I find

$$n_E = 13 \qquad \bar{x}_E = 1.231 \qquad s_E = 1.049$$
$$n_I = 20 \qquad \bar{x}_I = 1.550 \qquad s_I = 1.203$$

I have prior opinions about the value of experience, but I will suppress them and assume flat priors for both m's. Assume normal densities for the two populations. The posterior mean of m_E is the sample mean, 1.23, and the posterior standard deviation is $h_1 = h/\sqrt{n} = [1 + 20/13^2](1.049/\sqrt{13}) = .325$. The posterior mean of m_I is 1.55 and the posterior standard deviation is $[1 + 20/20^2](1.203/\sqrt{20}) = .282$. Since $1.231 - 1.550 = -.319$ and $.325^2 + .282^2 = .431^2$, the difference in population means, $d = m_E - m_I$, has a normal($-.319, .431$) density. The z-score for finding the probability of $d < 0$ (experienced analyst is better) is

$$z = \frac{0 - (-.319)}{.431} = .74$$

According to the Standard Normal Table, the probability that the experienced analyst is better ($d < 0$) is about .77.

A 95% posterior probability interval for d is

$$-.319 \pm 1.96(.431) = -.319 \pm .845, \text{ or from } -1.16 \text{ to } .53$$

This interval contains 0 and so the null hypothesis is supported.

Again, these are scores and so are difficult to interpret. If you want a probability interval that is convenient to interpret, choose a scoring system that corresponds more directly to the original scale. ◁

EXERCISES

13.10 The validity of the normal model assumption in Exercise 10.9 is questionable in view of the dot plot of Darwin's data that you made in Exercise 2.1. Instead of using normal models, group the data into three categories as follows: self-pollinated are much more vigorous, not much difference between self- and cross-pollinated, and cross-pollinated are much more vigorous. In particular, use these groups:

$$\begin{array}{ll}\text{N:} & \text{Difference less than } -20 \\ \text{M:} & \text{Difference between } -20 \text{ and } +20 \\ \text{H:} & \text{Difference greater than } +20\end{array}$$

(Recall that the data are in eighths of inches, so "20" means 2.5 inches.) The table at the top of page 436 repeats the difference row from Exercise 2.1 and also has a group row:

Pot	1	2	3	4	5	6	7	8	9	10	11	12	13	14	15
Difference	49	−67	8	16	6	23	28	41	14	27	56	24	75	60	−48
Group	H	N	M	M	M	H	H	H	M	H	H	H	H	H	N

Use the nine models and the prior probabilities from Example 13.3.

(a) Calculate the posterior probabilities of these nine population models.

(b) Find the predictive probabilities that the next observation falls in group N, group M, and group H.

13.11 Reconsider Darwin's data from Exercise 13.10. Now use the following categorization:

Category −2: Difference less than −32
Category −1: Difference from −32 to 0
Category 1:   Difference from 0 to 32
Category 2:   Difference greater than 32

(*Note:* You will find that category −1 is empty.) Assume a flat prior density for the (category) population mean m. (This would not have been Darwin's prior, but it has the advantage of being open-minded.)

(a) Find the posterior density of m.

(b) Find the posterior probability that m is greater than 0. (This is, for the categories we have assigned, the probability that cross-pollination produces more vigorous plants than self-pollination.)

13.12 Surgeons insert catheters into blocked coronary arteries to attempt to clear them of plaque. One type of catheter uses heat to melt and vaporize the plaque. Using it means that some of the tissue on the inside of the artery will be injured; the depth of injury should be minimized. Twenty-nine prototypes of an experimental catheter were used on cadaver tissue to assess the depth of injury. The following are the thicknesses (in thousandths of mm) of injured tissue:

```
105 182 125 130 104    91  63 112  34  61    112  45  72 153
178  84  38  45  90   118  54 112  89  78     77 359 157  92 147
```

(The 359 is an obvious outlier. However rare, such a deep injury may indicate a problem with the catheter's operation and the existence of this outlier may be the most important aspect of the experiment. Any report of these data should include a histogram—such as shown here—or some other clear indication of the existence of this outlier.)

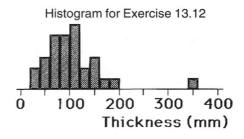

Histogram for Exercise 13.12

The sample mean is $\bar{x} = 107$ (in mm/1,000), and $s = 61.8$. The influence of the outlier, 359, is great: Excluding it changes these to $\bar{x} = 98.1$ and $s = 40.1$. The 35% decrease in standard deviation is especially indicative of the dominance of this observation in the analysis. Take a robust normal model approach to these data. Use the categories 0–99 (in mm/1,000), 100–149,

Sec. 13.3 / Categorizing Data 437

150–199, and 200 or greater, labeling them 1, 2, 3, and 4. Assume a flat prior distribution for the category population mean m and find the posterior distribution of m.

13.13 Exercise 13.9 gives results for three drugs designed to reduce numbers of premature heart beats—as shown in the table. Consider the difference, drug A minus drug C. Categorize the differences as follows:

Category −3: Difference less than −100
Category −2: Difference from −100 to −31
Category −1: Difference from −30 to −6
Category 0: Difference from −5 to 5
Category 1: Difference from 6 to 30
Category 2: Difference from 31 to 100
Category 3: Difference greater than 100

Patient Number	Premature Beats per Hour		
	Drug A	Drug B	Drug C
1	170	7	0
2	19	1.4	6
3	187	205	18
4	10	.3	1
5	216	.2	22
6	49	33	30
7	7	37	3
8	474	9	5
9	.4	.6	0
10	1.4	63	36
11	27	145	26
12	29	0	0

Assume a flat prior for the category population mean m and find the posterior probability that drug C is more effective than drug A ($m > 0$).

13.14 Repeat Exercise 13.13 for drug B vs drug C (instead of drug A vs drug C).

13.15 For the data in Exercise 13.13, associate the following categories with the actual numbers of premature beats per hour:

Category 0: From 0 to 3
Category 1: From 4 to 10
Category 2: From 11 to 30
Category 3: From 31 to 100
Category 4: Greater than 100

The categorization and the differences between drug A and drug C and between drug B and drug C are shown in the table at the top of page 438. Consider only the differences A − C. Assume a flat prior density for the (category) population mean m and a normal model for the population. Calculate the posterior probability that $m > 0$, that is, that drug C is more effective than drug A in eliminating premature beats.

| Patient | Scoring System | | | Difference in Scores | |
Number	Drug A	Drug B	Drug C	A − C	B − C
1	4	1	0	4	1
2	2	0	1	1	1
3	4	4	2	2	2
4	1	0	0	1	0
5	4	0	2	2	−2
6	3	3	2	1	1
7	1	3	0	1	3
8	4	1	1	3	−2
9	0	0	0	0	0
10	0	3	3	−3	0
11	2	4	2	0	2
12	2	0	0	2	0

13.16 Repeat Exercise 13.15 for B − C instead of A − C. (See table in Exercise 13.15.)

13.17 Reconsider Exercise 12.10. The improvements in oral hygiene index are repeated here:

Oxygen gel	10	15	6	10	11	3	8	8	3	13	10	9	8	9	8
	4	10	15	11	5	14	7	8	8	2	13	6	2	7	3
Placebo gel	5	6	4	3	3	5	6	4	4	2	0	7	0	3	2
	2	3	6	0	3	−3	1	6	6	8	2	12	24	5	3
	3	3	13	4											

A few improvements in the placebo groups were unusual—especially the value 24. These had the effect of increasing the sample standard deviation in that group and, in turn, that of the difference d. Now categorize the data as follows: improvement of 0 or less is category 1; 1–5 is category 2; 6–10 is category 3; and greater than 10 is category 4. Define d as before, $m_T - m_C$, but now interpret the m's as the means of the categorical populations. Assume flat prior densities for both m's.

(a) Find the posterior PdAL0 (compare it with your PdAL0 in Exercise 12.10).
(b) Find the 99% posterior probability interval for d.
(c) Convert this interval back to the original (index) scale, taking a unit increment in category to be an increment of 5 in the index scale.
(d) Test the null hypothesis that $d = 0$.

13.18 In Exercises 7.4 and 10.5, you considered the increases in miles walked while on an experimental drug as compared with the miles walked in the previous week while off the drug for 32 congestive heart-failure patients, as follows:

.00	+.56	+3.27	−2.55	+8.42	+1.07	−1.31	+3.19
−.59	+10.75	+11.73	−.05	+1.65	−3.42	+1.73	−1.44
+6.04	+12.21	+4.97	+1.68	+2.28	−6.57	−2.11	+.75
−.96	+1.68	+8.85	+7.45	−.59	+2.91	.00	+5.40

Exercise 9.13 gives data on a comparison group of 31 patients, who were given a placebo instead of the drug in the second week of the study:

+.69	+12.76	−.20	−1.09	−1.07	−4.51	−1.32	−1.78
+.08	+1.91	−1.81	−2.20	−1.41	−5.33	−2.44	−2.48
−.71	+1.44	−.84	−3.16	−.90	+.37	−2.86	+1.55
+.30	−1.23	−3.83	−6.31	−3.97	+.10	−.27	

Sec. 13.3 / Categorizing Data

Use the following categorization:

$$
\begin{array}{ll}
\text{Increases up to } -6.00: & -2 \\
\text{Increases from } -5.99 \text{ to } -2.00: & -1 \\
\text{Increases from } -1.99 \text{ to } +1.99: & 0 \\
\text{Increases from } +2.00 \text{ to } +5.99: & +1 \\
\text{Increases at least } +6.00: & +2
\end{array}
$$

Carry out a (robust) analysis using these category labels. Assume that the prior distributions of the population means are both flat.

(a) Find PdAL0, where $d > 0$ means the drug population has a greater mean increase in distance walked.

(b) Test the null hypothesis that $d = 0$.

13.19 A second half of the experiment described in Example 13.5 had 40 rats on a cancer "accelerator." Of these, 20 were randomly assigned to the "inhibitor." These were the results:

Survival Times in Inhibitor Group				Survival Times in Control Group			
18+	78	127	192+	37	43	51	62
19+	106	134	192+	38	43	51	66
40+	106	148	192+	42	43	55	69
56	106	186	192+	43+	48	57	86
64	127	192+	192+	43	49	59	177

Comparing the accelerator and nonaccelerator groups would be reasonable, both with and without inhibitor, but I want you to concentrate only on the data given here and—except in the prior densities as indicated below—to ignore the data in Example 13.5.

Use the following slight modification of the labeling scheme in the example:

Delete rats that died of other than cancer at 18 or 19 days.

Except for the above two, rats that died of other than cancer are category 4.

192+ is category 6.

Rats that had cancer at the end of the study are category 3 (there are none).

Rats that died of cancer:
    Between 150 and 192 days: category 2
    Between 100 and 150 days: category 1
    Before 100 days: category 0

Proceed as in the example to find the posterior PdAL0 in this new setting—with one exception. Now we have information from Example 13.5 to assess prior densities. We cannot simply use the posterior densities from Example 13.5 as prior densities here because observations from the two settings are not exchangeable (one group is with accelerator and the other is not). Since the accelerator is supposed to speed the onset of cancer, take the prior mean of m_T to be 2.50 instead of 3.79, which is the posterior mean from Example 13.5. In view of the additional uncertainty, quarter the prior precision from $1/.71^2 = 1.98$ to .50. So the prior standard deviation in the inhibitor group is doubled, to 1.42. (Quartering the precision is equivalent to observing one-quarter of the number of observations from a study with exchangeable observations, five instead of 20. So the prior is consistent with data from a previous study with the same standard deviation, five rats, and an average score that is about $\frac{2}{3}$ that of the actual one.) Thus, the prior density for m_T is normal(2.50, 1.42). Similarly, take the prior density of m_C to be normal(1.20, .70).

13.20 Reconsider Exercise 13.8 in which the numbers of trials required for the kittens to give at least 27 correct responses out of 30 were as follows:

Males (δ)	40	76	89	106	120	130	150	155	382
Females (φ)	66	69	94	103	117	391			

Use the following categorization:

Less than 100 trials:	1
From 100 to 200 trials:	2
More than 200 trials:	3

Use flat prior densities for both means (of categories).
(a) Find PdAL0, where $d = m_\delta - m_\varphi$.
(b) Find the posterior probability that d is contained in the interval from $-.5$ to $+.5$, that is, the posterior probability of a difference in means that is less than $\frac{1}{2}$ a categorical unit.

13.4 Ranking Data

Another transformation that leads to a robust analysis is **ranking.** The smallest observation in the set of interest is assigned rank 1, the second smallest has rank 2, and so on. For example, the numbers of trials of male kittens in Exercise 13.8 are 40, 76, 89, 106, 120, 130, 150, 155, 382. So their respective ranks are 1, 2, 3, 4, 5, 6, 7, 8, 9. Transforming 155 into 8 and 382 into 9 makes it clear that ranking shrinks outliers. Had two or more observations been tied, then each would have been assigned the average rank among those that are tied. For example, the ranks of the nine numbers 40, 40, 89, 106, 130, 130, 130, 155, 382 would be 1.5, 1.5, 3, 4, 6, 6, 6, 8, 9.

There are several possibilities for assigning ranks for two samples, say male and female kittens, depending on the set of interest. Ranking within samples changes

Males	40	76	89	106	120	130	150	155	382
Females	66	69	94	103	117	391			

into

Males	1	2	3	4	5	6	7	8	9
Females	1	2	3	4	5	6			

The result is not interesting for comparing populations: Regardless of the numbers of trials for the nine males and six females, their ranks will be exactly as shown (except when there are ties). So ranking within samples and comparing the populations on the basis of the ranks makes no sense.

Another possibility is to rank the entire data set:

Males	1	4	5	8	10	11	12	13	14
Females	2	3	6	7	9	15			

Sec. 13.4 / Ranking Data

This does make sense. For example, if the ranks in the combined groups were as follows:

Males	7	8	9	10	11	12	13	14	15
Females	1	2	3	4	5	6			

then every male took longer than every female, which clearly suggests a difference in the sexes. The question is whether the actual ranking suggests a difference.

This section considers combined rankings for both paired and independent samples. There is a complication when considering ranks. There are no populations from which the ranks are a sample. For example, if the sample size increases, then the ranks increase, as does the sample mean. However, in the paired-observation case, the "population" mean of differences between ranks would be 0 under the null hypothesis of no difference between treatment and control distributions. Similarly, in the case of independent samples, the difference between "population" means of ranks is 0—assuming the null hypothesis. The utility of taking ranks is therefore restricted to testing null hypotheses that the mean difference in ranks or the difference in mean ranks is 0. In addition, assessing prior information about ranks is difficult. So I will restrict consideration to flat priors in this section.

> **Consider ranks when testing hypotheses and when the data or other evidence suggest the existence of outliers in the population.**

First, consider an example of a paired analysis.

EXAMPLE 13.7
▷ **Cycles to failure of worsted yarn.** Examples 3.27 and 13.2 considered an experiment relating the number of cycles of loading to failure of worsted yarn to length of test specimen, amplitude of loading cycle, and load. As in Example 13.2, consider lengths of 250 mm and 300 mm where the other factors are used to form the various pairs:

Length (mm)	Cycles								
300	1,414	1,198	634	1,022	620	438	442	332	220
250	674	370	292	338	266	210	170	118	90

Now consider the ranks of the cycles (in the same order), and their differences:

Length (mm)	Ranks of Cycles								
300	18	17	13	16	14	11	12	8	5
250	15	10	7	9	6	4	3	2	1
Difference	3	7	6	7	8	7	9	6	4

Figure 13.16 Dot plot for Example 13.7: Comparing yarn lengths of 300 mm vs 250 mm

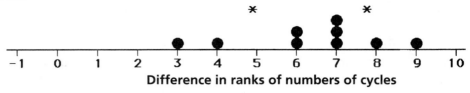

Difference in ranks of numbers of cycles

The dot plot shown in Figure 13.16 should be compared with those in Example 13.2. Just as when taking logs, the dots are closely packed, especially in comparison with their distances from 0. We proceed as usual with this sample of differences. The next table summarizes the data [where the small-n correction factor $k = (1 + 20/9^2) = 1.25)$*:

Differences Between Lengths of 300- and 250-mm yarn

	n	$\bar{x}$	s	sk	$sk/\sqrt{n}$
Ranks of Cycles	9	6.33	1.76	2.20	.733

As I suggested before this example, considering a sample of differences in ranks as a selection from a population is artificial. The population values depend on the sample size. But we are interested only in the "population" relative to 0 rather than in the actual values. In this example, we know in advance that the sample mean cannot be less than -9 nor greater than $+9$ and so the "population" mean must be similarly constrained. However, we will assume a flat prior density for the population mean, m. The fact that this allows for values outside the above range is not very important because the likelihood restricts essentially all the posterior probability to this range.

To test the hypothesis that $m = 0$ requires the z-score at 0:

$$z = \frac{0 - \bar{x}}{sk/\sqrt{n}} = -8.64$$

This is not as far from 0 as when taking logs (-14.44), but farther than when the data are not transformed at all (-4.16)—see Example 13.2. Since $z < -1.96$, the null hypothesis is not supported. Also, the 95% probability interval for m is as follows:

$$6.33 \pm (1.96)(.733) = 6.33 \pm 1.44, \text{ or from about 4.9 to 7.8}$$

which does not contain 0.

The lower and upper limits of this interval are indicated in the dot plot of Figure 13.16 using asterisks. Converting limits of ranks back to number of cycles or percentage change (taking "antiranks") is difficult. Interpreting ranks or their differences on the original scale of measurement is also difficult. Not having a good interpretation is

---

* Because the ranks and therefore also their differences are constrained—no difference can be bigger than 17 or less than -17, for example. This means that we have some information about the corresponding population standard deviation. But using the small-n correction factor and proceeding as usual is not unreasonable.

a disadvantage of ranking. The ranks of the original data go from 1 to 18 and the average difference in ranks is 6.33, or about one-third of this range. The maximum possible average difference is 9, which would happen only in cases such as this:

Treatment A	18	17	13	16	14	11	12	15	10
Treatment B	9	8	4	7	5	2	3	6	1
Difference	9	9	9	9	9	9	9	9	9

An average difference of 6.33 is relatively large, which we know from the size of the z-score. ◁

The data in the previous example are paired. In the next example, there are two independent samples. The example also shows that the presence of sample outliers is not necessary to take ranks.

EXAMPLE 13.8
▷ **Water purification.** An experiment[6] compared two methods of filtering water. The quantity of water filtered (m^3/m^2 of filter) by method A (20 runs) and method B (29 runs) are given, ordered from smallest to largest within groups:

Method A	81	98	105	119	148	157	172	182	182	194
	196	216	221	243	243	248	277	286	308	365
Method B	91	99	161	187	203	221	222	248	249	253
	258	271	280	282	294	296	305	307	308	311
	323	325	329	344	354	356	379	400	415	

There are no obvious outliers in these samples, but we can use ranks in any case. The overall ranks of the data (in the same order) are as follows:

Method A	1	3	5	6	7	8	10	11.5	11.5	14
	15	17	18.5	21.5	21.5	23.5	29	32	37	46
Method B	2	4	9	13	16	18.5	20	23.5	25	26
	27	28	30	31	33	34	35	36	38	39
	40	41	42	43	44	45	47	48	49	

The sample sizes, means, and standard deviations of the ranks for the two methods are as follows:

$$n_A = 20 \qquad \bar{x}_A = 16.90 \qquad s_A = 11.65$$
$$n_B = 29 \qquad \bar{x}_B = 30.59 \qquad s_B = 12.95$$

Just as in the single-sample case considered in the previous example, thinking of these two samples of ranks as selections from some population is artificial. The population values depend on the sample sizes. But we are interested in the populations relative to each other rather than in the actual values in the population. We know in advance of seeing the data that the sample means cannot be too small or too large (for example, $\bar{x}_A$ cannot be less than 10.5 nor more than 39.5) and so the "population" means are similarly constrained. In addition, the value of $\bar{x}_B$ is determined from that of $\bar{x}_A$ [the sum of all 49 ranks is $49(50)/2 = 1{,}225$ and therefore the sum of the ranks of the B's is $1{,}225 - 20(16.90) = 887$, which, divided by 29, gives $\bar{x}_B = 30.59$]. However, we will regard the samples as being independent and assume flat prior densities for the means, m_A and m_B. Just as in the single-sample case, the fact that this allows for values outside the possible range is not very important because the likelihoods will restrict essentially all the posterior probability to be in the range.

The posterior mean of m_A is the sample mean, 16.90. Using the small-n correction factor, the posterior standard deviation is $h_A = h/\sqrt{n_A} = (1 + 20/20^2)(11.65)/\sqrt{20} = 2.74$. For m_A, the posterior mean is 30.59; the standard deviation is $(1 + 20/29^2) \times (12.95)/\sqrt{29} = 2.46$. The difference in sample means is $16.90 - 30.59 = -13.69$. Since $2.74^2 + 2.46^2 = 3.68^2$, the difference in population means, $d = m_A - m_B$, has a normal(-13.69, 3.68) density. The z-score for finding the probability of $d < 0$ (method B filters more water on average) is

$$z = \frac{0 - (-13.69)}{3.68} = 3.72$$

According to the Standard Normal Table, the probability of $z < 3.72$ is .9999. The 95% posterior probability interval is $-13.69 \pm 1.96(3.68) = -13.69 \pm 7.21$, which does not contain 0, and so the null hypothesis is not supported.

To estimate how much more water method B filters on average (perhaps giving an interval), analyze the original, unranked data or take logs to assess the percentage increase. ◁

EXERCISES

13.21 Exercise 13.6 gives results of two experimental measurements of the amount of a blood factor present in each of 21 samples. The data are repeated here. Rank the results from 1 to 42 and calculate the differences in ranks for the 21 samples. Assume a flat prior for the mean of the population of differences in ranks. Find the posterior probability that the population mean is negative—that experiment 2 gives greater measurements on average than experiment 1 (which would indicate a persistent bias in one experiment relative to the other).

\multicolumn{6}{c}{Experiment}
1	2	1	2	1	2
12.77	13.16	3.98	4.12	3.90	4.15
6.54	7.73	2.56	2.92	5.36	4.69
6.03	6.79	3.19	3.83	2.27	2.49
12.60	16.85	2.24	2.79	1.57	2.25
7.69	9.43	3.11	3.23	.68	.95
5.69	5.18	16.20	17.45	2.18	1.90
1.23	1.74	5.44	6.33	8.85	8.42

13.22 In Exercise 13.18, you considered the additional miles walked while on a particular drug (T) for 32 congestive heart-failure patients:

Sec. 13.4 / Ranking Data

.00	+.56	+3.27	−2.55	+8.42	+1.07	−1.31	+3.19
−.59	+10.75	+11.73	−.05	+1.65	−3.42	+1.73	−1.44
+6.04	+12.21	+4.97	+1.68	+2.28	−6.57	−2.11	+.75
−.96	+1.68	+8.85	+7.45	−.59	+2.91	.00	+5.40

with the additional miles walked for 31 similar patients on placebo (C):

+.69	+12.76	−.20	−1.09	−1.07	−4.51	−1.32	−1.78
+.08	+1.91	−1.81	−2.20	−1.41	−5.33	−2.44	−2.48
−.71	+1.44	−.84	−3.16	−.90	+.37	−2.86	+1.55
+.30	−1.23	−3.83	−6.31	−3.97	+.10	−.27	

Rank the differences in distances walked from 1 to 63. Use flat priors for both population means.
    (a) Find the posterior probability that $d = m_T - m_C > 0$.
    (b) Test the null hypothesis $d = 0$.

13.23 Example 13.1 dealt with digoxin levels over time and demonstrated that the effects of outliers could be reduced by averaging measurements taken under similar circumstances. Exercise 13.3 showed how to reduce the effect still further by taking logs. Now instead of taking logs, take ranks (see table). Proceed as in Example 13.1: For each patient, average days C and D and subtract the average of the other 4 days from it. Assume a flat prior density for the population mean, m, of the differences (of ranks).

Plasma digoxin: Overall ranks

Patient	Day A	Day B	Day C	Day D	Day E	Day F
1	20.5	35	68.5	45	89	20.5
2	85	1	44	53	28.5	25
3	14.5	53	61.5	72	38.5	6
4	35	5	28.5	7	10	17.5
5	90	19	71	57.5	53	47.5
6	65	72	74	41.5	38.5	22
7	14.5	51	79	64	44	11.5
8	61.5	75.5	83	77	80	28.5
9	57.5	78	87	55	17.5	2
10	4	23.5	68.5	35	32	28.5
11	40	28.5	82	61.5	66	57.5
12	3	11.5	23.5	35	14.5	35
13	57.5	49.5	49.5	28.5	47.5	14.5
14	8	9	75.5	81	88	86
15	61.5	41.5	84	68.5	68.5	44

    (a) Calculate the posterior probability of $m > 0$.
    (b) Test the null hypothesis $m = 0$.

13.24 Repeat the analysis of Exercise 13.23, but use a different ranking system. Now rank the days within each patient, as shown in the table at the top of page 446. Assume a flat prior density for the population mean, m, of the differences (average ranks of days C and D compared with the average ranks of the other 4 days).

Plasma digoxin: Ranks within patients

Patient	Day A	Day B	Day C	Day D	Day E	Day F
1	1.5	3	5	4	6	1.5
2	6	1	4	5	3	2
3	2	4	5	6	3	1
4	6	1	5	2	3	4
5	6	1	5	4	3	2
6	4	5	6	3	2	1
7	2	4	6	5	3	1
8	2	3	6	4	5	1
9	4	5	6	3	2	1
10	1	2	6	5	4	3
11	2	1	6	4	5	3
12	1	2	4	5.5	3	5.5
13	6	4.5	4.5	2	3	1
14	1	2	3	4	6	5
15	3	1	6	4.5	4.5	2

(a) Calculate the posterior probability of $m > 0$.
(b) Test the null hypothesis $m = 0$.

13.25 Exercise 13.9 used logarithms to transform numbers of premature heartbeats on three drugs and Exercises 13.13–13.16 used categories. The results are repeated here. Now use ranks. Rank all 36 entries. Make two additional columns, one for the difference in ranks A − C, and the other for the difference in ranks B − C. Assume flat priors for the means of each difference, say, m_A for the first and m_B for the second. If m_A is greater than 0, then drug C is better than drug A in this comparison of ranks; similarly, if m_B is greater than 0, then drug C is better than drug B.

Patient Number	Premature beats per hour		
	Drug A	Drug B	Drug C
1	170	7	0
2	19	1.4	6
3	187	205	18
4	10	.3	1
5	216	.2	22
6	49	33	30
7	7	37	3
8	474	9	5
9	.4	.6	0
10	1.4	63	36
11	27	145	26
12	29	0	0

(a) Find the posterior probability that m_A is greater than 0.
(b) Test the null hypothesis $m_A = 0$.
(c) Find the posterior probability that m_B is greater than 0.
(d) Test the null hypothesis $m_B = 0$.

13.26 A company found that a substantial proportion of machine parts from its production lines had a defect that could not be discovered without actually using the part. It was not clear whether the defect hampered the machine's operation, but it concerned customers. The company wanted to fix the problem but did not know what caused it. It undertook a major testing program by simulating operating conditions and recording for each part whether the defect occurred. It made other measurements of the part's operating characteristics in hopes that one or more of the measurements would pinpoint the culprit. The data for a prime candidate (called "meas") are shown in the two tables for 297 tested parts. (The data and problem are real, but I will not give specifics.) I selected the 117 parts with the defect and listed these in the first table. Then I listed the 180 parts without the defect in the second table. Notice that the largest observation in the "defect" group is about 3 units larger than the next biggest observation. While the large sample sizes may lessen the effect of this outlier, taking ranks assures that its effect will be minimized.

I ranked all 297 parts and show each part's rank next to its meas in the tables. For example, there were two parts with meas = 4.26, one without and the other with the defect, so they appear in separate tables. These are the 22nd and 23rd smallest, and so they share the rank 22.5.

Parts with defect ($n = 117$) for Exercise 13.26

meas	rank	meas	rank	meas	rank	meas	rank
3.59	1	4.61	69	5.30	169	5.98	244
3.86	2	4.62	71	5.31	172	6.02	246
3.93	3	4.63	73	5.32	174	6.04	248
4.01	4	4.66	75	5.33	175.5	6.12	251
4.06	6	4.68	79.5	5.33	175.5	6.15	252
4.08	8	4.72	86.5	5.34	178.5	6.23	255
4.10	11	4.74	88	5.34	178.5	6.30	256
4.13	12	4.76	92	5.41	188	6.35	258
4.15	14	4.77	94.5	5.42	189	6.38	259.5
4.18	16	4.80	102	5.43	190.5	6.40	261.5
4.26	22.5	4.81	105	5.43	190.5	6.54	265
4.29	27.5	4.82	107.5	5.44	192.5	6.56	266
4.31	29	4.90	116	5.48	197.5	6.63	267
4.35	30.5	4.91	117.5	5.52	203	6.64	268.5
4.36	32	4.94	120	5.53	204	6.71	270.5
4.39	35	5.04	129	5.54	205	6.71	270.5
4.42	38.5	5.05	130.5	5.57	209.5	6.75	272
4.42	38.5	5.06	132.5	5.58	212.5	6.76	273
4.46	43.5	5.09	136	5.60	215	6.78	274
4.47	46	5.11	140	5.64	217.5	6.82	275
4.47	46	5.11	140	5.67	224	6.84	276
4.48	48.5	5.15	147	5.70	226.5	6.87	277.5
4.50	50.5	5.20	151	5.72	228	6.90	280.5
4.50	50.5	5.21	152.5	5.78	233	6.90	280.5
4.51	53	5.21	152.5	5.84	234.5	6.98	282
4.52	55	5.22	154	5.86	237	8.30	292
4.54	57.5	5.25	159.5	5.89	238	12.25	297
4.55	60	5.27	164	5.91	240		
4.57	64	5.27	164	5.96	242		
4.60	67	5.29	166.5	5.97	243		

Parts without defect ($n = 180$) for Exercise 13.26

meas	rank	meas	rank	meas	rank	meas	rank
4.04	5	4.68	79.5	5.12	142.5	5.60	215
4.07	7	4.69	82.5	5.12	142.5	5.60	215
4.09	9.5	4.69	82.5	5.14	144.5	5.64	217.5
4.09	9.5	4.70	84	5.14	144.5	5.65	220
4.14	13	4.71	85	5.15	147	5.65	220
4.16	15	4.72	86.5	5.15	147	5.65	220
4.19	17.5	4.75	89.5	5.16	149.5	5.66	222
4.19	17.5	4.75	89.5	5.16	149.5	5.67	224
4.23	19	4.76	92	5.23	155.5	5.67	224
4.24	20.5	4.76	92	5.23	155.5	5.70	226.5
4.24	20.5	4.77	94.5	5.24	157	5.74	229
4.26	22.5	4.78	97	5.25	159.5	5.75	230.5
4.27	24	4.78	97	5.25	159.5	5.75	230.5
4.28	25.5	4.78	97	5.25	159.5	5.76	232
4.28	25.5	4.79	99.5	5.26	162	5.84	234.5
4.29	27.5	4.79	99.5	5.27	164	5.85	236
4.35	30.5	4.80	102	5.29	166.5	5.90	239
4.38	33.5	4.80	102	5.30	169	5.95	241
4.38	33.5	4.81	105	5.30	169	5.99	245
4.41	36.5	4.81	105	5.31	172	6.03	247
4.41	36.5	4.82	107.5	5.31	172	6.07	249.5
4.43	40	4.85	109.5	5.34	178.5	6.07	249.5
4.44	41	4.85	109.5	5.34	178.5	6.19	253
4.45	42	4.86	111.5	5.35	181.5	6.22	254
4.46	43.5	4.86	111.5	5.35	181.5	6.32	257
4.47	46	4.87	113	5.38	183	6.38	259.5
4.48	48.5	4.89	114.5	5.40	185.5	6.40	261.5
4.51	53	4.89	114.5	5.40	185.5	6.45	263
4.51	53	4.91	117.5	5.40	185.5	6.50	264
4.53	56	4.92	119	5.40	185.5	6.64	268.5
4.54	57.5	4.97	121	5.44	192.5	6.87	277.5
4.55	60	4.98	122.5	5.45	194.5	6.88	279
4.55	60	4.98	122.5	5.45	194.5	7.02	283
4.56	62	4.99	125.5	5.46	196	7.07	284
4.57	64	4.99	125.5	5.48	197.5	7.08	285
4.57	64	4.99	125.5	5.49	199	7.11	286
4.58	66	4.99	125.5	5.50	200.5	7.39	287
4.61	69	5.03	128	5.50	200.5	7.55	288
4.61	69	5.05	130.5	5.51	202	7.60	289
4.63	73	5.06	132.5	5.55	206	7.74	290
4.63	73	5.08	134	5.56	207	7.87	291
4.67	76.5	5.09	136	5.57	209.5	8.50	293
4.67	76.5	5.09	136	5.57	209.5	8.65	294
4.68	79.5	5.10	138	5.57	209.5	9.21	295
4.68	79.5	5.11	140	5.58	212.5	9.56	296

The means and standard deviations of the ranks are given in the table below. Assume flat priors for both population means of ranks and test the null hypothesis that there is no difference between them (in which case, this measurement is unlikely to be implicated in the defect or to serve as a surrogate for its presence).

Ranks of meas for Exercise 13.26

Defect	With	Without
Mean	153.8	145.9
Std. dev.	91.0	82.0
n	117	180

13.5 Probabilities of Intervals and the Sign Test

The previous three chapters focused on the mean of a population and the mean of a sample. We made conclusions such as probability intervals for the population mean and for predicting future observations. In the previous sections of this chapter, we transformed sample observations (and by extension, unobserved population values) to enable stronger conclusions about the (transformed) population mean. In this section, we take a completely different tack. We will address probabilities of intervals directly, taking a simple-minded approach and using the sample proportion in the interval in question to make inferences about the corresponding population proportion. This is a straightforward application of the methods developed in Chapters 6 and 7, especially the latter. These inferences do not assume that the population distribution is normal or any other particular shape—hence, their inclusion in the present chapter.

We will focus on intervals that include all population values less than (or greater than) a particular value, but the methodology applies generally. In many problems, the value of interest is 0. In a paired experiment, for example, the two observations within each pair are subtracted from each other and the question of interest is whether the second is smaller than the first—that is, whether the difference is less than 0. So the problem is the same as that considered in Chapter 11 (especially Section 11.5). The difference between this section and Section 11.5 is that now we will not assume normal populations and we will not deal with means and standard deviations. Instead, we consider only the proportion of the observations less than 0 and use this proportion to draw conclusions about the proportion of the population that is less than 0. Addressing whether the population proportion equals $\frac{1}{2}$ is called a **sign test.** "Sign test" is a term used in classical statistics. The following development is a Bayesian version.

The next example is similar to that of Section 6.5: Observations are taken in pairs and the winning treatment within each pair is considered for analysis.

EXAMPLE 13.9
▷ **Is amiloride effective?** Example 2.7 and Exercise 11.2 considered a study of amiloride in patients suffering from cystic fibrosis. The patients' forced vital capacities

after 6 months of using an aerosol spray containing amiloride and after 6 months of using a spray without amiloride (vehicle only) are shown in Table 13.5. The table includes a column of signs of the differences, amiloride minus vehicle only; a + indicates a better response with the drug than without. (I will discuss the last column in the table shortly.) We will ignore the sizes of the differences and consider only their signs—a difference less than 0 means that the sign is −. In Exercise 11.2, you analyzed the column of differences. Because it is large relative to the other differences, the entry +550 for patient 3 had a substantial impact on the sample mean and therefore on the conclusion. In this example, it has the same effect as does any other positive entry.

Table 13.5

Patient Number	Forced Vital Capacity (in ml)			Sign of Diff	Sign of Diff − 100
	Vehicle	Amiloride	Amil − Veh		
1	2,925	2,760	−165	−	−
2	4,190	4,490	+300	+	+
3	5,067	5,617	+550	+	+
4	2,588	2,543	−45	−	−
5	3,934	3,810	−124	−	−
6	3,952	3,985	+33	+	−
7	2,547	2,392	−155	−	−
8	4,108	3,880	−228	−	−
9	2,646	2,732	+86	+	−
10	3,635	3,758	+123	+	+
11	2,890	2,960	+70	+	−
12	3,125	3,387	+262	+	+
13	3,805	4,048	+243	+	+
14	1,741	1,787	+46	+	−
# of +'s/# of −'s:				9/5	5/9

Of the 14 patients, 9 did better on amiloride and 5 did better on vehicle alone. Since + means amiloride was better, then, in the notation of Chapter 7, $s = 9$ and $f = 5$. Assume a beta(1, 1) prior density for the population proportion of +. This assumption means that the prior probability of some benefit for amiloride on a particular patient is $1/(1 + 1) = \frac{1}{2}$. The posterior density is beta(10, 6) and is shown in Figure 13.17. The predictive probability that the next patient will respond better on amiloride than on vehicle alone is $10/(10 + 6)$, about 63%.

With the posterior density in hand we can calculate areas as in Chapter 7. The calculations are easy but the logic and the language are cumbersome. Consider the area under the posterior density to the right of .5. The answer is 85%, a standard calculation from Chapter 7. This is the probability that more patients in the population would respond better on amiloride than on vehicle alone. This figure is easily confused with the 63% in the previous paragraph. The following two statements indicate the difference (1): "The probability that the next patient will respond better on amiloride than on vehicle alone is 63%." (2) "The probability that more patients respond better on amiloride is 85%." The first is a predictive statement about a particular patient, whereas the second refers to a population of patients.

Figure 13.17 Posterior density of population proportion of patients responding better on amiloride for Example 13.9.

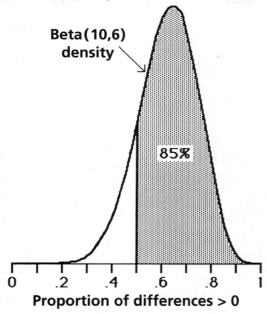

The null hypothesis of no treatment benefit is that the population proportion of + is equal to $\frac{1}{2}$. Recall from Chapter 7 that to test this hypothesis—called the *sign test*—means to decide whether $\frac{1}{2}$ is contained in the 95% probability interval, or equivalently, whether the z-score for proportion $\frac{1}{2}$ is between -1.96 and 1.96. To do this we need r, r^+, and t for the posterior density, that is, beta(10, 6):

$$r = \frac{a}{a+b} = \frac{10}{16} = .625 \qquad r^+ = \frac{a+1}{a+b+1} = \frac{11}{17} = .64706,$$

$$t = \sqrt{r(r^+ - r)} = \sqrt{\frac{10}{16}\left(\frac{11}{17} - \frac{10}{16}\right)} = .1174,$$

$$z = \frac{.5 - r}{t} = \frac{-.125}{.1174} = -1.06$$

Since this is between -1.96 and $+1.96$, the null hypothesis of no difference is supported. Equivalently, the 95% probability interval is $.625 \pm (1.96)(.1174) = .625 \pm .230$, or from about .395 to .855, and this interval contains $\frac{1}{2}$. (Minitab gives the exact 95% probability interval as .384 to .837, which is somewhat different from the approximate interval obtained using z-scores.) This interval is shown along with the beta(10, 6) density in Figure 13.18 (page 452).

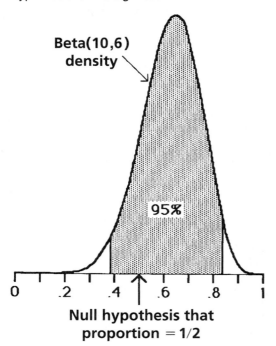

Figure 13.18 Beta(10, 6) density for Example 13.9 showing the 95% probability interval and the null hypothesis for the sign test

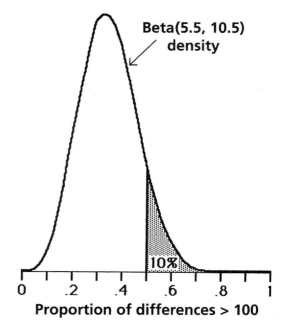

Figure 13.19 Posterior density for proportion having difference of 100 ml or more in Example 13.9

By testing for no benefit, the sign test addresses whether there is a benefit. A simple modification of the sign test addresses degree of benefit. It requires going back to the original data. Consider the proportion of patients in the population whose forced vital capacity would be at least 100 ml greater if given amiloride rather than vehicle alone. The last column in Table 13.5 at the beginning of this example indicates + for a patient who did respond at least that much better and a − for a patient who did not. Since there are 5 +'s and 9 −'s, now $s = 5$ and $f = 9$. Assuming a beta(1, 1) density is not consistent with the same assumption for the proportion of negatives: One's opinions about these two quantities could be the same only for someone who regards it impossible for the difference to be between 0 and 100 ml. To keep things consistent, take the prior density to be beta(.5, 1.5)—so the prior probability of a benefit greater than 100 is $.5/(.5 + 1.5) = \frac{1}{4}$. The posterior density is beta(5.5, 10.5). This is shown in Figure 13.19 and is nearly the mirror image of beta(10, 6). The predictive probability of a response at least 100 ml better on amiloride for the next patient in this population is $5.5/16 = .344$. The posterior probability that more patients will respond at least 100 ml better on amiloride than on vehicle alone is the area under the beta(5.5, 10.5) density, which is .098, or about 10%.

Other improvements besides 100 ml can be considered in the same way. ◁

Sec. 13.5 / Probabilities of Intervals and the Sign Test

> **Sign Test**
>
> Test the null hypothesis that the population proportion of $+$ is $\frac{1}{2}$ by deciding whether the 95% posterior probability interval contains $\frac{1}{2}$. If yes, then the null hypothesis is supported; otherwise, it is not.

> **Equivalent Sign Test**
>
> Find $z = (\frac{1}{2} - r)/t$ where r and t are calculated from the posterior beta density. If $-1.96 < z < 1.96$, then the null hypothesis is supported; otherwise, it is not.

EXAMPLE 13.10

▷ **Cycles to failure of worsted yarn (revisited).** Example 3.27 described a factorial experiment relating the number of cycles of loading to failure of worsted yarn to length of test specimen, amplitude of loading cycle, and load. In Example 13.2, we considered the following data, to which I have appended a row of signs of differences:

Length (mm)	Cycles								
300	1,414	1,198	634	1,022	620	438	442	332	220
250	674	370	292	338	266	210	170	118	90
Difference	740	828	342	684	354	228	272	214	130
Sign of difference	+	+	+	+	+	+	+	+	+

Consider the proportion of pairs in the population (with possibly different amplitudes and load) for which the 300-mm yarn lasts longer than the 250-mm yarn. Since for every observation the number of cycles is greater when length = 300 mm, $s = 9$ and $f = 0$. In view of prior information that longer yarns last longer, take the prior density to be beta(5, 1). It follows that the posterior density is beta(14, 1)—see Figure 13.20 on page 454. The predictive probability that the 300-mm yarn lasts longer for the next pair is $\frac{14}{15} = 93.3\%$. This may seem too small to you in view of the very clear evidence that 300-mm yarns last longer. But this is a predictive probability and not a statement about a population proportion. Considering the next pair selected, it would not be incredibly surprising if the 250-mm yarn lasted longer—especially for large amplitude and load—and a probability of 7% is not unreasonable for this event.

A sign test addresses a question about the population rather than about the next observation: Does the 300-mm yarn require more cycles to break in more than half of the population of 300-mm and 250-mm pairs? In view of the data, it *would* be surprising if the answer were "no." To find the z-score:

$$r = \frac{14}{15} = .933, \quad r^+ = \frac{15}{16}, \quad t = \sqrt{\frac{14}{15}\left(\frac{15}{16} - \frac{14}{15}\right)} = .06236$$

Then

$$z = \frac{.5 - .933}{.0624} = -6.95$$

Since this is not between -1.96 and 1.96, the null hypothesis is not supported. The corresponding 95% probability interval is $.933 \pm (1.96)(.0624) = .933 \pm .122$, or from .811 to over 1. This interval does not contain $\frac{1}{2}$. However, it does not make sense that the upper endpoint of the interval is greater than 1. The reason it is greater than 1 is that the beta(14, 1) density is not close to being symmetric, and this also makes the lower endpoint calculated using z-scores inaccurate as well. Using Minitab, the exact 95% probability interval is from .768 to .998.

Essentially all the probability (99.99%) under the beta(14, 1) density is to the right of .5—see Figure 13.20. So the evidence that 300-mm yarn requires more cycles to failure in more than half the population is very compelling.

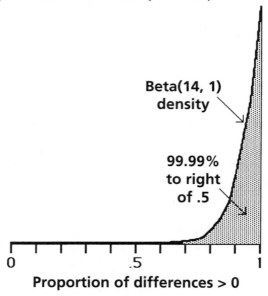

Figure 13.20 Posterior density for Example 13.10

◁

This chapter addresses changing data to minimize the influence of a small number of observations and to better fit the assumptions of analyses that you learned how to carry out. Manipulating data is dangerous. You might (knowingly or otherwise) choose a transformation that best makes a particular point. It may be that categorizing data as in Section 13.3 gives one answer, while ranking as in Section 13.4 gives a very different answer. Or two different categorizations might give rather different answers. In practice you should try different transformations—as well as no transformation. It will be reassuring if they give similar answers, and indeed this would suggest that a transformation was unnecessary. But if they do not, then try to understand why. If the null hypothesis is supported for the untransformed data and not for the log transformation, this suggests the presence of outliers in the original data. It also suggests that the transformation is appropriate. In any case, report the transformations you make, say why you made them, and give the results. Whenever possible, include the original data, perhaps in picture form (see Chapter 2).

EXERCISES

13.27 Exercise 13.4 describes an experiment in which two electrodes were placed on subjects and their resistances measured. The results are repeated in the accompanying table. I have added a column indicating the sign of the difference, electrode 1 minus electrode 2.

Resistance for two electrodes on 16 subjects in Exercise 13.27

| Electrode | | | Electrode | | | Electrode | | |
1	2	sign	1	2	sign	1	2	sign
500	400	+	27	84	−	15	45	−
660	600	+	100	50	+	160	200	−
250	370	−	105	180	−	250	400	−
72	140	−	90	180	−	170	310	−
135	300	−	200	290	−	66	1,000	−
						107	48	+

Assume a prior beta(1, 1) density for the population proportion of subjects for whom electrode 1 resistance is greater than that of electrode 2.

(a) Find the predictive probability that electrode 1 resistance will be greater than that of electrode 2 on the next subject.

(b) Carry out a sign test of the null hypothesis that electrode 1 resistance is greater than that of electrode 2 for half of the population.

(c) Find the posterior probability that electrode 1 resistance is greater than that of electrode 2 for more than one-third of the population. (Use the area to the right of proportion $\frac{1}{3}$ in your posterior density.)

13.28 Exercise 13.6 gives results of two experimental measurements of the amount of a blood factor present in each of 21 samples. The data are repeated here. In the third column, a positive (+) sign indicates that the larger reading was on experiment 2 and a negative (−) sign indicates that the larger reading was on experiment 1. The fourth column applies only to part (d). For parts (a), (b), and (c) assume a prior beta(2, 2) density for the population proportion of samples in which experiment 2 would give a bigger reading than experiment 1.

Results of experiments for Exercise 13.28

| Experiment | | | | Experiment | | | | Experiment | | | |
1	2	sign	ok?	1	2	sign	ok?	1	2	sign	ok?
12.77	13.16	+	y	3.98	4.12	+	y	3.90	4.15	+	y
6.54	7.73	+	y	2.56	2.92	+	y	5.36	4.69	−	y
6.03	6.79	+	y	3.19	3.83	+	y	2.27	2.49	+	y
12.60	16.85	+	y	2.24	2.79	+	y	1.57	2.25	+	y
7.69	9.43	+	y	3.11	3.23	+	y	.68	.95	+	y
5.69	5.18	−	y	16.20	17.45	+	n	2.18	1.90	−	y
1.23	1.74	+	y	5.44	6.33	+	y	8.85	8.42	−	y

(a) Find the probability that experiment 2's reading will be greater than that of experiment 1 on the next (22nd) sample.

(b) Carry out a sign test to judge whether experiment 2's reading is greater than that of experiment 1 for half of the population.

(c) Find the probability that experiment 2's reading is greater than that of experiment 1 for more than $\frac{3}{4}$ of the population.

(d) This laboratory would like to claim that these two experiments are measuring the same thing and that the two experiments agree in the population at least 90% of the time. Take "agreement" to mean that the measurements are within 1 unit of each other, regardless of which is bigger. Agreement is indicated by "y" in the "ok?" column of the table. Use a prior beta(1, 1) density for this proportion and find the posterior probability of at least 90% agreement.

13.29 In Exercises 10.5 and 11.15, you made some calculations for data involving 32 congestive heart failure patients. The numbers are increases in miles walked while on an experimental drug as compared with miles walked in the previous week while off drug. The entries below indicate the signs of the 32 differences and are given in the same order as in the previous examples:

```
0    +    +    −    +    +    −    +
−    +    +    −    +    −    +    −
+    +    +    +    +    −    −    +
−    +    +    +    −    +    0    +
```

Ignore the ties (the two 0's) and analyze only the 30 nonzeros. Assume a beta(3, 1) density for the population proportion of +'s (the drug results in longer walks).

(a) Find the predictive probability that the next patient will walk longer on drug.

(b) Carry out a sign test, deciding whether the data are consistent with the null hypothesis of no benefit for the drug based only on the table of signs.

13.30 In Exercises 11.13 and 13.1, we considered the following sample of ages in months at which 21 children spoke their first word:

15 26 10 9 15   20 18 11 8 20   7 9 10 11 11   10 12 42 17 11   10

Put these ages in order:

7 8 9 9 10   10 10 10 11 11   11 11 12 15 15   17 18 20 20 26   42

In the sample of 21 children, 16 spoke before they were 18 months old. Assume a beta(1, 1) prior density for the proportion of the population who speak before age 18 months. Find the posterior probability that more than half of the children in the population speak before 18 months.

13.31 Example 3.24 gives results for drugs formoterol (F) and salbutamol (S) in aerosol solutions compared with a placebo (P) solution for patients suffering from exercise-induced asthma. (See also Exercises 11.6–11.8, 14.4, and 14.23.) The accompanying table gives differences in responses between pairs of drugs along with columns indicating the signs of these differences. Assume a beta(1, 1) density in each case and using these signs,

(a) Find the posterior probability that formoterol will elicit a better response (i.e., bigger FEV) than salbutamol for the next patient from this population. [Ignore the tie (patient 8) and analyze only the 29 nonzeros.]

(b) Carry out a sign test for formoterol vs placebo, deciding whether the data are consistent with the null hypothesis of no benefit for formoterol.

Sec. 13.5 / Probabilities of Intervals and the Sign Test

(c) Find the posterior probability that salbutamol will elicit a better response than placebo in more than 80% of the patients in this population. [Ignore the tie (patient 12) and analyze only the 29 nonzeros.]

Differences in forced expiratory volume for drugs F, S, and P;
signs of differences in three right columns for Exercise 13.31

Patient Number	F − P	S − P	F − S	F − P	S − P	F − S
1	6	3	3	+	+	+
10	12	6	6	+	+	+
17	6	5	1	+	+	+
21	9	−1	10	+	−	+
23	12	6	6	+	+	+
2	13	6	7	+	+	+
11	12	6	6	+	+	+
14	15	−2	17	+	−	+
19	8	7	1	+	+	+
25	13	9	4	+	+	+
28	10	7	3	+	+	+
3	22	11	11	+	+	+
12	7	0	7	+	0	+
18	6	8	−2	+	+	−
24	22	21	1	+	+	+
27	11	8	3	+	+	+
4	15	11	4	+	+	+
8	8	8	0	+	+	0
16	17	7	10	+	+	+
6	3	2	1	+	+	+
9	10	11	−1	+	+	−
13	6	2	4	+	+	+
20	4	6	−2	+	+	−
26	9	7	2	+	+	+
31	11	8	3	+	+	+
5	20	10	10	+	+	+
7	5	11	−6	+	+	−
15	15	10	5	+	+	+
22	14	2	12	+	+	+
30	9	8	1	+	+	+
# of +'s/# of −'s:				30/0	27/2	25/4

Chapter Notes

1. G. W. Snedecor and W. G. Cochran, *Statistical Methods,* 6th ed. (Ames, Iowa: Iowa State University Press, 1967).

2. D. A. Berry, "Logarithmic transformations in ANOVA," *Biometrics* 43 (1987): 439–456.

3. E. J. Snell, *Applied Statistics* (London: Chapman and Hall, 1987), 153.

4. D. R. Cox and D. Oakes, *Analysis of Survival Data* (London: Chapman and Hall, 1984).

5. O. L. Davies, *Statistical Methods in Research and Production,* 2nd ed. (London: Oliver and Boyd, 1949). See also G. E. P. Box and G. C. Tiao, *Bayesian Inference in Statistical Analysis* (New York: Wiley, 1973), 208.

6. P. B. V. Vosloo, P. G. Williams, and R. G. Rademan, "Pilot and full-scale investigations on the use of combined dissolved-air flotation and filtration for water treatment," *Water Pollution Control* 85 (1986): 114–121.

14

Regression Analysis

CHAPTER 12 dealt with independent samples of a single type of measurement, such as an IQ test. This measurement is made or is otherwise available on each sample member. In this chapter, we consider a sample from a single population, but now there are two characteristics (an x-measurement and a y-measurement) available on each sample member. These characteristics may be of the same type, as in the paired samples of Chapter 11; for example, two IQ tests are given to the same individual at different times. But the characteristics may be of different types as well; for example, sample members take an IQ test and run a 100-meter dash. Our goal is to relate the two measurements.

Suppose the x-measurement takes only two values—high/low, female/male, and so on—and the y-measurement takes any number of values. So there are, in effect, two samples of y-measurements, one for each value of x. This is the setting of Chapter 12. The present chapter applies to that setting, but also to those in which there are more than two x-values—provided the x-values are numerical.

The first example considers two samples of the same measurement, but differing in time and treatment.

EXAMPLE 14.1
▷ **Treating asthma (revisited).** Example 3.24 and Exercises 11.6–11.8 and 13.31 dealt with a study comparing two drugs—formoterol (F) and salbutamol (S)—with placebo. The population consists of patients suffering from exercise-induced asthma. The measurements of interest are forced expiratory volumes (in ml) at different times. For the purposes of this example, consider only the measurements on F and on S. Also, ignore the order of administration. The results are shown in Table 14.1 (page 460), regrouped by patient number.

Figure 14.1 is a scatterplot of the data. There are 30 data points in the figure, one for each patient in the study. The four data points below the diagonal (labeled on the figure) are patients 7, 9, 18, and 20—the only ones who had a better response on S than on F.

The scatterplot suggests a relationship between FEVs (forced expiratory volumes) on the two drugs. Namely, if a patient's FEV on one drug is large, it will tend to be large on the other as well. It is precisely this relationship that makes a crossover design appealing: It minimizes the interpatient variability and makes for a more efficient treatment comparison. The relationship is not perfect for a number of reasons, including: (1) The relative benefit of the drugs may be different in different patients; and (2) these measurements were taken at different times and there are time-varying factors that affect FEV. Exercise 14.4 asks you to find a line that best fits these data.

Table 14.1
Forced expiratory volume (in ml) after F or S, in aerosol

Patient Number	F	S	Patient Number	F	S
1	35	32	16	34	24
2	31	24	17	23	22
3	32	21	18	14	16
4	26	22	19	23	22
5	29	19	20	13	15
6	25	24	21	23	13
7	20	26	22	38	26
8	28	28	23	30	24
9	32	33	24	32	31
10	34	28	25	30	26
11	28	22	26	26	24
12	23	16	27	31	28
13	14	10	28	31	28
14	31	14	30	28	27
15	27	22	31	25	22

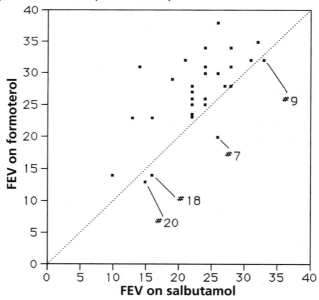

Figure 14.1 Scatterplot for Example 14.1

Various questions can be addressed regarding the previous example:

1. What will be the FEV of the next patient on salbutamol (or what is its probability distribution)? On formoterol?

2. What will be the difference between the next patient's FEV on formoterol as opposed to on salbutamol?
3. What is the mean of FEV values for patients on salbutamol in this population? On formoterol?
4. What will be the FEV of the next patient on formoterol if that patient's FEV on salbutamol is known to be 30? On salbutamol, if that patient's FEV on formoterol is known to be 30?
5. What is the mean of the population of FEV values of patients on formoterol whose FEVs on salbutamol are known to be 30? On salbutamol, if that patient's FEV on formoterol is known to be 30?

We addressed questions 1–3 in Chapters 10 and 11 using single-sample methods. Questions 4 and 5 are topics for this chapter.

To aid in addressing these questions, in the next section we will consider a purely mechanical problem: Find the line that best fits the points in a scatterplot. The solution depends on the interpretation of "best." Our interpretation is the standard one in statistical applications. The best fitting line will turn out to play an important role in addressing questions 4 and 5.

14.1 Least Squares Line

Two points determine a line. Alternatively, the slope and any one point (say, the y-intercept) determine a line. The equation of a line has this form:

$$y = a + bx$$

where x is one measurement and y is the other. As in Figure 14.2, think of x as the horizontal axis and y as the vertical axis. The y-intercept is a and the slope is b. In the figure, $b = \frac{1}{3}$ and $a = 2$. Setting $x = 0$ means $y = a + b(0) = a = 2$, the intercept.

Figure 14.2 Line $y = a + bx$ with y-intercept $a = 2$ and slope $b = \frac{1}{3}$

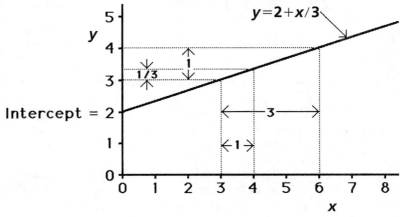

Increasing x by one unit—from 3 to 4, say—results in an increase in y of $b = \frac{1}{3}$—from 3 to $3\frac{1}{3}$. Increasing x by 3 units from 3 to 6 results in an increase of 3 times the slope ($3b = 1$) from 3 to 4.

The purpose of this section is to provide formulas for the slope and intercept of the line that best fits a set of data. I will develop these formulas in the context of the following example.

EXAMPLE 14.2

▷ **Relating log resistance on two electrodes.** Exercise 13.4 gives the resistance in kilo-ohms for two electrodes on 16 subjects. These data are repeated in Table 14.2 along with their natural logarithms. The scatterplot of the logs is also shown in Figure 14.3.

Table 14.2
Resistances and their logs for two electrodes, E1 and E2, on 16 subjects

Subject Number	Resistance		Log resistance	
	E1	E2	E1	E2
1	500	400	6.215	5.991
2	660	600	6.492	6.397
3	250	370	5.521	5.914
4	72	140	4.277	4.942
5	135	300	4.905	5.704
6	27	84	3.296	4.431
7	100	50	4.605	3.912
8	105	180	4.654	5.193
9	90	180	4.500	5.193
10	200	290	5.298	5.670
11	15	45	2.708	3.807
12	160	200	5.075	5.298
13	250	400	5.521	5.991
14	170	310	5.136	5.737
15	66	1,000	4.190	6.908
16	107	48	4.673	3.871

A way to think about these two measurements is that there is a line relating them and the observations are deviations about this line. Such a line is a hypothesis about the relationship. The dotted 45° line in the scatterplot is a candidate. The corresponding hypothesis is that the two measurements are equal, or equivalently, that their difference is 0, and that the observed deviations about the dotted line reflect randomness in the measuring process. These deviations are shown as vertical lines in the scatterplot in Figure 14.4. This figure indicates the heights of the vertical lines for subjects 1 and 15. The lengths of all the vertical lines (including signs, + for a data point above the line and − for below) are shown in Table 14.3 (page 464).

Sec. 14.1 / Least Squares Line 463

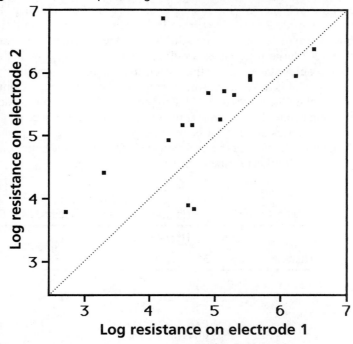

Figure 14.3 Scatterplot of logs of resistances for two electrodes

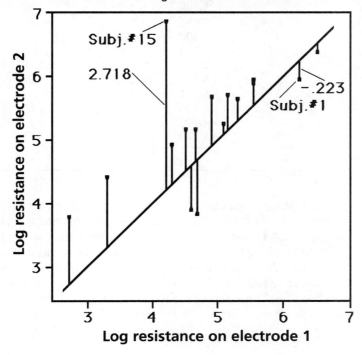

Figure 14.4 Vertical lines showing differences: Electrode 2 minus electrode 1

The differences in Table 14.3 indicate how closely electrodes 1 and 2 agree—that is, how well the 45° line fits the data. Most of the differences are positive, indicating that electrode 2 tended to give a larger reading. The mean difference in logs is .493, but we are really interested in the sizes of the differences irrespective of signs. Having two very large differences would suggest a poor fit, but if they had opposite signs they would cancel each other when averaging. A resolution would be to drop the signs—that is, take the absolute values of the differences. Conventional practice is to square the differences and add them up. Squares of numbers cannot be negative and so the sum cannot be negative. If the mean squared difference is 0, it would mean that all the differences are 0. That is, the line goes through all the points and so fits perfectly. Generally, the smaller the mean squared difference, the better the fit. The mean squared difference for the 45° line is .865 (see Table 14.3).

Table 14.3
Differences between log resistances for electrodes 1 (E1) and 2 (E2)

Subject Number	Log Resistance		Difference E2 − E1	Squared Difference
	E1	E2		
1	6.215	5.991	−.223	.050
2	6.492	6.397	−.095	.009
3	5.521	5.914	.392	.154
4	4.277	4.942	.665	.442
5	4.905	5.704	.799	.638
6	3.296	4.431	1.135	1.288
7	4.605	3.912	−.693	.480
8	4.654	5.193	.539	.291
9	4.500	5.193	.693	.480
10	5.298	5.670	.372	.138
11	2.708	3.807	1.099	1.207
12	5.075	5.298	.223	.050
13	5.521	5.991	.470	.221
14	5.136	5.737	.601	.361
15	4.190	6.908	2.718	7.388
16	4.673	3.871	−.802	.643
Mean			.493	.865

It seems possible to draw a line that fits better than the 45° line. I have drawn a candidate in the scatterplot in Figure 14.5. My method was simple: I drew a line connecting the leftmost dot (subject 11) with the rightmost dot (subject 2). This method will not usually give a good fit, but in this case it seems to do tolerably well. To decide how well, I have again drawn vertical lines representing differences between the line and the various points. These differences are called **residuals**—what is left over after the line is fit. The residuals are 0 for subjects 2 and 11, as they would be for any other points that happened to lie on the line. (*To be continued.*)

Figure 14.5 Scatterplot showing line connecting subjects 2 and 11, with residuals

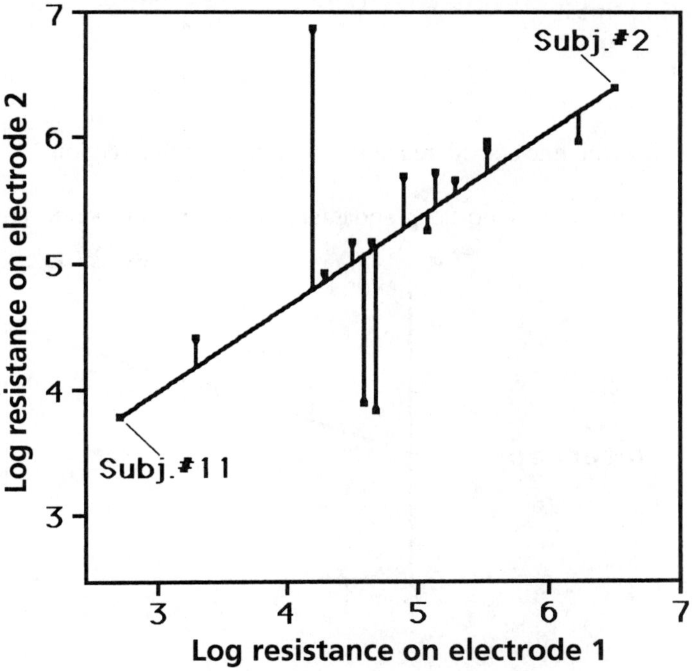

Fitting a Line and Finding Residuals

To find a residual, subtract the value of the line at each subject's value on electrode 1 (say, x) from the subject's reading on electrode 2 (say, y). To do this, we first have to find the line. Recall the equation of a line:

$$y = a + bx$$

where b is the slope and a is the y-intercept.

Figure 14.6 (page 466) shows an example of a line. The line goes through the two points $x = 3$, $y = 3$ and $x = 6$, $y = 4$. The slope is then

$$b = \frac{\text{increase in } y}{\text{increase in } x} = \frac{4 - 3}{6 - 3} = \frac{1}{3}$$

The y-intercept of a line is the value of y when $x = 0$, that is, the value of y where the line "intercepts" the y-axis: Setting $x = 0$ gives $y = a + bx = a + b(0) = a$. To find intercept a, extrapolate back to $x = 0$. Take either of the two points, say $x = 6$, $y = 4$. Multiply the slope by the x-value and subtract it from the y-value:

$$a = y - bx = 4 - (\tfrac{1}{3})6 = 4 - 2 = 2$$

This value on the y-axis is labeled in Figure 14.6. So $a = 2$, $b = \frac{1}{3}$, and the equation of the line is

$$y = 2 + \frac{x}{3}$$

This is the line considered at the beginning of this section.

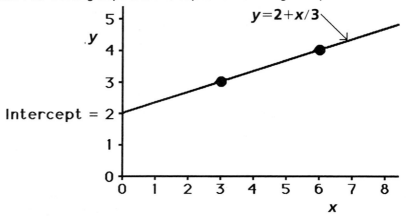

Figure 14.6 Finding slope and intercept of a line through two points

Now return to the electrode example.

EXAMPLE 14.3
▷ **Relating log resistance on two electrodes (continued).** The two points connected with a line in Example 14.2 are $x = 2.708$, $y = 3.807$ (subject 11) and $x = 6.492$, $y = 6.397$ (subject 2). So the slope of the line is

$$b = \frac{\text{increase in } y}{\text{increase in } x} = \frac{6.397 - 3.807}{6.492 - 2.708} = \frac{2.590}{3.784} = .684$$

Extrapolating from $x = 2.708$, $y = 3.807$ back to 0 gives the intercept:

$$a = 3.807 - 2.708(.6845) = 1.95$$

Therefore, the equation of the line connecting subjects 2 and 11 is

$$y = 1.95 + .684x$$

To evaluate how well this line fits the data, calculate the values of the line for the x values (log resistance on electrode 1) of each subject. The values on the line are called "fitted y" in Table 14.4. For example, for subject 1, the fitted value for electrode 2 is $1.953 + .6845(6.215) = 6.207$, and so the observed value for electrode 2 for this subject (5.991) is below the line.

Sec. 14.1 / Least Squares Line

Table 14.4
Fitted log resistances for electrode 2 (y) from electrode 1 (x), where $y = 1.95 + .684x$

Subject Number	Log Resistance		Fitted y	Residual	Squared Residual
	x	y			
1	6.215	5.991	6.207	−.216	.047
2	6.492	6.397	6.397	.000	.000
3	5.521	5.914	5.733	.181	.033
4	4.277	4.942	4.881	.061	.004
5	4.905	5.704	5.311	.393	.154
6	3.296	4.431	4.209	.221	.049
7	4.605	3.912	5.106	−1.194	1.425
8	4.654	5.193	5.139	.054	.003
9	4.500	5.193	5.033	.160	.025
10	5.298	5.670	5.580	.090	.008
11	2.708	3.807	3.807	.000	.000
12	5.075	5.298	5.427	−.129	.017
13	5.521	5.991	5.733	.259	.067
14	5.136	5.737	5.469	.268	.072
15	4.190	6.908	4.821	2.087	4.354
16	4.673	3.871	5.152	−1.281	1.640
Mean				.060	.494

As in the previous example, the residuals are the heights of the vertical lines in Figure 14.5, with the sign of the residual indicating whether the fitted value is above (+) or below (−) the observed value. The mean of the residuals is .060. Making the mean residual small is easy. Indeed, the horizontal line $y = 5.310$—the mean of the y-values—would make the mean residual equal to 0 because the positive residuals will exactly cancel the negative residuals. But a horizontal line fits the data poorly. Again, we will square the residuals—the last column in the table. The mean of the squares is .494, which compares with .865 for the 45° line of the previous example, that is, the line $y = x$. So this line fits better in the sense that its average of squares is smaller. (*To be continued.*) ◁

Equation of Least Squares Line

A line that makes the mean of the squared residuals as small as possible (or equivalently, the *sum* of the squares as small as possible) is called the **least squares line.** You could come close to finding it by trying various lines (that is, varying a and b) until you are satisfied that you are close enough to the best. However, a little algebra[1] provides easy formulas for the least squares values of a and b. I will call them A and B and spare you the details. However, I need some notation to give you the formulas. In earlier chapters, $\bar{x}$ indicates the mean of a sample of values and s indicates the sample standard deviation. In many settings of Chapter 11, we had two sets of observations, but we needed only a single notation because we subtracted one from the other and so reduced to a single sample. In Chapter 12, we had two $\bar{x}$'s and used subscripts derived from the population names. I will use a variant of this notation in the present chapter.

For the x-values, I will use $\bar{x}$ and s_x to stand for the sample mean and standard deviation. For the y-values, I will use $\bar{y}$ and s_y. Also, I need a notation for the average product of the x's and y's; I will use $\overline{xy}$ where the bar carries over both symbols. This notation is suggestive but it can be confusing—it does *not* mean the same thing as the product of $\bar{x}$ and $\bar{y}$, which I will write as $\bar{x}\bar{y}$. To calculate $\overline{xy}$, first multiply x times y and then average. To calculate $\bar{x}\bar{y}$, first average and then multiply.

> **The least squares line fits points in a scatterplot as closely as possible, in the sense that it minimizes the mean square of residuals.**

> **The least squares line has this slope and intercept:**
> $$B = \frac{\overline{xy} - \bar{x}\bar{y}}{s_x^2} \quad \text{and} \quad A = \bar{y} - B\bar{x}$$

The formula for B is complicated, but it is easy to remember the formula for the intercept, A. The formula for the line is $y = a + bx$. Rewrite it as $a = y - bx$. Putting bars on x and y and capitalizing a and b gives the formula for A. This relationship between the two formulas means that the least squares line goes through the point $x = \bar{x}$, $y = \bar{y}$.

To illustrate the calculations in the simplest of cases ($n = 2$), consider the points $x = 3$, $y = 3$ and $x = 6$, $y = 4$ from Figure 14.5. Our calculations must give $B = \frac{1}{3}$ and $A = 2$, for we know that the mean squared residual for the line through the two points is 0—no average of squares can be smaller than that and so this must be the least squares line.

Calculations for finding least squares line

x	y	x^2	xy
3	3	9	9
6	4	36	24
Average $\frac{9}{2}$	$\frac{7}{2}$	$\frac{45}{2}$	$\frac{33}{2}$

So

$$\bar{x} = \frac{9}{2} \quad \bar{y} = \frac{7}{2} \quad s_x^2 = \frac{45}{2} - \left(\frac{9}{2}\right)^2 = \frac{9}{4} \quad \overline{xy} = \frac{33}{2}$$

$$B = \frac{33/2 - (9/2)(7/2)}{9/4} = \frac{3/4}{9/4} = \frac{1}{3}$$

$$A = \frac{7}{2} - \left(\frac{1}{3}\right)\left(\frac{9}{2}\right) = 2$$

Now return to the electrode example.

EXAMPLE 14.4

▷ **Relating log resistance on two electrodes (revisited).** To find A and B, proceed as before, but now there are more calculations, as shown in Table 14.5. Using the results from this table:

$$\bar{x} = 4.817 \qquad \bar{y} = 5.310 \qquad s_x^2 = 24.06 - 4.817^2 = .8625 \qquad \overline{xy} = 26.09$$

$$B = \frac{26.09 - (4.817)(5.310)}{.8625} = .6003$$

$$A = 5.310 - (.6003)(4.817) = 2.419$$

(Remember that I carry more decimal places than I show. For example, if you use only the four-decimal accuracy given for the means, you will get $B = .5933$ rather than the more accurate .6003.) The least squares line, $y = 2.42 + .600x$, is shown in Figure 14.7 (page 470). Its slope is less than that of the previous line we considered ($y = 1.95 + .684x$).

Table 14.5
Calculating least squares line for electrodes experiment

Subject	x	y	x²	xy
1	6.215	5.991	38.62	37.23
2	6.492	6.397	42.15	41.53
3	5.521	5.914	30.49	32.65
4	4.277	4.942	18.29	21.13
5	4.905	5.704	24.06	27.98
6	3.296	4.431	10.86	14.60
7	4.605	3.912	21.21	18.02
8	4.654	5.193	21.66	24.17
9	4.500	5.193	20.25	23.37
10	5.298	5.670	28.07	30.04
11	2.708	3.807	7.33	10.31
12	5.075	5.298	25.76	26.89
13	5.521	5.991	30.49	33.08
14	5.136	5.737	26.38	29.46
15	4.190	6.908	17.55	28.94
16	4.673	3.871	21.84	18.09
Average	4.817	5.310	24.06	26.09

To sketch a line on a scatterplot or on any other graph, calculate the value of the line (fitted value) at two points and connect the dots. For example, the fitted value of y at $x = 3$ is $2.42 + .600(3) = 4.22$ and the value of y at $x = 6$ is $2.42 + .600(6) = 6.02$. These two points are shown as open circles on the line in the scatterplot.

I claimed that the least squares line minimizes the mean square of the residuals. In particular, it has to be smaller than the mean square of residuals (.494) for the line $y = 1.95 + .684x$. The calculations in Table 14.6 show that it does: .484 is smaller than .494, but not by much. Only a small amount of improvement is possible because the line connecting subjects 2 and 11 happens to fit the data quite well.

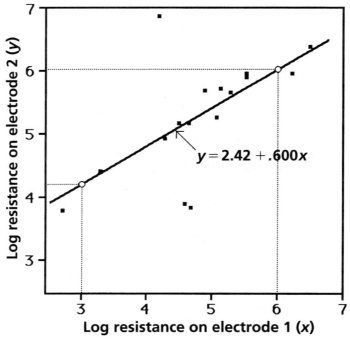

Figure 14.7 Least squares line in the electrode example

Table 14.6
Least squares line ($y = 2.42 + .600x$) for log resistances

Subject Number	Log Resistance		Least Squares		
	x	y	Fitted y	Residual	Squared Residual
1	6.215	5.991	6.150	−.158	.025
2	6.492	6.397	6.316	.081	.007
3	5.521	5.914	5.734	.180	.032
4	4.277	4.942	4.986	−.045	.002
5	4.905	5.704	5.364	.340	.116
6	3.296	4.431	4.397	.033	.001
7	4.605	3.912	5.183	−1.271	1.617
8	4.654	5.193	5.213	−.020	.000
9	4.500	5.193	5.120	.073	.005
10	5.298	5.670	5.600	.070	.005
11	2.708	3.807	4.045	−.238	.057
12	5.075	5.298	5.466	−.167	.028
13	5.521	5.991	5.734	.258	.067
14	5.136	5.737	5.502	.235	.055
15	4.190	6.908	4.934	1.974	3.896
16	4.673	3.871	5.224	−1.353	1.830
Mean				.000	.484

So far I have considered two measurements of the same type. However, least squares methods are commonly used for fitting lines to different types of measurements. The goal is to relate the two measurements. The next example is an illustration.

EXAMPLE 14.5
▷ **Chemical reaction versus temperature.** An experiment[2] was designed to relate the rate, K, of a chemical reaction to temperature. Assuming the Arrhenius law means that $\log K$ is linearly related to the (standardized) inverse of temperature, called x. Seven different temperatures were used and several replicate observations were made at each. The data are given here, along with the scatterplot in Figure 14.8.

x	$y = \log K$			
−3	−3.50	−3.40	−3.36	−3.50
−2	−3.51	−3.62	−3.53	
−1	−3.86	−3.70		
0	−3.99	−4.18		
1	−4.02	−4.23		
2	−4.44	−4.62	−4.34	
3	−4.74	−4.49	−4.61	−4.62

Figure 14.8 Scatterplot and least squares line for Example 14.5

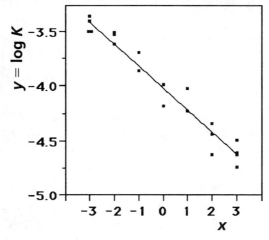

Calculating the slope and intercept of the least squares line (shown on the scatterplot) proceeds just as for the previous example—see Table 14.7 (page 472; I will not use the last column of this table until the next example). The fact that the mean of the x values is 0 makes the calculations rather easy:

$$\bar{x} = 0 \quad \bar{y} = -4.013 \quad s_x^2 = 5 - 0^2 = 5 \quad \overline{xy} = -1.0135$$

$$B = \frac{-1.014 - (0)(-4.013)}{5} = -.203$$

$$A = -4.013 - (.203)(0) = -4.013$$

Table 14.7
Calculating least squares line of reaction times for Example 14.5

Obs.	x	y	x^2	xy	y^2
1	−3	−3.50	9	10.50	12.25
2	−3	−3.40	9	10.20	11.56
3	−3	−3.36	9	10.08	11.29
4	−3	−3.50	9	10.50	12.25
5	−2	−3.51	4	7.02	12.32
6	−2	−3.62	4	7.24	13.10
7	−2	−3.53	4	7.06	12.46
8	−1	−3.86	1	3.86	14.90
9	−1	−3.70	1	3.70	13.69
10	0	−3.99	0	.00	15.92
11	0	−4.18	0	.00	17.47
12	1	−4.02	1	−4.02	16.16
13	1	−4.23	1	−4.23	17.89
14	2	−4.44	4	−8.88	19.71
15	2	−4.62	4	−9.24	21.34
16	2	−4.34	4	−8.68	18.84
17	3	−4.74	9	−14.22	22.47
18	3	−4.49	9	−13.47	20.16
19	3	−4.61	9	−13.83	21.25
20	3	−4.62	9	−13.86	21.34
Average	0	−4.013	5	−1.014	16.32

The least squares line is

$$y = -4.013 - .203x$$

In this example and generally, the slope of the least squares line is negative when there is an apparent inverse relationship between the measurements: A large value of one means a small value of the other. ◁

Correlation Coefficient

There is a slightly more convenient and intuitive way to write the slope of the least squares line. You know that slope is the "increase in y divided by the increase in x." Your experience with calculating z-scores means that you are used to measuring distances for variable quantities in terms of standard deviations. So if we write increases in terms of standard deviations, the slope becomes a proportion of ratios of standard deviations:

$$B = \frac{\text{increase in } y}{\text{increase in } x} = r\frac{s_y}{s_x}$$

The proportionality constant, r, is called the **correlation coefficient** between x and y. It measures how close the data points come to lying on a line. If all the data points lie on a line, then r is 1 or −1; otherwise, it is between these two extremes. In general, if

r is positive, then the least squares line has positive slope and if r is negative, then the least squares line has negative slope.

> **The correlation coefficient of measurements x and y is**
>
> $$r = \frac{\overline{xy} - \bar{x}\bar{y}}{s_x s_y}$$

> **Alternative Formulas for Least Squares Slope and Intercept**
>
> $$B = r\frac{s_y}{s_x} \quad \text{and} \quad A = \bar{y} - B\bar{x}$$

As an algebraic check, substitute the formula for r into this last formula for B; the result is the earlier formula for B.

EXAMPLE 14.6
▷ **Chemical reaction versus temperature (revisited).** Reconsider Example 14.5. In addition to the calculations from that example, finding the correlation coefficient requires the standard deviation of the y values, which is the purpose of the last column in Table 14.7 (page 472):

$$s_y^2 = 16.32 - (-4.013)^2 = .215$$
$$s_y = \sqrt{.215} = .464$$

Also,

$$s_x = \sqrt{5} = 2.24$$
$$\overline{xy} - \bar{x}\bar{y} = -1.014 - (0)(-4.013) = -1.014$$
$$r = \frac{-1.014}{(2.24)(.464)} = -.977$$

That r is so close to -1 indicates that the data points lie close to the least squares line.

Using the second formula for slope:

$$B = r\frac{s_y}{s_x} = -.977\,\frac{.464}{2.24} = -.203$$

as before. Of course, the intercept is the same as before: $A = -4.013 - (.203)(0) = -4.013$. ◁

The scatterplots in Figure 14.9 (page 474) are designed to build your intuition about correlation coefficients. Each shows a least squares line and gives the correlation coefficient. All plots show real data from industrial settings. The correlation coefficient does not depend on the number of data points. It does depend on how well the data points can be fit by a line.

Figure 14.9 Examples of least squares lines and correlation coefficients

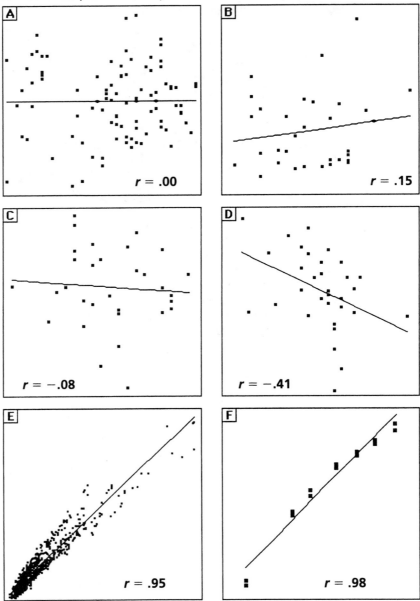

I have omitted the names of the variables and the scales because the correlation coefficient is unitless: It does not depend on the scales. The reason the scales are immaterial is that the units of both x and y appear in both numerator and denominator of r and so cancel. I have repeated plot F in Figure 14.10, but with the x-scale stretched out. In the stretched-out version, all the x values are multiplied by 2 and the y values are unchanged. The factor 2 appears in both numerator and denominator of the formula for r and so r is unchanged by the stretching—again it equals .98.

This section is unlike those in previous chapters. In particular, we have not calculated posterior probabilities. The next section will remedy that, with least squares calculations playing a central role.

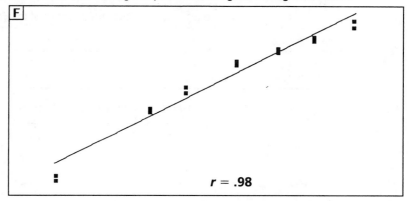

Figure 14.10 Scale change of plot F from Figure 14.9 gives same r

EXERCISES

Use graph paper or computer printouts for scatterplots; make your own graph paper if you do not have any.

14.1 A study[3] compared temperature (temp) of the air with the chirping frequency of the striped ground cricket. These are the observations:

Chirps/sec	20	16	20	18	17	16	15	17	15	16
Temp (°F)	89	72	93	84	81	75	70	82	69	83

(a) Find the correlation coefficient between $x =$ chirps/sec and $y =$ temp (°F).
(b) Find the least squares line relating y and x.
(c) Make a scatterplot of the data and sketch in the least squares line.

14.2 A chemistry student[4] made measurements and theoretical calculations of pH under six different conditions. Her results were as follows:

Condition	Calculated pH	Measured pH
Buffer solution	4.56	4.74
Buffer solution + .10 M HCl	4.12	4.37
Buffer solution + .10 M NaOH	4.94	5.11
Distilled water	5.50	7.00
Distilled water + .10 M HCl	1.74	1.54
Distilled water + .10 M NaOH	12.40	12.46

(a) Find the correlation coefficient between $x =$ Calculated pH and $y =$ Measured pH.
(b) Find the least squares line relating y and x.
(c) Make a scatterplot of the data and sketch in the least squares line.

14.3 Researchers captured eight squirrels at different times during the spring of 1994 in Minnesota, and assessed the osmolarity (solutes) of the squirrels' urine.[5] One question of interest was whether osmolarity depends on available moisture—it might be expected to decrease if the squirrel drinks more. Total rainfall over the previous 3 days and osmolarity are shown in the table and scatterplot that follow. Find the least squares line.

Osmolarity for eight squirrels and recent rainfall for Exercise 14.3

Squirrel Number	1	2	3	4	5	6	7	8
Rainfall (in.)	.45	0	0	.24	.12	0	.35	.44
Osmolarity (mg/l)	1,479	1,500	516	1,815	1,118	1,702	1,348	1,337

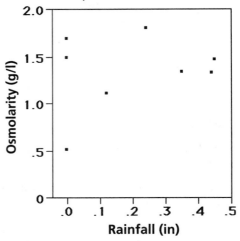

Scatterplot for Exercise 14.3

14.4 The study described in Examples 3.24 and 14.1 and Exercise 11.6 compared the forced expiratory volume (FEV) for the drugs formoterol (F) and salbutamol (S) in aerosol solutions with a placebo solution in 30 patients suffering from exercise-induced asthma. The following table repeats the data for F and S. The figure at the top of page 477 is the scatterplot ($x =$ FEV on S and $y =$ FEV on F) from Example 14.1 with the least squares line drawn in. (The dotted line on the figure is $y = x$.)

Data from crossover study of Example 3.24 for Exercise 14.4:
Forced expiratory volume (ml) for F and S in aerosol

F	S	F	S	F	S	F	S	F	S	F	S
35	32	31	24	31	28	31	28	32	33	29	19
34	28	28	22	32	21	26	22	14	10	20	26
23	22	31	14	23	16	28	28	13	15	27	22
23	13	23	22	14	16	34	24	26	24	38	26
30	24	30	26	32	31	25	24	25	22	28	27

Sec. 14.1 / Least Squares Line   477

Scatterplot and least squares line for Exercise 14.4

(a) Find the correlation coefficient between x and y.
(b) Find the equation of the least squares line.

14.5 Example 2.10 considered numbers of wins at home and away for the 1992 NBA teams. These are repeated in the table at the top of page 478. The scatterplot is also repeated from that example. In the example, I suggested that the number of away wins was about 60% of the number of home wins. Find the least squares estimate of number of away wins based on number of home wins. (The line I drew in the scatterplot went through the origin. The intercept of the least squares line is not exactly 0, but it is close.)

Scatterplot of home vs away wins for 1992 NBA season for Exercise 14.5

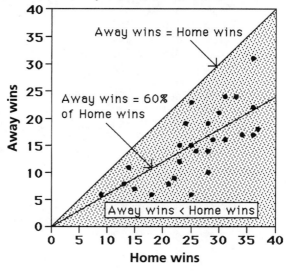

Numbers of home and away wins for 1992 NBA season for Exercise 14.5

Home	Away	Home	Away	Home	Away	Home	Away	Home	Away
23	15	18	6	24	19	30	21	31	16
34	17	25	23	28	10	13	8	28	19
22	9	31	24	25	6	23	12	21	8
36	31	28	14	9	6	36	17	37	18
36	22	26	14	25	15	33	24	14	11
15	7	29	16						

14.6 The following table duplicates that of Exercise 2.10 showing numbers of home and away wins for 1992 major league baseball teams (reading down the columns and then left to right gives the same order as in Exercise 2.10). Find the least squares line relating these numbers of wins, where x is home and y is away.

Home	Away	Home	Away	Home	Away	Home	Away	Home	Away
51	47	53	37	37	26	41	35	45	37
43	46	41	35	53	39	51	45	42	30
44	29	38	37	48	42	41	29	38	26
41	31	47	34	43	44	53	43	36	41
43	35	44	28	41	31	45	38	53	43
50	36								

14.7 Example 2.11 considers baseball winning percentage (y) and payroll (x) for the years 1988 and 1992. These are repeated here. Find the correlation coefficient between payroll and winning percentage and also the least squares line for:

(a) 1988
(b) 1992

1988				1992			
Pay*	Win%	Pay*	Win%	Pay*	Win%	Pay*	Win%
118	.338	125	.562	330	.605	274	.556
135	.335	97	.500	210	.549	159	.537
139	.546	153	.625	422	.451	445	.444
119	.463	194	.528	335	.444	345	.469
131	.475	101	.642	294	.481	397	.593
62	.441	138	.424	256	.531	238	.432
89	.540	60	.531	352	.556	326	.593
79	.461	129	.469	81	.469	266	.512
129	.543	93	.516	256	.463	274	.506
123	.506	124	.512	134	.500	325	.444
141	.532	73	.402	318	.444	222	.395
169	.584	54	.435	438	.389	282	.475
85	.537	121	.537	303	.568	427	.593

*Annual payroll in $100,000s

14.8 Exercise 2.9 considers two measures of retention of cannulae for 16 hearts, as listed at the top of page 479. Letting x = meas1 and y = meas2, find the least squares line and graph it on the scatterplot you made in Exercise 2.9.

meas1	meas2	meas1	meas2	meas1	meas2	meas1	meas2
.04	.06	.25	.21	.40	.10	.47	.58
.14	.00	.30	.47	.40	.39	.52	.67
.19	.22	.32	.56	.43	.60	.59	.45
.21	.17	.32	1.04	.44	1.17	.95	1.34

14.9 On a field trip[6] a scientist measured background radiation at 39 locations in the southwestern United States and related them to altitude. The following table gives Geiger counter counts per minute (Rad) for the various elevations (El) in thousands of feet above sea level. Find the least squares line for $x =$ El and $y =$ Rad and graph it on a scatterplot of the data.

Relating radiative background to elevation in southwestern U.S. for Exercise 14.9

El	Rad	El	Rad	El	Rad	El	Rad	El	Rad	El	Rad
.7	11.0	2.4	13.6	3.8	15.0	5.1	16.8	5.4	19.3	7.4	26.8
1.4	12.7	2.9	12.5	4.0	13.8	5.1	17.3	6.2	22.0	8.3	22.7
1.6	11.6	2.9	13.9	4.1	15.8	5.1	17.9	6.4	19.8	10.6	24.9
1.7	12.8	3.6	15.7	4.3	12.4	5.1	18.4	6.9	27.8	10.6	26.0
1.9	11.8	3.6	17.8	4.3	13.9	5.1	19.6	6.9	29.0		
2.1	11.5	3.7	17.2	4.8	18.0	5.1	20.1	7.1	20.4		
2.3	11.1	3.8	14.4	5.0	14.9	5.4	16.6	7.4	25.0		

14.2 Relating Two Measurements: Regression

This chapter considers relationships between two measurements that can be represented by a line. Such relationships are called *linear* and their models are called **regression models**. As usual, we deal with settings in which there is variability in measurement. Observations in different circumstances but with the same x may give different values of y. Perhaps the true line goes through the first value of y and perhaps it goes through the second—more likely, it goes through neither. Despite there being strength in numbers, any finite sample gives less than perfect information about the line. Since anything uncertain has a probability distribution, the line itself has a probability distribution. Lines are specified by slopes and intercepts and so probabilities for lines are specified by probabilities for b and a. A major focus of this section is on the slope, b, because it indicates the degree of the relationship between x and y: Increasing x by 1 unit means increasing y by b units. In particular, $b = 0$ means there is no relationship between x and y, at least not a linear relationship.

EXAMPLE 14.7
▷ **Chemical reaction versus temperature (revisited).** Examples 14.5 and 14.6 dealt with the relationship between the inverse (x) of temperature and the rate (y) of a chemical reaction. This is the least squares line:

$$y = -4.013 - .203x$$

This is only one of the many possibilities for the true relationship between x and y. The scatterplot in Figure 14.11 shows some others. All the indicated lines are possible, but some seem more plausible than others. We would like to associate probabilities with these lines in the usual way. Associating probabilities with a line means assigning prior probabilities to two unknowns—the line's slope and its intercept.

Figure 14.11 Scatterplot and possible lines for Example 14.7

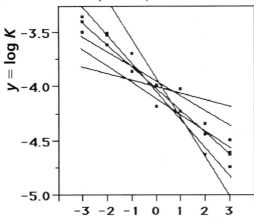

As always, assigning probabilities to models requires updating one's state of knowledge using Bayes' rule, which in turn requires finding likelihoods based on the data. Finding likelihoods means calculating the probability of each observation (that is, each point in the scatterplot) depending on the bowl model assumed. Models are now lines, or slope/intercept pairs. We will assume that observed differences from the true line—sometimes called *errors*—have a normal density. This density varies with x and we will assume that its mean is the value of the line at the corresponding value

Figure 14.12 Tilted scatterplot for Example 14.7

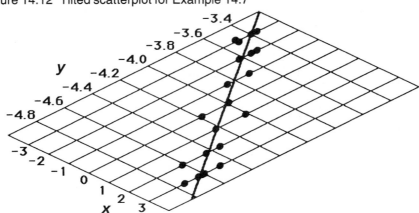

of x. The shape of the error density—as determined by its standard deviation—is unknown, but it is assumed to be the same for every value of x. To show a picture of error densities requires a third dimension. To facilitate showing the third dimension, I have tilted the scatterplot in Figure 14.12 (page 480), as well as added a grid. In addition, I have added a line to serve as a candidate model. This happens to be the least squares line, but it could be any line. Its function in the figure is merely to help in showing error densities and in explaining the calculation of the likelihood of a model (that is, of a line).

Figure 14.13 (page 482) shows two normal error densities on the tilted scatterplot—one each for $x = -1$ and $x = 2$. These are centered on the y-values of the line for the given values of x and both have standard deviation .2. Imagine such a density centered at the y-value of the line and oriented in the y-direction for each value of x in the sample, that is, for $x = -3, -2, -1, 0, 1, 2,$ and 3.

To calculate the likelihood of the model, proceed as in Chapter 10 and find the heights of the densities for all 20 data points at the corresponding values of x. The greatest height occurs for points on the line and the smallest height occurs for points farthest from the line—which, in this case, is the point $x = 2, y = -4.62$. Multiplying all 20 heights gives the likelihood for the line, together with the likelihood of the assumed error standard deviation. To calculate posterior probabilities, multiply likelihoods by the corresponding prior probabilities. As usual, for models with similar prior probabilities, bigger likelihoods mean bigger posterior probabilities.

A thorough analysis would be to evaluate likelihoods and assess prior probabilities for $a, b,$ and the error standard deviation in any particular model. Such an analysis is beyond our scope. Instead, in this chapter I will take the prior densities of a and b to be flat. Also, I will separate out inferences concerning the error standard deviation and use the sample data points to estimate this standard deviation in the manner of Chapter 11. If there is substantial prior information concerning a or b and the error standard deviation, then this could be used to adjust the estimate derived below, but I will not develop this possibility in this text.

While I will not apply Bayes' rule to the uncertainty about the error standard deviation, I will describe how its value affects the likelihoods. Imagine decreasing the error standard deviation of the normal densities from .2 in Figure 14.13(a). This would have the effect of squeezing the densities and raising the contribution to the likelihood for any data point that happens to lie near the line, but it decreases the likelihoods for points far from the line. This is shown for error standard deviation of .1 in Figure 14.13(b).

Squeezing changes the contributions to the likelihood by the individual data points—some increase and others decrease. The overall effect on the product of the 20 heights could be in either direction. However, as the standard deviation gets sufficiently small and the density highly peaked, the individual heights will start to decrease for any points not on the line. Therefore, their product will also start to decrease; eventually, the likelihood will become small and stay small. Now imagine increasing the standard deviation by spreading out the normal densities. The heights for the data points lying near the line will be decreased, but those far from the line will be increased. The product of all 20 heights may increase or decrease, but for large enough standard deviation the product will again start to decrease. The overall likelihood will

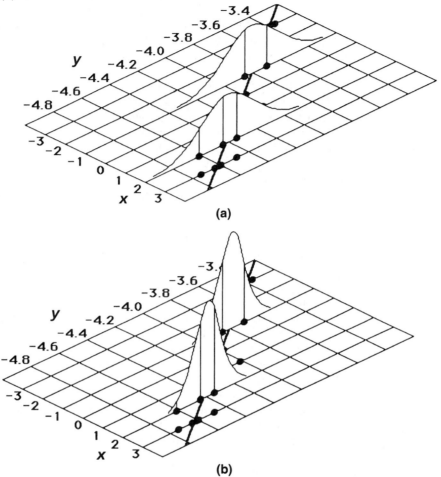

Figure 14.13 Scatterplot for Example 14.7: Normal densities are centered at the y-value of the line for two values of x (-1 and 2). (a) Error standard deviation is .2 (b) Error standard deviation is .1

reach a maximum somewhere between very large and very small error standard deviations. In this example, it turns out to be about .1 (see next calculation)—the one used in Figure 14.13(b).

In Chapter 11, we resolved the problem of having an unknown population standard deviation by estimating it from the sample. We based the estimate on the sample standard deviation s, multiplying s by a factor to account for small sample sizes. Now the error deviation is in the y-direction. But now we cannot use the sample standard deviation of the y-values (which I have called s_y) because the mean of the error densities is the value of y that lies on the line—that is, $a + bx$—and not $\bar{y}$. So instead of comparing each observation of y with the overall mean of the y values, we compare it with the least squares estimate of y evaluated at the corresponding x. These differences or residuals from the least squares line are estimated errors, or "deviations" from the true line; we seek a typical or standard deviation.

Table 14.8 shows the calculations. The first three columns duplicate the first three columns of Table 14.7 in Example 14.5. The fourth column gives the value of the least squares line at the corresponding value of x. The fifth column shows the residual for that observation—that is, it equals the third column value minus that in the fourth. The residual is what is left over when comparing the observed value of y with the least squares value of y. The last column is the squared residual. The sum of the last column divided by $n = 20$ is the mean squared residual: .0098. The square root of this is called the root mean squared error, or RMSE = .099.

Some algebra (which I will skip) gives a formula for the RMSE in terms of s_y, the sample standard deviation of the y values, and r, the correlation coefficient between x and y:

$$s_y\sqrt{1 - r^2} = .464\sqrt{1 - .977^2} = .099$$

as before. Use the small-n correction factor, $k = (1 + 20/n^2)$, to give the following estimate for the error standard deviation:

$$h = ks_y\sqrt{1 - r^2} = \left(1 + \frac{20}{n^2}\right)s_y\sqrt{1 - r^2} = (1.05)(.099) = .104$$

◁

Table 14.8
Calculating the mean of the squares of residuals for Example 14.7

Obs.	x	y	−4.013 − .203x	Residual	Residual²
1	−3	−3.50	−3.404	−.096	.0092
2	−3	−3.40	−3.404	+.004	.0000
3	−3	−3.36	−3.404	+.044	.0019
4	−3	−3.50	−3.404	−.096	.0092
5	−2	−3.51	−3.607	+.097	.0094
6	−2	−3.62	−3.607	−.013	.0002
7	−2	−3.53	−3.607	+.077	.0059
8	−1	−3.86	−3.810	−.050	.0025
9	−1	−3.70	−3.810	+.110	.0121
10	0	−3.99	−4.013	+.023	.0005
11	0	−4.18	−4.013	−.167	.0279
12	1	−4.02	−4.216	+.196	.0384
13	1	−4.23	−4.216	−.014	.0002
14	2	−4.44	−4.419	−.021	.0004
15	2	−4.62	−4.419	−.201	.0404
16	2	−4.34	−4.419	+.079	.0062
17	3	−4.74	−4.622	−.118	.0139
18	3	−4.49	−4.622	+.132	.0174
19	3	−4.61	−4.622	+.012	.0001
20	3	−4.62	−4.622	+.002	.0000
Average	0	−4.013	−4.013	.000	.0098

> **The estimated error standard deviation is the root mean squared error:**
> $$s_y\sqrt{1-r^2}$$
> **Adjusting for small samples:**
> $$h = s_y\sqrt{1-r^2}\left(1 + \frac{20}{n^2}\right)$$

Probabilities for a line are probabilities for an intercept a and a slope b. The intercept is the value of the line at $x = 0$. I will eventually discuss making inferences about intercepts and other y-values, but the first order of business is the slope. The slope is the change in y when the value of x increases by 1. So it is, in effect, a difference between two population means—one for $x = 0$, say, and the other for $x = 1$. This was the focus of our attention in Chapter 12. Just as in that chapter—see the rule of differences in Section 12.1—the posterior distribution of b is normal. Except for this appeal to similarity with Chapter 12, I will give little justification for the following results.

> **Updating Rule for Regression Models**
>
> For n observations, y, from normal$(a + bx, h)$ densities, where the values of x may differ for different observations, if the prior densities of both a and b are flat, then the posterior density of b is approximately normal$\left(B, \dfrac{h}{\sqrt{n}\,s_x}\right)$.

To calculate the z-score for any particular value of b, proceed as usual by first subtracting the mean of b, which is the least squares estimate $B = rs_y/s_x$, and then dividing by its standard deviation. An interesting value of b is 0, because it corresponds to no relationship. The z-score for $b = 0$ is

$$z = \frac{0 - B}{h/(\sqrt{n}\,s_x)}, \quad \text{where } h = s_y\sqrt{1-r^2}\left(1 + \frac{20}{n^2}\right)$$

To find a posterior probability interval for b, proceed just as in Sections 11.5 and 12.2, but with mean and standard deviation as given above. The following table is repeated from Section 11.5:

68% probability interval:	$z_{68} = 1.00$
80% probability interval:	$z_{80} = 1.28$
90% probability interval:	$z_{90} = 1.65$
95% probability interval:	$z_{95} = 1.96$
98% probability interval:	$z_{98} = 2.33$
99% probability interval:	$z_{99} = 2.58$

> A perc% probability interval for slope b is
>
> $$r\frac{s_y}{s_x} \pm z_{\text{perc}} \frac{h}{s_x\sqrt{n}}$$
>
> where z_{perc} is given in the preceding table.

The value $b = 0$ has a special interpretation. It means that the slope of the line is 0 and so there is no (linear) relationship between the two measurements—a null hypothesis. As in several earlier chapters, we cannot test null hypotheses by finding probabilities of particular values of b without a computer. As before, use the following alternative for testing $b = 0$:

> **Alternative Test of Null Hypothesis $b = 0$**
>
> Decide whether the 95% posterior probability interval for b contains 0. If yes, then the null hypothesis is supported; otherwise, it is not.

> **Equivalently, calculate**
>
> $$z = \frac{rs_y/s_x}{h/(s_x\sqrt{n})}$$
>
> If $-1.96 < z < 1.96$, then the null hypothesis is supported; otherwise, it is not.

Consider predicting the next observation y for a particular value of x. Perhaps x can be set by the experimenter or perhaps x precedes y in time. The predictive density again turns out to be normal. Its mean is just the least squares value of y: $A + Bx$. Its standard deviation is more complicated. The sources of variability are the uncertainty in the slope b and intercept a and the normal error that is present in every observation of y. The standard deviation of the normal error is estimated to be h, and so the predictive standard deviation cannot be less than h. The variability in y is smallest when x is in the middle of the x-values: $x = \bar{x}$. At this value, $A + Bx = A + B\bar{x} = \bar{y}$. At $x = \bar{x}$, the standard deviation of the next observation, y, is $h\sqrt{(n+1)/n}$, just as in Chapter 11 (except that h now has a slightly different interpretation).

If x differs from $\bar{x}$, then there is extra variability. Especially important is the uncertainty in the line's slope. A small change in the slope does not have much of an effect on y near the center of the data, but it results in a large change in y away from the center—see the possible lines in the scatterplot in Figure 14.11 on page 480. (Extrapolating to a value of x outside the range of the x-values in the data is even more problematic: The variability is greater because of the uncertainty in the slope, but prediction is also more sensitive to the assumption that the relationship is linear.) The general formula for the predictive standard deviation is as follows:

$$h\sqrt{\frac{n+1}{n} + \frac{(x-\bar{x})^2}{ns_x^2}} \quad \text{where } h = s_y\sqrt{1-r^2}\left(1 + \frac{20}{n^2}\right)$$

In the case of very large n, the predictive standard deviation reduces to h. Even if the line were perfectly known, the next observation is subject to normal error with standard deviation estimated as h.

> **Prediction for Regression Models**
>
> For n observations, y, from normal$(a+bx, h)$ densities for different values of x, if the prior densities of both a and b are flat, then the predictive density of the next observation y at a fixed x is approximately
>
> $$\text{normal}\left(A + Bx,\ h\sqrt{\frac{n+1}{n} + \frac{(x-\bar{x})^2}{ns_x^2}}\right)$$
>
> where A, B, and h are as defined previously.

EXAMPLE 14.8

▷ **Chemical reaction versus temperature (revisited).** Examples 14.5–14.7 dealt with the relationship between the inverse (x) of temperature and the rate (y) of a chemical reaction. These are values from the earlier examples:

$$n = 20 \quad s_x = \sqrt{5} = 2.24 \quad s_y = .464$$
$$A = -4.013 \quad B = -.203 \quad h = .104$$

The density of b is approximately normal$\left(B, \dfrac{h}{s_x\sqrt{n}}\right)$, or normal$(-.203, .0104)$. A 95% posterior probability interval for b is

$$B \pm z_{95}\frac{h}{s_x\sqrt{n}} = -.203 \pm 1.96(.0104)$$
$$= -.203 \pm .020, \text{ or from } -.223 \text{ to } -.183$$

This tight interval reflects very strong evidence that the slope is near $-.2$. Setting the temperature to increase x by 1 unit has the effect of slowing the rate of reaction by very close to .2, with high probability. Obviously, $b = 0$ is far from this interval and so the null hypothesis is not supported.

Suppose the next x is 0, the mean of the x values. The least squares line equals -4.013 at $x = 0$. (The two previous observations at $x = 0$ are -3.99 and -4.18, the mean being -4.085. The least squares estimate of -4.013 borrows strength from the other experimental results. It is a better estimate, provided the relationship between x and y is indeed linear, as we have been assuming.) A 95% probability interval for the next y at $x = 0$ is $-4.013 \pm 1.96h\sqrt{(n+1)/n}$, which is about $-4.01 \pm .21$, or from -4.22 to -3.80.

Suppose, instead, that the next x is 5, which is outside the range of the x values in the sample. The least squares estimate is $-4.013 - .203(5) = -5.03$. The predictive standard deviation at $x = 5$ is

$$h\sqrt{\frac{n+1}{n} + \frac{(x - \bar{x})^2}{ns_x^2}} = .104\sqrt{\frac{21}{20} + \frac{(5 - 0)^2}{20(5)}} = .104\sqrt{1.30} = .119$$

So a 95% probability interval for the next y at $x = 5$ is $-5.027 \pm 1.96(.119)$ or about $-5.03 \pm .23$, or from -5.26 to -4.80. This is somewhat wider than the corresponding interval at $x = 0$ because of the increased uncertainty in estimating y values for x values that are away from the center of the data. ◁

EXAMPLE 14.9

▷ **Relating log resistance on two electrodes (revisited).** Examples 14.2–14.4 considered logarithms of resistances on two electrodes. These are the relevant calculations, most of which are taken from the earlier examples:

$\bar{x} = 4.82$ $\bar{y} = 5.31$ $s_x = .929$ $s_y = .891$ $r = .625$
$h = .696(1 + 20/16^2) = .750$ $A = 2.42$ $B = .600$

A 95% posterior probability interval for b is $.600 \pm 1.96(.750)/(.929\sqrt{16}) = .600 \pm .396$, or from .204 to .996. This interval does not contain 0 and so the null hypothesis is not supported.

Predicting y at $x = 5$: The mean of the approximating normal density is $2.42 + .600(5) = 5.42$ and the standard deviation is

$$.750\sqrt{\frac{17}{16} + \frac{(5 - 4.82)^2}{.863(16)}} = .774$$

So a 95% probability interval for predicting y at $x = 5$ is $5.42 \pm 1.96(.774) = 5.42 \pm 1.52$, or from 3.90 to 6.94.

Predicting y at $x = 7$: The mean is 6.62 and the standard deviation is

$$.750\sqrt{\frac{17}{16} + \frac{(7 - 4.82)^2}{.863(16)}} = .890$$

A 95% probability interval for predicting y at $x = 7$ is $6.62 \pm 1.96(.890) = 6.62 \pm 1.74$, or from 4.88 to 8.36. ◁

The intervals in Example 14.9 seem wider than the corresponding intervals in Example 14.8. The reason is that the observations are more tightly packed about the least squares line in the earlier example, suggesting a much smaller error standard deviation and correspondingly greater accuracy in estimation and prediction.

EXERCISES

In these exercises, assume the prior probabilities considered in this section.

14.10 In Exercise 14.2, you found the least squares line for the following data:

Condition	x = Calculated pH	y = Measured pH
Buffer solution	4.56	4.74
Buffer solution + .10 M HCl	4.12	4.37
Buffer solution + .10 M NaOH	4.94	5.11
Distilled water	5.50	7.00
Distilled water + .10 M HCl	1.74	1.54
Distilled water + .10 M NaOH	12.40	12.46

A new condition has a calculated pH of 8.00. Find a 99% probability interval for the measured pH at that condition.

14.11 In Exercise 14.3, you found the least squares line relating osmolarity of squirrel urine to previous rainfall (table repeated here).

Squirrel Number	1	2	3	4	5	6	7	8
Rainfall (in.)	.45	0	0	.24	.12	0	.35	.44
Osmolarity (mg/l)	1,479	1,500	516	1,815	1,118	1,702	1,348	1,337

(a) Find a 95% posterior probability interval for the slope of the true line.

(b) Test the null hypothesis of no (linear) relationship between osmolarity and previous rainfall.

14.12 In Exercise 14.6, you found the least squares line relating the number of away wins to home wins for 1992 major league baseball teams (table duplicated here).

Home	Away	Home	Away	Home	Away	Home	Away	Home	Away
51	47	53	37	37	26	41	35	45	37
43	46	41	35	53	39	51	45	42	30
44	29	38	37	48	42	41	29	38	26
41	31	47	34	43	44	53	43	36	41
43	35	44	28	41	31	45	38	53	43
50	36								

(a) Test the null hypothesis of no (linear) relationship between away wins and home wins.

(b) Find the posterior probability that there is in fact a positive (linear) relationship between home and away wins (that is, a team that is good at home is also good away).

(c) Suppose that next year's teams are exchangeable with the teams in 1992. What is the predictive probability that a team wins 100 or more games in all if that team wins 55 home games?

14.13 A study[7] of reaction rate of a synthetase of a bovine lens suggests that this rate may be linearly related to the reciprocal (x) of the substrate concentration. Two observations were taken at each value of x, as given in the table here, for 14 observations in all. The scatterplot (with its least squares line) showing the rate vs x indicates that the relationship may be nonlinear. (The corre-

Sec. 14.2 / Relating Two Measurements: Regression   489

lation coefficient of .91 is high, but the rates at $x = 12, 16$, and 20 are well below the line and the rate at $x = 24$ is well above it.) Instead of relating rate with x, relate $y = $ log rate with x.

Reaction rate and $x =$ reciprocal of substrate concentration for Exercise 14.13

x	rate	log rate	x	rate	log rate
24	.429	−.85	24	.444	−.81
20	.293	−1.23	20	.293	−1.23
16	.251	−1.38	16	.268	−1.32
12	.207	−1.58	12	.216	−1.53
8	.239	−1.43	8	.218	−1.52
6	.180	−1.71	6	.199	−1.61
2	.156	−1.86	2	.167	−1.79

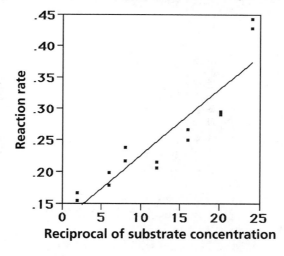

Scatterplot for Exercise 14.13 showing nonlinearity

(a) Find the least squares line for $y = $ log rate vs x.
(b) Find the 99% posterior probability interval for the slope of the true line.
(c) Test the hypothesis that the slope $b = 0$.

14.14 I do not know how the experiment described in Exercise 14.13 was conducted. However, the closeness of the reaction rates at the same x as opposed to the lack of fit of the data to a smooth curve—especially the anomaly at $x = 8$—suggests that the two rates having the same x were paired in the experiment and, perhaps, the experimental unit was the same. For example, these may have been replicate measurements of reaction rate on the same lens. If so, the errors within pairs are correlated and there are two separate sources of error in the experiment. This requires a more advanced treatment than is possible in this book. But since the within-pair error seems small, a reasonable approximation is to average the two rates at each x and carry out a regression analysis assuming a single y-value (the average of the two) at each x-value, for seven observations altogether. The revised data are given in the table at the top of page 490. The scatterplot shows log rate vs reciprocal of substrate concentration. Repeat the previous exercise.

Reaction rate
and x = reciprocal of
substrate concentration
for Exercise 14.14

x	Average rate	Log rate
24	.437	−.83
20	.293	−1.23
16	.259	−1.35
12	.211	−1.55
8	.228	−1.48
6	.190	−1.66
2	.162	−1.82

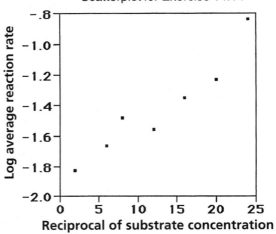

Scatterplot for Exercise 14.14

(a) Find the least squares line for y = log rate vs x.

(b) Find the 99% posterior probability interval for the slope of the true line. [Your answer to part (b) should be wider than in the previous exercise. This is appropriate since now there are only half as many data points.]

(c) Test the hypothesis that the slope $b = 0$.

14.15 Nine light-water reactor power plants were built in the United States (outside of the Northeast region) between 1967 and 1971. The following table gives power plant construction cost (millions of dollars), date of construction permit (for example, 69.33 means April 1969), and power plant net capacity.[8]

Plant	1	2	3	4	5	6	7	8	9
Cost	452.99	443.22	412.18	289.66	567.79	621.45	473.64	697.14	288.48
Date	67.33	69.33	68.42	68.42	68.75	69.67	70.42	71.8	67.17
Capacity	1,065	1,065	530	530	913	786	538	1,130	821

(a) How fast are costs going up? Answer by giving a 95% probability interval for the slope of cost y as it depends on date x (in years).

(b) Are costs related to net capacity? Answer by testing the null hypothesis that the slope is 0 for the line relating cost y to capacity x.

14.16 Exercise 2.14 gives data for an experiment to assess the percentage of dry matter and the amount of ascorbic acid in five varieties of lima beans. The scatterplot you made in Exercise 2.14 shows a bit of curvature. Taking the log of ascorbic acid shows a more nearly linear relationship. The following table gives the percentage of dry matter (DM) and the natural logarithm of ascorbic acid (AA). Combine all 25 observations into a single analysis for this exercise.

Variety 1		Variety 2		Variety 3		Variety 4		Variety 5	
DM	log AA	DM	log AA	DM	log AA	DM	log AA	DM	log AA
34.0	4.53	39.6	3.86	31.7	4.40	34.5	4.12	31.4	4.39
33.4	4.55	39.8	3.94	30.1	4.69	31.5	4.42	30.5	4.67
34.7	4.52	51.2	3.51	33.8	4.27	31.1	4.54	34.6	4.34
38.9	4.39	52.0	3.30	39.6	4.05	36.1	4.23	30.9	4.52
36.1	4.38	56.2	3.03	47.8	3.40	38.5	3.85	36.8	4.22

(a) Find the least squares line of y = DM on x = log AA.
(b) Find a 95% posterior probability interval for slope b.
(c) Test the hypothesis that $b = 0$.
(d) The next observation has ascorbic acid = 95 (the log is 4.55). Find a 90% predictive probability interval for the corresponding percentage of dry matter.

14.17 An experiment[9] used 13 blasts to assess the damage to residential structures depending on blast frequency (in cycles per second). The table shown here gives the frequencies used and the corresponding displacement of structures in inches. It also gives the natural logarithms of both quantities. The scatterplot of the unlogged quantities [page 492; figure (a)] shows nonlinearity. (The relationship cannot be linear over the whole range of x-values because the displacement cannot be negative.) Taking the log of displacement [figure (b)] improves linearity, but not as much as taking logs of both quantities [figure (c)].

Obs. No.	Frequency	Displacement	Log Frequency	Log Displacement
1	2.5	.390	.916	−.942
2	2.5	.250	.916	−1.386
3	2.5	.200	.916	−1.609
4	3.0	.360	1.099	−1.022
5	3.5	.250	1.253	−1.386
6	9.3	.077	2.230	−2.564
7	9.9	.140	2.293	−1.966
8	11.0	.180	2.398	−1.715
9	11.0	.093	2.398	−2.375
10	11.0	.080	2.398	−2.526
11	16.0	.052	2.773	−2.957
12	25.0	.051	3.219	−2.976
13	25.0	.150	3.219	−1.897

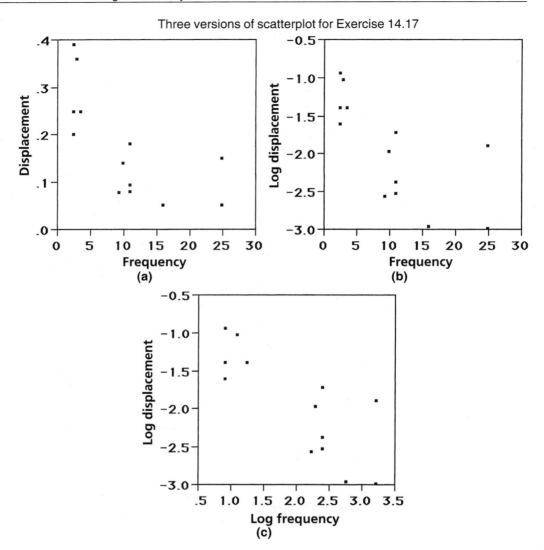

Three versions of scatterplot for Exercise 14.17

(a) Find the least squares line for $x =$ log frequency and $y =$ log displacement.

(b) According to your answer to part (a), what is the estimated change in displacement when frequency is doubled? (Doubling the frequency means increasing the log frequency $\log(2) = .693$. Applying part (a) will tell you the change in log displacement—use the antilog to find the corresponding change in displacement. Check your answer by referring to the scatterplots.)

(c) Consider a house built to withstand a displacement of .2 inch, but no more. What is the predictive probability that the house withstands a blast with a frequency of 10 cps? [*Note:* $\log(.2) = -1.609$ and $\log(10) = 2.303$.]

14.18 University of Minnesota English professor Donald Ross counted the total number of words and the number of different words in 156 literary texts. The results are shown in the following table and scatterplot. The scatterplot shows the least squares line as well.

For 156 texts, total number of words and number of different words for Exercise 14.18

Number words	Number different	Number words	Number different	Number words	Number different	Number words	Number different
129	82	330	207	415	237	489	211
145	76	334	203	416	247	489	219
173	111	336	158	419	233	491	249
175	112	337	205	421	238	494	322
180	97	338	177	425	241	499	268
182	102	343	169	430	217	501	196
198	125	351	185	430	237	504	260
202	116	354	185	432	236	505	262
205	90	354	196	434	233	505	313
209	120	356	199	434	241	507	280
217	139	362	184	438	211	508	239
222	140	362	190	438	248	511	238
225	152	363	188	439	250	516	286
226	115	363	204	443	252	525	208
232	108	364	176	445	271	531	235
241	120	364	209	447	255	540	305
241	128	365	210	449	240	543	302
243	131	365	221	449	261	547	272
247	139	366	210	452	245	552	241
251	112	370	221	454	238	555	212
255	153	373	205	455	270	560	243
274	152	378	202	456	253	560	330
276	143	380	216	458	232	564	277
278	188	381	227	459	242	566	287
286	176	382	222	460	223	575	240
290	175	384	241	460	243	575	256
295	173	386	231	460	247	577	257
297	174	387	212	462	223	592	214
300	195	387	241	463	241	592	327
304	145	400	225	463	340	600	346
306	190	400	225	465	244	633	348
311	174	401	213	470	220	651	347
317	181	402	192	470	280	655	304
318	168	402	212	472	273	707	406
318	184	402	227	473	240	720	337
319	173	406	211	476	223	902	413
326	200	408	233	478	275		
327	164	409	192	479	293		
328	198	413	200	482	226		
329	157	413	226	486	251		

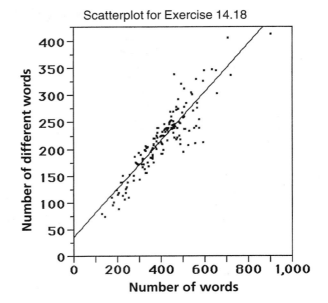

Scatterplot for Exercise 14.18

(a) Find the equation of the least squares line.
(b) Predict the number of different words in a new text of length 500 words from this population.
(c) Find a 95% predictive probability interval for your prediction in part (b).
[There are two aspects of the scatterplot that draw into question the assumptions we have made. First, any text with one word has one "different" word. So the true line surely goes through the point with number of words = number of different words = 1. The fact that the least squares line does not is a reflection of the lack of linearity of the true relationship between these two quantities. Indeed, for number of words less than 200, all the points in the scatterplot are below the least squares line. These points are consistent with a line through the origin. The least squares line is too high for a text having a small number of words because it is pulled down on the right by texts with a large number of words—a teeter-totter with the fulcrum near $x = 400$ and pulled down on the right will rise on the left. The second assumption that is questionable is that the standard deviation of the errors is the same for all values of x. The scatterplot suggests that the standard deviation is greater for texts with a larger number of words—especially those greater than 450. Both of these reservations are addressed in more advanced treatments of regression analysis.]

14.3 Regression Effect

The regression effect arises in every aspect of life and yet few people appreciate its existence. As a result, some people mistakenly attribute change to an ineffective intervention. A golfer has a terrible round and then takes a lesson—she does better the next time and praises her teacher. A baseball player is in a slump and changes bats—the next day he gets two hits and credits the bat. A woman with a backache goes to a specialist and her backache improves—she thinks the specialist is a genius. A man is feeling depressed and goes to a psychiatrist—a week later he is feeling better and signs up for more sessions.

Perhaps the golf instructor is good. The bat may be an improvement. Perhaps the back specialist is truly adept. The psychiatrist may be effective. But another explanation is possible and even likely in all these cases: the regression effect, also called regression to the mean. It is simple and ubiquitous: There are ups and downs in any sequence of exchangeable observations; drops tend to follow the ups and rises tend to follow the downs. (Exchangeability is not required for this effect—see Example 14.11 and Exercise 14.24—but it is the easiest case to understand.)

The first recorded observation of the regression effect was by Francis Galton. He found that seeds of mother sweetpea plants that happen to be larger than average tend to have daughters that are larger than average—but smaller than themselves. Symmetrically, mother seeds smaller than average tend to have daughters that are smaller than average—but larger than themselves. The comparison from one generation to the next is a reversion or regression to the mean. Similarly, Galton noticed that sons of tall men tend to be tall—but not as tall as their fathers, while sons of short men tend to be short—but not as short as their fathers. (Imagine what would happen to the population if the opposite were true and tall men tended to have sons taller than themselves and short men tended to have sons shorter than themselves!)

Sports statistics that show the regression effect are readily available. I used one newspaper for the data in the next example and also for that in Exercise 14.22.

EXAMPLE 14.10

▷ **PGA golf scores.** Consider the first 2 days' scores of the 1994 Professional Golf Association Championship at Tulsa, Oklahoma (August 11–12, 1994). These are shown in Table 14.9 (page 496) and the scatterplots of Figure 14.14. There is a positive correlation between the 2 days. [The correlation coefficient of .22 is driven largely by the player who scored 87 and 88—upper-right corner of Figure 14.14(a) scatterplot. This player's total score was 17 strokes higher than the second highest score. Without this player, the correlation coefficient drops to .12.] The correlation is positive because the player is the same on both days—the data are paired: A lower score on one day suggests that the player is better and, therefore, scores low on the other day as well. But not *as* low!

To see this, consider the differences between days. The third column of the table shows day 2 minus day 1. (The mean of this column is -1.2; apparently, the greens were watered between the two days and so were softer. It is easier to putt on a soft green and this accounts for the lower scores on day 2.) The companion scatterplot [Figure 14.14(b)] shows day 2 minus day 1 on the vertical scale vs day 1 on the horizontal scale. If a player's day 1 score is low, this difference tends to be positive. (While I have not shown the relevant scatterplot, symmetrically: If a player's day 2 score is low, this difference tends to be negative.) That these are negatively correlated is an example of the regression effect ($r = -.57$; without the highest scoring player, $r = -.61$). At this level of competition, the players have comparable abilities (except for the player with the highest score who was either ill or out of his league), and the overall mean of 72.5 is a better estimate of a player's second day's score than is his own first

Table 14.9
Two days' golf scores of 151 players (ordered by D1) for Example 14.10

D1	D2	D1 − D2	D1	D2	D1 − D2	D1	D2	D1 − D2	D1	D2	D1 − D2
67	65	−2	72	68	−4	74	67	−7	77	66	−11
67	76	9	72	68	−4	74	69	−5	77	67	−10
68	71	3	72	70	−2	74	69	−5	77	69	−8
68	71	3	72	70	−2	74	69	−5	77	72	−5
68	72	4	72	71	−1	74	70	−4	77	73	−4
68	74	6	72	71	−1	74	71	−3	77	74	−3
69	71	2	72	71	−1	74	71	−3	77	74	−3
69	72	3	72	71	−1	74	72	−2	77	74	−3
69	72	3	72	72	0	74	73	−1	77	75	−2
69	73	4	72	73	1	74	73	−1	77	76	−1
69	73	4	72	73	1	74	74	0	77	77	0
69	73	4	72	74	2	74	75	1	77	80	3
69	75	6	72	74	2	74	76	2	77	81	4
69	76	7	72	75	3	74	77	3	78	67	−11
70	67	−3	72	75	3	74	77	3	78	70	−8
70	67	−3	72	75	3	75	68	−7	78	73	−5
70	69	−1	72	77	5	75	68	−7	78	74	−4
70	69	−1	72	78	6	75	69	−6	78	75	−3
70	70	0	72	78	6	75	69	−6	78	76	−2
70	71	1	72	78	6	75	70	−5	78	76	−2
70	72	2	73	67	−6	75	71	−4	78	79	1
70	72	2	73	68	−5	75	71	−4	79	71	−8
70	73	3	73	70	−3	75	73	−2	79	73	−6
70	74	4	73	70	−3	75	73	−2	79	74	−5
70	78	8	73	71	−2	75	74	−1	79	74	−5
70	79	9	73	71	−2	75	74	−1	79	79	0
71	66	−5	73	71	−2	75	75	0	81	69	−12
71	67	−4	73	71	−2	75	76	1	81	74	−7
71	68	−3	73	71	−2	75	78	3	82	76	−6
71	69	−2	73	72	−1	75	79	4	83	74	−9
71	69	−2	73	72	−1	76	69	−7	87	88	1
71	69	−2	73	73	0	76	69	−7			
71	71	0	73	73	0	76	69	−7			
71	73	2	73	73	0	76	70	−6			
71	76	5	73	74	1	76	71	−5			
71	76	5	73	74	1	76	73	−3			
71	79	8	73	74	1	76	74	−2			
71	79	8	73	77	4	76	76	0			
72	66	−6	73	77	4	76	77	1			
72	67	−5	74	64	−10	77	65	−12			

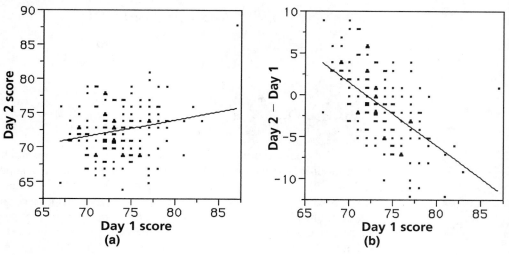

Figure 14.14 Scatterplots with least squares lines of first 2 days' scores in 1994 PGA Championship for Example 14.10

day's score. The mean of squared differences is about 14 in the first case and 21 in the second. This conflicts with most people's intuitions, including intuitions of professional golfers.

Of course, a combination of the two is better than either extreme. The least squares line shown in the scatterplot of Figure 14.14(a) is

$$\text{Day2} = 54.4 + .2454 \times \text{Day1}$$

The minimum mean of squared deviations for this best-fitting line is 13.5—not very much better than estimating 72.5 for everyone. This equation indicates that we would expect someone whose day 1 score was 67 to score 70.9 on day 2 and someone whose day 1 score was 87 to score 75.8 on day 2—barring prior information about a particular player's abilities. In this comparison, a range of 20 strokes on day 1 reduces to a predicted range of 5 on day 2. That does not mean there is less variability on day 2. The two standard deviations are essentially the same (3.42 on day 1 vs 3.74 on day 2). Moreover, exchanging the roles of days 1 and 2 would give the same result, with 20 strokes on day 2 reducing to about 5 strokes on day 1. The regression effect indicates a tendency of the extremes to come together, but there is still variability in the extremes and in the middle of the distribution as well.

Nick Price, who is listed first in the table (and also first in the table for Exercise 14.21), won this tournament by six strokes over the 4 days. There was some evidence in 1994 that he was the best player in the world. He was the only player who scored under 70 on day 1 to buck the regression effect and score even lower on day 2. Some people who score 67 are 70-plus golfers who get a little lucky. For Price at this stage of his career, shooting 67 was more typical than lucky. ◁

> **Regression Effect (Regression to the Mean)**
>
> For two correlated measurements, x and y, of the same type, when x is larger or smaller than its mean, y tends to be less extreme than x and therefore closer to its own mean.

> To show the regression effect, plot $y - x$ vs x and show that they are negatively correlated.

In the previous example, both x and y are on the same scale and have roughly the same variability. In that case, the formula for the slope of the least squares line reduces from $B = rs_y/s_x$ to r. We know that r can never be larger than 1 (or smaller than -1). So, except when $r = +1$ or -1 (in which case, the linear fit is perfect), the least squares line shrinks or "regresses" the x-value by the factor r. This same phenomenon occurs and is easy to see on scatterplots quite generally, but the effect is only easy to see from the formula $B = rs_y/s_x$ when the two standard deviations cancel. This shrinkage gives rise to the "regression effect" and also to the name of the analysis of linear relationships that is the subject of this chapter. [A way to see the regression effect when measurements x and y have different units or when their standard deviations are different is to divide each measurement by its standard deviation and plot $(y/s_y) - (x/s_x)$ vs (x/s_x).]

Astute analysts (such as *Baseball Abstracts* author Bill James) recognize the regression effect in baseball and call it the law of competitive balance. Teams who do very well one year tend to do worse in the next—better than most of the league but worse than their own previous high. The same is true for players. For example, Rookies of the Year tend to do worse in their sophomore years. Correspondingly, teams or players who do poorly in one year tend to do better in the next, although usually worse than average. As I suggested in the previous example, to do very well in a particular year requires a combination of skill and luck. Most teams that do well have both. In the following year, the skill may still be there (barring extensive personnel changes), but the same degree of luck is apt not to be. So, in the following year, the team will tend to do well (the skill part), but not *as* well (the luck part).

A quick and dirty method of predicting a player's batting average next year is to average his last year's batting average with last year's average for the entire league. This formula suggests that skill and luck contribute in equal measure. While they do not contribute equally, the formula works quite well. It certainly outdoes using the player's last year's batting average to predict next year's.

The next example is one that you have dealt with before. It is an example in which observations within pairs are not exchangeable because of a treatment effect. Also, it shows that the regression effect occurs in settings other than sports.

EXAMPLE 14.11
▷ **Playing around makes rats brainy (revisited).** Exercise 11.24 compared cortex weights of rats that lived in a communal setting with an ample supply of toys (treat-

Sec. 14.3 / Regression Effect

ment) with those of littermates that lived in isolation and without toys (control). Treatment vs control cortex weight is shown in the scatterplot of Figure 14.15(a), in which the symbol plotted is the experiment number. The correlation coefficient is .48. The standard deviations are comparable in the two groups and so the slope is about .5, which is less than 1 and shows the regression effect.

Table 14.10
Cortex weights (mg) of control and % increase for treatment for Example 14.11

				Experiment					
1		2		3		4		5	
C	%	C	%	C	%	C	%	C	%
657	4.9	669	5.7	668	3.3	662	5.7	641	−.2
623	5.3	650	13.8	667	5.1	705	1.8	589	11.2
652	2.5	651	14.4	647	5.9	656	3.5	603	3.5
654	.9	627	4.0	693	8.4	652	13.8	642	6.2
658	3.2	656	−1.1	635	1.9	578	26.0	612	12.3
646	2.6	642	5.3	644	.5	678	−.1	603	8.3
600	10.7	698	.1	665	8.3	670	3.9	593	10.1
640	1.1	648	7.4	689	4.2	647	9.9	672	−1.8
605	14.7	676	5.3	642	11.8	632	6.0	612	9.2
635	−.3	657	7.8	673	3.4	661	−1.5	678	.1
642	1.7	692	8.2	675	−2.5	670	6.1	593	7.6
		621	11.1	641	6.1	694	2.3	602	7.8

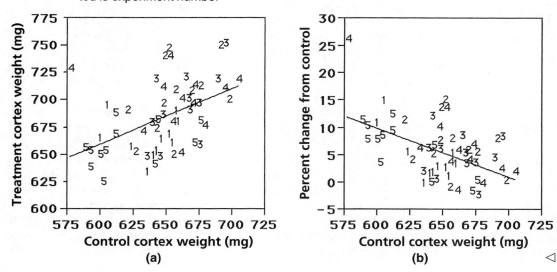

Figure 14.15 Scatterplots showing regression effect for Example 14.11; symbol plotted is experiment number

The earlier exercise focused on the percentage increase in cortex weight of treatment over control. This measure is shown for each pair (along with the control weight)

in Table 14.10 and in the scatterplot of Figure 14.15(b). Again, the symbol plotted in the scatterplot is the experiment number. Clearly, the control values depend on the experiment, but the regression effect shows through the effect of the experiment. The negative correlation evinced in this scatterplot ($r = -.52$) is an instance of the regression effect. [There is not very much variability in the control weights (compared with zero) and so the scatterplot of percentage increase vs control looks very much like the scatterplot of weight increase (in mg) vs control.] This effect does not change the estimated increase in cortex weight of 5.72% for the treatment group over control. But it does mean that the estimated increase tends to be less for a control cortex that happened to be large and it tends to be greater than 5.72% for a control cortex that happened to be small.

Daniel Kahneman was teaching a course in the psychology of training to air force flight instructors at Hebrew University in the 1960s.[10] He cited animal studies, some with pigeons, showing that rewards are more effective than punishment. One of the students objected: "I have often praised people warmly for beautifully executed maneuvers, and the next time they almost always do worse. And I have screamed at people for badly executed maneuvers, and by and large the next time they improve. Do not tell me that reward works and punishment does not. My experience contradicts it." Other students agreed.

Kahneman said, "I suddenly realized that this was an example of the statistical principle of regression to the mean, and that nobody else had ever seen this before. I think this was one of the most exciting moments of my career . . . Once you become sensitized to it, you see regression everywhere." For example, great movies have disappointing sequels and disastrous presidents have better successors.

One place the regression effect shows up is in clinical trials. To be admitted to a clinical trial, a patient has to be sick. For example, in a trial evaluating the effects of an antihypertensive drug, only those with high blood pressure are included. Blood pressure measurements vary within each individual. Just as in baseball, where a result is a combination of skill and luck, a diagnosis of high blood pressure is a combination of real hypertension and blood pressure that just happened to be high when it was measured. The real hypertension persists, but the "just happened to be high" component does not. So patients' blood pressures tend to decrease—even if the drug has no effect at all!

To see this, reconsider Example 14.10 in terms of a clinical trial. Suppose only those golfers with "high golf pressure"—say 75 strokes or greater on day 1—are continued in the "trial." These are shown in Figure 14.16, with the excluded golfers shown shaded. There are 56 players with a score of 75 or more on day 1; their mean was 77.1. Their scores dropped on the second day to a mean of 73.2. (The former mean is shown by the pointer on the vertical axis in the figure and the latter is shown by the pointer on the horizontal axis.) The difference of 3.9 strokes might be attributed to a "drug" taken between the two days, but most of it is the regression effect. Watering the greens, as mentioned in Example 14.10, plays the role of a "golf pressure lowering drug" and accounts for 1.2 strokes, but the remaining 2.7 is the regression effect.

Regardless of the entry criteria, an ineffective treatment will "work" best on patients having the highest blood pressures among those in the study. More generally, the illusion of the regression effect makes some treatments seem to work for the most

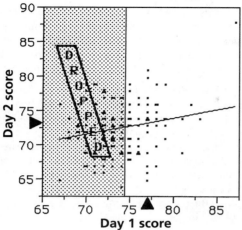

Figure 14.16 Restricting inclusion to day 1 scores of 75 and over artificially lowers day 2 scores; shaded scores are dropped; pointers show means of those not dropped

seriously ill patients (except for terminal diseases, such as many types of cancer and AIDS). These are patients who by virtue of their status have the most room for improvement. Such observations have duped unwary researchers into thinking that a worthless treatment is effective for certain types of patients.

There are protections against the regression effect. One is to separate criteria for admission and analysis. For example, a patient's blood pressure can be taken several days after being admitted to a trial and this will serve as a baseline for comparison with pressures taken after administering treatment. This baseline is used even if it indicates that the patient does not have high blood pressure. A better resolution is to include a randomized control group. This group would also show the regression effect and would allow for separating out this effect from any real treatment benefit. [However, in studies with placebo controls and restricted entry criteria, some investigators and others have marveled at dramatic regression-effected improvements in the control group and have credited them to a placebo effect (see Example 3.1) or simply to "the benefits of being in a study"!]

The regression effect applies to exchangeable observations and those having some type of central tendency. But it does not apply to those in which there is drift, at least not without adjusting for the drift. For example, the fact that a country's population has just had a big increase does not mean that it is likely to drop. Stock market prices (measured by the Dow Jones Industrial Average, say) reaching an all-time high today do not mean a drop is likely tomorrow. However, daily increases in the Dow index are approximately exchangeable (at least over a 1- or 2-month period); so if today's increase is relatively large, tomorrow's will probably be smaller.

EXERCISES

14.19 Since Roger Maris's major league baseball record of 61 home runs in a 162-game season set in 1961, several players have been on a pace to break Maris's record. That is, if they continued to hit home runs at their early season rate, they would have more than 61 by the season's end. Why is it not surprising that these would-be record holders have come up short?

14.20 In baseball, teams in first place at the "All-Star Break"—roughly halfway through the season—usually have a worse record in the second half of the year. Similarly, teams in last place at this break tend to have a better record in the second half of the year. Explain this phenomenon. If you were to compare second halves of seasons of teams in first place at the break with those in last place at the break, which records would tend to be better and why?

14.21 The scores on days 3 and 4 of the 1994 PGA Championship (cf. Example 14.10) should also show the regression effect. These scores are shown in the following table. Only these 76 golfers made the "cut"—after the second day, only those having a total score of 145 or less on the first 2 days continued to compete. Make a scatterplot of the day 3 scores and the difference, day 4 minus day 3. Calculate the correlation coefficient between these two and argue that the regression effect holds. (The cut has no impact on the regression effect for days 3 and 4. It serves to define an even more homogeneous set of golfers. However, the cut means that the scores on the last 2 days will tend to be greater than they were on the first 2 days *for these same golfers*. This too shows the regression effect since these golfers were selected precisely because their first 2 days' scores were low. Making the cut has an element of luck as well as of skill; the skill stays but the luck evens out.)

Third and fourth rounds of 1994 PGA Championship (ordered by position at finish) for Exercise 14.21

D3	D4	D4 − D3	D3	D4	D4 − D3	D3	D4	D4 − D3	D3	D4	D4 − D3
70	67	−3	71	68	−3	75	69	−6	72	75	3
69	69	0	72	70	−2	71	70	−1	72	73	1
67	70	3	68	70	2	68	74	6	73	75	2
71	66	−5	68	70	2	72	70	−2	73	72	−1
67	70	3	68	72	4	74	71	−3	74	72	−2
69	70	1	67	72	5	73	72	−1	73	74	1
66	69	3	69	73	4	74	75	1	74	74	0
70	70	0	69	75	6	70	75	5	72	74	2
69	70	1	66	75	9	72	76	4	73	76	3
67	71	4	67	73	6	70	76	6	71	77	6
67	71	4	71	70	−1	72	73	1	70	77	7
70	72	2	73	66	−7	71	73	2	76	73	−3
73	66	−7	70	71	1	74	73	−1	76	76	0
68	75	7	69	71	2	78	68	−10	73	80	7
71	70	−1	69	71	2	69	77	8	78	78	0
67	70	3	70	72	2	75	71	−4	74	79	5
70	71	1	72	71	−1	72	72	0			
70	72	2	75	72	−3	72	72	0			
70	73	3	68	74	6	75	73	−2			
74	68	−6	73	74	1	70	74	4			

14.22 The 1994 First of America Classic was held in Ada, Michigan concurrently with the 1994 PGA Championship. The scores for the first and second nine holes of the tournament (on August 12, 1994) are shown in the following table. Make a scatterplot of the first nine-hole scores and the difference, second nine minus first nine. Calculate the correlation coefficient between these two and argue that the regression effect holds for comparing first and second nine-hole scores.

First and second nine-hole scores (ordered by first nine) for Exercise 14.22

1st	2nd	Diff.	1st	2nd	Diff.	1st	2nd	Diff.	1st	2nd	Diff.
32	34	2	36	33	−3	37	35	−2	38	36	−2
33	33	0	36	34	−2	37	35	−2	38	37	−1
33	33	0	36	34	−2	37	35	−2	38	37	−1
33	34	1	36	34	−2	37	36	−1	38	38	0
33	34	1	36	34	−2	37	36	−1	38	42	4
33	35	2	36	35	−1	37	36	−1	39	33	−6
33	39	6	36	35	−1	37	36	−1	39	37	−2
34	32	−2	36	35	−1	37	36	−1	39	37	−2
34	33	−1	36	35	−1	37	37	0	39	38	−1
34	34	0	36	35	−1	37	38	1	39	38	−1
34	34	0	36	36	0	37	39	2	39	38	−1
34	35	1	36	36	0	37	40	3	39	39	0
35	33	−2	36	36	0	38	34	−4	39	39	0
35	34	−1	36	36	0	38	34	−4	40	35	−5
35	35	0	36	37	1	38	35	−3	40	36	−4
35	35	0	36	38	2	38	35	−3	42	38	−4
35	35	0	36	39	3	38	35	−3	43	42	−1
35	35	0	37	33	−4	38	36	−2	46	47	1
35	37	2	37	35	−2	38	36	−2			
35	38	3	37	35	−2	38	36	−2			

14.23 In Exercise 14.4, you considered a study described in earlier examples comparing FEV for the drugs formoterol and salbutamol. The data are very convincing that formoterol is more effective, with an average benefit over salbutamol of more than 4 ml. But take another look. The accompanying table repeats the data for F = FEV on formoterol and also gives the difference, S − F. (This is the negative of the F − S column from the table for Exercise 11.6.) Most of these values are negative and those that are positive have rather small values of F. Find and plot the least squares line for $x = F$ and $y = S - F$. Show that if F is less than 16.3 ml, then the estimate of S − F is positive and, therefore, salbutamol is estimated to be better than formoterol for such patients. Does this mean that salbutamol should be given to patients who do poorly on formoterol in the sense that their FEV is low on formoterol? Explain.

Data from crossover study of Example 3.24 for Exercise 14.23:
Forced expiratory volume (ml) after F and S, in aerosol

F	S − F	F	S − F	F	S − F	F	S − F	F	S − F	F	S − F
35	−3	31	−7	31	−3	31	−3	32	+1	29	−10
34	−6	28	−6	32	−11	26	−4	14	−4	20	+6
23	−1	31	−17	23	−7	28	0	13	+2	27	−5
23	−10	23	−1	14	+2	34	−10	26	−2	38	−12
30	−6	30	−4	32	−1	25	−1	25	−3	28	−1

14.24 Scores on the verbal (V) and analytical (A) parts of the Graduate Record Examinations for 26 applicants to a graduate program in statistics are given in the following table. Take x to be the verbal score and y to be the analytical score. (These measure different characteristics and are not exchangeable.)

Graduate Record Exam scores on verbal and analytical parts for Exercise 14.24

V	A	V	A	V	A	V	A	V	A	V	A
240	540	370	560	430	580	490	550	670	670	790	650
270	500	380	660	450	470	540	690	710	680		
290	420	390	520	450	600	600	690	780	710		
290	420	420	550	470	670	610	710	780	730		
330	560	420	670	470	670	640	730	790	620		

(a) Find the least squares line.
(b) Suppose the program requires an analytical score of 700, but a student who scored 500 on the verbal part did not take the analytical part. Using the prior probabilities assumed in Section 14.2, what is the probability that this student would have scored at least 700 on the analytical?
(c) Explain the regression effect for this problem and argue that it exists in this data set.

14.25 The numbers of quarterback sacks for the 28 National Football League teams for each of the years 1991, 1992, and 1993 are shown in the accompanying table. (A team earns a "sack" when the opposing team's quarterback is attempting to throw the ball to a team member and is tackled before he can do so.) Take x to be the number of sacks in 1991 and y to be the number in 1992.

Numbers of quarterback sacks for each NFL team in each of 3 years for Exercises 14.25–14.28

Team	1991	1992	1993	Team	1991	1992	1993
Arizona	25	27	34	LA Rams	17	31	35
Atlanta	29	31	27	Miami	35	36	29
Buffalo	31	44	37	Minnesota	33	51	45
Chicago	40	43	46	New England	25	20	34
Cincinnati	21	45	22	New Orleans	50	57	51
Cleveland	35	48	48	NY Giants	34	25	41
Dallas	23	44	34	NY Jets	35	36	32
Denver	52	50	46	Philadelphia	55	55	36
Detroit	30	29	43	Pittsburgh	38	36	42
Green Bay	45	34	46	San Diego	28	51	32
Houston	45	50	52	San Francisco	31	41	44
Indianapolis	29	39	21	Seattle	36	46	31
Kansas City	39	50	35	Tampa Bay	39	36	29
LA Raiders	42	46	45	Washington	50	39	31

(a) Find the least squares line.
(b) Assume that these 28 pairs of observations represent a sample from some larger population. Suppose a team sacks the quarterback 20 times in the first year. What is the probability that the team gets more than 20 sacks in the next year? (Use the prior probabilities assumed in Section 14.2.)
(c) Explain the regression effect for this problem and argue that it exists for pairs x and y.

14.26 Repeat Exercise 14.25, but take x to be the number of sacks in 1992 and y to be the number in 1993.

(a) Find the least squares line.

(b) Assume that these 28 pairs of observations represent a sample from some larger population. Suppose a team sacks the quarterback 20 times in the first year. What is the probability that the team gets more than 20 sacks in the next year? (Use the prior probabilities assumed in Section 14.2.) [Your answer should be similar to your answer to Exercise 14.25(b).]

(c) Explain the regression effect for this problem and argue that it exists for pairs x and y.

14.27 Repeat Exercise 14.25, but take x to be the number of sacks in 1991 and y to be the number in 1993.

(a) Find the least squares line.

(b) Assume that these 28 pairs of observations represent a sample from some larger population. Suppose a team sacks the quarterback 20 times in one year. What is the probability that the team gets more than 20 sacks 2 years hence? (Use the prior probabilities assumed in Section 14.2.) [Your answer should be similar to your answers to Exercises 14.25(b) and 14.26(b). But since a team changes more in 2 years than in 1—and therefore has more opportunity to regress to the mean—it seems reasonable that this answer will be larger than the previous two.]

(c) Explain the regression effect for this problem and argue that it exists for pairs x and y.

14.28 In parts (c) of the preceding three problems, you found a clear negative relationship between the number of sacks in one year and the increase in number of sacks from that year to a subsequent year. Consider the correlation between sacks in 1991 and the increase in sacks from 1992 to 1993. Would you expect this to be negative, positive, or not clearly one or the other? Is this comparison an example of the regression effect?

14.29 Report any data (sports records, medical measurements, examination grades, and so on) you have collected that address the regression effect. If it demonstrates this effect, then explain why it does and if it does not show the regression effect, then explain that as well.

14.30 In late 1994, the Merrill Lynch stock brokerage firm reported that there was a "silent bear market."[11] That is to say, the Dow Jones Industrial Average and the Standard & Poor's 500 show little change, but individual stocks drop. The report claimed that between January 1993 and November 1994, 98% of all stocks fell at least 10% from high to low and that more than half fell at least 30%. This seems much more pessimistic than the drops of 10% and 9% in the Dow and S&P 500 over the same period—hence, "silent bear." Disagreeing with the report's conclusion, Katherine Hensel of the Lehman Brothers brokerage firm said that she would "bet you that in the exact same period, 98% of stocks also *rose* 10% from their *lows*." (The point is well taken, but since this was a period of overall decline, 98% is too large—90% would be a more reasonable bet.) Explain Ms. Hensel's remark in terms of the regression effect.

Appendix: Using Minitab for Regression Analysis

The Minitab program **'lin_reg'** can be used to make inferences about the slope and to predict future y values for regression models. Consider Example 14.5, where one is relating the rate, K, of a chemical reaction with temperature. We believe that the logarithm of the rate, y, is linearly related to the inverse of the temprature, x. We are interested in the strength of this relationship. In addition, we are interested in predicting the logarithm of the rate when x is equal to 0 and 5.

We first place the *x* and *y* values in Minitab columns 'x' and 'y.' To find the least squares line, the Minitab command **regress** can be used:

regress 'y' 1 'x'

```
MTB > name c1 'x' c2 'y'
MTB > set 'x'
DATA> -3 -3 -3 -3 -2 -2 -2 -1 -1 0 0
DATA> 1 1 2 2 2 3 3 3 3
DATA> end
MTB > set 'y'
DATA> -3.50 -3.40 -3.36 -3.50 -3.51 -3.62 -3.53 -3.86 -3.70
DATA> -3.99 -4.18 -4.02 -4.23 -4.44 -4.62 -4.34 -4.74 -4.49
       -4.61 -4.62
DATA> end

MTB > regress 'y' 1 'x'

The regression equation is
y = - 4.01 - 0.203 x
```

To test hypotheses about the slope and to predict *y*-values, type

exec 'lin_reg'

The program asks for the numbers of the columns that contain the *x* and *y* data. The output is the mean and standard deviation for the slope of the least squares line. Using these numbers and the prior distributions discussed in Section 14.2, a 95% posterior probability interval for *b* is −2.03 ± 1.96 (.010), or from −2.05 to −2.01. Since this interval does not contain 0, the null hypothesis of no relationship between *x* and *y* is not supported. The program next asks whether we wish to predict future values of *y*. We input 'y' and on the 'DATA' line input the two *x* values of interest, 0 and 5. The program gives the mean and standard deviation for the future *y* value at each *x* value. Finally, the program displays a scatterplot of the data in a separate graph window. On the scatterplot, the least squares line and 95% prediction bounds are drawn for the entire range of *x* values assuming the prior distributions of Section 14.2. These bounds are slightly wider on the left and right than they are in the middle of the graph.

```
MTB > exec 'lin_reg'

INPUT COLUMN NUMBERS OF X AND Y DATA:
DATA> 1 2

THE POSTERIOR DENSITY FOR B IS NORMAL
WITH MEAN AND STANDARD DEVIATION:
K52        -0.202700
K56         0.0103956

Input 'y' and 'return' to obtain prediction intervals:
y

NOTE:
   INPUT X VALUES OF INTEREST:
```

```
DATA> 0 5
DATA> end

NOTE:
    THE PREDICTIVE DENSITY OF THE NEXT OBSERVATION FOR DIFFERENT
    VALUES OF X IS NORMAL WITH MEANS AND STANDARD DEVIATIONS GIVEN
    BELOW:

    Row      xo      MEAN_Y        STD_Y

     1        0      -4.0130      0.106523
     2        5      -5.0265      0.118528
```

Chapter Notes

1. For example: D. A. Berry and B. W. Lindgren, *Statistics: Theory and Methods* (Pacific Grove, Calif.: Brooks/Cole, 1990), 635.

2. G. E. P. Box and G. C. Tiao, *Bayesian Inference in Statistical Analysis* (New York: Wiley, 1973), 178ff.

3. G. W. Pierce, *The Songs of Insects* (Cambridge, Mass.: Harvard University Press, 1949), 12–21.

4. I thank Erin Berry for these data.

5. I thank Jennifer Berry-Graber for these data.

6. F. M. Mims, III, "Field trip report," *Science Probe!* (November 1990): 61–68.

7. D. A. Berry and B. W. Lindgren, *Statistics: Theory and Methods* (Pacific Grove, Calif.: Brooks/Cole, 1990) 669–670.

8. W. E. Mooz, "Cost analysis of light water reactor power plants," Report R-2304-DOE (Santa Monica, Calif.: Rand Corporation, June 1978).

9. H. R. Nichols, C. F. Johnson, and W. I. Duvall, "Blasting vibrations and their effects on structures," (U.S. Dept. of the Interior: Bureau of Mines Bulletin 656, 1971): 17.

10. As described in *Discover* (June 1985): 22–31.

11. D. Kadlec, "Silent bear market stirs loud discussion," *USA Today* 13 (December 2, 1994): 3B.

Appendix—Using the Bayesian Minitab Macros

A set of Minitab macros for performing Bayesian computations is contained in a directory named 'bayes' on the disk that is included with the text. These programs are designed to work on Release 10 of Minitab for Windows or the Student Edition of Minitab for Windows. Some of the graphics commands used in the macros are incompatible with older releases of Minitab. A second version of the programs is available, which uses the older style of character graphics, and will work for Minitab Versions 7 and greater. These programs are available at no cost by sending a 3½-inch disk to Jim Albert, Department of Mathematics and Statistics, Bowling Green State University, Bowling Green, Ohio 43403.

To use these programs on Windows, first define a new directory on the hard disk. Copy all of the programs (each ending with MTB) into this directory. Then start Minitab by double-clicking on any program in this directory. Now you can run the programs by the use of 'exec' commands typed in the session window. For example, to run the program 'bayes_se' described in Chapter 5, type the command

```
exec 'bayes_se'
```

The Minitab macros are illustrated by examples in the appendix material in Chapters 5, 6, 7, 8, 9, 10, 11, 12, and 14. For convenience, all of the macros are summarized below. Under each program name, a short description of the program and basic instructions for running the macro are given.

CHAPTER 5: Conditional Probability and Bayes' Rule

bayes_se   Sets up models, prior probabilities, and likelihoods for Bayes' rule
To run: Type exec 'bayes_se'. You input the number of models, the prior probabilities, and the likelihoods.

bayes   Implements Bayes' rule for an independent sequence of outcomes
To run: Type exec 'bayes'. (The program 'bayes_se' must be run first.) You input the sequence of independent observations (1 is the first possible outcome, 2 is the second outcome, etc.) and Bayes' rule is implemented one observation at a time.

_bayes, _bayes2, _bayes3   Macros used by 'bayes'. You do not run these programs.

CHAPTER 6: Models for Proportions

p_disc Computes and graphs the posterior distribution for p when there is a finite set of models

To run: Name two Minitab columns 'p' and 'prior'. Place the values of p in 'p' and the prior probabilities in 'prior'. Then type exec 'p_disc'. The posterior probabilities are stored in the column 'post'.

p_disc_p Computes and graphs predictive probabilities for the number of successes in a future set of observations

To run: The values of p are contained in a column 'p' and the probabilities are contained either in the column 'prior' or 'post'. Type exec 'p_disc_p'. You will input the number of trials in the future experiment, if you want to use prior or posterior probabilities for p, and the range of number of successes that you are interested in. The predictive distribution is stored in the columns 'succ' and 'pred'.

CHAPTER 7: Densities for Proportions

p_beta Summarizes beta density for p

To run: Type exec 'p_beta'. After inputting a and b of the beta(a, b) density, you can see a graph of the density, compute areas under the density for a list of values of interest, and compute percentiles of the density.

p_beta_p For beta prior or posterior, computes predictive probabilities for number of successes in future experiment.

To run: Type exec 'p_beta_p'. You input the beta parameters, the number of trials, and the range of numbers of successes that you are interested in. The values of number of successes and corresponding predictive probabilities are stored in the columns 'succ' and 'pred'.

CHAPTER 8: Comparing Two Proportions

pp_disc Finds posterior probabilities for (p1, p2) when a uniform prior on a set of possible models is used

To run: Type exec 'pp_disc'. You input the low and high values of each proportion, the number of models, and the number of successes and failures for each sample. The values of p1 and p2 are stored in the columns 'p1' and 'p2' and the values of the prior and posterior distribution are stored in the columns 'prior' and 'post', respectively.

pp_discm Finds probabilities for (p1, p2) when an informative prior is given in a table form

To run: Place the values of p1 in column C1 (say), the values of p2 in column C2, and the table of prior probabilities in columns C3–CN (each row corresponds to the particular value of P1 and all values of P2). Type exec 'pp_discm'. You tell the program where the prior is located and input the number of successes and failures for each sample. The values of p1 and p2 are stored in the columns 'p1' and 'p2' and the values of the prior and posterior distribution are stored in the columns 'prior' and 'post', respectively.

CHAPTER 9: Densities for Two Proportions

pp_beta Using simulation, summarizes the distribution of p2 − p1 when proportions have independent beta densities.

To run: Type exec 'pp_beta'. Input *a* and *b* of the beta(*a, b*) density for each proportion and the number of values to simulate. Program gives a dotplot of the posterior distribution of p2 − p1 and computes probabilities that p2 − p1 exceeds each of a list of values that you input.

CHAPTER 10: General Samples and Population Means

m_disc Computes the posterior distribution for *m* when there is a finite set of models

To run: Name two Minitab columns 'm' and 'prior'. Place the value of *m* in 'm' and the prior probabilities in 'prior'. Then type exec 'm_disc'. The posterior probabilities of *m* are stored in the column 'post'.

CHAPTER 11: Densities for Means

m_cont Summarizes posterior distribution for a normal mean, *m,* with a normal prior.

To run: If data are in raw form, then place them into a particular column. Type exec 'm_cont'. You input the mean and standard deviation of the normal prior distribution and the column number of the data. The output is the mean and standard deviation of the approximate normal marginal posterior density for *m*.

CHAPTER 12: Comparing Two or More Means

mm_cont Summarizes the difference of means, m2 − m1, when means have independent normal distributions

To run: Type exec 'mm_cont'. You input the mean and standard deviation for the normal distribution for each of the two means, m1 and m2, and the program gives the mean and standard deviation for the normal distribution of the difference in means m2 − m1.

CHAPTER 13: Data Transformations and Nonparametric Methods

Use programs indicated in either Chapter 11 or Chapter 12, depending on whether there are one or two means. Also, Section 13.5 uses beta densities—see Chapter 7.

CHAPTER 14: Regression Analysis

linreg Gives inference about slope and future *y*-values for linear regression model

To run: Place the *x* and *y* data in two columns. Type exec 'lin_reg'. You input the numbers of the columns of the data. The output is the mean and standard deviation of the regression slope. In addition, for each value of *x* that you input, the program gives the mean and standard deviation of the predicted *y*-value.

Short Answers to Selected Odd-Numbered Exercises

Complete solutions to all the odd-numbered exercises are given in the student solutions manual, sold separately.

CHAPTER 2
2.21. (a) $\bar{x} = 141.8$, (b) $s = 5.57$, (c) median = 140, (d) first quartile = 139.5, third quartile = 146.5
2.23. (b) median = 23.3, (c) $\bar{x} = 21.6$, $s = 4.05$
2.25. (a) $\bar{x} = 2.41$, (b) $s = 4.44$, (c) median = 2
2.27. (a) $\bar{x} = .731$, (b) $s = .388$, (c) median = .645, (d) first quartile = .43, third quartile = .95
2.29. For group 0: 289.4, 171.9, 210. For group 1–2: 211.7, 137.9, 210. For group 3–5: 318.5, 226.8, 260. For group 6+: 289.4, 167.7, 260.

CHAPTER 3
3.17. One way: Use 30 separate pots, one plant per pot, and assign them to cross- or self-pollination (randomly).
3.21. Randomly choose 4 of the 24 to be assigned to each one of these six sequences: ABC, ACB, BAC, CAB, BCA, CBA.
3.23. Assign 2 plots to each of the 6 possible combinations of fertilizer and corn variety.
3.25. Assign 200 patients each to the 4 possible treatment combinations.

CHAPTER 4
4.1. (a) {1, 3, 5}, (b) {5, 6}, (c) {3, 4, 5}, (d) {5}, (e) {1, 2, 3, 4, 6}, (f) {1, 3, 4, 5}, (g) {5, 6}, (h) {2, 4, 6}, (i) {2, 6}
4.3. (a) 1/2, (b) 5/6, (c) 2/3, (d) 2/3, (e) 1/2
4.5. (a) 1/3, (b) 1/3, (c) 1/3, (d) 1/6, (e) 0
4.7. 125/216   **4.9.** 4/47
4.11. (a) 4/8, (b) 3/8, (c) 6/8, (d) 5/8   **4.13.** 1/4
4.15. .15   **4.19.** $4.60, $186,600

CHAPTER 5
5.1. $19.05, $5.95   **5.3.** (a) 3/5, (b) 1/4, (c) 7/10, (d) Joint probabilities are products of individual probabilities.
5.5. (a) 1/4, (b) 2/9   **5.7.** .961   **5.9.** 1/2
5.11. (a) Joint probabilities are not products of individual probabilities. (b) All are 1/2. (c) 1, (d) 1   **5.13.** 1/3
5.15. .194   **5.17.** (a) 3/10, (b) 2/3   **5.19.** 2.5
5.21. 5/12   **5.23.** (a) 2/3, (b) 1/8, (c) 9/20
5.25. 91/216   **5.27.** 0, 1/7, 6/7   **5.29.** 8/9
5.31. (a) 1/5, (b) 1/5, (c) 2/5, (d) 2/5
5.33. .000099   **5.35.** You: 2/3, Luke: 1/3.   **5.37.** 9/11
5.39. (a) 100,000, (b) .5000, .9091, .9901, .99996, .99999, .9999999   **5.41.** .46

CHAPTER 6
6.3. .5725
6.7. (a) .203, (b) .016, (c) .794, (d) .666
6.9. (a) .315, (b) .683   **6.11.** (b) .117   **6.13.** .213
6.15. .115

CHAPTER 7
7.1. (a) .036, (b) .912   **7.3.** (a) beta(4, 25), (b) 4/29
7.5. .55   **7.9.** 1.000   **7.11.** 1.000   **7.13.** .75
7.15. (a) 11.8% to 20.3%, (b) Null hypothesis is not supported. (c) .997   **7.17.** 13.3% to 18.9%
7.19. (a) 39.4% to 46.5%, (b) 2
7.21. 1.5% to 22.5% (Using Minitab: 3.5% to 24%)
7.23. 31.9% to 40.2%   **7.25.** 40.2% to 50.4%
7.27. (a) 1/3, (b) 4/15, (c) 5/7   **7.29.** .911

CHAPTER 8
8.1. (c) .998, .759, .0499   **8.3.** (b) .999, .874, .378, .044
8.5. .599   **8.7.** (a) .079, (b) .082 − .003 = .079
8.9. .975   **8.11.** (a) .047, (b) .951
8.13. (b) .356, (c) .644, .553

CHAPTER 9
9.1. (a) .994, (b) 1.000
9.3. (a) beta(29, 2), beta(7, 5), (b) .993, .855, .371, .042
9.7. (a) .0041 to .0341, (b) Null is not supported.
9.9. (a) −25.1% to −8.9%, (b) −23.4% to −9.6%, (c) −23.5% to −10.1%
9.11. (a) 17.1% to 60.3%, (b) Null is not supported.
9.13. .996, .959, .785, .436, .133, .020, .001, .000, .000, .000, .000
9.15. (a) .767, .421, .129, (b) −9.8% to 25.6%, (c) Null is supported.
9.17. (a) 8.2% to 30.2%, (b) Null is not supported.
9.19. (a) .994, (b) .994, .618, .027, (c) −.5% to 68.7%, (d) Null is not supported.

513

9.21. (a) Evidence (plus prior information) does not support null. (b) Evidence supports null.
9.23. (a) $-.5\%$ to 10.3%, (b) .932, .488, .061
9.25. -26.5% to 19.8%
9.27. (a) 1.000, .925, .271, (b) .716, .072, .000, (c) 1.000, .900, .221 **9.29.** Null is not supported.
9.31. Null is supported.
9.33. (a) Null is not supported. (b) Null is not supported. (c) Null is supported. (d) Null is supported.
9.35. Null is not supported. **9.37.** Null is supported.
9.39. (a) Null is not supported. (b) Null is not supported.

CHAPTER 10
10.1. .003, .178, .001, .818 **10.5.** 3% and 97%
10.7. .004, .234, .676, .086, .000
10.9. (a) For $m = -5$ to 45 (the others are $< .001$): .001, .007, .031, .095, .192, .256, .224, .130, .049, .012, .002
10.11. Smith: 98.9%

CHAPTER 11
11.1. (a) .933, (b) .610 **11.3.** Normal(77.6, 41.8), 4%
11.5. (a) .009, (b) .017
11.7. (a) 1.000, (b) 1.000, 1.000, 1.000, .999
11.9. 6, 5.77
11.11. Every normal density gives positive probability to negative values. **11.13.** (a) Normal(14.19, 1.769), (b) 11.8%
11.15. 70% **11.17.** (a) 14%, (b) 3.1%, (c) 22%
11.19. (a) 17.52 to 18.98, (b) 19.17 to 23.11
11.21. 18.4 to 24.8 **11.23.** .304 to .540
11.25. (a) 7.012 to 7.666, (b) 7.008 to 7.666
11.27. (a) Null is not supported. (b) Null is supported.

CHAPTER 12
12.1. .994
12.3. (a) .946, (b) -5.12 to .06, (c) Null is supported.
12.5. .997, .926, .560, .125, .007
12.7. (a) Normal(23.0, 19.2), (b) .885, .752, (c) -14.7 to 60.7, (d) Null is supported. **12.9.** .140 to .590
12.11. $-.0152$ to .0162
12.13. (a) 1.000, (b) .36 to .52, (c) Null is not supported.
12.15. Null is supported.
12.17. (a) 2.01 to 5.17, (b) Null is not supported.
12.19. (a) .0095 to .0118, (b) Null is not supported.
12.21. (a) Null is not supported. (b) Null is supported.
12.23. (a) .028, (b) $-.97$ to $-.07$, (c) Null is supported.
12.25. (a) 99.1%, (b) 7.5 to 42.1, (c) Null is not supported.
12.27. (a) Essentially 1, (b) Null is not supported.
12.29. (a) 99.6%, (b) 99.98%
12.31. (a) Null is supported. (b) Null is not supported. (c) Null is not supported.

CHAPTER 13
13.1. (a) Normal(2.56, .096), (b) 8.5%
13.3. (a) Essentially 1, (b) 10% to 44%, (c) Null is not supported. **13.5.** 11.3%
13.7. (a) 99.9%, (b) Null is not supported.
13.9. (a) 99.2%, (b) 94.8%, (c) 80.8%, (d) 45.6%
13.11. (a) Normal(.933, .348), (b) .996 **13.13.** 99.0%
13.15. 98.3%
13.17. (a) 99.99%, (b) .255 to 1.22, (c) 1.3 to 6.1, (d) Null is not supported. **13.19.** 99.99%
13.21. 99.99%
13.23. (a) Essentially 1, (b) Null is not supported.
13.25. (a) 99.3%, (b) Null is not supported. (c) 90.1%, (d) Null is supported.
13.27. (a) 27.8%, (b) Null is not supported. (c) 29.5% (Using Minitab 28.1%)
13.29. (a) 67.6%, (b) Null is not supported—barely.
13.31. (a) 83.7%, (b) Null is not supported. (c) 97.6% (Using Minitab: 95.6%)

CHAPTER 14
14.1. (a) .9173, (b) $y = .4878 + .2069x$
14.3. $y = 1,254 + 489x$
14.5. Away wins $= -1.569 + .6434$(Home wins)
14.7. (a) .235, Winning% $= .4402 + .000516$(Payroll), (b) .0188, Winning% $= .4961 + .000013$(Payroll)
14.9. Rad $= 8.55 + 1.89$(El)
14.11. (a) $-1,310$ to 2,290, (b) Null is supported.
14.13. (a) $y = -1.91 + .039x$, (b) .029 to .049, (c) Null is not supported.
14.15. (a) 18.2 to 121 million dollars per year, (b) Null is supported.
14.17. (a) $y = -.680 - .633x$, (b) 35.5%, (c) 88%
14.21. Correlation coefficient is $-.63$.
14.23. Least squares line for $y = S - F$ is $y = 6.34 - .389F$.
14.25. (a) $y = 23.7 + .481x$, (b) 93%
14.27. (a) $y = 23.1 + .406x$, (b) 92%

Index

Abortion drug as morning-after contraceptive, 180–181
Acorn crop, effect on deer population, 401–402
Age of children when speak first word, 348, 355, 422–423, 456
Aging and oxygen use, 403–404
AIDS, 289–290, 293–295, 295–296, 301
Alcohol and caffeine consumption, 11–13, 15, 31–33, 35, 53, 55–57, 392
Alcohol use and family history, 298
Alcoholism, genetic component of, 304
Ali, Muhammed, 120
Amiloride in treating cystic fibrosis, 18–20, 61–62, 344–345, 363, 449–452
Ammonia lost in making nitric acid, 345
Analysis of variance (ANOVA), 395
Angioplasty, 192
Antilog, 417, 419
Aphids, kill rates of, 268–269
Area plot, 31–33
Argon, discovery of, 390
Aristotle, 2
Arm preference for carrying babies, 192
Arrowhead fracture, 15–16
Assay of carbon, 434–435
Assessing prior information
  about a population mean, 347–348
  about a proportion, 210–216
Attraction of opposites, 225, 299–300
Audit of accounts receivable, 225
Average value, 173
Averaging replicate measurements, 414–416
AZT
  treatment for AIDS, 289–290
  preventing transmission of AIDS to newborns, 295–296

Baird, Zoë, 225
Bar chart, 14–17, 45–46
  showing standard deviation, 46
Baseball, home run record in, 501–502
Baseball team performance
  first half vs second half, 502
  home vs away, 29, 478, 488
  vs payroll, 25–26, 30, 53–54, 478
Basketball team performance, home vs away, 23–25, 477–478
Bayes factor, 149–152
Bayes' rule, 147–157, 179, 313
  generalized, 152–157
Bayes, Thomas, 147
Behavioral intervention for distress, 99–100, 265, 382–383
Bell curve (*see* Normal density)
Belmont Stakes, 119–120
Beta density, 200–205
  and the sign test, 450–455
  choosing as a prior, 210–216
  construction of, 202

Beta density (*continued*)
  finding areas for a, 217–220
  mean of a, 204–205
  median of a, 221
  percentiles of a, 221
    using Minitab, 235
  prediction for a, 203–204
  standard deviation of a, 219
  updating rule for a, 205–209
  using Minitab for a, 233–236, 305–308
  z-score for a, 217
Betting odds, 116–119
Betting on baseball, 118–119, 125, 128
Bingo, 113, 139–140
Blinding, 67
Blood pressure decreases on drug, 353–355, 359–360, 362, 377–378
Bovine lens synthetase, reaction rate vs substrate concentration, 488–490
Breakaway bases and injury rates in baseball, 192–193, 227
Breast cancer, 129, 131, 146, 158, 233, 265–266, 299
  adriamycin for, 103
  and DDT and PCBs, 366
  and dietary fat, 84
  heritability of, 182
  taxol for, 103
Breast implants, 267–268
Burrowing owls, 72, 191, 210, 220, 238, 297

c (*see* Precision in a sample)
c_0 (*see* Precision, prior)
c_1 (*see* Precision, posterior)
Caffeine and miscarriage, 70
Calcium and blood pressure, 385–387
Calibration experiment, 121
Cancer, risk of, and exposure to phenols, 266–267, 300
Cancer, volume of, 44
Cannulae, pressure to remove, 14, 18, 29, 53, 384, 478–479
Cardiac Arrythmia Suppression Trial, 289
Case–control study, 70, 78
Categorizing data, 410, 426–435
Catheters to vaporize plaque, 436–437
Cause–effect, 66
Central limit theorem, 355–356, 431
Certain event, 107–108
Chance, 106
Changeover study (*see* Crossover study)
Charon, density of, 390–391
Chemical reaction rate vs temperature, 471–473, 479–483, 486–487, 505, 507
Chicks' diet and weight gain, 100–101
Chip-from-bowl
  experiment, 121–122, 153–156
  model, 153–156, 165
Chip model (*see* Chip-from-bowl model)
Chlorine in glacial moraines, 345, 383

Cholera, 79
Classical music and intelligence, 91–92
Cocaine use and fetal health problems, 296
Coin tossing, 168
Combining measurements (*see* Averaging replicate measurements; Categorizing data)
Combining samples, 394–399
Communal living and cortex weight, 364–366, 404–405, 498–499
Comparing means, 370–408
  when not appropriate, 399–400
Comparing samples in pairs, 394–399
Comparing several means, 392–400
Comparison group, 63
Competitive balance, law of, 498
Complement of an event, 107
Conditional probability, 130–137
Condoms, effectiveness of, in preventing AIDS transmission, 293–295, 301
Confidence interval, 361
Construction of nuclear power plants, cost of vs date of, 490–491
Control, 61–62
  group, 63
Controlled study, 60–62
Corn yield increases, 41–43, 47–48
Correlation coefficient, 472–475
Cosmopolitan sex survey, 73
Counts, 417
Court-martial rate of Gulf War trophy takers dependent on rank, 303
"Craps," 143–145
Cricket chirping frequency vs temperature, 475
Cross- vs self-pollination, 13, 17, 29, 52, 86–88, 333, 410–411, 435–436
Crossover study, 91–95
Cycles to failure of worsted yarn, 101–102, 420–422, 423, 441–443, 453–454
Cystic fibrosis (*see* Amiloride in treating cystic fibrosis)

d (*see* Difference of two means; Difference of two proportions)
Darwin, Charles, 13, 17, 29, 52, 86–88, 833, 410–411, 435–436
Data-fixer problem, 60
Data histogram (*see* Histogram, sample)
Data transformation, 409–458
Decision making, 4
Defect in manufactured part, 446–449
Degree of belief, 120–122
Density
  beta (*see* Beta density)
  flat prior (*see* Flat priors)
  for a difference in two means, 371–377
  for a difference in two proportions, 277–309
  for a mean, 336–369

515

Density (*continued*)
  for a proportion, 198–200
  predictive (*see* Normal density, predictive)
  normal (*see* Normal density)
  with heavy tails, 411–413
  with light tails, 412
Deprenyl, 103
Designing experiments, 3, 59–103
  deviations in, 59–60
Dietary fat and breast cancer, 84
Difference
  in ranks, 441–444
  of two means, 374
  of two proportions, 252
Differences, rule of, 374
Digoxin levels in blood, 27–28, 414–416, 423–424, 445–446
DiMaggio, Joe, 288–289, 295
Disjoint events, 125
Distances walked using congestive heart failure drug, 13, 17, 39, 52, 95–96, 210, 297, 332–333, 355, 438–439, 444–445, 456
DNA fragments, molecular weights of, 48–50, 356–357
  in criminal case, 333
Dot diagram (*see* Dot plot)
Dot plot, 11–13
  using Minitab for, 54, 56
Double-blind study, 67

Earthquake, San Francisco Bay Area, 125
ECMO, 256–257, 289
Electrode resistances, 424, 455, 462–467, 469–470, 487
Engineer test scores 50–51, 358–359
Equally likely outcomes, 108–109, 121
Error standard deviation in regression, 480–484
  estimating (*see* Root mean squared error)
Errors of measurement in regression, 480–484
ESP, 165–166, 189–190
Event, 107
Exchangeability of observations, 136, 167, 170, 183
  adjustment for lack of, 216
  and the regression effect, 501
Exchangeable experiments, 134–137
Expected value, 173
Experiment, 2, 105–106
Experimental evidence, 1
Experimental unit, 59
Extracorporeal membrane oxygenation (*see* ECMO)
Eyewitnesses, accuracy of, 89, 290, 371, 377

Factorial design, 96–102
Fair bet, 117
Fair coin, 146
Fair die, 108
Fair odds, 117
First quartile (*see* Quartiles)
Fish oil and duration of pregnancy, 1, 392–393, 395–398, 400

Flat priors, 337–347
  in regression analysis, 481–485
  principle of, 354
  with ranks, 441–444
Flawed data, 60
Football, quarterback sacks in, 504–505
Formoterol for asthma, 93–95, 346–347, 383, 456–457, 459–461, 476–477, 503
Free-throw success in basketball by attempt, 299
Frequency distribution (*see* Histogram)

Galton, Francis, 495
Gambling, philosophy of, 119
Gastric freezing in treated ulcers, 68–69, 299
Gay friends, proportion having, 226
Gays in the military, proportion favoring, 223
Germ warfare testing and birth defects, 83–84
Glass breakage, cause of, 227–229, 232
Golf, putting success in, 7, 78–79, 80–82
Golf scores, 495–497, 501–503
Graduate Record Examinations, verbal vs analytical, 504

h (*see* Standard deviation of a density)
h_0 (*see* Standard deviation of a population mean, prior)
h_1 (*see* Standard deviation of a population mean, posterior)
Half-life, 424
Harvard graduates, literacy of, 226
Hawking, Stephen, 2, 105, 120, 124
Health insurance and employment, 130
Heart pumps
  performance of, 20–21, 85
  pressure generated by, 29–30, 33, 40, 384
Heavy tails (*see* Density with heavy tails)
Helms, Jesse, 78
Hensel, Katherine, 505
Histogram
  population, 319
  sample, 36–39
Honey and hemoglobin, 86, 190–191, 197–198, 210, 220
"Hot hand" in basketball, 258–264
Hyperactivity and vocation, 303
Hypothesis, 166

Impossible event, 108
Independent events, 127–128, 133
Independent experiments, 128
Interaction, 97
Intersection of events, 125–128
Interval
  confidence (*see* Confidence interval)
  of population values and sign test 449–455
  probability (*see* Probability interval)
Inverse probabilities 148

James, Bill, 498
Joint probability, 125–128

k (*see* Small-n correction factor)
Kahneman, Daniel, 500
Kaiser Family Fund, 130

Large numbers, law of 168–169, 321, 340, 413–414
Law of competitive balance (*see* Competitive balance, law of)
Law of large numbers (*see* Large numbers, law of)
Lead, exposure to, and graduation rate, 233, 289, 295
League of Women Voters, 130
Learning, 4
Least squares line, 461–475
  equation of, 467–475
Left-handedness and being accident prone, 225
"Let's Make a Deal," 156–157
Leukemia, treatment of 16–17, 90, 182–189, 207–209, 220, 223–224
Light tails (*see* Density with light tails)
Likelihood, 106, 149, 313
  for a normal model, 324–325, 329–330
  for a proportion, 174–181
  for pairs of proportions, 239–244, 277–281
  for regression, 480–484
Lima beans, ascorbic acid vs dry matter 30–31, 491
Line
  equation of a, 461
  finding residuals for a, 465–467
  of least squares (*see* Least squares line)
  plot, 27–28, 101
  slope of a, 461–462
  y-intercept of a, 461–462
Literary Digest poll, 73–74
Log (*see* Logarithm)
Logarithm, 401, 416–426
Long-run frequency, 114–115
Lotteries, 139

m (*see* Mean of a density)
m_0 (*see* Mean of a population mean, prior)
m_1 (*see* Mean of a population mean, posterior)
Macaques, social ties among, and immune function, 303–304
Malaria and sickle cells, 6–7, 88–89, 239–255, 277–286
Mammograms, benefit of, 70–71
Maris, Roger, 501–502
Marital therapy, effectiveness of, 269–270
Marriage and mental health, 65–66
Matched case–control study (*see* Case–control study)
Mean
  as a center of gravity, 43
  of a bar chart, 173
  of a beta density, 204–205
  of a density, 321, 336
  of a population (*see* Mean of a density)
  of a population mean
    posterior, 349–355
    prior, 347–348
  of a sample, 40–46
    using Minitab, 55
Means, rule of, 349–355
Median of a beta density (*see* Beta density, median of)